JURASSIC WEST

Life of the Past   James O. Farlow and Thomas R. Holtz, editors

The Dinosaurs of
the Morrison
Formation and
Their World

SECOND EDITION

JURASSIC
WEST

JOHN FOSTER

Indiana University Press

This book is a publication of

Indiana University Press
Office of Scholarly Publishing
Herman B Wells Library 350
1320 East 10th Street
Bloomington, Indiana 47405 USA

iupress.indiana.edu

Manufactured in the
United States of America

Third printing in 2023

Library of Congress
Cataloging-in-Publication Data

Names: Foster, John Russell, author.
Title: Jurassic West : the dinosaurs
    of the Morrison Formation and
    their world / John Foster.
Description: Second edition. | Bloomington,
    Indiana : Indiana University Press,
    [2020] | Series: Life of the past | Includes
    bibliographical references and index.
Identifiers: LCCN 2020015322 (print) |
    LCCN 2020015323 (ebook) | ISBN
    9780253051578 (hardback) |
    ISBN 9780253051585 (ebook)
Subjects: LCSH: Vertebrates, Fossil—
    Southwestern States. | Paleontology—
    Jurassic. | Paleontology—Southwestern
    States. | Dinosaurs—Southwestern
    States. | Paleoecology—Jurassic. |
    Paleoecology—Southwestern
    States. | Morrison Formation.
Classification: LCC QE733 .F68 2020
    (print) | LCC QE733 (ebook) |
    DDC 567.90978—dc23
LC record available at https://
    lccn.loc.gov/2020015322
LC ebook record available at https://
    lccn.loc.gov/2020015323

In memory of
My mother, Lucille Foster,
who would have loved the Late Jurassic climate

And of Jack McIntosh, who deserves much more credit
for what we know about the Morrison Formation
than he would have ever accepted

And for Harrison, who may tire of the Morrison Formation
by the time he can read this

The seasons came and went, and seemingly everything changed and nothing changed. There were images of sauropod necks and backs glistening in monsoonal rains. Of lush savannahs, clouds of insects, and streams of migrating giants.

Dale Russell, *The Dinosaurs of North America*, 1989

Contents

Foreword

We are about to visit a certain region that is precisely located in space and time. Our objective's position in space is easily attainable—the Western High Plains of midlatitude North America. However, the region is separated from us by almost two million human life spans of elapsed time and cannot be reached directly. Only a fraction of a lifespan is available, and time can be traversed in only one direction—toward the future.

Yet we have found in the author a guide who can take us to this region as it used to be. First, we must assimilate the fact that the region was then just as immediately real as our present surroundings are now. Tangible indications of its former actuality are preserved in softly multi-colored "badlands" where the accumulated sands, silts, and clays of that time are exposed. Our guide knows how to reconstruct landscapes from ancient sediments, from the shapes of long-vanished plants and animals from carbon impressions, and from petrified bones. Experience, logic, and analogy translate these images into words, which pass into our minds as though they were sensory perceptions. These images can easily be manipulated by our imaginations, making it possible for us to actualize, in a dreamlike manner, an otherwise impossible safari that will take us two million lifetimes into the past.

A preliminary briefing is necessary. We will find ourselves in a world where the geography, the climate, and even the atmosphere differ from the one into which we were born. Many of the animals we will encounter are gentle, quadruped giants, and some are much larger than any land animals we have ever seen. Their bodies are curiously shaped, for all bear long and heavy tails, and most possess elongated necks. Their habits are also peculiar, in that these often elephantine creatures lay their eggs in nests on the ground. Our lives could be imperiled by large, flesh-eating bipeds that might mistake us for their normal prey. In order to identify the animals, we are provided with a handbook containing their names, illustrations, lengths, weights, probable habits, and abundance. Inasmuch as it can be charted today, a map of this seemingly limitless region is included with the handbook, together with general notes on the vegetation and weather conditions to be anticipated.

It is difficult to visualize, in detail, once-living communities of organisms that are so distant from us in time. We must be prepared for surprises, not all of which may be pleasant. Hardships may be difficult to endure except by very positive thinking. They may be alleviated by moments of surpassing natural beauty, accompanied by feelings of humility and gratefulness for the opportunity of witnessing the grandeur of the effects of vast time. The perspective these occasions may impart could be analogous to those experienced by astronauts beholding the spheres of the earth and moon from far beyond the maternal biosphere of our planet. It may change our lives, too, and deepen our appreciation for the congeniality of the modern world, where we are supported by all the organisms we have domesticated, and from which we will be separated, if only for a few months. The emotions that will haunt our memories of the adventure after our return are also unpredictable but likely to be powerful.

Before our jump-off into the outback of time, we would be well advised to prepare for the voyage. A vigorous exercise and weight-loss program should be a high priority. It should be accompanied by an assessment of how well we know ourselves. This is best carried out under conditions most likely to resemble those prevailing in our objective in the past. Appropriate modern analogues include tropical or subtropical forests and savannas inhabited by large animals. How much sun, heat, thirst, cold, hunger, sleeplessness, and fatigue can we withstand without compromising our health? Are we alert to wildlife? How well do we support other members of a team? Are we able to strike fires to cook our food? How do we boil a one-gallon egg? Can we tolerate such foods as may be readily available, such as insect larvae or salamanders? If, in the course of our voyage in time, we are bitten by insects that adapted to feeding on pachyderms, how should the bites be treated?

Some intellectual preparation would be useful as well. It would be a good idea to visit a museum and reflect on the skeletal anatomy of large land animals, noting particularly the structure of their teeth and claws and examining their limb bones for indications of speed and agility. Recall that living animals always appear to be larger than their skeletons. Peruse journals of those who have explored the equatorial regions of the Old and New Worlds. Be mindful that, unlike in these accounts, we will not be assisted by—or even encounter representatives of—indigenous human cultures. Our journey will take place long before the distant ancestors of our species appeared. Libraries can be searched for information on the behavior of large animals, particularly when hungry or defending their young. Large herbivores strongly affect the appearance of the vegetation on which they browse and can transform pristine forests into open prairies. All of this information should provide us with a basis for comparison in our assessment of the characteristics of the communities of organisms we will encounter. A notebook and a light digital camera must not be left behind.

Our objective is not an ordinary, random sample of the past. It is arguably the most exotic, most extensive, most spectacular, and most completely preserved biotic community ever known to have existed in the history of life on land. Only a few plants and animals typical of its time still survive today in Australia and on the islands of New Zealand and New Caledonia in the southwestern Pacific. Much is also obscured by the fog of time, so that many uncertainties persist and many interesting observations lie before us. Some of them may relate to the notion that this vast community of organisms somehow possessed the biotic potential, over eons, to change with inconceivable slowness into the modern terrestrial biosphere.

Discourses in the pages ahead on geology, anatomy, the history of collection, faunal composition, and ecology will further prepare us for our intellectual voyage. Can one not already feel the hot wind and sense fetid odors of desiccating ponds mingled with aromas of resins exuding from broken conifer branches? Will the savannas be sparsely or heavily populated with animals? Will movement increase as heat dissipates with the setting of the sun? Will we hear cacophonies of shrieks, grunts, and snapping twigs, or will landscapes harbor only silent breezes wafting through the lengthening shadows of passing giants?

Our guide awaits us in Laramie, Wyoming.

Dale Russell, North Carolina Museum of Sciences

Preface

About 100 km (60 mi) northwest of Laramie, Wyoming, there is a turnoff for a dirt road that heads north from an east-west-trending stretch of Highway 287. From this turnoff, one travels 6.4 km (4 mi) of gravel across a rolling high plain, with little stretching out to the horizon on a summer day but waves of golden prairie grass. Elk Mountain, the northern anchor of the Medicine Bow Mountains, which extend down into Colorado, is visible to the southwest, but otherwise, there are few noticeable topographical features in this land. After those 6.4 km, however, the gravel road drops suddenly into a shallow valley hidden from the main highway by low south-facing slopes. What you find on the other side of those slopes, on the side facing north into the valley, are the multicolored mudstone deposits left by ancient rivers that flowed through the region 150 million years ago. In the spring of 1877, these red, green, and gray hillsides at Como Bluff began yielding to scientists (and other characters) fossilized bones of extinct vertebrate animals both small and almost unimaginably huge. That spring, the ancient life of western North America during the latest part of the Jurassic period began to be revealed, a picture of a past world that has become, over the last 140-plus years, one of the best of any ancient vertebrate ecosystem in the world.

Hundreds of expeditions to Morrison Formation rocks have, over the years, packed museum collection rooms and exhibit halls around the country and the world with dinosaur and other vertebrate remains, including some of the most familiar dinosaurs known by the public: *Stegosaurus*, *Allosaurus*, and *Apatosaurus* (*Brontosaurus*). But more than 100 types of vertebrates are now known from the formation, and barely a third of these are dinosaurs. The formation contains bones and other fossils from a surprising variety of life-forms, from crocodiles and turtles, frogs and salamanders, dinosaurs and mammals, to clams and snails, to ginkgoes, ferns, and conifers. Among the dinosaurs are familiar animals as well as those with more exotic-sounding names such as *Torvosaurus*, *Tanycolagreus*, and *Ornitholestes*.

Fossils have been collected from the formation by a large cast of collectors for more than 100 years, from the early years in Colorado and Wyoming, when O. C. Marsh of Yale College and E. D. Cope from Philadelphia squared off in bids to outdo each other in the naming of paleontological specimens, through the early twentieth century, when large quarries were identified in Utah and northern Wyoming, to recently when, surprisingly, totally new types of dinosaurs have been found of which there was no trace in the first 100 years of exploration. The fact that new things can be learned and new animals found in a formation as extensively explored as the Morrison only points out how relatively little we know of what was around at the time and of how the animals lived. This is still one of the times and areas in Earth history, however, that we do know the best.

This book is based on several years of research both in the field at many of the Morrison Formation vertebrate fossil localities and in most of the museum collections of these vertebrates from around the country. A study of the Morrison Formation is enjoyable duty. The rocks are exposed throughout the Rocky Mountain region of the western United States, particularly in some of the most spectacular outcrops of the red rock region of the Colorado Plateau country in the Southwest. It is beautiful terrain in which to work, and when the field season is over, the museum work takes one to cities such as New York, Chicago, and Washington, DC. There is nearly as much excitement to be had digging through museum collections of Morrison material (which are surprisingly consistently housed in poorly lit basements) as there is in being out hiking the formation itself. Even unspectacular vertebrate fossils are beautiful things, and coming across a well-preserved toothed jaw of a *Dryosaurus* while looking through a museum cabinet drawer, for example, does elicit a certain thrill, even when the specimen is no longer in the original rock and is 1,500 miles from where it was collected some 120 years ago. There are times while doing paleontological research when one is suddenly reminded that we are dealing with what used to be living animals in all their rather unusual glory and that what we hold in our hands is concrete evidence of a world so uniquely different from our own that it almost seems imaginary. And yet it was real, and that is what makes paleontology so enjoyable.

In chapters 1 and 2, I introduce background information on geology and paleontology that should be helpful for the rest of the book. I am not assuming that the reader has a background in geology or biology, just an interest in paleontology, and these first sections introduce what geology and biology are necessary as a foundation for understanding the Morrison Formation. I also introduce a bit of the history of vertebrates leading up to the Late Jurassic and some of the background geology of the Morrison Formation itself, and how that helps us understand the Late Jurassic world in North America. Chapters 3 and 4 cover the history of excavations of dinosaurs and other vertebrates in the Morrison Formation and introduce some of the animals these projects revealed. Chapters 5 and 6 are the longest and possibly the most important sections, in that they cover the morphology and ecology of each type of vertebrate from the Morrison Formation. Groups of species are sometimes introduced first, and when the ecologies of the species in a group were likely similar, they are sometimes covered in the group introduction section rather than under the species or genus heading. I have also gone into a fair amount of background material as it relates to debates in previous studies regarding individual groups and their ecology. Chapter 7 introduces some newer data and conclusions that are based on an overall survey of the Morrison

Formation vertebrate fauna and the relative abundances of individual taxa. Chapter 8 is a presentation of one view of the Morrison world and its animals in the form of a mental exercise imagining a journey across the Late Jurassic floodplain and what we might have encountered. In the epilogue I briefly summarize what the Morrison Formation vertebrate fauna may have to say about the history of vertebrate animals.

John Foster
Grand Junction, Colorado
February 2006

Preface to the Second Edition

There was a time not that long ago when the dinosaur renaissance that started in the 1970s was not all that distant in the past and when research on dinosaurs and the Morrison Formation was slowly picking up a bit of steam. Though hardly as slow as dinosaur research had been midcentury in North America, there still were relatively few workers dedicated to the Morrison Formation. Jack McIntosh was one of very few in this country truly obsessed with sauropods. Most theropod workers seemed to truly care only about the Cretaceous. (Okay, that's still true to some degree.) Papers on aspects of the Morrison came out relatively infrequently. Few and far between the publications were, or at least that's how it seemed to some graduate students. Things developed on a pace suiting geologic time. As the renaissance permeated more of the science and as more researchers dove into the field, it became inevitable that more would investigate the Morrison Formation. That has resulted in an increasing pace of publications and an increasing rate of surprises for those of us studying the formation. Things have changed so much in the decade since the first edition of *Jurassic West* that updating it seemed important if it were to remain of any use as a general reference about the terrestrial Late Jurassic of western North America.

New finds and reanalysis of existing collections (which happily are often now stored in better conditions), along with development of new localities and reinvestigation of lost sites, has resulted in entirely new taxa being described, existing taxa being reassigned taxonomically, whole data sets being analyzed anew, and conclusions being reassessed. As always, the project seemed saner at the outset than it proved to be in the depths of revision, incorporating literally mounds of papers with new data and conclusions and creating almost all new figures.

Here, you will find a significant reworking of the 2007 version of *Jurassic West* with, just as a few examples, more information on the plants and invertebrates of the Morrison; illustrations of and information on new taxa of sauropods and theropods; name changes and classification adjustments to taxa of mammals, pterosaurs, and crocodylomorphs; several recently named new nondinosaurian taxa; discussions of analysis techniques and new technologies; and almost every illustration expanded and redone, with numerous entirely new figures. I didn't hold back on citations of where information originated (as you will see; there are well more than 150 new references), and unlike the previous edition, I put these citations right in the text to make it easier to reference results of original research. Make use of the text citations and the bibliographies in the back. I say this with the hope that nonspecialists will give a shot at diving into scientific literature (more of which is now available as open access) to see the data in raw form. Although as a graduate student you do have an introduction to unfamiliar terms in classes and from advisers and classmates, the truth is, as a young student starting out, a significant percentage of what you pick up you also get just by braving the literature and learning terminology as you go. So give it a shot! It will take a while, but you will make progress. And don't be afraid to ask for help from the authors (whose email addresses are usually right on the papers). That's what we're here for. Alternatively, if you're not that obsessed, I've tried to gather in one place as much information as possible about the Morrison Formation, so pull up a recliner!

Hopefully this edition of *Jurassic West* will prove to be both very up to date and significantly more enlightening to readers than it was a decade ago and will be a valuable resource for current and future students of the Morrison at all levels.

John Foster
Vernal, Utah
February 2020

Acknowledgments

Many people have helped me during this study, which goes back to my early days in graduate school in South Dakota, when my adviser, Jim Martin of the School of Mines, suggested I "go check out those outcrops along Interstate 90." I have learned much from discussions and correspondence with many fellow students of the Morrison, and their discoveries and work have all added to this book. Some have provided advice, manuscripts, references, further input about their work, tours of their quarries and collections, parts of figures, and many other donations of time and effort that have helped immensely; a number we were pleased to host as backyard BBQ guests (always a good venue for discussions). These colleagues include (in totally random order) Luis Chiappe, Brent Breithaupt, Jim Farlow, Dan Chure, Jack McIntosh, Scott Madsen, Christine Turner, Zhe-Xi Luo, Bob Bakker, George Callison, Kirk Johnson, Tom Holtz, Scott Sampson, Jim Clark, Rod Scheetz, Peter Dodson, Jim Madsen, Sid Ash, Kelli Trujillo, Carole Gee, Brooks Britt, Jim Kirkland, Cliff Miles, Spencer Lucas, Ken Carpenter, Susan Evans, George Engelmann, Nick Fraser, Joe Peterson, Emanuel Tschopp, Scott Hartman, Adrienne Mayor, Brian Curtice, Matt Carrano, Brian Davis, John Damuth, Steven Salisbury, Takehito Ikejiri, Susie Maidment, Mike Flynn, Bhart-Anjan Bhullar, Carol Hotton, Mark Norell, Chris Noto, Ray Wilhite, Harley Armstrong, Sue Ann Bilbey, Tony Fiorillo, Adrian Hunt, Peter Galton, Ken Stadtman, Heinrich Mallison, Martin Sander, Steve Hasiotis, Kirby Siber, Octavio Mateus, Jerry Harris, Marc Jones, Randy Nydam, Bruce Schumacher, Matt Lamanna, Rich Cifelli, Eric Dewar, Randy Irmis, Dave Gillette, Joe Sertich, Sharon McMullen, Andy Heckert, Oliver Rauhut, Mark Loewen, Dave Burnham, Pat Monaco, Donna Engard, Melissa Connely, Dale Russell, Jason Pardo, Pedro Mocho, Josh Mathews, Matt Bonnan, ReBecca Hunt-Foster, Mark Gorman, Tom Rich, J. P. Cavigelli, Matt Wedel, Steve Sroka, Ken Stadtman, Cary Woodruff, Craig Sundell, Kent Stevens, Scott Williams, Kevin Madalena, Julia McHugh, Larry Martin, Mike Taylor, Katie Tremaine, Alison Mims, Brian Engh, Don DeBlieux, John Whitlock, Thomas Adams, Forest Frost, Michael Caldwell, Bill Wahl, Gabe Bever, Walter Joyce, Guillermo Rougier, Scott Sampson, John Flynn, Rob Gay, Darrin Pagnac, Tracy Ford, Greg Paul, Jordan Mallon, Dave Lovelace, Gerard Gierliński, and others.

The early stages of my work in the Morrison were guided by my graduate advisers and committee members, particularly Jim Martin, Richard

Stucky, Martin Lockley, Fred Peterson, David Armstrong, Mary Kraus, Paul Weimer, and Phil Bjork.

Thanks also to the dozens of curators and collections managers that facilitated my work in the various museums housing vertebrate material from the Morrison Formation. I would like to thank all of them, particularly the Carnegie Museum, American Museum of Natural History, Yale Peabody Museum, Natural History Museum at the University of Kansas, Natural History Museum of Los Angeles County, Denver Museum of Nature and Science, Museums of Western Colorado, South Dakota School of Mines and Technology Museum of Geology, and National Museum of Natural History, for allowing me access to their collections and for granting me permission to use my photographs of their specimens in this book. Unless otherwise noted, photographs are my own. Numerous archivists also assisted with the location and use of historic photographs.

My fieldwork in the Morrison Formation has been made much easier thanks to the many volunteers and fellow staff of the Museum of Western Colorado, the South Dakota School of Mines and Technology, and the Museum of Moab, particularly my field assistants over the years, including Greg Goeser, Don Chaffin, Josh Smith, Vaia Barkas, Lorin King, Alex Morrow, Zach Cooper, Krista Brundridge, Lucia Herrero, Nehali Dave, and Tom Temme. In recent years at the *Dystrophaeus* site, I owe huge thanks for many exhausting ascents and much hard work to Natural History Museum of Utah staff and volunteers, especially Tylor Birthesel, Carrie Levitt-Bussian, and Riley Black.

A big thanks to the artists that have contributed their skills to this book, including Brian Engh, Thomas Adams, Todd Marshall, Donna Braginetz, Karen Foster-Wells, Matt Celeskey, Mark Witton, William Berry, and Laura Cunningham. Their work is what really brings the Morrison world back to life. Mark Klingler and Donna Sloan also drew several reconstructions that were used. Amy Henrici, Zhe-Xi Luo, Kelli Trujillo, Kirk Johnson, Andy Heckert, Rick Adleman, Richard Peirce, Alan Titus, Anjan Bhullar, and Ken Carpenter graciously provided several of the figures. Dan Chure, Logan Ivy, Darrin Pagnac, and Richard Stucky made available equipment for some of the microphotography. Help with some of the figures and associated computer glitches was provided by Nita Kroninger and Zeb Miracle. Use of some computer equipment and software was generously allowed by Don Montoya and the Board of the Museum of Moab. Brent Breithaupt, Beth Southwell, Richard Peirce, and Ray Bley read and helped edit parts of the early versions of the manuscript. Many of the specimens illustrated in the book have been collected from land administered by the Bureau of Land Management, National Park Service, and USDA Forest Service.

Many thanks to Gary Dunham, David Hulsey, Jim Farlow, Bob Sloan, Nancy Lightfoot, David Miller, Dawn Ollila, Miki Bird, Elisabeth Marsh, Peggy Solic, and Anne Teillard-Clemmer at Indiana University Press, as well as copyeditors Jamie Armstrong and Karen Hellekson, for their help on the two editions of this project.

I've always felt that I owe quite a bit to Don Prothero and David Archibald, both of whom introduced me to field paleontology early on in my training and thus allowed me to discover just how much fun it is. And finally, thanks to my family and friends for support during the projects. Now, quit sneaking up behind me, Ruby.

Abbreviations

AMNH	American Museum of Natural History, New York, New York
BMNH	Natural History Museum, London, United Kingdom
BYU	Earth Sciences Museum, Brigham Young University, Provo, Utah
CM	Carnegie Museum of Natural History, Pittsburgh, Pennsylvania
CMNH	Cleveland Museum of Natural History, Cleveland, Ohio
DINO	Dinosaur National Monument, Jensen, Utah
DMNH	Denver Museum of Nature and Science, Denver, Colorado
FHPR	Utah Field House of the Natural History State Park Museum, Vernal, Utah
FMNH	Field Museum of Natural History, Chicago, Illinois
GPDM	Great Plains Dinosaur Museum, Malta, Montana
KU	University of Kansas, Lawrence, Kansas
LACM	Natural History Museum of Los Angeles County, Los Angeles, California
MOR	Museum of the Rockies, Bozeman, Montana
MWC	Museums of Western Colorado / Dinosaur Journey, Fruita, Colorado
NMMNH	New Mexico Museum of Natural History and Science, Albuquerque, New Mexico
OMNH	Sam Noble Oklahoma Museum of Natural History, Norman, Oklahoma
SDSM	South Dakota School of Mines and Technology, Rapid City, South Dakota
SMA	Saurier Museum Aathal, Aathal, Switzerland
SMM	Science Museum of Minnesota, St. Paul, Minnesota
SUSA	Museum of Moab, Southeastern Utah Society of Arts and Sciences, Moab, Utah
TM	Tate Geological Museum, Casper, Wyoming

TPII	Thanksgiving Point Institute, Inc. (North American Museum of Ancient Life), Lehi, Utah
UCMP	Museum of Paleontology, University of California, Berkeley, California
UMNH	Utah Museum of Natural History, Salt Lake City, Utah
USNM	National Museum of Natural History, Washington, DC
UW	University of Wyoming Collection of Fossil Vertebrates, Laramie, Wyoming
YPM	Yale Peabody Museum, New Haven, Connecticut

JURASSIC WEST

1.1. Typical outcrops of the Morrison Formation. A mix of sandstones and colored mudstones is characteristic of the formation in most areas; this is in eastern Utah.

The natural world of the Late Jurassic was drastically different from the one we know today. In many ways, it would have seemed to us an almost alien world, although the evidence for that world is sometimes found literally in our backyards. That juxtaposition of the strange and familiar, of the ancient and the modern, as seen in the discovery of bizarre animals of inconceivable antiquity among our recognizable modern landscapes, is paralleled by the **vertebrate*** faunas found in rocks of Late Jurassic age. There is an interesting mixture of animals with no modern relatives and almost no modern morphologic or physiologic equivalents, but then in the same rocks are found bones of animals that, if we saw them alive, we would have a difficult time distinguishing from their modern descendants. Mixed in with 24-ton herbivorous reptilian behemoths, with skulls the size of a horse's, are remains of frogs and salamanders that would have looked much like those we might see today in a neighborhood pond. Earth during the Late Jurassic was filled with animal groups we know well today—sharks, horseshoe crabs, shrimp, beetles, crayfish, lizards, frogs, salamanders, dragonflies, crocodiles, turtles, and mammals. The mammals of the time were all small, many had unusually large numbers of teeth, and many retained some "reptilian" characters in their lower jaws, but otherwise, most of these animal groups were even then quite similar to those we know in modern ecosystems.

It is the dinosaurs of the time that are truly strange and would have given us the strongest sense of being in an alien world. However, the physical environment was different enough from what we know now to have also made us feel lost on our own planet. For example, sea levels were higher than they are today, and overall sea surface temperatures were higher, so there were essentially no polar ice caps; the ocean temperatures above the Arctic Circle, where Alaska is today, would have been about the same as those of the modern central California coast; atmospheric carbon dioxide levels were several times higher; global land surface temperatures were generally higher and differed less extremely between low and high latitudes; the continents were in slightly different positions, with the Atlantic Ocean being only about 1,600 km (1,000 mi) wide; and the topography and physical geography of the continents were different, with what is now the Rocky Mountain region of North America nearly flat and close to sea level. Europe was mostly under a shallow sea, with the only exposed land being largely subtropical islands. Even the **terrestrial** plant

Bolded words are the first occurrences of words in the glossary.

life was different, with no grass (which was not to originate for some 80 million years) and no flowers or other angiosperms. The main plant types of the time were relatives of the modern ferns, araucarian conifers (the Norfolk Island pine is one), ginkgoes, and cycads (such as the sago palm, which is not a true palm at all, palms being angiosperms).

The evidence of one small corner of this Late Jurassic world is preserved in the red, gray, and green **mudstone** badlands and the tan-brown **sandstone** benches of the Morrison Formation in the western United States, mostly in the Rocky Mountain region (fig. 1.1). The **formation** occurs in an area encompassing some 1 million square kilometers and is exposed in Arizona, New Mexico, Oklahoma, Utah, Colorado, Wyoming, South Dakota, and Montana. The rocks of the Morrison Formation are exposed in some areas totally devoid of vegetation, and here they often form spectacularly colored moonscape hills of brutally steep popcorn-textured mudstone. In many areas, the formation is partially covered by rockfall debris, piñon pine, and juniper trees, and in some places, such as Como Bluff, Wyoming, the exposure rises from the grass-covered plains. The mudstone units preserve the deposits of millions of years of floods from rivers flowing out of mountains that lay to the west and south; the rivers then flowed east and northeast across the **floodplain**. Mixed in with the muds are altered ash and rare pure ash deposits from volcanoes in those same mountains. From analysis of the ashes, we get some idea of the numerical age of the formation. The sandstone units in the Morrison Formation preserve the ancient river channels. In some cases, the sandstone units are largely composed of rounded pebbles, indicating that this stretch of the river was gravelly and had a higher discharge rate than most of the sandy channels. In these sandstones, we often find the remains of dinosaurs whose bones got entrained in the currents before getting buried and preserved. The mudstones also contain bones of the animals that lived at the time, and these are often preserved in areas that were low, wet parts of the floodplain.

Only rarely will you walk far in the Morrison Formation and not see some sign of ancient life. Clamshell casts, invertebrate traces, small scraps of bone, and petrified wood fragments are not uncommon, but don't let a paleontologist's somewhat warped perspective fool you—it isn't easy to find identifiable vertebrate remains in the Morrison Formation, and hiking for several days straight with no luck is to be expected. Still, the number and **diversity** of vertebrate fossils that have been found in the formation over the last 135 years are impressive, and it remains one of the best-known Mesozoic vertebrate ecosystems in the world.

Road-Tripping to the Late Jurassic

The Morrison Formation was deposited during the Late Jurassic, the most recent part of the Jurassic period of geologic time. The Jurassic period covers the time from approximately 201 to 145 million years ago (mya), although these numeric dates have been assigned to it since its original designation, which was based on extinction patterns of marine

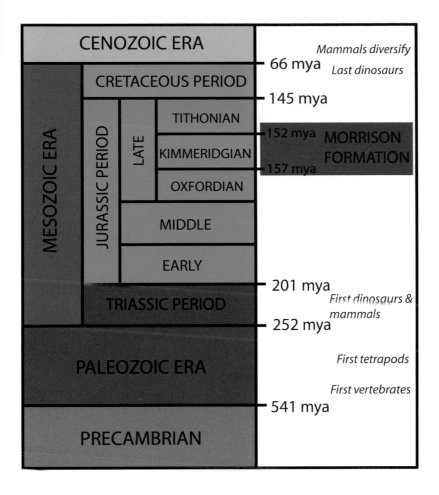

1.2. Generalized geologic timescale showing approximate ages of period boundaries in the Mesozoic era. Also noted are the position of the Morrison Formation and major events in vertebrate evolution.

invertebrates in western Europe. The numeric dates have changed over the years as the dating methods of the rock contacts have been refined, and these 201 to 145 mya estimates are current as of 2016. The Late Jurassic covers the time from approximately 163 to 145 mya. But where does this, the time of the Morrison Formation and its animals, fit into the overall history of the earth and its life? Was this early or late, both in time and in vertebrate development? To get an idea of this, it helps to have a better understanding of the geologic timescale and the relative length of geologic time itself.

The history of the earth is divided into a number of subdivisions, beginning with the earliest years after the planet's formation about 4.6 billion years ago. The time from this first formation up to around 541 mya, when the first animals with hard skeletal parts appear, is all included in the Precambrian[1] (fig. 1.2). The time from 541 to 252 mya is known as the Paleozoic era (referring to "old life"). This is the era during which trilobites lived, the first vertebrates appeared, and plants and animals invaded the land. The Mesozoic era (for "middle life") extended from 252 to 66 mya, and the modern Cenozoic era (the effective Age of Mammals) from 66 mya to the present. The Mesozoic is the time of the dinosaurs and is further divided into the Triassic, Jurassic, and Cretaceous periods. The Triassic is the time of emergence of the dinosaurs and mammals. During the early part of the Jurassic, the dinosaurs essentially took over the world. The dinosaurs reached their peak during the Cretaceous and then disappeared at the end of this period.

The Late Jurassic world of the Morrison Formation was close to the middle of the reign of the dinosaurs. This was a time of peak diversity of the very large sauropod dinosaurs (**Sauropoda**). But just how long ago was this? On a human scale, 150 million years is a nearly inconceivably long time, but geologically speaking, it wasn't really all that long ago. Still, put into everyday terms, 150 million years is a rather fantastic concept and equal to about 6 million human generations, roughly 7.8 billion workweeks, 1.95 billion full moons, and all of *Homo sapiens*'s (modern humans') tenure about 500 times over. It's difficult to deal in hundreds of millions of years. We need to put a reasonable human measurement of time into perspective and relate it to the geologic timescale by visualizing time as interchangeable with distance.

We need to travel in time, and what more appropriate, modern way to travel than in a car? For our trip, we are going to say that 1 mm (about 0.04 in.) is equal to 1 year. Thus, a human generation is about 2.5 cm (1 in.), and most human lifetimes fit in a space of 10 cm (4 in.). A millennium is contained in 1 m (just more than 1 yd). Now pretend that, starting in our driveway, we will be traveling back in time along the Road of the Geologic Timescale to the age of the Morrison Formation in the Late Jurassic, 150 mya. For most of the trip, once we are out of the driveway and on the open road, we will travel at about 60 mph (97 kph). As we travel the road, remember, each millimeter (0.04 in.) represents a single year with its 365 sunrises and sunsets; 13 lunar cycles; spring, summer, fall, and winter seasons; weather; waves, earthquakes, and volcanic eruptions—everything we observe today in a single year. (These numbers change as we go back in time far enough: during our backward journey, the days become shorter and the moon comes closer to the earth. With shorter days, there may have been more days per year, perhaps a 23.5-hour day and 370 days in a year.) For each hour of driving, as we cruise along at 60 mph, we travel back in time 97 million years. So we start the trip, and what would we see as we traveled the road? Well, we'd pass the end of the most recent glacial cycle of the current ice age 10,000 years ago before we got out of the driveway, and we would pass the time of origin of modern humans somewhere down the next block. The beginning of the Pleistocene ice ages would be a little over a mile into the trip, and the point of origin of our own **genus**, *Homo*, would be just beyond that. As we got on the highway, accelerated to 60 mph (97 kph), and settled in for the journey, it would take close to 20 minutes into the expedition to reach the time of the Oligocene mammals seen in the rocks of Badlands National Park in South Dakota. About 40 minutes (40 mi or 65 km) into the trip, we would finally reach the time of the extinction of the dinosaurs—the time of *Tyrannosaurus* and *Triceratops* and the Hell Creek Formation of

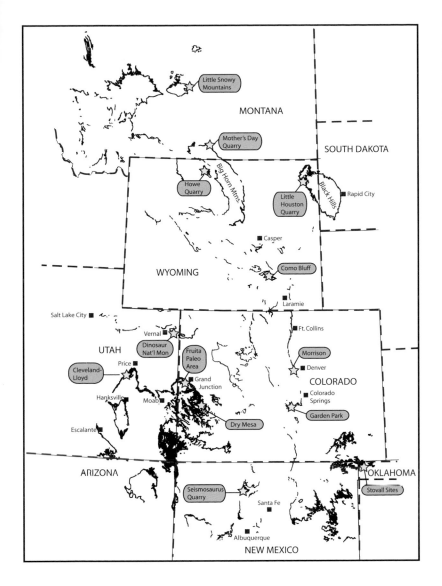

1.3. Map of outcrops of the Morrison Formation in the Rocky Mountain region with some major quarry sites indicated. Outcrops are based on compilations by and courtesy of Kelli Trujillo and Ken Carpenter.

Montana and North and South Dakota. Then, for the next 53 minutes or so, we would watch as the later part of dinosaur time flew by—the time of duckbilled dinosaurs, ceratopsians, and early tyrannosaurids, of inland seaways, marine reptiles, South American titanosaurs, the feathered dinosaurs of China, and the large dromaeosaur *Utahraptor*. For nearly an hour, we would watch as we traveled back in time and the years flew by at a rate of more than 260 centuries every second. Finally, after we'd been on the road for more than an hour and a half, we would arrive at the world of the Late Jurassic, having traveled the Road of the Geologic Timescale for about 93 mi (150 km). Although the Morrison Formation is a relatively thin rock unit and does not seem at first to represent much time, it in fact spans about 7 million years, and it would take us about four minutes to roar down the highway through all of Morrison times. Indeed, by our human perspective, the Late Jurassic was a very long time ago, and deposition of the Morrison Formation was a long process. The time of the Morrison Formation alone was 23 times longer than the current tenure of *Homo sapiens*.

If we continued our trip past the Morrison, through the time of large sand dunes in western North America, and through the time of the carnivorous dinosaur *Dilophosaurus*, it would take us an additional 46 minutes to get back to the time of the origins of dinosaurs and mammals in the Late Triassic. It would be nearly four hours into the trip before we got to the time when the first vertebrates began moving about on land, another half hour beyond that to the time of the first major land plants, and another hour to the time of origin of vertebrates during the Early Cambrian period. To get back to the time of the first fossils, we would now, 5.5 hours into the trip, have to settle in for a 30-hour drive along the Road of the Geologic Timescale—30 hours through the Precambrian at almost 100 million years for each hour of driving. If we started the trip in Los Angeles and headed east, we would get to the time of the Morrison Formation somewhere in the Mojave Desert, but we would need to drive a little beyond Chicago to get to the time of the first fossils, with a year going by with each millimeter of road we traveled. From this perspective, the Late Jurassic was, geologically, "yesterday." On a human scale, however, the depth of geologic time is beyond comprehension.

The Morrison Formation is one of the most widely distributed rock units in North America. Whereas many Mesozoic-age formations are restricted to one or a few US states or have equivalents in other regions that go by different names, the Morrison Formation occurs over an area of 1 million square kilometers and is exposed in eight states in the Rocky Mountain region. In addition to the surface outcrops found in the states mentioned previously, wells drilled in Kansas, Nebraska, Texas, and North Dakota show the formation in the subsurface under these states. Vertebrate fossils are known from the Morrison Formation in each of the eight states in which the unit is exposed on the surface (fig. 1.3).

Wild World of the Jurassic: The Setting

Globally, the Late Jurassic was a time of transition. The continents had, some 100 million years earlier, coalesced into the supercontinent Pangaea. All the continents were joined together in one landmass on one side of the planet and stretched from near the North Pole to close to the South Pole. The rest of the earth was ocean. Sometime in the Late Triassic, this landmass began to break up, and by the Late Jurassic, some of the continents had begun to separate from each other with narrow oceans forming in between (fig. 1.4). In the south, during Late Jurassic time, Antarctica, Australia, India, and Madagascar were together in one mass centered near 60° south latitude and were attached to Africa, which stretched north across the equator. Africa itself was still mostly connected to South America to the west, although the southern Atlantic Ocean was just beginning to form and to separate the two large continents. North of

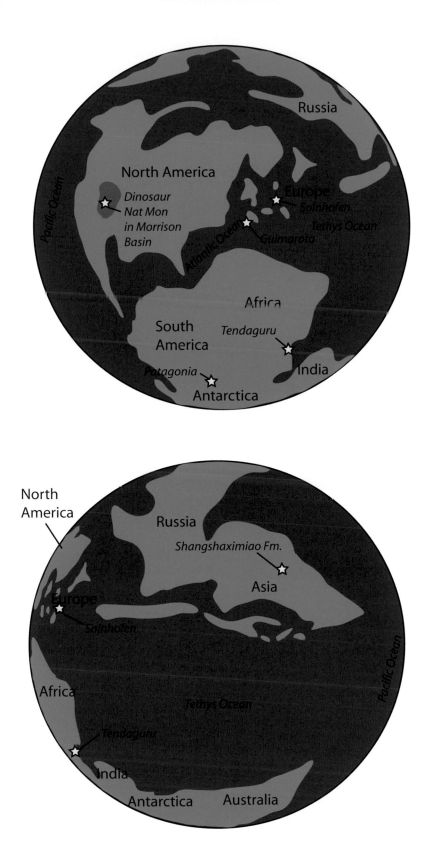

1.4. Globe as it appeared during the Late Jurassic. The Morrison Formation was deposited on a North American continent that was only a few hundred miles from Africa and South America. Europe was a group of islands in the tropical sea. The southern continents were all connected into one landmass. Other Late Jurassic vertebrate faunas indicated include the Tendaguru from Tanzania, the Solnhofen in Germany, those of Patagonia, the Guimarota and others in Portugal, and the Shangshaximiao Formation in China.

Africa, a narrow ocean separated that continent from Europe and Asia, except western Europe, which was still close to Africa. The northern Atlantic Ocean was less than one-third as wide as it is now, at 1,600 km (1,000 mi) across; today the Atlantic is approximately 5,700 km (3,500 mi) wide. Because the Atlantic was so narrow and most continents were still close together, the Pacific Ocean covered an entire half of the planet during the Late Jurassic. The region that is now northern Europe and the United Kingdom contacted North America at Greenland during the Late Jurassic, and as mentioned above, western Europe and northern Africa were nearly in contact as well, so the similarity of faunas from the Morrison, Europe, and to some degree Africa is not surprising. On the basis of geologic evidence, it appears that North and South America may have nearly been in contact as well. But Late Jurassic faunas from South America do not yet appear to be particularly similar to those from North America, Africa, or Europe (although they are not yet as well known as in some other areas), so there may have been some type of geographic barrier separating the animals in South America from other areas.

In North America, the vertebrates of the Morrison Formation were then living between about 30° and 40° north latitude, on a floodplain basin between highlands immediately to the west and far to the east. North America was about 648 km (400 mi) farther south than it is today. To the north, in what is now Canada, there was a shallow continental seaway, and to the southeast was the Late Jurassic equivalent of the Gulf of Mexico, in approximately its present position. Far to the east were the eroding Appalachian Mountains, formed by the initial collision of North America and Africa near the end of the Paleozoic and now overlooking the relatively newly created shoreline of the young North Atlantic. The eastern part of the Morrison Formation appears to have been eroded off sometime during the earliest Cretaceous period, so we know little about what was happening in that region, but the Morrison basin probably continued some distance to the east, because ancient stream channels in the southern part of the Morrison Formation seem to indicate flow right across the outcrop area and continue eastward. If sediments from the east were being washed into the northern Morrison basin, they likely were a minor component relative to that coming in from the west, because stream directions and source areas in the Morrison indicate nearly all the influence coming from the south and west. There are some **clasts** of Pikes Peak Granite in the Morrison along the Front Range of Colorado that indicate small local sources within the basin, but these are minor.

To the west and southwest of the Morrison Formation depositional area, mountains lined a wide swath of the western edge of North America. During the Late Jurassic, as today, the west coast of North America formed what is known in geology as a **convergent boundary**.[2] This simply means that two plates of the Earth's crust are moving toward each other. When the two plates happen to be continental pieces, the result is a collision and uplift of large mountains, as seen in the Himalayas today with the continued pushing of India into Asia. Two plates of oceanic crust can also collide, but the result is often one plate diving under the other. In many cases, however, a convergent margin brings oceanic crust up against continental crust, and in this situation, the denser, cooler oceanic crust almost always will be pushed downward into the Earth's mantle and underneath the overriding continental plate. This is going on today on the west coast of South America and all the way around through the west coast of North America, along Alaska and Russia, and down to Japan. The pushing of the ocean plate under the North American continent during the Late Jurassic (as the two plates moved toward each other) resulted in the formation of mountains along the west coast and for a ways east, and among these mountains were a number of volcanoes. We can see evidence of the volcanoes now only in the ash layers and the altered volcanic ash in muds of the Morrison Formation.[3]

We also see evidence of the volcanoes in some of the granitic rocks of today's Sierra Nevada in California as well as ranges across Nevada, which represent the cooled and now lithified magma that had supplied the volcanoes during the Jurassic (E. Christiansen et al. 2015). The west coast of North America was slightly farther east of where it is now because some of far western California, for example, simply was not there yet during the Late Jurassic. Much of the west coast we see today was actually added on to the continent as volcanic islands (perhaps similar to Japan and Indonesia) that rode the oceanic plate "crashed" into North America. This has been going on along the west coast for some time, but some of the far western edges of the states of California, Oregon, and Washington may well have been added to North America since the Late Jurassic. We do know that at least some of the rocks in the hills along either side of California's great Central Valley farming region were deposited in shallow marine waters just offshore of the coastal mountains (represented by the Sierra Nevada batholith). West of this, many Jurassic rocks of the Coast Ranges seem to derive from oceanic crust that was pressurized, ground up, and scraped off as the Pacific plate slid under the North American plate. Other rocks along the coast, such as those in the hills above the north end of San Francisco's Golden Gate Bridge, appear to be mudstones representing deep ocean bottoms just offshore. From this convergent margin traveling east (during the Late Jurassic), then, there appears to have been a shallow marine environment, then the coast itself, then the western mountains, and then the Morrison basin.

The low but extensive western mountains appear to have had some rain-out effect on the Morrison, in part causing its semiarid climate. Systems moving off the Pacific Ocean from the west had hundreds of miles of low, rolling mountains and foothills to cross before coming to the Morrison basin, and most of the precipitation appears to have been forced out of the storms by the mountains before getting to the Morrison. Most of the water that did come to the Morrison basin was by way of groundwater and surface rivers from the mountains, and most rain that did fall directly on the Morrison seasonally was either from the paleo–Gulf of Mexico or from larger storms off the Pacific. Water draining to the east out of the western mountains carried eroded sediments out onto the Morrison floodplain. The rock types included as pebbles in the Morrison Formation give us an indication of what rocks were exposed in these western mountains and were washing down to the floodplain; most appear to be older **sedimentary rocks** and some metamorphic rocks. Although they were wide, the western mountains were not high in elevation, and it probably did not snow on even the highest peaks.

Compared with today, the most unusual aspect of the Morrison floodplain, as it existed between the mountains and volcanoes to the west and the possibly lower hills much farther to the east, was that it was essentially flat for hundreds of miles in all directions and was also close in elevation to sea level. The only anomaly in this was an area of very low hills in central Colorado that were the last remnants of the Ancestral Rockies. Some granite-derived sediments eroded from these hills were carried by east-flowing streams of the time. Most of the Colorado Plateau, Wyoming, and eastern Colorado (including what is now the Colorado Front Range) was then nearly flat, low-elevation floodplain sloping gently to the east and north. This is drastically different from today, when the Morrison Formation is exposed in the Rocky Mountain region and at rather high elevation. It is important to keep in mind, given the rough similarity of the Late Jurassic west coast of North America to today's west coast, that what is now the Rocky Mountain region was then in almost every respect unlike what we see now.

The geographic and tectonic situation in North America was essentially the same during most of the Jurassic, so that formations representing the early and middle parts of the period in the Colorado Plateau were deposited under basically the same geologic conditions, but climatically, conditions changed drastically from 201 to 145 mya. In fact, the world of the Morrison was very different from that of the first half of the Jurassic. During the Early and Middle Jurassic, the continents of the world were in roughly similar positions, and the west coast of North America was already a convergent margin. The North Atlantic was younger and a little narrower than it was during the Morrison, but otherwise there were few major differences tectonically. Because the climate was rather drier than later, however, the Early Jurassic started with much of the Colorado Plateau covered in large sand dunes (Baars 2000; Fillmore 2011). These are now lithified in the orange cliffs of the Wingate Sandstone, which overlies the Late Triassic Chinle Group. To the southwest of this dune field were sandy **braided streams** represented by the Moenave Formation. After this came a period of more braided streams and mud and the theropod dinosaur *Dilophosaurus* (among other vertebrates), followed by more widespread sand dunes (the Kayenta Formation and then the Navajo Sandstone, which makes up the cliffs of Zion National Park in Utah). After the Early Jurassic had been dominated by sand dunes, the Middle Jurassic was a time of several shallow continental seaways, and these are represented by the Carmel, Swift, and Sundance Formations. These shallow seas came in mostly from the north and northwest and flooded the flats of what was later to become the Morrison basin in the modern Rocky Mountain region. The Entrada Sandstone and overlying Curtis Formation, famous for dinosaur footprints in the Moab, Utah, area, represent a last major deposit of sand dunes followed by a shallow seaway during the Jurassic. To the northwest of the Entrada sand dunes, however, there still existed a shallow continental seaway. Deposition of the Morrison Formation began after the final withdrawal of the interior seaway represented in Wyoming by the Sundance Formation. As the Sundance Sea retreated to the north, it drained the Morrison basin area for the last time in the Jurassic. A shallow continental seaway would not return to the Rocky Mountain region for nearly another 50 million years (Baars 2000).

The retreat of the sea and the beginning of deposition of river **channel sandstones** and floodplain mudstones, as represented by the Morrison Formation, also represented the first appearance of such extensive fluvial deposits in the region since the rivers and floodplains of the Chinle Group gave way in the Late Triassic to the sand dunes of the Wingate Sandstone, a period of nearly 55 million years. The Chinle can look a lot like the Morrison in outcrop exposures, with multicolored, layered mudstones dominating parts of the section. On close inspection of the muds and especially the sandstones, however, the two are quite distinguishable, and although both were deposited by streams, floodplains, and lakes, some environmental differences are apparent. The Southwest during Late Triassic times was still under strong monsoon conditions, and the Chinle plain was crossed by rivers and dotted by lakes much like the Morrison, but the characteristics of the rivers and lakes were different from those of the Morrison, and the source area was to the southeast and flow to the northwest in the Chinle, unlike the Morrison. Still, the onset of Morrison Formation deposition represented a return to landscape conditions most like those of the Late Triassic, which had not been seen on such an extensive scale for tens of millions of years. During that time, Sahara-like sand dune deserts and shallow seaways larger than any seen today had prevailed. As the Sundance Sea retreated and exposed the new Morrison plain, terrestrial animals immediately moved in and set the stage for the next 7 million years in the region.

Stratigraphic Setting

The Morrison Formation is one of many in the Mesozoic section of the Rocky Mountain region. Formations are mappable units of sedimentary rocks that occur in vertical sequences in any one area. Unfortunately for many students new to an area, the names of formations in any one sequence will change as you travel laterally, usually because formations' geographic distributions vary, and thus the number of names involved can become confusing. Stratigraphy is a subdiscipline of geology that deals with identifying and laterally correlating formations and other sedimentary rock units. This section is an introduction to the stratigraphy of the Rocky Mountain region as it relates to the Morrison Formation.

Figure 1.5 presents a general stratigraphy of the Mesozoic section of the Colorado Plateau (Utah and western Colorado) and some surrounding areas (eastern Colorado, Wyoming, and up into Montana), based in part on Imlay (1980) and other sources. It is not complete but gives a good overall picture of the rock units that one sees above and below the Morrison in most areas (and where time is missing from the rock record). The Triassic period saw deposition of shallow marine and intertidal sediments of the Moenkopi Formation on the Colorado Plateau and of the Spearfish Formation in the Black Hills, followed by the river, lake, and floodplain deposits of the Chinle Group, which yielded the small dinosaurs of Ghost Ranch, New Mexico (*Coelophysis*), and the large fossil logs of Petrified Forest National Park in Arizona. By the Early Jurassic, deposition had switched to sand dunes and braided streams, as represented by the Wingate Sandstone and Moenave Formation. More braided streams and floodplains are represented by the Kayenta Formation, famous for producing the type specimen of the carnivorous dinosaur *Dilophosaurus*. Sand dunes returned to the Colorado Plateau with deposition of the Navajo Sandstone, the unit that forms the high cliffs of Zion National Park in Utah. The Navajo Sandstone also includes layers of **limestone** that represent oasis deposits between the sand dunes, and both the sand dune and oasis deposits contain dinosaur and other vertebrate tracks. There are even burrows made by small vertebrates in the Navajo sands. The limestones also yield fossil stumps near Moab, Utah, indicating that the oases were well vegetated.[4] As we move upward into the Middle Jurassic, the Carmel Formation of the Colorado Plateau and the Gypsum Spring Formation of northern Wyoming and South Dakota both represent shallow marine deposition—evidence of yet another shallow continental seaway invading the area. Above the Carmel on the Colorado Plateau is the Entrada Sandstone, another sand dune and marginal marine unit that has produced many thousands of dinosaur footprints around Moab and near Escalante, both in Utah. Interestingly, of all these trackways, only one (possibly two) is not of a carnivorous dinosaur. Above the Gypsum Spring in Wyoming and South Dakota is the Sundance Formation, a dark, silty, and sandy shallow marine unit that has yielded many fossils of oysters and the squidlike animal *Pachyteuthis* from sites near the Bighorn Mountains, Como Bluff, and Devils Tower National Monument, all in Wyoming. In these areas to the north, the Sundance underlies the Morrison Formation. On the Colorado Plateau, there is a thin, shallow marine unit between the Entrada and the Morrison, usually known as the Summerville, Wanakah, or Curtis Formation. In Montana and southeastern Colorado, the formations directly under the Morrison Formation are most commonly the Swift and Bell Ranch Formations, respectively. The Morrison Formation itself is divided into several subunits called **members**, but these are described later. Above the Morrison, the overlying formations are Early Cretaceous in age and are separated from the Morrison by an **unconformity**, a surface between rock layers that represents nondeposition or erosion. Unconformities usually

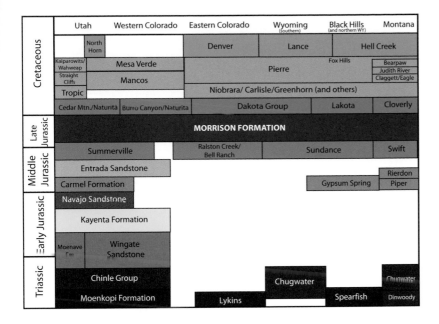

1.5. Mesozoic stratigraphy of the Rocky Mountain region showing the position of the Morrison Formation relative to other formations. Positions of formations are relative and are not exact in age or alignment of boundaries; similarly, not all formations are represented in each section.

also represent significant time gaps in the rock record. That is, several years up to entire periods of Earth history elapse between the end of deposition of the underlying layer and the beginning of deposition of the upper one (or somewhere in that time, any sediment that was deposited was washed away before deposition of the overlying layer). In any case, the amount of missing time represented by the unconformity between the Morrison Formation and the overlying Lower Cretaceous formations is significant and is much more than the small gap between the Morrison and the shallow marine deposits below it. In fact, in many areas, there may be as many as 20 million years missing between the Morrison and the Lower Cretaceous units.

The formations above the Morrison include the Cedar Mountain Formation in Utah (which has yielded a significant fauna in recent years, including *Utahraptor*); the Naturita (formerly Dakota) Formation, Sandstone, or Group (depending on the area) in Arizona, New Mexico, and eastern Colorado; the Cloverly Formation in Wyoming and Montana (which produced the remains of the small dromaeosaur *Deinonychus*); the Lakota Formation in Wyoming and South Dakota; and the Burro Canyon Formation in western Colorado. Most of these formations represent river, lake, and floodplain deposits that are not that different from the Morrison Formation. The exception is the Naturita, which, in parts of southern Utah, includes both coastal terrestrial and shallow marine deposits.

Above these formations are most commonly found the shallow marine mud (or chalk) units of the Upper Cretaceous. These formations include the famous Pierre Shale of the Great Plains states, a gold mine for marine reptiles, large **pterosaurs**, and ammonoids (similar to the chambered nautilus); the underlying Niobrara Formation, also of the

Great Plains; the Mancos Shale of western Colorado and eastern Utah; the Tropic Shale of southern Utah; and the Mowry and Thermopolis Shales of Wyoming. These units represent large continental seaways that covered much of the central part of North America from the Gulf of Mexico to the Arctic. Similar in age to these units but generally along the western edge and representing deltaic deposits much like those currently being deposited along the Gulf Coast of the United States, are the Straight Cliffs Formation of southern Utah and the Mesa Verde Group of western Colorado.

The last stage of Mesozoic deposition in the Rocky Mountain region is represented by the terrestrial floodplain, river, and swampy sediments of the Hell Creek Formation of Montana, North Dakota, and South Dakota; the Lance Formation of Wyoming; the Denver Formation of eastern Colorado; the North Horn Formation of central Utah; the Kaiparowits Formation of southern Utah; and the Kirtland Formation of New Mexico. Broadly speaking, the latest Cretaceous was the time of *Tyrannosaurus*, *Triceratops*, and *Edmontosaurus* and was the final stage of dinosaurian ecosystems. After this time came the Cenozoic and the Age of Mammals.

So the Morrison Formation sits nearly midway through the 160-million-year history of the dinosaurs, approximately 75 million years after the origin of the group but a full 85 million years before the last **species** would eventually become extinct. Think of all the individual animals that lived in all that time!

Common Rock Types

The Morrison Formation consists of a variety of rock types, but most of the section is composed of mudstone and sandstone. Below are short descriptions of the most common rocks found in the formation.

Green-Gray Mudstone

Much of the Morrison comprises layers of gray, green, and green-gray mudstone (fig. 1.6). North of Moab, Utah, the green mudstones of the upper Morrison are particularly bright, while up in Wyoming, the mudstones are more often gray. The muds contain fine-grained sedimentary particles mixed with clay minerals derived from the weathering of different rock types or from the alteration of volcanic ash that fell after eruptions many miles away. The drab colors have been taken as evidence of relatively damp local conditions on the floodplain. A number of vertebrate fossil quarries are known from these mudstones, including the site that yielded the sauropod dinosaur *Brontosaurus* that is now mounted at the Yale Peabody Museum in New Haven, Connecticut.

Red Mudstone

Red mudstone is often one of the most notable rock types seen in the Morrison Formation (fig. 1.6). It can be bright, and its banding in the

1.6. Outcrops of red mudstones and gray-green mudstones in the Morrison Formation of Capitol Reef National Park.

outcrop is visible from afar. The red, maroon, or brownish-red outcrops contain similar mud particles and clay minerals derived from similar sources as the green-gray mudstones, but their reddish color suggests better draining and thus drier, more oxidizing conditions on the floodplain. Many of the bright-red mudstone deposits contain **paleosols**, the preserved remains of ancient soils (Demko et al. 2004; Dodson et al. 1980; Tanner et al. 2014). These paleosol deposits are actually common in the Brushy Basin Member of the Morrison Formation and are sometimes distinguished by abundant root stains, red or green mottling, calcium carbonate nodules, or root casts in their fine-grained sediments.

Fossil quarries are less common in the red mudstones than in the drab mud deposits, but some of the first and largest dinosaurs found in the Morrison Formation came from the red mudstones of Colorado, and there are many other sites. Recent excavations have also uncovered an articulated sauropod skeleton in a mottled red mudstone immediately overlain by a **crevasse-splay** sandstone in the lower Brushy Basin Member of the formation.

Channel Sandstones

Morrison Formation **channel sandstones** are often tan layers 3 m (10 ft) or more in thickness and can be traced laterally for a mile or more. Other channel sands are thin and more laterally restricted. In the southwest outcrop area, the sandstones can be medium to coarse grained or conglomeratic and of a range of mineralogies, whereas sandstones in the northeastern part of Morrison deposition are generally fine grained and with few grains other than quartz. In areas further north and east, the sandstones are also generally thinner and more laterally restricted than in, for example, the Colorado Plateau region.

1.7. Channel sandstone (Cs) in the Salt Wash Member of the Morrison Formation in (A) Utah and (B) western Colorado. These thick, crossbedded sandstone units overlie and often downcut into red, green, and gray mudstone layers (Ms) and may be hundreds of meters to several kilometers wide. Day pack for scale in A (just below Cs label); Josh Smith for scale in B. The arrow in B indicates the approximate flow direction suggested by crossbeds.

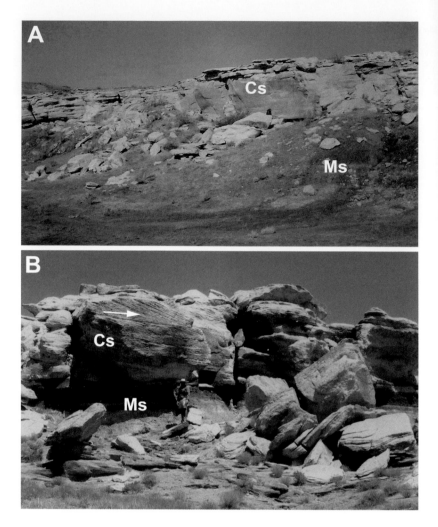

These sandstone units in the Morrison Formation represent ancient river channels (fig. 1.7). Depending on the internal characteristics of the channels in the outcrop, we can determine whether a given layer represents a **meandering river**, an **anastomosing channel**, or a **braided stream**. All three types are present in Morrison Formation deposits. As a general model, the meandering rivers of the Morrison Formation were similar to the modern Brazos River in Texas, and the braided streams were like the South Saskatchewan River in Canada. Meandering rivers tend to consist of single channels that snake back and forth across a floodplain over time. Anastomosing channels are interconnected but often vegetated on the banks and do not migrate laterally. Braided streams are generally straighter but consist of many interconnecting channels that "braid" around sandbars, which are covered during high water flow but otherwise are mostly exposed (Miall 2010).

The thick channel sandstones in the Morrison Formation are commonly **crossbedded** (fig. 1.8). This sedimentary structure type is formed

1.8. Crossbedding (*upper right*) in a pebbly channel sandstone in the Morrison Formation. Crossbeds form as sand grains migrate continuously downstream onto steep, leeward sides of large ripples in river channels, forming multiple sets of diagonally striated rock, like that seen here. Note also the interbedding of sandy and pebbly layers. Scale is 10 cm.

by flow characteristics of the ancient river and can be used to determine the direction of flow. The crossbeds are visible when natural erosion cuts a cross section through diagonal beds in a sandstone, and these diagonal beds were formed by the downcurrent migration of large sand ripples. Most of the channels in two study areas in the Brushy Basin Member of western Colorado appear to have been low gradient and anastomosing and with perennial water flow (Galli 2014), while channels in other areas appear to have been meandering (Derr 1974).

Many important quarries are known from channel sandstones, including the Carnegie Quarry at Dinosaur National Monument, Utah; the Bone Cabin Quarry north of Como Bluff, Wyoming; the Dry Mesa Quarry near Delta, Colorado; the Hanksville-Burpee Quarry in southern Utah; and the Marsh-Felch Quarry at Garden Park, Colorado. These quarries tend to contain many associated to articulated skeletons of a number of different dinosaur taxa.

Limestone

Most limestones in the Morrison Formation are light gray and usually 15 cm (6 in.) to 1 m (3.3 ft) thick. Most of these limestones represent wetlands and lakes on the floodplain, and some contain sponges, stromatolites, fish, and complete or fragmentary freshwater snail fossils, indicative of quiet water. Freshwater and marine limestones are both formed by the biogenic precipitation of calcium carbonate from the water. Some of the thicker limestone beds, in southeastern Colorado and in the northwestern Black Hills, for example, contain many small ooids, rounded concretionary grains of calcium carbonate formed by the oscillating action of shoreline waves. Vertebrate body fossils are not common in the limestones, but they do occur. Finds have included sauropod elements and fish near La Junta, Colorado; sauropods north of Sundance, Wyoming; and crocodilian elements at Garden Park, Colorado.

We have just recently come to understand that some of the limestones in the Morrison Formation represent wetland deposits and that not all were permanent lakes (Dunagan and Turner 2004). The distinction here is that wetlands are seasonally flooded or dry, and their water is derived from groundwater sources more than from the surface. The lake deposit limestones were formed in more perennial bodies of water also supplied by rain and surface runoff. Distinguishing between the two is difficult but can be accomplished through close study of the outcrops and geochemistry of the rocks. The lake limestones should have associated river and delta deposits, whereas those of wetlands would lack these elements.

Dark-Gray Mudstone

In some areas, the top levels of the Morrison Formation contain unusual beds that are dark gray to black (fig. 1.9). These mudstone units represent wet floodplain deposits. Their dark color is caused by high carbon content. The reducing environments of the wet floodplains preserved the carbon of some of the plant material that fell to the ground over the millennia, and the carbon mixed with mud grains and clay minerals, forming the dark mudstone. These beds preserve vertebrate fossils at a few localities, including at Morrison, Colorado, where the type specimen of *Apatosaurus* was found, and at Breakfast Bench, Wyoming, where some of the microvertebrate material came from this lithology. Bones are also found in this dark-gray mudstone on the Colorado Plateau near Blanding, Utah (fig. 1.9). Some of the best plant fossils in the Morrison Formation come from dark-gray mudstone and **coal** deposits near Belt, Montana (see below).

1.9. Dark-gray mudstone in the Morrison Formation. Jim Kirkland, Don DeBlieux, and Scott Madsen (*left to right*) work a dark-gray carbonaceous layer containing bone at a site in the upper Morrison southwest of Blanding, Utah.

Thin Sheet Sandstones

Some sandstone beds in the Morrison are relatively thin and can be restricted laterally. These include crevasse-splay sandstones. Crevasse splays are thin, fan-shaped deposits of sand that form when a river breaks out of its levee and floods the surrounding plains. Although the river may be over its banks and levees all along the flooded stretch, the main flow can cut through and erode a narrow opening in the levee itself (under water), and sandy **bedload** material can spill out in a fan across the muds of the floodplain next to the channel. In the rock record, these crevasse-splay deposits are characteristically thin, laterally restricted, and convex on top, and they may contain current ripple marks. Unlike channel sandstones, they generally are not convex on the bottom (fig. 1.10). Some vertebrate

1.10. Relatively thin (1 m [3.3 ft]), flat-bottomed sandstone bed (ts) in the Brushy Basin Member of the Morrison Formation in western Colorado. Channel sandstones in this region tend to be thicker, coarser grained, and convex bottomed.

1.11. Crossbedded eolian sandstone in the Bluff Sandstone Member of the Morrison Formation near Bluff, Utah. Crossbed sets are approximately 3 m (10 ft) thick.

fossils occur in these deposits, but large quarries are rare. Many of the microvertebrate specimens from the Fruita Paleontological Area are associated with crevasse splays.

Eolian Sandstone

Sand dune deposits occur in the lower part of the Morrison Formation in Wyoming and South Dakota, and in the Salt Wash Member in the Four Corners region. The thickest sand dune deposit in the Morrison is the Bluff Sandstone Member in southeastern Utah. These beds are relatively rare and have not produced any fossils, but they are characterized by having rounded, fine-sized sand grains composed mostly of quartz (with few of the softer or heavier minerals often found in water deposits). The **eolian** sandstones also often have large, meter-scale crossbeds, indicative of the migration of the sand dunes (fig. 1.11).

Pedocal and Paleosol Carbonates

Paleosol carbonates (formerly **caliche**) and **pedocal** soils are calcium carbonate deposits formed below the ground's surface, often by the evaporation of groundwater. Modern pedocal soils are characterized by a layer of calcium carbonate or calcium carbonate nodules below the ground surface. Caliches are solid and hard layers of calcium carbonate. Both are characteristic of semiarid climates. The paleosol carbonates in the Morrison are generally hard, blocky, and gray, and some have been used as marker beds to try to identify time-equivalent horizons in the formation. A number of nodular calcium carbonate layers also exist in the formation, and these may be equivalent to modern pedocal soils. Because of the presumed role of evaporation in their origin, paleosol carbonates and pedocals of the Morrison Formation are thought to be indicative of semiarid climatic conditions, although their formation is thought to involve more the evaporation of groundwater than the drying of soils wetted by rain.

Paleosols in the Morrison indicate a semiarid to tropical wet-dry (monsoonal) climate with fluctuating groundwater conditions and increasing humidity through time (Demko et al. 2004; Tanner et al. 2014).

Chert

Cherts are hard, often colorful, glassy-looking rocks composed of solid silicon dioxide. They can be formed as a result of biochemical processes or other accumulation (often in oceans) or by diagenetic (postdepositional) alteration of limestone. Limestone-derived chert beds are rare in the Morrison Formation, but they do occur. More commonly, chert is seen as reworked clasts in Morrison sandstones.

Coal

Coal is extremely rare in the Morrison Formation and occurs only in one outcrop belt in central Montana, where plant material is common. It is believed that this coal was formed by accumulation of plant matter on the low coastal plain near the northern extent of Morrison deposition, close to the northern seaway into which many Morrison streams drained. Much of the rock in this outcrop area is dark mudstone, and some is coal; both appear to have been formed as a result of slightly cooler temperatures and lower evaporation rates, resulting in wetter soil than further south. This was not exactly a swamp, but it was wetter (at least in the soil) than most areas of the Morrison.

Ash

Most ash that fell on the Morrison floodplain mixed with the muds of the floodplain and was later modified to types of clay by diagenetic processes. But some unaltered volcanic ashes do occur, and these can be composed of small, glassy grains that give the bed an overall white to gray appearance. These beds are rare, and no fossils have been found in them other than a few plants. Although most beds are at least partly decomposed, one section in Utah that is 110 m (361 ft) thick was found to have at least 35 individual ash-derived beds ranging a mere 2.2 million years (E. Christiansen et al. 2015). This would suggest significant eruptions on average every 63,000 years or so.

Laminated Shale and Limestone

Although most mud rocks in the Morrison are soft mudstones, there are some layers that are much harder and are distinctly bedded. These hard **shales** (and rare limestones) are rare and often represent lake deposits, but they have produced fossils. For example, a purple hard **shale** layer in western Colorado (fig. 1.12) has yielded several complete fish skeletons

1.12. Laminated rocks of the Morrison. (A) 40 cm (16 in) thick hard laminated shale lacustrine unit (hls) in the Morrison Formation in Rabbit Valley, Colorado. This shale unit (hls) has yielded fish *Morrolepis*, cf. *Leptolepis*, and "*Hulettia*." (B) Laminated bedding in a shale sample from the unit in A. (C) Laminated bedding from a calcareous lacustrine shale sample from the Temple Canyon sites in Colorado.

of previously unknown varieties, and other laminated layers near Cañon City, Colorado, have produced a lungfish skull and several fish skeletons, plants, and invertebrates. The Brushy Creek plant site in Utah is in a soft, distinctly and finely laminated shale and produces numerous specimens of ferns and *Czekanowskia*.

1.13. Coarse conglomerates from the base of the Morrison Formation at Temple Canyon, Colorado. (A) Basal contact with conglomerate composed of fist-sized, subangular cobbles. (B) Bone fragment preserved in the basal conglomerate. Rock hammer and lens cap for scale in A and B, respectively.

Conglomerate

Although pebbly conglomeratic channel sandstones occur commonly in the Brushy Basin and Salt Wash members of the Morrison Formation, conglomerates with the largest clasts I've seen in the formation are at the base of the Morrison in the Temple Canyon area, near Cañon City, Colorado (fig. 1.13). These conglomerates at the base of the formation include clasts of Pikes Peak granite and other Precambrian rocks that are up to fist size and larger, and in many cases the clasts are rather angular. This particular deposit represents an accumulation of coarse material next to an exposure (perhaps in low hills) of Precambrian rock that was eroding during Morrison times.

Members

1.14. Stratigraphy of members of the Morrison Formation in different areas of the Rocky Mountain region. Inset shows approximate map lines of the color-coded fence sections. Based on Turner and Peterson (2004), Kirkland et al. (2020), and the author's field observations.

If you drive to various outcrops of the Morrison Formation in states of the Rocky Mountain region, you will always see something a little bit different. In some areas, the formation is nearly 300 m (990 ft) thick, while in areas to the north, it thins considerably and can be less than 30 m (98 ft) thick. Similarly, there are varying amounts of gray limestones and of fine-grained sandy to very gravelly sandstone units, all of differing thickness. The distinctive bright-red, green, and gray mudstones, so characteristic of the Morrison, occur everywhere but almost always in different thicknesses and in different orders in each place. These subtle but important differences in the character of the formation in different regions are the basis for dividing the formation into members. **Members** of a formation are usually composed of one or more distinctive lithologic facies (rock types representing specific environments). Several members have been proposed for the Morrison Formation for different areas (fig. 1.14); some have been accepted by nearly everyone, while others have not. The main members in the western and northern Colorado Plateau region of Arizona, Utah, and Colorado are, in ascending order, the Tidwell, Salt Wash, and Brushy Basin. In southeastern Utah, there is a geographically

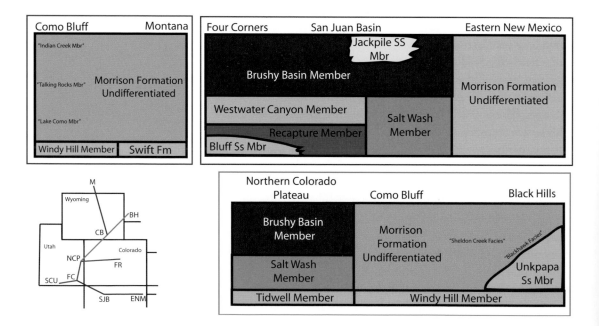

restricted unit at the base of the formation called the Bluff Sandstone Member, and in south-central Utah, near the town of Escalante, there is another geographically restricted unit forming the upper half of the Morrison, known as the Fiftymile Member. The Morrison Formation of the San Juan Basin in northwestern New Mexico contains several members that are exposed only there and, in some cases, just over the state borders of Arizona, Utah, and Colorado. These include the Recapture Member at the base of the formation, the Westwater Canyon Member in the middle of the formation, and the Jackpile Sandstone Member, which is above the Brushy Basin Member in some areas.

In areas further north and east, there are fewer established and named members. In fact, the Morrison Formation at the type area along the Front Range of Colorado is undivided, except into informal units by some workers. The upper beds of formations underlying the Morrison have recently been labeled as being essentially in the Morrison by some workers as a result of a major unconformity identified at their bases. These include the Windy Hill Member of the Sundance Formation in Wyoming and South Dakota and the Ralston Creek Formation in the Colorado Front Range. In central Wyoming, the main Morrison rocks have been divided into the Lake Como, Talking Rocks, and Indian Fort Members, and in the Black Hills, a geographically restricted unit known as the Unkpapa Sandstone has been included in the Morrison Formation. Figure 1.14 summarizes the members present in different major outcrop areas of the Morrison Formation.

Tidwell Member

The Tidwell Member consists of red and gray mudstone, gray and light-brown sandstone, gray limestone, and some white gypsum. It is exposed at the bottom of the Morrison mainly in the western and northern Colorado Plateau region (fig. 1.15). It is most often 8 to 23 m (25 to 75 ft) thick, although it can be much thicker in areas where it fills in paleovalleys that existed on the pre-Morrison plain. The bottom contact of the member is marked by colorful chert pebbles or a zone of chert concretions, by a white gypsum marker bed, or by an angular unconformity in which the horizontal beds of the Tidwell overlie slightly tilted and truncated beds of the underlying Summerville Formation. The upper contact between the Tidwell and the overlying Salt Wash Member is gradual but can be placed at the lowest laterally continuous tan sandstone bed, indicating the lowest unit in the Salt Wash. The gradual nature of the upper contact is a result of interfingering of the Salt Wash channel sandstones and the mudstones of the Tidwell. This can be seen at several locations, including south of Moab, Utah. On the west side of the San Rafael Swell, south of Ferron, Utah, and within sight of Interstate 70, the Tidwell is particularly thick and is directly overlain by the Brushy Basin Member (with the Salt Wash missing).

1.15. Exposures of the Tidwell Member of the Morrison Formation with the overlying Salt Wash Member and underlying formations at (A) the west slope of La Sal Mountains, Utah; and (B) Colorado National Monument near Grand Junction, Colorado.

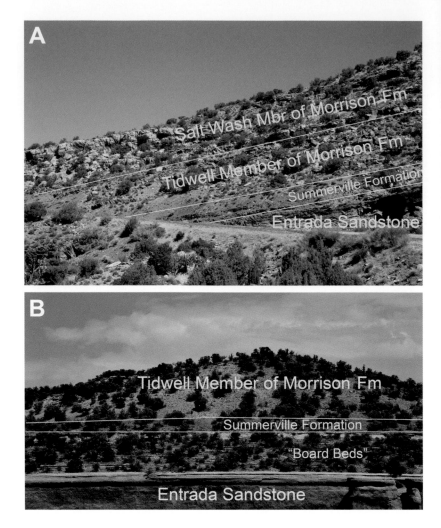

The Tidwell Member represents mudflats, lakes, some eolian dunes, and narrow river channels of the lateral and distal facies of the large fan of rivers later represented by the Salt Wash Member (with rare marine beds near the base) (F. Peterson and Turner-Peterson 1987). As water and sediments washed out of western mountains, the muds and lake limestones of the Tidwell were deposited ahead of and between the main, large river channel complexes, which later brought coarse-grained sand and gravels to the Colorado Plateau area. The Tidwell may be roughly equivalent to the Ralston Creek Formation, which lies below the undifferentiated Morrison along the Front Range of Colorado.

Fossils are not common in the Tidwell Member, but several track sites are known, indicating the presence of pterosaurs and sauropods. Similarly, the few body fossils include the first North American sauropod (*Dystrophaeus*) and several elements of the **pterosaur** *Utahdactylus*, both from near Moab, Utah. Recently, a lungfish **tooth plate** was found in what has been mapped as the Tidwell Member in Colorado National Monument near Grand Junction.

Windy Hill Member

The Windy Hill Member is a shallow marginal marine unit and was originally described as the uppermost member of the Sundance Formation, which underlies the Morrison Formation in most of Wyoming, South Dakota, and north-central Colorado. Some workers have identified an unconformity (essentially a time break) at its lower boundary and thus separated it from the Sundance and put it in with the Morrison Formation (F. Peterson and Turner 1998). The sea represented by this unit was to the north of the main Morrison basin, and the Windy Hill is roughly equivalent in time to the Tidwell Member.

Bluff Sandstone Member

The Bluff Sandstone is best exposed near the town of Bluff, Utah (fig. 1.16), where it is up to 92 m (302 ft) thick (O'Sullivan 2010). It thins to the north near Blanding and extends south into northeastern Arizona. In southwestern Colorado, the Bluff is equivalent to the Junction Creek Sandstone. The Bluff Sandstone is up to 105 m (344 ft) thick and consists of fine-grained sandstone with abundant, large crossbeds, although some rare parts of the Bluff are flat bedded. The Bluff intertongues with the Tidwell and Salt Wash Members to the west, and in northeastern Arizona, the Bluff rests on an angular unconformity that cuts across warped bedding of the underlying Summerville Formation. The Bluff Sandstone Member represents sand dune deposits with some interdune sand flats. The dune field occupied what is now the southeastern corner of Utah and formed downwind of a low-lying structural high on the early Morrison alluvial fan complex to the west. Rivers probably flowed around this high and left it dry to deflate by wind action and have its sand blown eastward into the Bluff dune field (F. Peterson and Turner-Peterson 1987). No vertebrate body fossils have yet been found in the Bluff Sandstone

1.16. Eolian Bluff Sandstone Member of the Morrison Formation sitting on the Summerville Formation near Bluff, Utah. Early in Morrison times, this area was covered in sand dunes.

Member, but tracks of several theropod dinosaurs are known from just 2 m (6.5 ft) above the unit west of Blanding, Utah.

Salt Wash Member

Approximately the lower half of the Morrison Formation in much of the Colorado Plateau region is composed of the Salt Wash Member. This unit comprises rocks of three major alluvial complexes and is distinguished from the overlying Brushy Basin Member in most areas by its distinctly higher percentage of sandstone. In the 1950s, 1960s, and 1970s, the Salt Wash Member was actively mined for its uranium and vanadium as the demand for nuclear weapons and power plant material expanded. The area around the Four Corners was particularly heavily worked, and most geologic members of the Morrison were involved. The Salt Wash Member alone produced more than 45 million kilograms (99 million pounds) of uranium between 1947 and 1982 (Chenoweth 1998). One result of this work was that jeep trails were put into many otherwise inaccessible areas, and these trails are also useful for paleontological work. Calling some of the 60-year-old roads "accessible" may be somewhat of a stretch, however, and they can make for some interesting driving: the roads often go up and over the characteristic large sandstone benches of the unit through piñon and juniper country. I once got airborne (at low speed, no less) and nearly ran over a field assistant going over one of these benches in a four-wheel-drive vehicle on Deerneck Mesa in Utah. The same sandstone that yielded the uranium still produces dinosaur tracks and bones in this region, but the member can be particularly difficult to work precisely because of the thick, hard sandstone benches. Far more quarries are known from the Brushy Basin Member and its equivalents, but there are likely plenty still undiscovered in the Salt Wash. Recent work in the Salt Wash has suggested that fossil bones are actually relatively abundant in the upper half of the Salt Wash, at least in Utah, but because the material is almost always in cliffs and benches of sandstone it is very difficult to remove.

The Salt Wash Member is up to 122 m (400 ft) thick in the southwestern part of the Morrison outcrop in southern Utah, near one of its source areas. To the northeast in northwestern Colorado, it is much thinner—about 30 m (98 ft). The member comprises tan and white, cliff-forming sandstone and conglomerate (fig. 1.17), often weathered brown or black and accumulating a coating of desert varnish, which consists of manganese and clay minerals precipitated and attached to the rock surface by bacteria. Between the layers of sandstone are thinner layers of red, green, and gray mudstone with some thin siltstones. The sandstone is generally medium to coarse grained, and the conglomerates can include thick, pebbly layers within each bed set. Most sandstone units also contain many crossbeds, which are generally seen in the outcrop as repeated diagonal lines in the sandstone bed, highlighted by textural or mineral differences in the rock.

Crossbeds represent the preserved migration of sandbars or large or small ripples—in this case within the ancient rivers. The crossbeds in the sandstones of the Salt Wash Member are several centimeters to several meters in height. The characteristics of the sand and conglomerate and the size, type, and orientation of the crossbeds give clues to the types of rivers that existed at the time.

The Salt Wash Member represents a distributive fluvial (river) system that flowed generally east across the Morrison Formation floodplain (Tyler and Ethridge 1983; F. Peterson and Turner-Peterson 1987; Weissmann et al. 2013). There were many source areas for the water and sediment that washed down from the mountains and formed the Salt Wash streams and rivers. At the time, there were two major mountainous regions to the west and southwest: the Sevier Highlands in what is now Nevada and western Utah, and the Mogollon Highlands in southern Arizona. Water runoff from these mountains flowed down and spilled out from several major outlets onto the Morrison plain, fanning out into a complex of connected smaller rivers as they went. At one point during deposition of the Salt Wash, there was significant outflow from an area near modern Lake Powell that fanned sediment and rivers out into much of southern Utah; most of these rivers flowed east and northeast. Another alluvial fan system flowed east from an area of what is now western Utah. In some areas of Utah even today, the sinuous channels of the Salt Wash Member can be seen from above in aerial or satellite photos (Heller et al. 2015).

These alluvial systems flowing out of the western and southern mountains contained a variety of river types. Some river channels contained mostly sand bottoms; some were mostly gravelly. Many were a mixture. There were river channels that were more or less straight, contained within leveed banks but with variable flow. These rivers would have, at low water, a single meandering channel surrounded by sandbars within the confines of the levees. During increased runoff, the rivers would fill the whole channel and sometimes overflow the levees in a major flood. Sometimes the rivers would reroute themselves across a different part of the floodplain when they overflowed the levees and the main channel of flow directed itself across a new track of the floodplain, a process known as **avulsion**. These major river channels are usually represented by medium to coarse sandstones and conglomerates that have large crossbeds and are up to 20 m (66 ft) thick and laterally continuous for hundreds of meters.

The Salt Wash also contains thinner sandstone units that have smaller crossbeds. These crossbeds indicate flow that is much more variable in direction. The sizes of the sand grains and of the crossbeds in these units become smaller going upward through the bed. These units represent meandering rivers that perhaps were tributaries to the larger channels described above (Robinson and McCabe 1997). These meandering rivers probably had a degree of variable discharge like the main channels but had much higher sinuosity.

There were also, in the Salt Wash alluvial system, many sandy braided streams that had relatively low sinuosity and flowed across much of the middle part of the floodplain. These rivers were shallow and wide and less confined within leveed banks. The relatively straight track of these rivers flowed most of the time as a complex braid of connected channels separated by sandbars, only filling the whole expanse of the river during floods.

A significant percentage of the surface runoff from the western mountains sank in along the western edge of the Morrison basin and fed the groundwater aquifer under the plain. The depth to groundwater was not great in most areas of the floodplain, and groundwater flow was to the east. There were numerous perennial river channels, mainly those draining large areas. The streams of smaller drainage areas were likely seasonal, but because of the high water table, holes dug in the dry streambeds would likely have produced water.

Thin lenses or sheets of sandstone in the Salt Wash represent crevasse-splay deposits. Crevasse splays occur when a flooding river overflows its banks or levees and a thin sheet of coarser sediment (usually sand) flows out onto the mud of the floodplain. These fan-shaped wedges of sediment appear in the outcrop as thin sandstone units, often with a flat bottom surface that contrasts with the scoured surface created by river channels.

In the distal parts of the alluvial system to the north and east (away from the source areas), most of the streams were meandering, like the tributaries further southwest. The mudstones between sandstone beds in the Salt Wash are red, green, and gray. These muds represent well-drained and poorly drained floodplain deposits, ponds and small lakes, and abandoned river channel fills. In some places, carbonaceous plant material occurs in the mudstones.

The mountains to the west and southwest appear to have been composed of a significant amount of sedimentary rock themselves. The sand grains and clay in the Salt Wash sandstones, along with clasts in these units such as chert and quartzite pebbles, demonstrate that the rocks eroding up in those mountains and washing down onto the Morrison floodplain were even more ancient lithified sediments. Some clasts in the Salt Wash Member have been identified as eroded fragments of Paleozoic-age carbonates, and some of these contain fossils—fossils that would have been such even to the animals of Morrison times.

Characteristics of some of the sand grains in the Salt Wash suggest that Lower Jurassic sand dune deposits may also have been eroding to the west and having their reworked sand carried into the Salt Wash fluvial system of the Late Jurassic. The Navajo Sandstone and Aztec Sandstone of Nevada and California, which were both already 40 million years old by the Late Jurassic, perhaps were eroded and had their sand washed into the Morrison basin. Interestingly, there are small eolian deposits in the Salt Wash that indicate a prevailing wind from the west. Although the Lower Jurassic material has not actually been found in these eolian units in the Morrison, it is entirely possible that sand from the Early Jurassic dunes of the Navajo or Aztec was eroded from the west, washed into the Morrison system, and reincorporated into a "new," Late Jurassic–age sand dune.

Bentonite, altered volcanic ash, is common in many Morrison mudstones, including some of those of the Salt Wash (although it is much more abundant in the overlying Brushy Basin Member) and indicates that there were also volcanoes in those mountain ranges to the southwest and west. Indeed, some of the intrusive igneous rocks of the eastern Sierra Nevada range and of Death Valley in California, as well as in a number of ranges across Nevada, are Late Jurassic in age and provide additional indirect evidence of the presence of volcanoes to the west of the Morrison floodplain (E. Christiansen et al. 2015).

There are only a handful of fossil quarries in the Salt Wash Member (and equivalent, unnamed levels), compared with the number that occurs in overlying units. A few important collections are known, however, including the type and referred specimen of a new species of the carnivorous dinosaur *Allosaurus* and several specimens of the sauropod *Haplocanthosaurus*. These two species seem to be some of the only dinosaurs in the Morrison Formation that are restricted to the lower levels of the unit, as we will see later. The Jensen-Jensen Quarry near Dinosaur National Monument in Utah has produced parts of a *Brachiosaurus* and is in the Salt Wash Member. An important quarry for *Stegosaurus* and *Camptosaurus* remains, Quarry 13 at Como Bluff, is not in the Salt Wash but is from a roughly equivalent level.

What the Salt Wash really has in abundance is footprint localities— dozens of them. The record of footprints in the Morrison Formation is reasonably extensive, mostly because of the number of sites in the Salt Wash of the Colorado Plateau. Whereas the amount of sandstone in the unit causes trouble for finding and excavating dinosaur skeletal material (because of cliffs, hard overburden, talus, and vegetation), it makes prime territory for finding tracks. Extensive bedding surfaces of sandstone preserve track impressions and many trackways, and sandstone overhangs above mudstones preserve abundant natural casts of sauropod and other tracks. Even mine ceilings are known to preserve tracks within sandstone units. There are also traces of possible termite nests and crayfish burrows. The Salt Wash Member should someday yield important skeletal material, but in the meantime, it forms the core of our knowledge of Morrison Formation vertebrate (and invertebrate) ichnology.

Brushy Basin Member

Overlying the Salt Wash Member throughout much of the Colorado Plateau is the other widespread member of the Morrison Formation, the Brushy Basin. This is a slope-forming unit consisting of a large percentage of mudstone (fig. 1.18)—far more mudstone and less sandstone than in the underlying Salt Wash. The Brushy Basin mudstones are red, grayish purple, maroon, gray green, green, and light to dark gray in color, and at many levels, the rock weathers to a texture that can only be described as

1.18. Mudstone outcrops of the Brushy Basin Member of the Morrison Formation. (A) Near Colorado National Monument and Grand Junction. (B) High above the Colorado River in Horsethief Canyon, Colorado.

mud popcorn. Traversing these units can be like walking on ball bearings when they are dry. The mudstone becomes singularly slick and sticky when wet. The reason for this strange and sometimes dangerous texture is that much of the mudstone in the Brushy Basin Member is composed of bentonite, a type of altered volcanic ash that has, as its main constituent, a clay mineral called montmorillonite. Among the four groups of clay minerals, montmorillonite is the main member of the smectite group, a collection of similar minerals containing aluminum, magnesium, silicon, oxygen, and hydrogen. Clay minerals consist of two- or three-layer sheets of silicates weakly bonded together, and montmorillonite and other smectites are three layered. These clays have the ability to absorb large amounts of water, which, along with their sheetlike microstructure, makes for the slick mud that the bentonitic parts of the Brushy Basin become after a heavy rain. The clay minerals also expand when they absorb water, and this leads to the popcorn texture of many of the Brushy Basin mudstone outcrops (fig. 1.19). There are some layers, particularly low in the member, that are not composed of bentonite (and thus lack montmorillonite and its expanding character).

The Brushy Basin is not all mudstone, however. Thin gray limestones are rare but can be seen in the member; and laterally relatively restricted—but in places fairly thick—tan channel sandstones are not uncommon. Most of these sandstones are medium-fine to coarse grained, but some can be very gravelly and conglomeratic. In the San Juan Basin of New Mexico, the Brushy Basin Member can be up to 165 m (540 ft) thick, although in most areas of the Colorado Plateau, it is about 91 m (300 ft) thick. A study of the characteristics of these river channel sandstones in the Brushy Basin and the Salt Wash found no real differences between them (other than how much there was in one versus the other). So there doesn't appear to be much change in the river systems from Salt Wash to Brushy times; the difference seems to be an increase in the input of volcanic ash to the floodplain muds of the Brushy Basin Member (Heller et al. 2015).

Most of the deposits of the Brushy Basin Member represent floodplain muds mixed with volcanic ash. These muds were the flat parts of the plain, away from or along the river channels and small lakes represented by the sandstones and limestones, respectively. The mudstones were deposited during floods when fine sedimentary particles settled out of the quiet waters away from the channel. As water levels dropped and the river returned to the confines of the channel, the newly deposited mud was added to the floodplain. Some of these muds sat in low areas of the floodplain that were kept wet by high groundwater tables or by occasional rains, and other areas of the floodplain were better drained and stayed dry more of the time. Drier parts of the plain may have been away from rivers because the water table there would not have been as elevated as it was close to flowing surface water. The former types of mud tend to be preserved as green and gray mudstones, and the latter as reddish mudstones

1.19. Popcorn weathering texture of mudstones with a high smectite content in the Brushy Basin Member of the Morrison Formation near Moab, Utah (A), and at the Riggs Quarry 13 area, Grand Junction, Colorado (B).

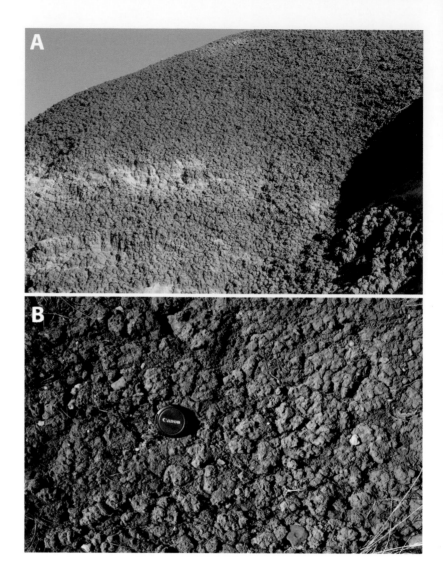

in the Brushy Basin. Some of the better-drained muds often began to form soil horizons, which are preserved in the rock record as well.

The sandstones of the Brushy Basin Member represent mostly sandy to gravelly meandering river deposits (high sinuosity), although some braided streams existed too. One study of these meandering river units suggested that the discharge of the rivers could have ranged from 8 to 120 m³/s (300 to 4,300 cfs), and the rivers' gradients may have been about 38 cm/km (2 ft/mi) (Derr 1974). These flow characteristics are roughly equivalent to those of low flow on the Colorado River near the Colorado–Utah border today.[5] As with the Salt Wash Member, there are places in Utah where the paths of channels of these Brushy Basin rivers can be seen in aerial or satellite photos (Williams et al. 2011). The meandering and braided streams of the Brushy may have given way to anastomosing rivers downstream to the north (Cooley and Schmitt 1998). An anastomosing river consists of multiple stable channels that are interconnected and do not migrate laterally like those of a meandering river.

The Brushy Basin Member appears to represent an interval of at least 2 million years, from 152 or more mya to approximately 150 mya (Trujillo and Kowallis 2015); some of the uncertainty in the older age stems from the fact that we have few good samples from low in the member. Part of the uncertainty in the ages of the members of the Morrison Formation also results from the fact that few individual sections have more than one sample taken from them (often one from the base or one from the top, for example); a recently analyzed section with three dates preliminarily suggests that the upper Westwater Canyon through upper Brushy Basin Members range close to 4 million years.

During the time of deposition of the upper part of the Brushy Basin Member, there existed a large, shallow basin in which evaporation was particularly high. The normal eastward flow of groundwater in this area was blocked by a low upwarp of basement rock near the western edge of what is now Colorado's Rocky Mountains. If this upwarp broke the surface, it did not do so significantly, so surface-water flow was unimpeded, but the high water table caused by the blocked groundwater resulted in high evaporation up through the soil in the basin. (As mentioned above, there seem to have been a couple of isolated high points just east of this basin area where basement rock was eroding; the Morrison Formation near Cañon City, Colorado, contains large cobbles of the nearby and very ancient Pikes Peak granite [pers. obs. and C. Turner, pers. comm., 2004].) Because of the semiarid climate of the time, volcanic ash input, and the high evaporation rates, this basin did not fill up into a large lake but rather became an alkaline/saline, seasonally wet-dry open wetland plain, crossed by intermittent streams and dotted with ponds that dried up completely during long dry times of the year (Dunagan and Turner 2004). This area was in what is now northwestern New Mexico and southwestern Colorado and also included some of Arizona and Utah. It has been named the T'oo'dichi basin, from a Navajo word for "bitter water." The basin was identified on the basis of the alteration of the clays within it (caused by evaporation) versus those around the perimeter. The moisture content of the mud in the basin was influenced seasonally by intermittent stream runoff, but most often it was controlled by groundwater. The surface would have been a wetland much of the time but was likely dry for long periods. Most of the river waters that flowed in either evaporated or sank into the ground, because deltaic deposits (where a river would have drained into a standing pond) are rare. During the dry season, there may have been, in the center of the basin, both damp, muddy areas still fed by groundwater and small, better-drained, dry lake deposits similar to but smaller than those seen in the Great Basin deserts of the western United States today. Because of the evaporation and the chemistry of the clays in the basin, the standing surface water, like that in the soil, may have been saline and of high pH. There was a constant input of volcanic ash to the basin, blown in on winds from the west, contributing to the alkaline/saline conditions (F. Peterson and Turner-Peterson 1987; Turner and Fishman 1991). The wetlands of the Brushy Basin Member and its equivalent levels were an important part of the Morrison Formation landscape. Those of T'oo'dichi were similar to others in the eastern part of the Morrison outcrop area (near the modern Front Range, for example) but were generally drier with more alkaline/saline water. Authors of some other studies on the geochemistry and paleosols of the Morrison in this region, however, have been unconvinced of the evidence for the Lake T'oo'dichi system (Tanner et al. 2014; Galli 2014), so the topic remains a bit contested.

Modern prairie playas of the Great Plains, or possibly vernal pools, may provide a good smaller-scale model of how the seasonal wetlands in the center of the Morrison's T'oo'dichi basin and other areas may have appeared. The stretch of plains from west-central Texas up through eastern Colorado and into Nebraska today is dotted with circular wetlands that are shallow and temporarily hold surface water from a small, internally drained basin. Rainwater drains to the shallow ponds and then both evaporates into the atmosphere and sinks into the ground, where it joins the local aquifers (L. Smith 2003). Although Morrison Formation wetlands were structurally similar, they were hydrologically different in that their water levels were determined mostly by groundwater fluctuation and less by precipitation. In modern wetlands of the Great Plains, rainwater floods the small basin and then drains down to a deep water table. In the Morrison system, by contrast, standing water was usually present seasonally as a result of some rain but predominantly because of a high water table, and this water would evaporate.

Although the small wetlands of the modern Great Plains are dry during much of the year and then flooded for periods, vegetation grows all around and within the pond areas. Only small areas in the center are clear of plants, so that when dry, the wetland looks much like a meadow. When flooded, the pond has some **aquatic** plants that rise above the surface. Most of the plants growing in the pond area are necessarily adapted to wet-dry cycles and thus are different from those growing on the surrounding plains outside the pond area. Invertebrate animals are common in the playas but mainly appear when the pond has standing water, and because of the ephemeral nature of the pond, the playas generally lack fish. Frogs and salamanders, however, are relatively common because these **amphibians** can remain dormant in the dried mud between wet periods. Other vertebrates living in or near the modern playas of the Great Plains include turtles, lizards, and small mammals.

The area of west-central Texas, where modern playas are so common, currently receives approximately the same amount of annual precipitation that has been estimated for the Morrison basin 150 mya (based on computer modeling). These modern playas may indeed be useful for picturing the environment for the wetlands of the Brushy Basin Member. The plants, the hydrologic mechanisms, and some of the animals were

Cedar Mountain Formation
Poison Strip Mbr
upper Yellowcat Mbr
lower Yellowcat Mbr
Brushy Basin Member
Morrison Formation

1.20. Upper Jurassic–Lower Cretaceous contact near Cisco, Utah. The Morrison Formation slopes that are visible comprise only the Brushy Basin Member. A thick calcium carbonate nodular layer occurs in the lower Cedar Mountain (left arrow, between the upper and lower Yellowcat), and contact (thick yellow line) occurs below this calcium carbonate layer, just above the well-developed paleosol at the top of the Morrison Formation (right arrow).

different, but the structure of the wetlands was likely much the same. We probably would have seen a huge, flat expanse, mostly covered in low-growing vegetation, that was dotted with flooded and plant-choked ponds during the wet season. In the T'oo'dichi basin, the wetlands would have been a bit drier and less vegetated than in other areas of the Morrison.

The lateral equivalents of the Brushy Basin Member to the north and east are also mostly mudstone, and most quarries in the Morrison are in either the Brushy Basin or similar levels. Thus, most of what we know of the vertebrates of the Morrison Formation comes from this member. Dinosaur National Monument's Carnegie Quarry, the Dry Mesa Quarry, the Cleveland-Lloyd Quarry, and the Fruita Paleontological Area micro-vertebrate quarries all occur in the Brushy Basin Member, and the Cope quarries at Garden Park and many of the quarries at Como Bluff, including Quarry 9, come from levels probably equivalent to the Brushy Basin.

The upper contact of the Brushy Basin Member (and thus the Morrison Formation) is with the Cedar Mountain Formation (Lower Cretaceous) in most of the Colorado Plateau (fig. 1.20). The Cedar Mountain Formation has proved to host its own rather impressive faunas of dinosaurs and other animals, although these are quite distinct from that of the Morrison.

Fiftymile Member

The Fiftymile Member is restricted to southern Utah and is best exposed south of the town of Escalante (F. Peterson and Turner-Peterson 1987). It consists of interbedded sandstone and mudstone and is up to 107 m (351 ft) thick. The unit is laterally equivalent to the Brushy Basin Member and sits above the Salt Wash Member. It contains many more and laterally more continuous sandstone beds than the Brushy Basin, but it has slightly less sandstone than the Salt Wash. It is also less cliff forming than the

1.21. The Recapture and Westwater Canyon Members of the Morrison Formation northwest of Bluff, Utah.

Salt Wash. This member represents an alluvial complex, similar to but smaller than the Salt Wash one, that existed during Brushy Basin times. The source area of this complex was probably, like that of the Salt Wash, to the southwest. No significant vertebrate specimens have yet been found in the Fiftymile Member.

Recapture Member

The Recapture Member is restricted to the San Juan Basin of northwestern New Mexico and adjacent parts of the Four Corners (Condon and Peterson 1986). The rocks exposed in this unit are a mix of interbedded sandstone, siltstone, limestone, and mudstone, generally tan, reddish brown, green, or gray in color (fig. 1.21). The two types of sandstone in the member represent river and sand dune deposits, and these facies are interbedded in the western part of the San Juan Basin. Elsewhere, the dune deposits lie underneath those of the rivers. The lower part of the member is equivalent to the Salt Wash Member, and the Recapture includes generally less sandstone than the overlying Westwater Canyon Member. The Recapture reaches a maximum thickness of 152 m (498 ft). Few significant vertebrate fossils are known from this member.

Westwater Canyon Member

The Westwater Canyon Member is also recognized mainly in the San Juan Basin of New Mexico up into southeastern Utah and consists of cliffs of crossbedded sandstones (fig. 1.21; Turner-Peterson 1986). The member is up to 110 m (360 ft) thick, and the sandstones are reddish brown and fine to medium grained (although local conglomerates do occur). The reddish-brown color of the sandstones is due to postdepositional alterations in the chemistry of the rock. The Westwater Canyon Member also includes a minor component of red and green mudstone between many of the sandstone layers. Some of these units have thin limestone beds as well, and the mudstone facies probably represent **overbank** and pond deposits. This member lies below the Brushy Basin Member in many areas, but it has been shown to interfinger with the Brushy Basin on its northwestern edge or to be laterally equivalent to the Salt Wash Member (O. Anderson and Lucas 1998). The Westwater Canyon Member represents the deposits of sandy braided streams that flowed east and northeast across the floodplain, and the source area mountains to the southwest appear to have contained a variety of rock types. On the basis of studies of pebbles found in the ancient stream deposits, it is apparent that sedimentary, metamorphic, and both intrusive and extrusive igneous rocks were eroding from the mountains, carried by the rivers. Some of the sedimentary rocks that were exposed in the mountains, and whose fragments were being washed out into the Morrison basin, were Paleozoic limestones containing crinoid and coral fossils. Late Jurassic fossils in the Westwater Canyon Member include logs and fragmentary dinosaur bones in New Mexico and southeastern Utah, although no quarries are known.

Jackpile Sandstone Member

The Jackpile Sandstone Member was named for a uranium mine and is exposed mostly in the San Juan Basin of northwestern New Mexico (Owen et al. 1984), although it also occurs in sections to the east in the state, out toward the Great Plains (Lucas 2018). It can be up to 91 m (300 ft) thick, although in most areas it is 30 m (100 ft). It consists of off-white to yellowish-tan, medium- to coarse-grained, crossbedded sandstone representing northeast-flowing, low-sinuosity braided streams on a distal alluvial fan. There are some minor interbedded thin lenses of pale-green to red mudstone, and crossbeds in the sandstones are up to 1 m (3.3 ft) thick. This member is at the very top of the Morrison Formation, locally above the Brushy Basin and interfingered with it to the northwest. The only fossils known from the Jackpile Sandstone are some invertebrate trace fossils and dinosaur bone scraps.

Lake Como, Talking Rocks, and Indian Fort Members

Unfortunately, not even the most extensive members of the Colorado Plateau region, the Salt Wash and the Brushy Basin, can be correlated to the Morrison Formation in Wyoming or farther north or east. The Morrison along the Front Range of Colorado (and to the southeast in Oklahoma) remains undivided into members, although the Brushy Basin of the Colorado Plateau is likely equivalent to the upper Morrison along the Front Range. In part, the problem is that in eastern Colorado, there is no distinctive equivalent of the abundant sandstone beds of the

Salt Wash Member. A similar problem exists in Wyoming, where there also is no Salt Wash equivalent. Work by Robert Bakker and Al Allen in southeastern and central Wyoming has, however, led them to propose three members for the Morrison in this area. In ascending order, these are the Lake Como, Talking Rocks, and Indian Fort Members (A. Allen 1996). The Lake Como Member is characterized by lacustrine limestones and crevasse-splay sandstones in the Como Bluff area and similar beds plus eolian sandstones to the north. The top of the unit is defined by the "boundary caliche layer," which is described as a datum plane caused by a regional aridity event that produced evaporative calcium carbonate deposits. An important quarry in this member at Como Bluff is Quarry 13, which produced a number of *Stegosaurus* and *Camptosaurus* skeletons in the 1800s. The overlying Talking Rocks Member is composed of red and gray floodplain mudstones and fluvial sandstones; the mudstones are largely smectitic at Como Bluff. The Indian Fort Member, the uppermost unit, comprises dark-gray mudstone with some limestones and channel sandstones; the mudstones are mostly nonsmectitic. This is the level of the famous Breakfast Bench quarries at Como Bluff.

Unkpapa Sandstone Member

In the Black Hills of South Dakota and Wyoming, the Morrison Formation is thinner than in most of the Colorado Plateau. In some parts of the Black Hills, it is only 24 m (80 ft) thick. For many years, the only rocks included in the Morrison Formation in this region were fluvial and floodplain in origin: floodplain mudstones, limestones deposited in small ponds, and river sandstones. In 1981, however, South Dakota School of Mines and Technology geologists George Szigeti and Jim Fox proposed that a local sand dune deposit, the Unkpapa Sandstone, which directly underlies the traditional Morrison rocks in the eastern and southern Black Hills, should properly be included as a basal member of the Morrison in this region (Szigeti and Fox 1981). If true, this suggests that in the early years of Morrison deposition in the area, there was a field of sand dunes blowing in from what is now western Nebraska, with lake, river, and floodplain muds surrounding it. The source of the sand may have been uplifted elements of the Sundance Formation, whose sand grains eroded and were blown north.

The Morrison Formation in the Black Hills is actually a productive area for vertebrate fossils (Foster and Martin 1994; Foster 1996a, 1996b, 2001a), but none has come from the eolian Unkpapa Sandstone Member. The more typical red, green, and gray mudstones, gray limestones, and tan sandstones of the rest of the Morrison are what have yielded the fossils. None of these more traditional Morrison rock types occurs in one part of the southern Black Hills, where the overlying Lower Cretaceous Lakota Formation rests directly on top of the thickest section of the Unkpapa. As the Unkpapa thins both west and north along the east flank of the Black Hills, the normal Morrison Formation becomes thicker and always lies between the Unkpapa and the Lakota. The Morrison consists entirely of the regular mudstones, limestones, and sandstones on the western and northern rim of the Black Hills, from northwest of Edgemont, South Dakota, all the way around through Wyoming and down to Sturgis, north of Rapid City. These rocks include nonsmectitic mudstones similar in color to those of Wyoming and the Colorado Plateau (reds, greens, and grays) and tan sandstones that are generally thinner, finer grained, more quartz rich, and more laterally restricted than in areas to the southwest (Como Bluff and the Colorado Plateau). There are thin gray limestones in this unit as well, and many contain gastropod shells, indicating quiet-water ponds; a few thicker limestones contain ooids, indicating lacustrine wave action. Where the undifferentiated Morrison occurs above the northward-thinning Unkpapa Member (mainly along the east edge of the Black Hills from Hot Springs to just south of Sturgis), it often contains a lower unit of light-gray to white siltstone with interbedded limestones and an upper unit of gray to green mudstone. The lower siltstone sections are interpreted by Szigeti and Fox (1981) as being wet interdune areas formed near the end of local eolian deposition. To some degree, the undifferentiated Morrison in the Black Hills can be described as being of two informal types or laterally adjacent megafacies: the light-colored (gray and white) siltstone and green mudstone units overlying the Unkpapa Sandstone Member in the eastern Black Hills comprising a "Blackhawk facies," which is best exposed near the town of Blackhawk, South Dakota, north of Rapid City; and the red, green, and gray mudstones with gray limestones and tan channel sandstones comprising a "Sheldon Creek facies," which occurs away from where the Unkpapa Sandstone is present and which is best exposed on ranches south of Sundance, Wyoming (particularly on the Sheldon Creek 7.5-minute quadrangle map), and along Interstate 90 west of Sundance. Most of the Sheldon Creek facies is time equivalent to parts of the Unkpapa Sandstone Member and to the Blackhawk facies, but it occurs more to the north and west in the Black Hills. In the northwestern Black Hills, where no Unkpapa Sandstone or Blackhawk rocks occur, the Sheldon Creek lithologies comprise the entire Morrison Formation (Foster 1992, 1996b). These three divisions existed simultaneously during Morrison deposition but represented different environments. The sand dunes of the Unkpapa Sandstone Member in the southeastern part of today's Black Hills graded into the wet, silty interdune areas represented by the Blackhawk facies, which graded into the better-drained floodplains, ponds, and small rivers of the Sheldon Creek facies (which, of the three, most resembles the Morrison Formation in other areas).

The most productive site for vertebrates in the Blackhawk facies is the Wonderland Quarry near Blackhawk, where elements of a *Barosaurus* and other animals have been found. The overall most diverse and productive site in the Morrison of the Black Hills is in the Sheldon Creek facies at the Little Houston Quarry west of Sundance, Wyoming. This site has yielded the remains of many dinosaurs and mammals.

Although most Morrison Formation researchers probably subscribe to the subdivision of the formation as outlined above, several others maintain that the situation is much simpler, at least in the Colorado Plateau region. Under this scheme, there are only two members of the Morrison, the Salt Wash and the Brushy Basin; the Tidwell Member is part of the underlying marine Summerville Formation; the Bluff Sandstone is its own formation, and the Recapture Member is part of it; and the Westwater Canyon Member is the same as the Salt Wash Member (O. Anderson and Lucas 1995, 1996, 1998). The main proponents of this subdivision stress comparisons with the type section of the Morrison Formation west of Denver, where the rocks include nothing similar to the Salt Wash, the Bluff Sandstone, or the several other members of the formation from the Colorado Plateau. Although they are likely equivalent to more than just the upper member of the Morrison of the Colorado Plateau, the rocks along the Front Range are similar to the Brushy Basin overall. Orin Anderson and Spencer Lucas include the Salt Wash in the Morrison, despite its dissimilarity to the type section, because the member interfingers with the overlying Brushy Basin and thus was depositionally transitional with it. They also put the Bluff Sandstone under the Morrison because at the type section of the San Rafael Group (the handful of formations underlying the Morrison in the Colorado Plateau region) in Utah, the group is partly defined by having eolian deposits, which the Bluff is. Two of the main current proponents of the opposing view, Christine Turner and Fred Peterson of the United States Geological Survey, have, along with their predecessors, included a number of units with eolian beds in the Morrison Formation.

A major source of disagreement here is the placement of the J-5 unconformity, the erosional surface or break in deposition that represents a significant time gap between rock layers and the last major unconformity of the Jurassic period in the western United States. Major unconformities such as this are sometimes used as formational boundaries. Because there seems to have been such a surface formed sometime between the last full extent of the region's Middle Jurassic continental seaway and deposition of the Morrison's fluvial floodplains, the J-5 is generally agreed to define the base of the Morrison. Many workers place the J-5 unconformity below the Tidwell, Bluff, or Recapture Member and thus include those in the Morrison. Those favoring the simplified, two-member Morrison see the J-5 as being at the base of the Salt Wash Member and so include the Tidwell, Bluff, and Recapture in San Rafael Group formations.

Because we are mainly concerned with vertebrate fossils here, and because more than 99% of the identifiable fossil material in the Morrison comes from the Salt Wash and Brushy Basin Members or their equivalents, the fact that these two camps see things so differently regarding other members is not much of an issue for our look at the animals and their interactions. Only scraps of dinosaurs are known from the Recapture and Westwater Canyon, and there is nothing but tracks from the Bluff. Several important specimens (*Dystrophaeus*, a lungfish, and dinosaur tracks) are known from the Tidwell, however, and the proper placement of this member is therefore important. I have seen several outcrops in Utah and Arizona where there appears to be an unambiguous unconformity below the Tidwell, and, in the field, I have more than once even mistaken the Tidwell for a lower muddy facies of the Salt Wash, so I tend to include the Tidwell Member in the Morrison Formation. Most of the analysis of the vertebrate fauna of the Morrison Formation presented here, however, is based on data from the Salt Wash–Brushy Basin level.

Age

The age range of the Morrison Formation has been argued almost from the beginning. Workers in the past have proposed a Jurassic, Cretaceous, or Jurassic-Cretaceous age, but it is generally agreed today that at least most of the formation is Late Jurassic (Kowallis et al. 1998; Turner and Peterson 2004; Trujillo and Kowallis 2015). For example, recent isotopic and micropaleontological work suggests a Late Jurassic age for the formation. A few relatively recent studies, however, have hinted that some of the very top beds of the formation may be Early Cretaceous. Although the very base and the very top of the formation, in some areas, may be Oxfordian or Tithonian in age, respectively, most of the Morrison appears to be within the Kimmeridgian age of the Late Jurassic.

Clay minerals in altered volcanic ash in mudstones of the formation, isotopically dated by uranium-lead measurements on zircon crystals or recalibrated argon-argon ages, indicate that the formation ranges in age from about 157 to 150 mya. Interestingly, several quarries to the east (Quarry 9 at Como Bluff, the Fruita Paleo Area microsites, the McKinsey-REX pipeline near Laramie, and the Mygatt-Moore Quarry in western Colorado) are each about 152 million years in age and approximately the same age as the top of the Salt Wash Member in areas farther west (Trujillo et al. 2014; Trujillo and Kowallis 2015; Foster et al. 2017), even though the Mygatt-Moore Quarry at least is near the middle of the Brushy Basin Member. The top of the Morrison is also of somewhat different age in different areas, suggesting that the top of the formation was scoured to varying degrees before deposition of the overlying formations.

The age estimates indicate that the Morrison was deposited during the middle part of the Late Jurassic, a conclusion also suggested by paleomagnetic studies and by the microfossils. Plant pollen samples recovered from the Morrison indicate a Late Jurassic age for the formation (specifically Kimmeridgian-Tithonian). Spores and pollen from the lowest beds that may be part of the formation suggest a Kimmeridgian age but could be latest Oxfordian in age (early part of the Late Jurassic). Stratigraphically higher levels of the formation have yielded plant pollen that clearly indicates a Kimmeridgian and early Tithonian age for the formation. Small fossils known as charophytes and ostracods (which represent, respectively, the reproductive structures of algae and the shells of small aquatic arthropods) indicate a similar age for the Morrison, although some data from recent pollen samples suggest that the very top

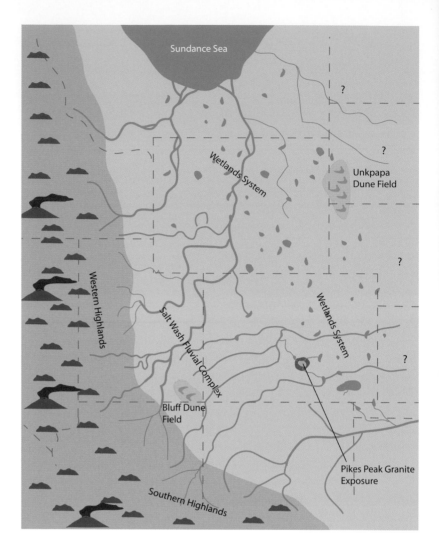

1.22. Generalized paleo-geography of the Rocky Mountain region during the early stages of Morrison times (Salt Wash Member and equivalents). Based in part on Turner and Peterson (2004) and the author's own work. To some degree the data are time averaged so that not all paleoenvironmental features necessarily co-occurred; this was done simply so that fewer time maps were required.

of the formation may be Lower Cretaceous, at least in some areas. These results need to be further tested, however.

Paleoenvironments and Climates

In most areas, the Morrison Formation is interpreted to consist of continental sediments deposited over much of western North America after the retreat of a shallow continental seaway during the early Late Jurassic. The source area for these siliciclastic and volcanic sediments was to the west and south (E. Christiansen et al. 2015), and this highland source area was probably part of the Andean-type range of mountains and volcanoes that existed along the western margin of North America at the time (Santos and Turner-Peterson 1986). Rivers from these western and southern mountains flowed east and north across the nearly flat Morrison floodplain, which was scattered with lakes, marshes, and other wetlands, the rivers becoming smaller to the north (figs. 1.22 and 1.23).

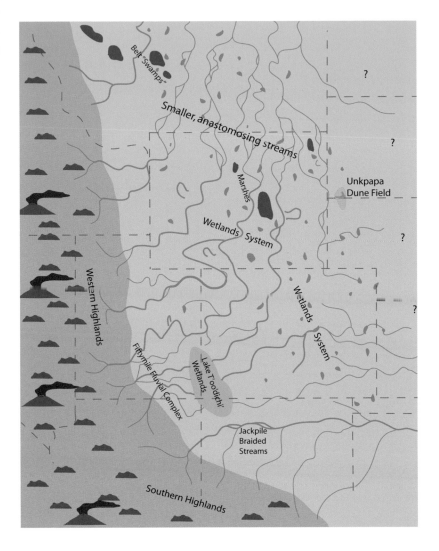

Belt "Swamps"

Smaller, anastomosing streams

Marshes

Wetlands System

Western Highlands

Fiftymile Fluvial Complex

Lake T'oo'dichi' Wetlands

Wetlands System

Unkpapa Dune Field

?

?

?

?

Jackpile Braided Streams

Southern Highlands

1.23. Generalized paleogeography of the Rocky Mountain region during the later stages of Morrison times (Brushy Basin Member and equivalents). Based in part on Turner and Peterson (2004) and the author's own work. To some degree the data are time averaged so that not all paleoenvironmental features necessarily co-occurred; this was done simply so that fewer time maps were required.

As we have seen, the environments represented by the sedimentary rocks of the Morrison Formation are generally thought to include sandy meandering and braided streams, lakes, ponds, marshes and other wetlands, and well-drained floodplains. Different workers have envisioned the landscape differently through the years, however. In the early part of the last century, some pictured an environment much like the present Amazon Basin, with abundant tropical vegetation and many rivers and lakes. Others envisioned an alluvial plain with an interconnected system of rivers, lakes, and swamps with abundant vegetation. Still others felt that the lakes on the alluvial plain were only seasonally present, and that the apparent lack of abundant plant remains in the Morrison Formation and the occurrence of some eolian deposits indicated that semidesert conditions existed. Geologist Ralph Moberly proposed the Gran Chaco Plain of northern Argentina as a modern analog for the Morrison Formation. The Gran Chaco is a vast, 2,200-by-1,100-km-wide alluvial plain, well vegetated with patches of forest scattered over savannas and between lakes and freshwater swamps. A high average rainfall occurs, but most precipitation is during the rainy season in the spring and summer, when many areas of the plain become flooded. During the winter dry season, many of the low-lying areas that are underwater in the summer become baked mud flats. In their classic study of the Morrison Formation, Peter Dodson, Kay Behrensmeyer, Robert Bakker, and Jack McIntosh concluded that although the absence of coals, apparent scarcity of small aquatic vertebrates, abundance of oxidized sediments, and presence of calcretes indicated an occasionally dry environment, plant productivity had to have been high enough to support large **herbivores** (Dodson et al. 1980). The current view of the Morrison is that it represents a semi-arid savanna with abundant vegetation that was thickest and tallest near surface water (Turner and Peterson 2004). The plant cover out on the plains, away from surface water, was likely much shorter, but overall, the amount of vegetation was probably greater than might be expected on the basis of the atmospheric dryness, largely because groundwater and a high water table were so important to the system. The Morrison is now known to contain more small aquatic vertebrates and more plant fossils than were recognized 35 years ago. In fact, there is increasing evidence of abundant, and in places rather tall, vegetation existing in the Morrison basin for significant periods of time.

Sea level, which is feared today to be rising with global warming, has in fact been both significantly higher and significantly lower in the past. During recent ice ages, for example, the amount of water locked up in continental glaciers dropped sea level some 100 m (330 ft) lower than it is today.[6] In the early Paleozoic, sea level got as high as 152 m (500 ft) above today's level. At many times during the Triassic, Jurassic, and Cretaceous periods, sea level was close to 100 m (330 ft) higher than today, and the seas were generally warmer. Global sea level during the Late Jurassic was somewhat higher than the present level, but not as high as the apparent maximum it reached during the Late Cretaceous, near the end of the time of dinosaurs. Global sea level rose throughout the Oxfordian and Kimmeridgian ages of the Late Jurassic, then dropped slightly and rose throughout the latest Jurassic and into the Early Cretaceous (Haq et al. 1987). Regional sea levels in western North America, though, appear to have dropped overall during the Late Jurassic, as indicated by the withdrawal of the Sundance Sea during the Oxfordian. This was likely the result of the interplay of global sea-level changes and regional tectonics.

The different paleoenvironments represented in the Morrison Formation have also influenced interpretations of the climate of the Late Jurassic in North America. The rocks in much of the formation seem to indicate semiarid and seasonal climates (Demko et al. 2004; Turner and Peterson 2004), but evidence from plant and small vertebrate fossils suggests a moderately humid to humid-semiarid climate with limited water stress (Tidwell 1990; Parrish et al. 2004; Foster and Heckert 2011; Tanner et al. 2014; Gee 2016), even in more southern areas. Paleoclimatologist

Anthony Hallam noted that globally, the Jurassic was warmer and latitudinally more equable than the present day, with seasonal rainfall and no (or very little) polar ice (Hallam 1994). Although occurring seasonally (probably during the northern winter), rainfall during the Late Jurassic in North America may have been a little lower compared with what it had been earlier in the Mesozoic. During the Triassic period, when all the continents were fused into the pole-to-pole supercontinent of Pangaea, the land areas were under a monsoonal climate, and total rainfall was quite high but occurred mostly during half the year. By the Late Jurassic, the fact that the two main components of the supercontinent, Laurasia and Gondwana, were separated by the equatorial and east-west-trending ocean Tethys meant that the monsoon had broken down. North America was now most likely subhumid to semiarid, with less-seasonal and lower total rainfall than during the Triassic (Parrish 1993; Parrish et al. 2004; Tanner et al. 2014).

Computer simulations of average surface temperatures for the Late Jurassic suggest that western North America, in the area of Morrison deposition, was warmer than it is now. One study predicted an average temperature of 4°–20°C (39°–68°F) in the northern winter and 24°–36°C (75°–97°F) in the summer in one model, and 0°–16°C (32°–61°F) in the winter and 20°–36°C (68°–97°F) in the summer in another (Valdes and Sellwood 1992; Valdes 1994). The differences between the results arose from assigning different sea-surface temperature estimates as part of the model parameters. Atmospheric carbon dioxide levels during the Late Jurassic may have been three to four times higher than modern preindustrial levels. Temperature estimates for the area of the Morrison Formation that account for these higher CO_2 levels suggest even hotter days for Late Jurassic dinosaurs, perhaps as high as 20°C (68°F) in winter and 40°–45°C (104°–113°F) in summer, on average.

Some precipitation estimates for the Morrison outcrop region have called for semiarid conditions with less than 500 mm (20 in.) per year. Other estimates include approximately 1 to 2 mm per day for the winter months and less than 1 mm per day for the summer months, approximately 45 cm (18 in.) over the course of the year (Valdes 1994). These estimates were based on computer models. In other studies, based on paleosols of the formation, the results were an average precipitation of about 1.6–2.5 mm per day (~58–90 cm or 23–35 in. per year) (Retallack 1997) and, in a recent study, between 80 cm (22 in.) and 110 cm (43 in.) of precipitation per year (Myers et al. 2014). This amount of precipitation is approximately equivalent to that seen today in central to southern Texas. But other evidence (see below) suggests the lower end of these estimates especially may be rather low. The fossil wood and soil studies also suggest some waterlogging of the soils in the Morrison. All this evidence indicates that conditions were not exactly humid or junglelike but that things were much wetter than the region of Morrison outcrop sees today.

One argument for higher precipitation in the Morrison basin is the fact that the formation contains a fair amount of petrified wood and, in a number of cases, large petrified logs, not the kind of trees that could get by on limited water. There are fossil conifer logs at sites near Escalante and Capitol Reef in Utah, for example, with diameters that indicate trees pushing 30–52 m (98–170 ft) in height (fig. 1.32; see below). (These estimates are based on comparisons with Oregon old-growth forest trees in Givnish et al. [2014], fig. 5, using trunk diameter; and on my own measurements of trunk circumference and height in several species of modern araucarians, cupressaceans, and pinaceans, fig. 1.32.) The Morrison fossils are trees in some cases as big as you see in the Chinle Group in Petrified Forest National Park (though not in as large numbers), and you do not get trees that big in a desert. The Morrison logs may even suggest a precipitation/evaporation ratio during the Late Jurassic that was close to 1.0, as some modern trees' maximum height is correlated with this relative amount of precipitation (Givnish et al. 2014, although that study was admittedly focused on the angiosperm genus *Eucalyptus*). The maximum height ranges apparently reached by some Morrison conifers are suggestive of a small net gain in precipitation relative to evaporation (40–50 m = ~1.1 precipitation/evaporation; see fig. 1.32 and Givnish et al. 2014, their fig. 4). And some of the biggest fossil logs in the Morrison are in southern Utah in a region that we believe to have been relatively drier than the more northern reaches of the basin.

Although the Morrison has a reputation for not having much petrified wood (e.g., Parrish et al. 2004), I've seen a lot of it in the formation over the years and all over the region, and others have noted this abundance too (Gee 2016). The abundance and size of what is preserved in terms of fossil wood, along with numerous specimens showing little water stress (no growth rings; e.g., Tidwell et al. 1998), suggests that things were not as dry back then as we have been thinking. Many of the logs in the formation indicate trees up to at least 45 m (150 ft) and more in height (see fig. 1.32; Gee et al. 2019); giant trees of this size today require annual precipitation of at least 150 cm (60 in.) (Scheffer et al. 2018).

It has also been suggested that the presence of sand dune deposits in some Morrison units indicates aridity, but of course, all you need for sand dunes is a source of sand and wind—it doesn't have to be dry. Beaches even in humid areas commonly have sand dunes along them, after all.

The apparent contradiction between the semiarid climates indicated by the rocks (Demko and Parrish 1998; Parrish et al. 2004; Engelmann et al. 2004) and the wetter setting suggested by the plant, wood, fish, and **semiaquatic** animal fossils (Tidwell 1990; Ash and Tidwell 1998; Foster and Heckert 2011; Foster and McMullen 2017; Gee et al. 2019) may be at least partially explainable, however. The geochemistry of the rocks of the Morrison Formation suggests a semiarid climate, whereas the abundance of large herbivorous dinosaurs, aquatic and semiaquatic vertebrates, and relatively large plants indicates a significant supply of water. But not all water needs to come from the sky in the form of humidity and rain. Rather, the air mass over the Morrison basin may have been fairly dry, and evaporation was possibly high. Much of the water in the basin could have been groundwater flowing from the mountains to the west. Surface runoff from the mountains also supplied a number of perennial streams, and rainfall into the basin itself provided an additional, seasonal component to the dampness of the soil and the levels of the rivers and ponds. In fact, a system of shallow ponds and lakes (some of them perennial) that existed in what is now the eastern half of Colorado was ostensibly groundwater fed and resulted from a reemergence of the water table caused by downslope flow of groundwater coming out of the western mountains (Turner and Peterson 2004). It is also possible that the contradiction between paleosols and large logs and other indications of abundant precipitation reflect long-term, regional wet-dry cyclicity in the climate that we do not have the fine-scale stratigraphic resolution to distinguish in a formation representing about seven million years.

The airflow during the Late Jurassic in North America was west to east, and storms carrying moisture off the paleo-Pacific had to cross a significant stretch of mountainous terrain before reaching the Morrison basin. What is now central California east all the way to western Utah was, during the Late Jurassic, covered by low hills and mountains with some volcanoes. These mountains and hills caused rain to fall from the weather systems, and because the mountains were so wide west to east, by the time systems got to the Morrison basin, relatively little moisture may have been left to rain on the basin. This is different from a rain-shadow effect, which occurs immediately in the lee of a tall (but not necessarily wide) mountain range. Moisture that might have fallen on the Morrison basin was possibly "rained out" of systems as they crossed the wide expanse of low mountains and hills to the west. The effect was essentially the same, if not quite as severe, as those in modern rain-shadow deserts: the air was relatively dry, rainfall was moderate at best, and evaporation rates may have been high. It was possibly the groundwater supply and surface runoff in the rivers more than rainfall on the basin itself that made life possible for plants and larger animals in the Morrison basin. When the rains did come, however, they were a welcome source of regeneration for many elements of the biota.

The dinosaurs may be the flashy part of the story, but a lot of fossils other than vertebrates are known from the Morrison. Invertebrate fossils known from the Morrison Formation include insects, crayfish from the Mygatt-Moore Quarry and Fruita Paleontological Area, gill- and lung-breathing snails, freshwater bivalves, and two major groups of tiny, shelled, aquatic arthropods (Chure et al. 1998a, 2006).

Freshwater unionid mussels are often found in river channel sandstones, and freshwater snails are more often associated with limestones and other quiet-water pond deposits. Several types of unionids are known from the formation (3 genera and 17 species, though they may be over-split), and those found commonly in the Morrison Formation (fig. 1.24) indicate perennial water in that their larval stage depends on the gills of

Invertebrate Paleontology of the Morrison

1.24. Mollusks of the Morrison Formation. (A) Unionid bivalves *Vetulonaia faberi* from Wyoming. (B) Unionid bivalves *Vetulonaia mayoworthensis* from Wyoming. (C) Unidentified unionid bivalves from the Fruita Paleo Area, Colorado. (D) Gastropods *Viviparus reesidei* from Wyoming. (E) Gastropods *Amplovalvata scabrida* from Wyoming. (F) Gastropods *Viviparus* sp. (*top*) and *Amplovalvata* sp. (*bottom*) from Wyoming. (G) Gastropods *Gyraulus veturnus* from Wyoming. (H) Gastropods *Lymnaea atavuncula* from Wyoming. (I) Unidentified unionid bivalve from the Little Houston Quarry, Wyoming. (J) Unidentified elongate unionid bivalve from the Little Houston Quarry, Wyoming. Images in A–H courtesy of Emmett Evanoff. All scale bars = 1 cm, except C, which is in cm.

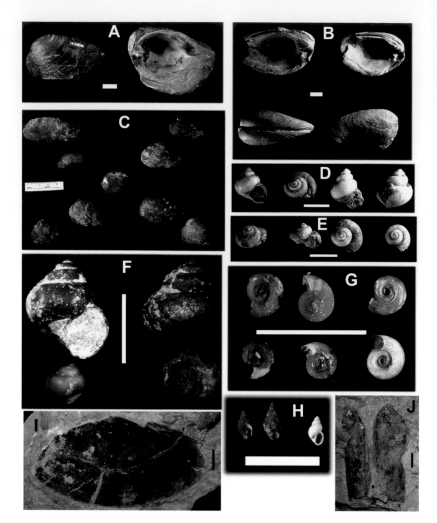

ray-finned fish as hosts. Thus, the presence of unionids in many Morrison channel sandstones also demonstrates that fish lived in the rivers too. Unionids are widespread geographically and stratigraphically in the Morrison, and the biology of their modern forms indicates that they preferred low-turbidity, well-oxygenated, medium-pH water that was perennial, shallow, warm, and lime rich (Good 2004). They also required abundant plankton and fish (as mentioned above) and may have served as food for lungfish. Gastropods are also reasonably common in Morrison Formation rocks (fig. 1.24).

Among the arthropods known from the Morrison Formation are an unnamed crayfish (fig. 1.25) and several species of phyllopods (previously called conchostracans), a group of small freshwater branchiopod crustaceans that look more or less like a tiny shrimp with a thin clam shell draped over its back. Usually only the shell is preserved (fig. 1.25), but such fossils are common at some sites in the formation.

The Temple Canyon site near Cañon City, Colorado, recently produced *Parapleurites morrisonensis*, a fossil orthopteran wing from a

Arthropods

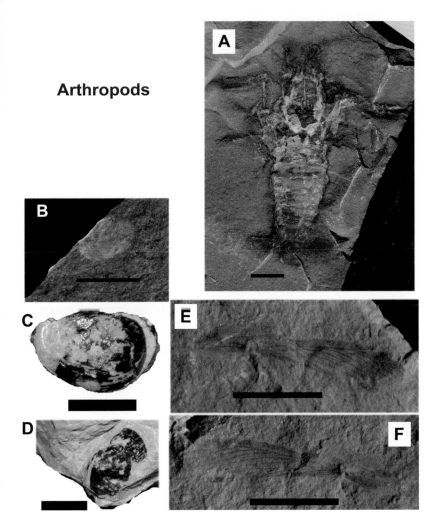

laminated lake-bed deposit of the Morrison Formation. This is the first insect body fossil from the Morrison Formation (fig. 1.25), and it represents an order that today includes grasshoppers, crickets, and katydids. This specimen, *P. morrisonensis*, appears to be a member of the orthopteran family Locustopsidae (D. M. Smith et al. 2011). The newly named nepomorph hemipteran insect *Morrisonnepa jurassica* is large and consists of a forewing, parts of the abdomen, and a possible head and was found recently in the Brushy Basin Member of the Morrison in southeastern Utah (Lara et al. 2020).

Trace fossils of invertebrates are present in the formation, and these appear to indicate the presence of beetles, ants, termites, bees, horseshoe crabs, snails, and crayfish (Hasiotis and Demko 1996a, 1996b; Hasiotis 2004; E. Smith et al. 2016), although some of these interpretations of the traces have been challenged (Bromley et al. 2007). The traces are particularly common in sandstone units in the formation throughout the outcrop area. Trace fossils from Utah indicate that although we may not be able to attribute some of these relatively large nests directly to ants or termites, they do at least seem to indicate that some type of social insect was responsible for their construction (E. Smith et al. 2016; E. Smith 2017). Burrows of possible termites and crayfish indicate that the water table depth in the Morrison was consistently less than 4 m (13 ft).

Small, circular pits on the bones of dinosaurs are common at some sites in the Morrison Formation. I've pulled sauropod bones out of a quarry in the Black Hills that look as if someone has spent hours with a power tool drilling hundreds of millimeter-deep holes all over more than half the outer surface of some of the bones—this bone modification isn't subtle; you can't miss it when it's present. While the bones of the Little Houston Quarry and other sites have this modification often in abundance, some sites have no sign of it at all (Mygatt-Moore, for example, which, like Little Houston, I worked for more than a decade and had plenty of opportunity to see it if it were present). These strange circular pits have been interpreted as being caused by low pH soil conditions or by excavation of the pupation chambers of dermestid beetles (or some similar unknown arthropod), making them essentially insect trace fossils on dinosaur bone (Fiorillo 1998a; Britt et al. 2008; Bader et al. 2009). These pupation chambers would have been used by the beetles during decomposition of a dry carcass that was subaerially exposed. So even the seemingly simplest and strangest modifications to bone can provide us with a rather vivid image of the past and show us glimpses of the ancient ecosystem at the same time.

Paleobotany of the Morrison

Although it is better known for its remains of dinosaurs and other vertebrates, the Morrison Formation has produced one of the most diverse Mesozoic **paleofloras** in the world. Plant fossils are less common than vertebrate fossils but are present at more than 100 localities in the formation (fig. 1.26), and some plant remains are even mixed in with vertebrate material in the same quarries (J. T. Brown 1972, 1975; Tidwell et al. 1998). The plants from the Morrison Formation include genera of algae, mosses, quillworts, ferns, horsetails, tree ferns, cycadophytes, seed ferns, ginkgoes, and conifers (fig. 1.27) from sites near Belt, Montana (fig. 1.28); Montezuma, Utah; Brushy Creek, Utah (fig. 1.29); Como Bluff, Wyoming; and the Mygatt-Moore Quarry and Temple Canyon sites in Colorado (fig. 1.30; Tidwell 1990; Ash and Tidwell 1998; Tidwell et al. 1998, 2006; Gorman et al. 2008) and from numerous palynomorph and spore localities (Hotton and Baghai-Riding 2010, 2016). The approximately 35 known genera include 77 species of megafloral remains (leaves, etc., as opposed to pollen). Pollen and other plant microfossils (spores) occur at a number of different localities throughout the Rocky Mountain region and have been assigned to more than 200 microfloral species. They indicate a significant diversity of ferns, many more than are seen in the macrofossils, and an abundance of conifers (Hotton and Baghai-Riding 2010). Silicified petrified wood is not uncommonly found as one hikes

1.26. Some plant localities in the Morrison Formation. (A) One of the Temple Canyon, Colorado, quarries low in the Morrison. (B) Scott Madsen working the Brushy Basin Member in the Brushy Creek area of Utah.

1.27. Some plant fossils from the Morrison Formation (boxes) with generalized reconstructions of how whole plants may have appeared (to scale; height of *Podozamites* ~20 m [65 ft]). Fossils based on illustrations in Tidwell (1990), Ash and Tidwell (1998), and Tidwell et al. (1998); full plant reconstructions based on the work of Kirk Johnson, Tidwell et al. (1998), and the author's observations.

Structural Variety of Some Morrison Formation Plants

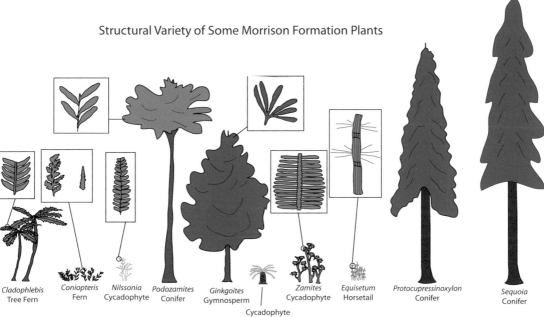

Cladophlebis
Tree Fern

Coniopteris
Fern

Nilssonia
Cycadophyte

Podozamites
Conifer

Ginkgoites
Gymnosperm

Cycadophyte

Zamites
Cycadophyte

Equisetum
Horsetail

Protocupressinoxylon
Conifer

Sequoia
Conifer

1.28. Plant fossils of the Morrison Formation from Belt, Montana. (A) Fern *Coniopteris*, DMNH 6061. (B) Fern *Coniopteris*, DMNH 6064. (C) Conifer cone, DMNH 6062. (D) Conifer *Podozamites*, DMNH 6066. (E) Unidentified conifer, DMNH 6063. (F) Cycadophyte *Zamites*, DMNH 6065. (G) Fern, DMNH 6070. (H) Slab with fern pieces, DMNH 6067. (I) Fern *Cladophlebis*, DMNH specimen. All scale bars = 1 cm, except G and H = 5 cm.

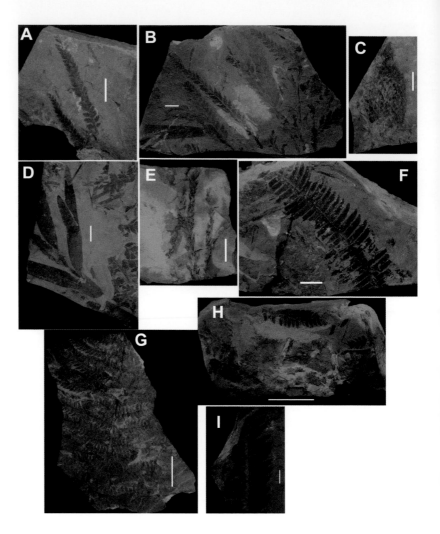

the Morrison Formation (Tidwell and Medlyn 1993), but good, identifiable plant macrofossils occur at only a handful of sites. Whether these sites indicate rare preservation of more widespread vegetation patterns or whether the sites represent local conditions that truly were limited in space and time (Engelmann et al. 2004) is not entirely clear. Sites with fragmentary, unidentifiable but abundant plant remains are actually quite common in the formation, however. Some of the best of the well-preserved macrofossils are near Belt, Montana, not far from Great Falls and north of most of the good vertebrate material in the formation (although there are more recent finds of vertebrates in this northern area). Some well-preserved plant fossils have also been found in southeastern Utah, Colorado, and Wyoming. Conifers, ferns, and cycadophytes are the most common, diverse, and wide-ranging elements of the flora. At one recently discovered site in Utah, the wet-adapted *Czekanowskia* is particularly abundant, possibly indicating humid climatic conditions (Sun et al. 2015). The data suggest a diversity of conifer trees (araucarians,

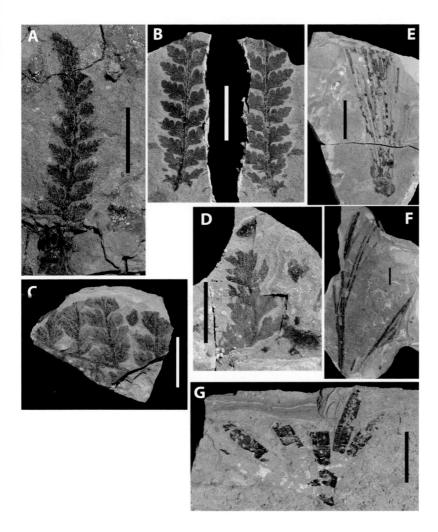

1.29. Plant fossils of the Morrison Formation from Brushy Creek, Utah. (A–D) Various forms of the fern *Coniopteris hymenophylloides*. (E–F) Specimens of the wet-adapted *Czekanowskia* sp. (G) Leaf of *Ginkgo* sp. All SUSA specimens now at FHPR. All scale bars = 1 cm.

podocarpacreans, cheirolepidiaceans, and cupressaceans) with an understory of ferns and some open fern "meadows" (Hotton and Baghai-Riding 2010).

The Morrison Formation paleoenvironment appears to have had significant groundwater and seasonal rain with no evidence of freezing. Plant **biomass** was higher than expected given the apparent amount of rain. There may have been high rates of evaporation relative to precipitation, so groundwater and the surface flow from the western mountains were important to plants growing in the basin. Conifers were probably concentrated near rivers, ponds, and lakes with areas of open forest out on the floodplain, and there may have been a lower story of cycadophytes and tree ferns. However, the abundance of fossil wood, as well as logs (fig. 1.31) that indicate tree heights of at least 25 m (82 ft) and commonly higher, suggests the presence of extensive mixed conifer forests, at least in some regions (Gee 2016). Details of the trees preserved from these forests suggest that there was little seasonality in temperature or precipitation

1.30. Plant fossils of the Morrison Formation from Temple Canyon and the Mygatt-Moore Quarry, both Colorado, and the Little Houston Quarry, northeastern Wyoming. (A) Horsetail *Equisetum* sp., Temple Canyon (DMNH 6057). (B) Horsetail *Equisetum* sp., Little Houston Quarry. (C) Unidentified seed(?), Little Houston Quarry. (D) Fern *Coniopteris*(?), Temple Canyon (DMNH 6058). (E) *Czekanowskia* sp., Temple Canyon (DMNH 6060). (F) Fern, Temple Canyon (DMNH specimen). (G) Conifer *Brachyphyllum* sp., Mygatt-Moore Quarry (MWC specimen). (H) Fern *Sphenopteris*?, Mygatt-Moore (MWC specimen). (I) Indeterminate, Mygatt-Moore (MWC specimen). (J) Fern *Coniopteris* sp., Mygatt-Moore (MWC 6000). (K) Unidentified conifer, Mygatt-Moore (MWC 2190). Scale bar C = 1 mm; all others = 1 cm.

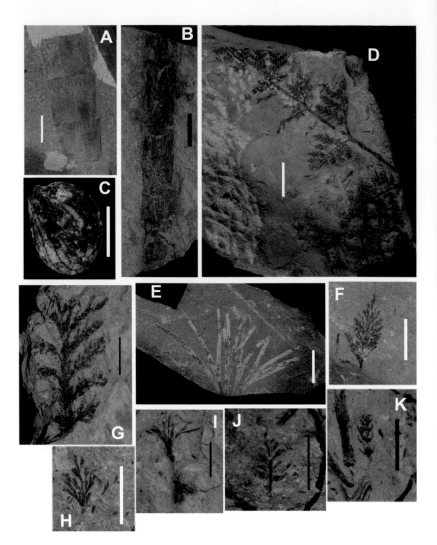

and that the climate was not arid (Gee et al. 2019). Fossilized seed cones are known from at least Utah, Colorado, and Wyoming in the Morrison Formation, with cones of five types of conifers, including araucarians, pines, and cheirolepidiaceaens, having been reported just from Utah (Gee et al. 2014).

There are relatively few studies that have attempted to estimate the original heights of fossilized trees (Mosbrugger et al. 1994), but my comparisons of trunk circumference and total height in some modern araucarian, cupressacean, and pinacean conifers with circumferences of some of the larger fossil logs out of the Morrison Formation (fig. 1.32) suggest that maximum tree heights of these ancient conifers may have been up to 52 m (170 ft) and commonly were over 30 m (98 ft) even in what is now the southern Colorado Plateau region. Estimates of the heights of the same specimens using the calculation of Gee et al. (2019) (based on Mosbrugger et al. 1994) give roughly similar heights. And estimates of heights of recently discovered logs in the Dinosaur National Monument area are up

1.31. Examples of fossil wood and petrified logs from the Morrison Formation. (A) Brian Engh with a petrified log from the Salt Wash Member near Capitol Reef National Park. (B) Large petrified log in the Brushy Basin Member at Escalante Petrified Forest State Park, southern Utah. This log was from a tree that may have been up to 41 m (135 ft) tall (see fig. 1.32). (C) Petrified log in the Salt Wash Member near Capitol Reef. (D) Fossil wood fragments from the Mygatt-Moore Quarry in western Colorado; scale bar = 5 cm. (E) Petrified log section from near the site in A; scale in centimeters.

to 48 m (158 ft). As we saw above, giant trees in these height ranges today require at least 1500 mm (60 in.) mean annual precipitation and at least a moderate density of growth (Scheffer et al. 2018), which suggests that some areas and/or times in the Morrison basin must have had open forests of very large trees. This also would have promoted growth of diverse and abundant plants below this canopy, including shade-tolerant ferns.

The forest understory was probably composed of these ferns plus small cycadophytes (C. Miller 1987; Turner and Peterson 2004). Ginkgo trees may have been scattered among the conifers, and in some plant localities in the Morrison ginkgophytes are quite common. Petrified wood of conifers in the Morrison Formation is of types either containing growth rings or often lacking them. The growth rings indicate seasonal growth patterns, and the lack of rings indicates continuous growth and a lack of water stress, the latter of which suggests mesic conditions (moderate moisture) in many areas (Gee 2016). These two styles may represent trees growing away from and close to perennial water sources, respectively. In the absence of grass or any other type of angiosperm, ferns and low-growing herbaceous cycadophytes may have formed the main ground cover during the Jurassic, particularly in areas away from other vegetation. Laterally, the open Morrison floodplains were probably covered with more low-growing ferns and herbaceous cycadophytes. Closer to the rivers and ponds were probably open woodlands with tree ferns, seed ferns, cycads, and smaller conifers and ginkgoes. The watercourses were likely lined with denser woodlands of large conifers, ginkgoes, and tree ferns with an understory of other plants also seen in other areas. Ponds and rivers were probably lined with horsetails and may have had some

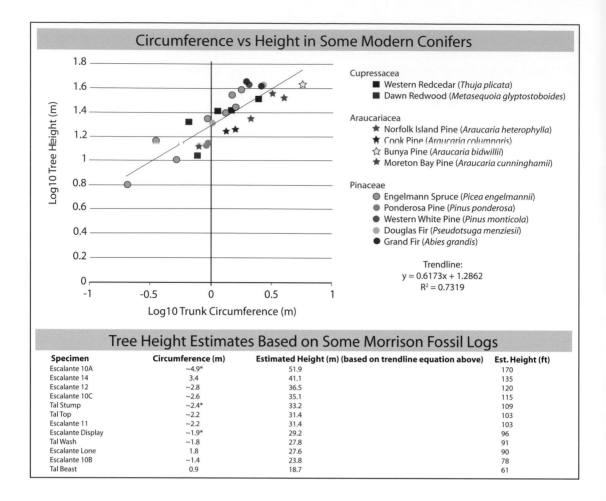

Circumference vs Height in Some Modern Conifers

Cupressacea
- ■ Western Redcedar (*Thuja plicata*)
- ■ Dawn Redwood (*Metasequoia glyptostoboides*)

Araucariacea
- ☆ Norfolk Island Pine (*Araucaria heterophylla*)
- ★ Cook Pine (*Araucaria columnaris*)
- ☆ Bunya Pine (*Araucaria bidwillii*)
- ★ Moreton Bay Pine (*Araucaria cunninghamii*)

Pinaceae
- ● Engelmann Spruce (*Picea engelmannii*)
- ● Ponderosa Pine (*Pinus ponderosa*)
- ● Western White Pine (*Pinus monticola*)
- ● Douglas Fir (*Pseudotsuga menziesii*)
- ● Grand Fir (*Abies grandis*)

Trendline:
$y = 0.6173x + 1.2862$
$R^2 = 0.7319$

Tree Height Estimates Based on Some Morrison Fossil Logs

Specimen	Circumference (m)	Estimated Height (m) (based on trendline equation above)	Est. Height (ft)
Escalante 10A	~4.9*	51.9	170
Escalante 14	3.4	41.1	135
Escalante 12	~2.8	36.5	120
Escalante 10C	~2.6	35.1	115
Tal Stump	~2.4*	33.2	109
Tal Top	~2.2	31.4	103
Escalante 11	~2.2	31.4	103
Escalante Display	~1.9*	29.2	96
Tal Wash	~1.8	27.8	91
Escalante Lone	1.8	27.6	90
Escalante 10B	~1.4	23.8	78
Tal Beast	0.9	18.7	61

1.32. Log-transformed measurements of the trunk circumference and estimated full tree height for 11 species in modern conifer families that have ancient relatives identified in the Morrison Formation. Morrison trees seem to have reached heights easily ranging from 20 to 50 m (65–164 ft). These data were used to generate a trend-line equation from which to estimate the height of trees represented by Morrison petrified logs, based on circumference. Modern trees' circumference was measured at ~1.5 m above the ground with tape; height was estimated by sighting with Brunton to the treetop at a 45° angle and measuring the distance to the trunk center with metric tape, then adding 1.75 m for eye height. Morrison fossil logs were measured with tape at Escalante Petrified Forest State Park (Brushy Basin Member) and the Tal Site (Salt Wash Member) near Fremont River, both in southern Utah. Tildes (~) indicate estimations necessitated by partial measurements due to incomplete preservation or burial. Asterisks (*) indicate measurements taken near the basal root flare. Circumference measurements lacking the ~ symbol preserved full cross sections. Circumference (and thus height) estimates in most cases are minimums due to the lack of bark preservation in nearly all fossil specimens (modern specimens were measured with bark) and to the assumption that the maximum measureable circumference is from the base of the trunk (i.e., if the thicker end of an isolated fossil log section was in fact from several meters up the trunk of the living tree, the estimated height for the fossil will be low).

charophyte-producing green algae in the water itself, especially in areas east and north of the modern Colorado Plateau.

Some of the conifers found in the Morrison Formation include *Sequoia, Agathoxylon, Xenoxylon, Brachyphyllum, Pagiophyllum, Araucarioxylon,* and *Podozamites.* These are similar to modern redwoods and Norfolk Island pines, and *Podozamites* is similar to the kauri tree of New Zealand and to *Nageia* from Japan. Among the conifer families present in the Morrison Formation were Araucariaceae, Pinaceae, Cupressaceae, and the drier-adapted Cheirolepidaceae (Ash and Tidwell 1998; Tidwell et al. 1998; Gee 2016), the latter of which is more abundant as palynomorphs in some southern outcrop areas such as New Mexico (Hotton and Baghai-Riding 2016).

Morrison ferns include *Coniopteris* and *Cladophlebis*, but there are almost 80 other types of ferns known from the formation based on palynomorphs. The fern *Sphenopteris* was recently identified from the Morrison Formation of Utah. The Morrison cycad *Zamites* may have grown like a branching bush (even more bush-like than the modern *Zamia*; K. Johnson, pers. comm., 2006) rather than like a typical cycad such as the modern *Cycas*. There undoubtedly were cycads in the Morrison that grew similar to *Cycas* and probably others that grew like the modern *Zamia* (fig. 1.27).

Burying the Bones

We would not know anything about the animals that came to live on the Morrison Formation floodplain if they were not lucky enough to have become fossilized and preserved for 150 million years. Without burial, preservation, and subsequent uplift and exposure of the animals' skeletons (fig. 1.33), the Late Jurassic in North America would be yet another gap in our understanding of the history of vertebrate life. As it is, we lucked out with the Morrison Formation. Conditions were right then and the landscape is right now for the preservation of impressive amounts of information about that time and its fauna.

All of the fossils found in the Morrison Formation are preserved in sedimentary rocks, which are simply rocks composed either of particles derived through erosion from preexisting rocks or minerals, or of minerals precipitated from freshwater or saltwater through chemical or biochemical processes. **Sedimentation** is the process by which these particles or minerals accumulate. Most of the sediments of the Morrison Formation accumulated through the actions of streams and their related systems. Rivers and their floodplains, ponds, lakes, and marshes and other wetlands all contributed to the deposition of sediment. Most of the sediments were composed of detrital grains, meaning particles of previously existing rocks that were eroded from mountains or low hills and washed down streams and rivers to be deposited on the floodplain. Particles in sedimentary rocks can be of a variety of sizes, including sand grains, mud particles, and larger clasts like gravel, pebbles, and cobbles. In a sense, the bones of the vertebrates that are preserved in the formation are a type of clast, although very large ones. The sedimentation process that deposits the sands and other grains of sedimentary rocks is critical for the preservation of the bones because the skeletal remains must be buried by these processes in order to be preserved.

One element critical for the deposition and preservation of sediments is net **subsidence**. This simply means that there must be a place for the sand and mud grains to be deposited and accumulate. If the sediments

1.33. Dinosaur fossils in their natural environment in the Morrison Formation, as found. (A) Sauropod dinosaur bones (arrows) in a channel sandstone in the Brushy Basin Member near Moab, Utah. (B) *Allosaurus* sacrum and ilia in sandstone at the Poison Creek area near Buffalo, Wyoming. (C) Small sauropod humerus and partial radius on the underside of a sandstone outcrop also near Buffalo, Wyoming. Brian Flynn's rock hammer and forearm for scale.

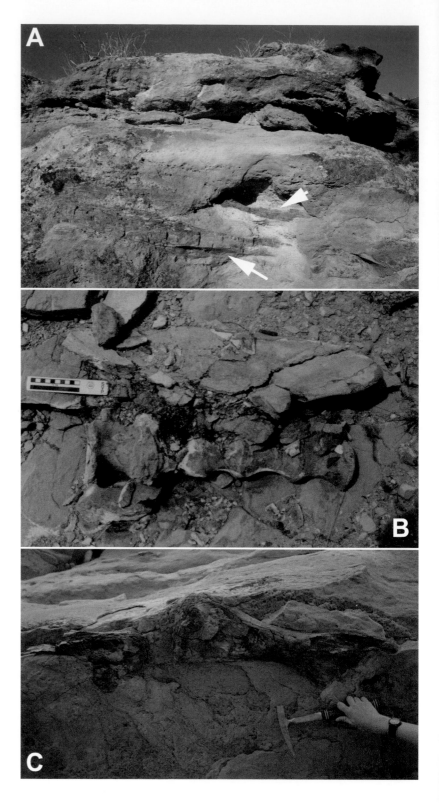

eroded out of the mountains were simply washed across the land and out into the deep ocean, few terrestrial vertebrate fossils of the time would be preserved. The key to the preservation of Morrison sediments and fossils is the fact that the Morrison basin (the land east of the mountains of the time) was slowly sinking, or subsiding, at a rate approximately equal to that at which the western mountains were eroding. This created **accommodation space**, room in the basin that the sediments could fill in. If not for the relative subsidence creating accommodation space, the sediments would have kept right on going across the plain and drained into the ocean. As it was, much of the sediment was deposited on the Morrison floodplain by the rivers, muds out on the plains, and sands in the channels. These sediments subsequently sunk and were overlain by more muds and sands. The process continued for nearly 7 million years, until several hundred feet of mud and sand had accumulated, forming what was later named the Morrison Formation. Net subsidence means that the basin was subsiding at a high rate relative to the levels of the rivers and the rate of erosion. The bottom of the Morrison basin may have sunk and been filled in, as it appears, but the same effect would be seen if a basin were simply uplifted at a slower rate than the mountains. In that case, although all areas would be uplifted, as long as the difference in uplift between the basin and the mountains created an equal or greater volume than was eroded from the mountains, accommodation space would result.

The main rivers of Morrison time that flowed out of the mountains fanned out as they opened onto the floodplain. The primary channels that flowed across the Morrison floodplain were large and had numerous tributaries within the basin. The main large channels were perennial, but other smaller ones were likely ephemeral. Sandstones occur throughout the Morrison, so streams must have been abundant, and these were natural gathering places for the vertebrates looking for shade, water, and vegetation. Most rivers were perhaps no more than 200 m (660 ft) across, particularly farther from the source areas to the southwest. Some of the larger rivers may have been up to 500 m (1,650 ft) wide.

Sediments in rivers can move as either bed load or **suspended load**. Usually sand grains move along the bottom bed load of the stream through **saltation**, a process of hopping along the bottom. Larger clasts may roll along the bottom, sometimes only in stronger currents. Mud most often moves as suspended load, continuously floating in the water and often contributing to a river's brown or red color. This mud will slowly settle out of the water and to the bottom when flow slows down enough or stops. Bones of dinosaurs most likely moved down river channels as bed load and were eventually buried in sand or gravel. Sometimes bones ended up in ponds where the muddy waters were still, and the mud slowly settled to the bottom and buried the bones in what would turn into fine-grained mudstone. Some bones were buried in distal floodplain muds that turned into soils, others in sandy crevasse splays deposited where floodwaters broke through river levees, and still others in flash-flood deposits.

Animals that died near river courses were most likely to be preserved. The skeleton of such an animal would become entrained in the river waters and carried for some distance by the current before settling to the bottom, where the continued flow would cover the bones with sand. This process would continue, and the skeleton would be buried deeper and deeper as the Morrison basin subsided slowly. Perhaps eventually the river would begin a new course. Then mud would be deposited on top of the skeleton instead of sand.

Other water sources such as lakes are good for preserving skeletons, although sedimentation rates are lower there than in rivers and thus the skeletons get buried more slowly, which leaves them susceptible to damage before they are safely under the sediment. Bones as large as those of some dinosaurs are generally safe at the bottom of a lake, though, so these shallow pond deposits in the Morrison are a common place to find preserved vertebrates. The least likely place of preservation for bones in the Morrison was the drier parts of the floodplain, where the bones could be damaged or destroyed by predators, scavengers, insects, and the elements as they sat exposed to the air and sun for years. If soil forms, it will often but not always destroy bones too, and the longer bones sit anywhere, the more likely they will be damaged. Overall, the faster a bone is safely encased in sediment, the more likely it will be fossilized, and this happens most quickly in water. Thus, even terrestrial animals' skeletons are most often found preserved as fossils in freshwater deposits like river channel sandstones and pond mudstones.

Over millions of years of seaways and rivers, and with new types of animals living on the land above, a buried skeleton would become compressed somewhat and slightly distorted. As it became buried deep enough, the sand would begin to turn to rock, and groundwater would deposit minerals in the pore spaces of the bones, turning the bone essentially to rock. Many more millions of years would go by, and eventually the area would become uplifted to some 1,524 m (5,000 ft) above sea level, and erosion would begin to strip away the rock. The sandstone of the old river channel would become exposed. After a few hundred years, perhaps, the first elements of the skeleton would become exposed and would be visible, to be found by a hiker or paleontologist.

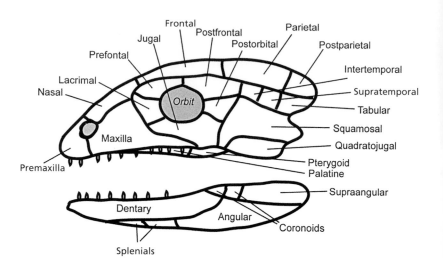

2.1. Skull of a generalized early tetrapod showing most of the external cranial bones. Some of these bones are lost in later terrestrial vertebrates, but most are retained. Based on Hildebrand (1995).

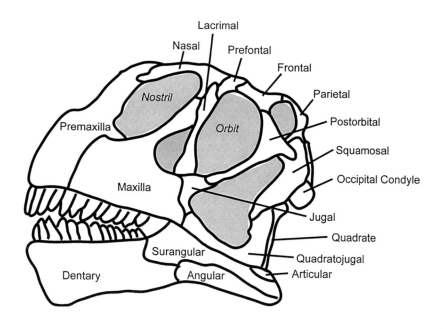

2.2. Skull of the Morrison sauropod dinosaur *Camarasaurus* showing the external bones. Note the openings in the skull (shaded) both behind and in front of the orbit. Nostril = external naris of text.

Setting the Stage: Vertebrates and the Jurassic World

2

Because this book focuses on vertebrates, it will be helpful to review some of the group's evolution up to the time of the Late Jurassic. First, we must review the basic vertebrate skeleton so that the changes that led to the osteology of Late Jurassic animals make sense. The vertebrate skeleton is divided into several sections. The **cranial skeleton** consists of the bones of the skull, and the **postcranial skeleton** is all bones posterior to (behind) the skull. The postcranial skeleton consists of **axial** and **appendicular** elements; axial are the vertebrae of the neck, backbone, pelvis, and tail, and appendicular are the elements of the limbs and pelvic and pectoral girdles. Throughout the book the terms *anterior* and *posterior* refer to directions toward the front (skull or tip of snout) and back (tip of tail) of the animal, respectively. *Dorsal* refers to the direction toward the top of the spinal column and *ventral* to that toward the bottom of the neck, torso, or tail.

Some of the more important skeletal terms for land vertebrates are as follows. Cranial elements comprise the skull. A representation of an early terrestrial vertebrate shows most of the external bones of the skull (fig. 2.1); not all of these show up in all later groups. Figure 2.2 shows most (but not all) of the bones of the dinosaur skull, demonstrated here by a side view of the most abundant dinosaur in the Morrison Formation, *Camarasaurus*. Cranial elements include the tooth-bearing **premaxilla** and **maxilla** of the upper jaw. The upper edge and ascending process of the premaxilla form the front edge of the **external naris**, the skeletal opening for the fleshy nostrils. The top of the skull is formed mostly by the **nasal, frontal, parietal,** and **squamosal** bones, and the **orbit** (or eye socket) is enclosed by the **jugal, postorbital,** frontal, **prefrontal,** and **lacrimal** bones. (The "orbital horns" of the dinosaur *Allosaurus* are formed by the top edges of the lacrimals.) The back part of the skull includes the **quadratojugal,** the **quadrate** (which in reptiles articulates with the lower jaw), and the **occipital condyle,** which is the process on which the skull pivots relative to the neck. There are also many bones on the interior of the skull, including the **vomer** and **palatine,** which together help form the roof of the mouth; and the **pterygoid,** ectopterygoid, basioccipital, and basisphenoid, the latter two of which, together with several other bones, form part of the braincase.

The lower jaw includes the tooth-bearing **dentary** bone, plus (in reptiles) postdentary bones such as the **angular, splenial, surangular, articular, coronoid,** and **supra-angular**. Most mammals lack the postdentary

The Vertebrate Skeleton

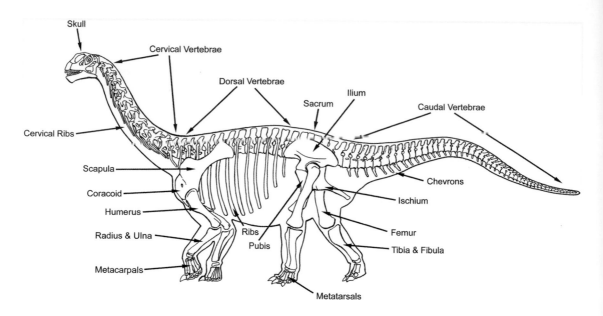

Skull

Cervical Vertebrae

Dorsal Vertebrae

Ilium

Sacrum

Caudal Vertebrae

Cervical Ribs

Scapula

Coracoid

Humerus

Radius & Ulna

Metacarpals

Ribs

Pubis

Metatarsals

Chevrons

Ischium

Femur

Tibia & Fibula

2.3. Skeleton of the Morrison sauropod dinosaur *Camarasaurus* showing the postcranial bones.

bones and have only the dentary, although primitive mammals retain some of the postdentary bones in very reduced form.

Postcranial elements are shown by a *Camarasaurus* skeleton (fig. 2.3). **Cervical vertebrae** are the bones of the neck. Vertebrae are composed of four main elements: the **centrum,** or the spool-shaped main body of the vertebra; the **neural arch,** which attaches to the top of the centrum and surrounds the spinal cord; the **neural spine,** which projects upward from the arch and provides an attachment for muscles and ligaments of the vertebral series; and the **zygapophyses,** which are just above the spinal cord at the base of the neural spine, each set of which articulates with the zygapophyses of the vertebra in front or behind.

Dorsal vertebrae are the bones of the vertebral column between the shoulder blades and the pelvis. **Sacral vertebrae** are the fused vertebrae of the pelvis. **Caudal vertebrae** are the bones of the vertebral column in the tail.

Cervical ribs are short to long bones extending backward along the sides of cervical vertebrae; these sometimes fuse onto the dual articulations with the vertebrae but often are detached. **Dorsal ribs** are the typical large thoracic ribs that one thinks of in a vertebrate skeleton. They are longer than cervical ribs and project downward perpendicular to the vertebral series, rather than parallel to it as in cervical ribs, and also form the rib cage. **Chevrons,** or **hemal arches,** are short, forked bones that extend downward from between the centra of the caudal vertebra series, most of the way along the tail. In life, they enclosed the main artery that extended down the base of the tail.

Transverse process is a term most often used to describe lateral projections of the dorsal or caudal vertebrae. In dorsal vertebrae, these processes mostly include attachment points for the dorsal ribs. In caudal vertebrae, the transverse processes are sometimes called **caudal ribs** and do not contain any other bone attachment point but rather form an attachment for and division between the upper and lower muscle masses of the tail.

The **scapula** and the **coracoid** are the main elements of the shoulder blade in vertebrates of the Morrison Formation. The **humerus** is the bone of the upper forelimb; the **radius** and **ulna** form the lower forelimb. The **femur** is the bone of the upper hind limb; the **tibia** and the **fibula** are the bones of the lower hind limb. The **astragalus** and **calcaneum** are the main bones in the reduced, simplified ankle of most dinosaurs (and the only ones, in fact, in sauropod dinosaurs, as far as we know); there are a number of other bones in most vertebrate ankles, including those of humans. **Carpals** are the bones of the wrist, and **metacarpals** and **metatarsals** the bones of the hand and foot, respectively, between the wrist or ankle and the knuckles. The fingers and toes are composed of phalanges (singular **phalanx**), individual bones that articulate in series. The human thumb has two phalanges, and each of the other fingers has three. The claws of many vertebrates, including dinosaurs, are known as **unguals**. The bones of the vertebrate pelvis include the **ilium,** which is attached to the sides of the vertebrae of the sacrum, and the **pubis** and **ischium,** which project downward from the ilium on both sides.

How did the vertebrates of the Morrison Formation come to be living in that area 150 million years ago (mya)? We need to understand a little of the history of their ancestors. Vertebrates as we know them today include a variety of animals, from horses and pigs and cats and dolphins to frogs, lizards, crocodiles, birds, fish, and lampreys. The key aspect of being a vertebrate is having, among other things, a backbone (which is not always actually made of bone but may be, as in the case of sharks, **cartilaginous**—that is, made of **cartilage,** the same material that gives structure to your ears and nose). All vertebrates are part of a slightly larger phylum of animal life known as the **Chordata**. Chordata comprise animals that possess a small suite of characters, including a dorsal hollow nerve cord (forming the spinal cord and brain), which contrasts with the solid and ventral nerve cord of most other animal groups; a notochord (a stiffening rod between the digestive tract and the nerve cord); and a muscular postanal tail (nonchordates have a digestive tract that extends nearly the full length of the body) (Campbell 1993; Hildebrand 1995). All vertebrates are chordates, but not all chordates are vertebrates. Vertebrates retain the **chordate** characters in a slightly modified form; the notochord is replaced by cartilaginous or bony vertebrae, and the dorsal hollow nerve cord expands anteriorly into a more complex brain (Carroll 1997).

The nonvertebrate chordates include sea squirts (also known as urochordates or tunicates) and the marine lancelets (cephalochordates; *Branchiostoma*). Although most adult sea squirts are stationary animals that attach themselves to stones or docks, for example, and filter food out of seawater that they circulate through their hollow bodies, both larval sea

Vertebrate Evolution as a Prologue to the Morrison

squirts and the lancelets are capable of swimming. Lancelets settle into the sand on the bottom of the ocean and also filter feed but can move from place to place; the larval sea squirt eventually attaches to one spot and develops into an adult form. In sea squirts, the chordate characters of a notochord, dorsal hollow nerve cord, and postanal tail are obvious only in the larvae and are not readily recognizable in the eggplant-shaped adults.

True vertebrates may have developed in the Early Cambrian from the free-swimming larvae of sea squirt–like (urochordate-like) animals. How could this happen? In a phenomenon known as **paedomorphosis,** juvenile characteristics are retained in adults. This happens because sexual maturity can be reached at a stage of development that was previously immature in the species, either by speeding up the development of sexual maturity or by slowing down other development in the individual. Thus, the larvae of an early urochordate may have developed the ability to reproduce while still in the swimming larval stage before becoming stationary adult filter feeders. Because this process was successful, the metamorphosis from the free-swimming to sessile life modes may have been stopped altogether, and the first fishlike chordate species would have originated. I hinted above that not all sea squirt adults are stationary, and in fact there are a few that are "larval" throughout life. Recent work on these free-swimming animals suggests that they are among the more primitive of sea squirts, which indicates that, contrary to the paedomorphosis idea of fish origins, the sessile adult stage for the sea squirts may rather be a specialized condition (instead of the free-swimming form being the new invention).

Chordates are known from the fossil record starting in the Cambrian period more than 500 mya. On a high ridge in the mountains of British Columbia, Canada, there are well-preserved fossils of a variety of animals from the 515-million-year-old Burgess Shale, and in these deposits have been found 3.8-cm-long (1.5 in.) specimens of *Pikaia*, a small, early chordate that lived among trilobites and the large and rather unusual marine predatory invertebrate *Anomalocaris*. *Pikaia* is not necessarily a chordate ancestral to modern vertebrates, however, because recently in China, paleontologists have found well-preserved true vertebrate fossils in Early Cambrian rocks of about 530 mya. These animals, known as *Haikouichthyes*, are less than 5 cm (2 in.) long. In overall appearance, they are similar to modern lancelets, but they have the distinctive vertebrate character of vertebrae along their body length. And more recently it was discovered, based on new specimens, that the Burgess Shale form *Metaspriggina*, previously recognized as a possible second chordate from the formation, was in fact a full-blown vertebrate (Conway Morris and Caron 2014). From the chordate stage, all that had been required for true vertebrate status was to develop the vertebrae and the enlarged brain and sense organs of the head. These Early to Middle Cambrian vertebrates such as *Metaspriggina* and *Haikouichthyes* eventually evolved into the first diverse group of vertebrates, the **Agnatha,** or jawless fish (Briggs et al. 1994; Shu et al. 1999).

In the earliest part of the Paleozoic era, almost all known vertebrates were agnathan fish. These were animals with cartilaginous vertebrae like the modern lamprey; they also shared with the lamprey the lack of a bony jaw or real teeth. Unlike lampreys, however, most of the Paleozoic forms had at least the front half of their bodies covered with a series of small bony plates in the skin that probably provided more defensive protection than a structural base. This was the only bone in their bodies; the rest of the skeleton was cartilage. The agnathans were essentially cartilaginous fish swimming the seas in a coating of bony mail armor. Most agnathans were also small and lacked paired fins. Many likely fed on organic material that they filtered out from the water or sediment.

For a long time, fossils known as conodonts were biostratigraphic superstars, although neither the paleontologists that studied them nor the geologists that relied on them for dating rocks had any idea what kind of animals they represented. Recent studies indicate that the strange, pointy, multicusped teeth of the conodonts are the only hard parts of otherwise rather elongate, worm-shaped vertebrate "jawless" fish.

Jawed fish became common near the middle of the Paleozoic era. The development of jaws was accomplished by a modification of the previously existing supports for the gill slits. The gill regions had been used for some time both as filter-feeding devices and for gas exchange, as in modern fish, but the incorporation of two sets of the gill slit supports into the mouths of fish was a major event that allowed more predation and a greater diversity of individual ecologies of the fish. Shortly after the development of jaws, true teeth originated as well (twice, in fact).

Jawed fish also developed bony internal skeletons, as seen in almost all modern fish. (Modern lampreys and hagfish are modified descendants of the agnathans, and sharks, with their jaws and teeth but a skeleton composed of cartilage, are probably modified descendants of the bony ancestors.) These bony fish are known as the **Osteichthyes** and comprise two subgroups: the **Actinopterygii,** including the trout and tuna and most of today's other freshwater and marine fish; and the **Sarcopterygii,** including lungfish and the great living-fossil coelacanths, long thought to have gone extinct at the end of the Cretaceous period but found living deep in the waters off Madagascar in the twentieth century. It is from the sarcopterygians that land vertebrates developed in the Devonian period about 370 mya. Sarcopterygians had elements in their fins that ended up being turned into the elements of the limbs of tetrapods (vertebrates with digits and legs). Unlike actinopterygians, which have fins made of skin and thin, fanned-out rays, the bases of sarcopterygian fins are composed of muscular, scale-covered flesh and are supported by one proximal and two distal bones. These were then followed distally by more bones fanning out in a manner roughly similar to those of actinopterygians. This fleshy base is the structure that inspired the group's name, Sarcopterygii ("flesh-fin") and is the precursor to all terrestrial vertebrate limbs. All tetrapods have a single proximal limb element (the humerus and femur of forelimbs and hind limbs, respectively) and two more distal ones (the radius and ulna, and the tibia and fibula). The bones of the sarcopterygian distal radials developed into the wrists, ankles, fingers, and toes of the tetrapod **autopod,** and the fin rays were lost (Amaral and Schneider 2017). The first tetrapods were essentially sarcopterygians that had developed their fins into walking limbs. However, some of these retained a fishlike tail and were largely aquatic.

The neat, simple stories of major transitions in vertebrate history almost invariably break down as soon as we start learning more about them. The transition from cynodont **synapsids** to true mammals during the Triassic is one such case. As more and better fossils were found, we ended up with animals with such mixed suites of mammalian and "mammal-like reptile"–grade characters that the line defining what is a mammal and what isn't has blurred. Likewise, there was a time when, the story went, a sarcopterygian named *Eusthenopteron* (or an animal very much like it) gave rise to the first **tetrapod** to crawl out of the water (*Ichthyostega*), and these first tetrapods eventually led to primitive amphibians like *Eryops*. All this fell apart when paleontologists started finding sarcopterygians more closely related to tetrapods than was *Eusthenopteron*, and when they began to find new taxa and more and better-preserved specimens of some known tetrapods. What they found were tetrapods that not only retained fishlike tails but also had seven or eight digits on some of the limbs and, in the case of *Ichthyostega*, had large, weight-bearing forelimbs and small, paddle-like hind limbs best suited for steering in the water. It may have spent some time in the water and some time hauling itself around on land like a sea lion. *Acanthostega*, a tetrapod and contemporary of *Ichthyostega*, appears to have had paddle-like limbs and is not likely to have left the water much at all, but there is some question as to whether it is a tetrapod that had not left the water yet or whether, perhaps, it was a secondarily aquatic tetrapod (from a terrestrial ancestor) even this early in the group's history. *Tiktaalik* was described fairly recently, and it appears to be transitional between the "front-wheel drive" forms like *Ichthyostega* with weight-bearing forelimbs and later full tetrapods in that its hind appendages (pelvic girdle and fins) were more developed (Shubin et al. 2014). One aspect of the problem is that, whereas in the past we assumed that limbs developed in order to move about on land and thus appeared tetrapods, in reality there are fish today that have what are essentially limbs but that almost never leave the bottom of the ocean; and conversely, there are fish that move around on land just fine with regular fins. And just a few decades ago, we would not have guessed that front and back limbs would have developed out of synch. So the origin of limbs and the origin of terrestriality may be two totally different stories. And thus we find strange vertebrates like *Tiktaalik*, *Ichthyostega*, and *Acanthostega*, none fully terrestrial, among some of the first tetrapods.

Another common theme of this and other transitions (such as that leading to mammals) is the concept of **mosaic evolution,** which is the term applied when different parts of the skeleton in a group of animals change more quickly or slowly—that is, when the entire animal does not progress into the next mode as a unit (as we saw above). In the mammalian transition, for example, lower jawbones left over from the nonmammalian synapsid grade are retained in species otherwise defined as mammals. In the early tyrannosaurid dinosaur *Dilong*, the skull seems to have changed toward an advanced tyrannosaur condition, more so than the postcranial skeleton. Among the first tetrapods, the same lurching evolution also occurred in different parts of the skeleton. Most have reduced the number of skull bones and shortened the skull in general, as in later tetrapods, and limbs have developed, but the animals retain fishlike tails.

The occurrence of seven and eight digits in the feet of some early tetrapods is interesting. The five fingers and toes that humans have is the long-established primitive condition for modern tetrapods. Some species have reduced this number to varying degrees (most carnivorous dinosaurs have only three fingers in the hand, horses only one), but none has more than five (house cats and humans as species have five even though there are the occasional six-toed individuals). Thus, the recent discovery of seven to eight toes in the earliest tetrapods indicates that we terrestrial vertebrates probably inherited our five-toe body plan from slightly younger descendants of *Acanthostega* and *Ichthyostega*, after next-generation tetrapods had begun moving about somewhat on land and adopted an amphibious lifestyle, and the number of fingers and toes, after early experimentation with a greater finlike number, had been settled at five (Clack 2002).

The amphibious tetrapods of the end of the Devonian and into the early Carboniferous period (about 350 mya) were the first terrestrial vertebrates, but they still depended on water for the reproductive cycle, laying eggs in water or in very damp areas. They were unlike amphibians we know today, such as frogs and salamanders. For one thing, these ancient vertebrates were much larger, some of them 1.7 m (5 to 6 ft) long. At some point early in the Carboniferous, one group of tetrapods developed something different from all others that allowed the animals to lay eggs away from water sources. This was the **amnion,** a fluid-filled sac formed around the embryo and within the egg (or in the uterus in mammals). This solved the tetrapods' previous dependence on water for their eggs, and although amphibians continued as before, the new group (the amniotes) developed hard eggshells and began laying eggs far from water sources. All modern and ancient birds, reptiles, and mammals are amniotes. Eutherian (placental) and metatherian (marsupial) mammals have retained the amnion but lost the shelled eggs of their ancestors (monotremes still lay eggs).

Sometime before the end of the Carboniferous, before about 300 mya, the **amniote** line itself split in two. One group eventually led to

2.4. Skulls of tetrapods showing the basal design (anapsids) and the major split among most amniotes into diapsids with two temporal fenestrae and synapsids with just one. There were many other synapsids before mammals (including the sail-backed Permian *Dimetrodon*), but our furred group is all that remains today.

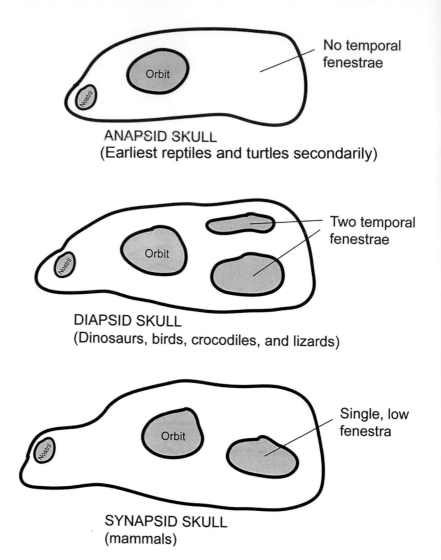

No temporal fenestrae

Orbit

Nostril

ANAPSID SKULL
(Earliest reptiles and turtles secondarily)

Two temporal fenestrae

Orbit

Nostril

DIAPSID SKULL
(Dinosaurs, birds, crocodiles, and lizards)

Single, low fenestra

Orbit

Nostril

SYNAPSID SKULL
(mammals)

today's reptiles and birds (**anapsids** and **diapsids**), as well as a wide diversity of ancient "reptile" groups. The other line led through many ancient forms and eventually to mammals (synapsids). The division at this branching is based mainly on the structure of the skull: the mammal line developed a single opening low on each side of the skull behind the eyes. In humans, this opening, first developed at least 75 million years before the appearance of dinosaurs, is expressed by the gap between the cheekbones and the braincase of the skull just in front of each ear. At the time, most other amniotes had no openings in their skulls behind the eyes (fig. 2.4). As the line leading to mammals continued, developing a wide range of species (including *Dimetrodon*, the sail-backed, four-legged animal of children's plastic "dinosaur" sets), the reptile-bird line also diversified. Most of the early forms on this branch still had no openings behind the eyes, but later, some developed two openings behind the eyes (diapsids, e.g., *Petrolacosaurus*), and from these ancestors developed many familiar

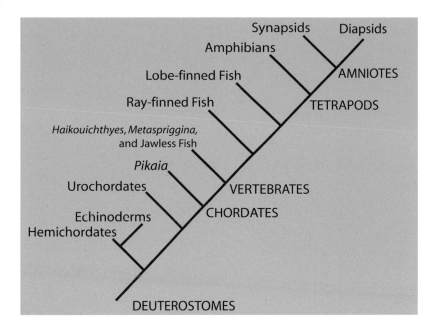

2.5. Relationships of vertebrate groups as discussed in the text. The groups closest to the urochordates and chordates appear to be the echinoderms, which include modern starfish and sea urchins, and the hemichordates (acorn worms). Chordates and vertebrates first appear during the Cambrian period. Among amniotes, synapsids include mammals and diapsids include crocodiles, birds, dinosaurs, and lizards. Note that lobe-finned fish (lungfish and coelacanths) are closer to tetrapods than they are to other fish.

ancient and modern groups, including lizards, snakes, crocodiles, birds, pterosaurs, and dinosaurs. Turtles have no openings in the skull behind the eyes, but they appear to be modified diapsids that secondarily closed the two openings.

Figure 2.5 summarizes the relationships of chordates and vertebrates as outlined above. Although the fauna of the Morrison Formation includes some ray-finned fish, lungfish, and amphibians, most of the vertebrate taxa are amniotes. During the Triassic period, about 225 mya, dinosaurs and mammals appeared on the scene at nearly the same time. In western North America, this is the time of the Chinle Group, a time dominated by large reptiles, but only a few of them were dinosaurs: *Coelophysis*, the small carnivorous theropod made famous by the Ghost Ranch Quarry in New Mexico; some other small theropods; and unidentified prosauropod dinosaurs known only from fragmentary remains and footprints. There are also ancient forms of groups later known from the Morrison around at this time (mammals, pterosaurs, sphenodontids, and crocodylomorphs), but dinosaurs are rare. Most of the large reptiles of Chinle times were archosaurs, a particular group of diapsid reptiles that includes dinosaurs, crocodiles, and birds (but not lizards or snakes). Among the nondinosaur **archosaurs** of the Late Triassic were low, tank-like, plant-eating aetosaurs; large, carnivorous, terrestrial rauisuchids; and carnivorous, semiaquatic, crocodile-like phytosaurs. There were large, herbivorous synapsids too, including *Placerias*. So the world of the Late Triassic had some of the same groups as we see in the Morrison Formation but overall was rather different. A major change in the terrestrial vertebrate fauna occurred at the Triassic–Jurassic boundary, however, and by the beginning of the Early Jurassic 201 mya, the stage was set for the Morrison Formation's cast of characters.

During the Early Jurassic, there were still sphenodontids, mammals, and pterosaurs, but now there were turtles, more modern crocodylomorphs, and a diversity of dinosaurs as well. The first large, long-necked, long-tailed sauropod dinosaurs appear at this time, and the bipedal, carnivorous theropod dinosaurs diversify. The Early Jurassic–age Kayenta Formation is exposed throughout much of the Four Corners region of the US Southwest, and in most of this outcrop area, it is a deposit consisting mainly of sandy braided stream sediments. But in northern Arizona, there is a muddy facies of the Kayenta that preserves a number of vertebrate body fossils. (Little is known from the Kayenta in other areas except for tracks.)

Crews working in the muddy facies have found the carnivorous dinosaurs *Dilophosaurus* and *Coelophysis*(?) (*Syntarsus*).[1] *Dilophosaurus* everyone will recognize from the 1993 movie *Jurassic Park*, although in reality there is no evidence at all that this animal had a neck frill or was venomous, and it was in fact quite a bit larger than portrayed in the film, being about 6 m (20 ft) long. (*Dilophosaurus* did, however, have a pair of parallel crests on the top of its skull.) Other animals that have turned up in the muddy facies of the Kayenta include pterosaurs, frogs, turtles, mammals, and the herbivorous dinosaurs *Massospondylus* and *Scutellosaurus*. As we move into the Middle Jurassic in North America, there are few fossils other than some footprints, and we thus have an incomplete picture of the faunas. The records from Asia and Europe, however, are more complete, and these indicate that most of the major groups of animals known from the Morrison Formation had appeared by the middle part of the Jurassic, including spiked stegosaur dinosaurs (**Stegosauria**), armored **ankylosaur** dinosaurs, and lizards. In fact, some of the same vertebrate genera from the Morrison Formation first appear in the Middle Jurassic of Europe.

Dinosaurs evolved from small, carnivorous, bipedal archosaurs about 230 mya. There are several types of archosaurs that are very dinosaur-like but that are not by definition included within the **Dinosauria**. The best-known and most diverse group of archosaurs that is closely related to the dinosaurs is the pterosaurs, but several obscure taxa of archosaurs are closer and are even more dinosaur-like. Dinosaurs are defined by specific characteristics, mainly of the skull and ankle, which all species of dinosaurs share, from the oldest and most primitive to the very last.

The origin of dinosaurs in the Triassic was an important event of the Mesozoic, although it did not have immediate significance. Dinosaurs appeared some 30 million years before they rose to any kind of dominance of the terrestrial ecosystems (after the end of the Triassic), and many were small compared with other archosaurs that were around. During Chinle Group time, for example, phytosaurs and rauisuchids were **carnivores** many times heavier in body weight than the carnivorous dinosaur *Coelophysis*. The herbivorous archosaurs were smaller than other nondinosaurian herbivores such as *Placerias* and the aetosaurs (although diets among these groups likely were quite different), and the poorly known prosauropod dinosaurs of the Chinle were barely larger than their nondinosaurian counterparts. There were also plenty of taxa around that were superficially rather similar to dinosaurs but were not actually members of Dinosauria (e.g., silesaurids), and true dinosaurs may have come to dominate different parts of the globe at different times (Irmis et al. 2007). Ecologically, then, the Late Triassic was not dinosaur dominated; that would come only with the onset of the Jurassic period.

So the stage was now set for the Morrison Formation and its vertebrate cast. These vertebrates included ray-finned fish and lungfish; frogs; salamanders; turtles; lizards; lizard-like sphenodontians; one of the earliest and smallest members of the extinct, semiaquatic archosauromorphs called choristoderes; several terrestrial and semiaquatic crocodylomorphs; the flying pterosaurs; the bipedal, meat-eating theropods; the giant, plant-eating sauropods; the plated and spiked stegosaurs; the heavily armored ankylosaurs; the bipedal, plant-eating ornithopods; and mammals (table 2.1).

Table 2.1. Fossil vertebrates from Morrison Formation. Updated from Foster (2003a) and Chure et al. (2006)

Fish	Hoplosuchus	Haplocanthosaurus
"Hulettia"	Fruitachampsa	Camarasaurus
Morrolepis	New shartegosuchid	Brachiosaurus
cf. Leptolepis	Theriosuchus	Armored dinosaurs
Amiiformes	Amphicotylus	Stegosaurus
Pycnodontoidea	Eutretauranosuchus	Hesperosaurus
Pholidophoriformes	Pterosaurs	Alcovasaurus
Lungfish	Mesadactylus	Mymoorapelta
Potamoceratodus	Dermodactylus	Gargoyleosaurus
Ceratodus	Kepodactylus	Heterodontosaurid
Frogs	Comodactylus	Fruitadens
Enneabatrachus	Harpactognathus	Basal neornithischians
Rhadinosteus	Utahdactylus	Nanosaurus
Pelobatidae	Theropod dinosaurs	Ornithopod dinosaurs
Salamanders	Ceratosaurus	Dryosaurus
Iridotriton	Fosterovenator	Camptosaurus
Caudata B	Torvosaurus	Mammals
Turtles	Allosaurus	Docodon
Glyptops	Saurophaganax	Fruitafossor
Dinochelys	Marshosaurus	Ctenacodon
Uluops	Stokesosaurus	Psalodon
"Dorsetochelys"	Aviatyrannus-like form	Glirodon
Sphenodontids	Ornitholestes	Zofiabaatar
Opisthias	Coelurus	Morrisonodon
Theretairus	Tanycolagreus	Triconolestes
Eilenodon	Koparion	Aploconodon
Lizards	Hesperornithoides	Comodon
Dorsetisaurus	Sauropod dinosaurs	Priacodon
Paramacellodus	Dystrophaeus	Trioracodon
Saurillodon	Suuwassea	Amphidon
Schillerosaurus	Maraapunisaurus	Tinodon
Snakes	Diplodocus	Foxraptor
Diablophis	Barosaurus	Paurodon
Choristoderes	Kaatedocus	Comotherium
Cteniogenys	Galeamopus	Tathiodon
Crocodylomorphs	Supersaurus	Amblotherium
Hallopus	Apatosaurus	Dryolestes
Macelognathus	Brontosaurus	Laolestes

Spanning the Globe: Jurassic Vertebrate Communities

Morrison Formation vertebrates did not exist in isolation. Other areas have fossils of the same age that illustrate what the world was like and how the Morrison was similar and in other ways unique. As we will see, there were some genera and at least one species that occur both in the Morrison and on other continents.

Europe

In Europe, the Late Jurassic was a warm time, and much of the area was covered by a shallow continental sea along the northern margin of an ocean that formed as the European continent and northern Asia split away from Africa and the other southern continents. What is now southern Germany was then a shallow tropical sea with scattered islands, some the size of modern-day Cuba, and lagoons of this archipelago were quiet and undisturbed by heavy currents. The bottom may well have been very low in oxygen, and these conditions proved to be excellent for the preservation of complete and nearly unscavenged skeletons of all types of animals. These fossils are preserved in the thin Solnhofen Plattenkalk limestones that formed in the deeper basins between sponge and algal mounds near coral reefs and are from the Tithonian age of the Late Jurassic. The Solnhofen deposits, from near the Danube Valley north of Munich, include one of the best-known faunas in the world and record Late Jurassic life complementary to that of the Morrison. Preserved in these thin rocks that look like stacks of plywood are sponges, jellyfish, corals, worms, clams, marine snails, squids, ammonoids, barnacles, shrimp, lobsters, horseshoe crabs, dragonflies, cockroaches, water skaters, beetles, wasps, flies, starfish, sea urchins, sharks, rays, bony fish, turtles, ichthyosaurs, plesiosaurs, lizards, crocodiles, pterosaurs, the dinosaur *Compsognathus*, and the first bird, *Archaeopteryx* (fig. 2.6; Barthel et al. 1990). This diversity of preserved animals gives some indication of what would have been around in the seas of North America during deposition of the nearly contemporaneous Morrison Formation. *Archaeopteryx* is certainly the most famous of the fossils found in the Solnhofen Plattenkalk. The Berlin specimen in particular is one of the most beautiful vertebrate skeletons ever uncovered, preserving exquisite details of the bones and feathers, and its importance as the first bird is well known (Shipman 1998). But the sheer diversity of other vertebrates and invertebrates from this deposit stands out as well. Although they are not preserved in or near the Morrison, sharks, shrimp, starfish, and other animals likely lived in the seas and oceans surrounding the North America of the time, and the types of insects preserved in the Solnhofen deposits probably had related species that served as food for insectivorous vertebrate species of the Morrison. We do know that the Sundance Sea, which immediately preceded Morrison deposition in the northern part of the Rocky Mountain region, was home to belemnoids, plesiosaurs, and crinoids, among other types of animals.

2.6. Late Jurassic bird, *Archaeopteryx*, from the Solnhofen limestones of southern Germany. Animals like this may have existed at the same time in the area of the Morrison Formation, but we have not yet found evidence of them.

Another taxon from the Late Jurassic of Germany, *Europasaurus*, is a dwarf sauropod that was found to be among the camarasauromorphs phylogenetically (Carballido and Sander 2014).

To the northwest of the Solnhofen deposits, in what is now England, several terrestrial deposits of Late Jurassic age have yielded a number of vertebrates. These rocks represent floodplain deposits near the coast of the seaway that existed to the southeast. The Jurassic part of the Purbeck Formation actually contains several small vertebrates assigned to the same genera as animals found in the Morrison: the sphenodontian *Opisthias*, the lizards *Paramacellodus* and *Dorsetisaurus*, and the mammals *Ctenacodon* and *Amblotherium*. There are not many dinosaurs known from the Purbeck, but one, *Echinodon*, is closely related to a form found in the Morrison Formation.

The Kimmeridge Clay is a Late Jurassic–age formation in England and contains several dinosaurs, including the brachiosaurid *Bothriospondylus*, the stegosaur *Dacentrurus*, and the ornithopod *Camptosaurus*. *Camptosaurus* is a bipedal plant eater also known from the Morrison

Formation and is particularly common at Quarry 13 at Como Bluff. The Kimmeridge also recently yielded a new species of the theropod *Stokesosaurus*, a genus originally found in the Morrison Formation (Benson 2008). These faunal similarities between the Upper Jurassic deposits of Great Britain and those of North America attest to the wide geographic range of many of the groups of vertebrates that existed at the time, and as we will see, parallels between these two particular regions are just the beginning.

Near the town of Leiria in western Portugal, in 1959, a coal mine began yielding what became one of the most important windows into the world of the Late Jurassic, ultimately resulting in a collection of some 800 skulls and jaws and nearly 7,000 isolated teeth of mammals, along with many other organisms. The Guimarota Mine is in the coal and limestone deposits of the Alcobaça Formation, and the rocks at the site represent a coastal swamp environment similar in overall appearance (although not in actual plant types) to the mangrove swamps of the Florida Everglades. The fossils here are of the Kimmeridgian age of the Late Jurassic, correlative in time to most of the Morrison Formation but slightly older than the Solnhofen Plattenkalk. Among the vertebrate specimens preserved in the Guimarota Mine are sharks, ray-finned fish, nearly 9,000 **amphibian** bones (including frogs, salamanders, and albanerpetontids), turtles, lizards, crocodiles, pterosaurs, mammals, and several types of dinosaurs (T. Martin and Krebs 2000). Dinosaurs are represented mainly by teeth and include two types of neornithischians (the hypsilophodontid *Phyllodon* and an iguanodontian), sauropods, and several types of theropods (Rauhut 2001). Genera in common between Guimarota and the Morrison Formation include the lizards *Saurillodon*, *Paramacellodus*, and *Dorsetisaurus*; the choristodere *Cteniogenys*; goniopholidid crocodyliforms; and the mammal *Dryolestes*. Tiny teeth referred to *Archaeopteryx* have also been found in the mine.

Other Upper Jurassic sites in Portugal have produced a camarasaurid sauropod, *Lourinhasaurus*; the diplodocine sauropod *Dinheirosaurus* (possibly a species of *Supersaurus*); and the sauropod *Lusotitan*, a possible brachiosaurid; plus the theropods *Allosaurus*, *Torvosaurus*, and *Ceratosaurus*, all three also known from the Morrison Formation (Pérez-Moreno et al. 1999; Mateus and Antunes 2000a, 2000b; Mateus et al. 2006; Mannion et al. 2012; Mannion et al. 2013; Mocho et al. 2014a, 2014b; Mateus et al. 2017). The new theropod *Aviatyrannus* is known from Portugal, and a similar form may be represented in the Morrison Formation (Rauhut 2003). The long-necked ("sauropod-mimic") stegosaur *Miragaia* was recently described from the Late Jurassic of Portugal also (Mateus et al. 2009). There are also dryosaurid and possible camptosaurid ornithopod dinosaurs reported from Upper Jurassic deposits in Portugal (Mateus and Atunes 2001; Escaso et al. 2014; Rotatori et al. 2020).

Africa

In the early 1900s, German paleontologists Eberhard Fraas, Werner Janensch, and Edwin Hennig began leading excavations in the Tendaguru Beds in what is now Tanzania in east Africa. The Tendaguru excavations involved a number of individual quarries near the Mbemkuru River in Upper Jurassic rocks. The site had been found in 1906 by B. W. Sattler, who managed a local garnet mine, and the subsequent fieldwork was carried out by local crews led by German researchers from 1907 to 1912 (Maier 2003). After World War I, British crews worked Tendaguru for many years, and ultimately these deposits produced a dinosaur fauna roughly contemporaneous to the Morrison Formation and including several similar genera: *Giraffatitan*, a close relative of *Brachiosaurus*; *Dysalotosaurus*, which is similar to *Dryosaurus*; *Elaphrosaurus*, a cerato-saurian perhaps related to an unidentified Morrison form; the diplodocid sauropods *Tornieria* and ?*Barosaurus*; the dicraeosaurid *Dicraeosaurus*; and possibly *Allosaurus*. The purported diplodocid sauropod *Australodocus* was described recently (Remes 2007), although this specimen has also been argued to be a possible brachiosaurid (Whitlock 2011a). Tendaguru also has the new carcharodontosaurid theropod *Veterupristisaurus* (Rauhut 2011). The Tendaguru deposit represents a terrestrial paleoenvironment near a coastal shoreline; shallow marine and freshwater units are interbedded at the site (Russell et al. 1980).

The Upper Jurassic Kadzi Formation of northern Zimbabwe has yielded fragmentary dinosaur remains, and among these are the Morrison genera ?*Barosaurus* and ?*Camarasaurus*, a brachiosaurid, plus the Tendaguru genera *Dicraeosaurus* and *Tornieria* (Raath and McIntosh 1987). Recent work has also identified indeterminate theropod and ornithopod dinosaurs, plus the crocodilian *Goniopholis*, in Upper Jurassic deposits in Ethiopia (Goodwin et al. 1999).

Asia

In Asia, the similarities with the Morrison Formation seem to lessen, but recently some related taxa have been identified. The Shangshaximiao Formation, in China, is early Late Jurassic in age and contains a possible allosaurid (*Szechuanosaurus*), the unusual nonneosauropods *Omeisaurus* and *Mamenchisaurus*, and two types of stegosaur (*Tuojiangosaurus* and *Chungkingosaurus*). The Suining Formation, also Late Jurassic in China, has the mamenchisaurid sauropods *Memenchisaurus* and *Qijianglong* (Xing et al. 2015a). The dinosaurs from China are generally more distantly related to those of Europe, Africa, and North America, which among them share some common genera. Most elements of the Chinese fauna, however, are essentially endemic or, among sauropods, more basal. There have been reports of an Early Cretaceous diplodocid sauropod in Asia (Upchurch and Mannion 2009), although this claim has been contested

(Whitlock et al. 2011), and in fact diplodocids may have never existed in Asia (Xing et al. 2015b).

The Late Jurassic Shar Teg locality in southwestern Mongolia has yielded a variety of animals including many insects, fish, turtles, crocodylomorphs, dinosaurs, a mammal, and labyrinthodont amphibians, the latter of which are not found in the Morrison Formation. Several of the shartegosuchid terrestrial crocodylomorphs, such as *Shartegosuchus* and *Nominosuchus*, are closely related to forms in the Morrison Formation (Clark 2011).

Although it is Early Cretaceous in age, the stegosaur *Stegosaurus homheni* from the Lianmuqin Formation of China once appeared most closely related to *Hesperosaurus mjosi* from the Morrison Formation (Maidment 2010), although now it appears to be closest (perhaps not surprisingly) to *S. stenops* (Raven and Maidment 2017). In either case, it seems to represent one of the Morrison's closer ties with Asia among dinosaurian taxa.

South America

The Upper Jurassic Cañadón Cálcareo Formation in the Patagonia region of Argentina yielded the camarasauromorph sauropod *Tehuelchesaurus* and has just recently produced a dicraeosaurid sauropod dinosaur named *Brachytrachelopan* (Novas 2009). This latter sauropod was rather small by sauropod standards, with a total length of less than 10 m (33 ft). It also had a short neck and may have filled the same ecological role, as a low-browsing **herbivore,** that was filled by large ornithopod dinosaurs such as *Camptosaurus* in North America (Rauhut et al. 2005). The formation has also recently yielded an indeterminate brachiosaurid sauropod close to *Brachiosaurus* (Rauhut 2006) and a coccolepidid fish, *Condorlepis*, that is in the same family as the Morrison's *Morrolepis* and several other forms from Asia and Europe as well (López-Arbarello et al. 2013). The Cañadón Cálcareo Formation is just about the same age range as the Morrison Formation, Kimmeridgian-Tithonian. Late Jurassic dinosaurs have been found in other parts of South America as well. Among these are a sauropod vertebra from northern Colombia and dinosaur tracks from Chile (Novas 2009).

The Start of It All: The Morrison Vertebrates Come to Light

3

Paleontology is a seasonal profession. More so than in some lines of work, paleontologists have a cyclical schedule that involves months of winter laboratory preparation, administrative responsibilities, teaching, and collections research, for example, followed by the outdoor summer field season. How long that field season is often depends on the climate in one's research area. In western Colorado and eastern Utah, one can generally depend on decent weather for most of the period from April through October. In Wyoming, however, the reliable part of summer is only June, July, and August (and sometimes not even June). In either case, paleontologists are usually out somewhere at the first sign of warm weather in the spring. In addition to a myriad of other individual reasons for choosing this particular career path, one nearly universal characteristic of our group is a love of being outdoors, and few waste an opportunity to do so. The reward of an important find after days of fruitless prospecting, or of getting a large field jacket safely removed from a quarry after weeks of work, is another draw. And in all of the work, each find, no matter how mundane, is a surprise and draws you on to keep looking.

Paleontological fieldwork, particularly quarrying, can thus be a particularly meditative experience. In the heat of the summer sun, uncovering a small dinosaur bone, splitting shale in search of trilobites, or working on part of the trench around a large sauropod bone, the world narrows down to just one's work space, and it becomes easy to ignore outside happenings and sounds. A sense of obsessive concentration prevails for hours on end, and the work becomes, almost counterintuitively, relaxing. Perhaps the simplicity of the task and the patience required add to the effect. It is work that few of us ever get bored with.

There are also experiences resulting from the remote nature of fieldwork that, although not unique among outdoor enthusiasts, are certainly unusual as a regular part of a job (fig. 3.1). In the olden days, excavating year-round in cold and driving snow were not uncommon, at least at Como Bluff in Wyoming, and weather and animals have always been interesting factors. Securing food or hauling enough in to a remote camp were important factors to consider a hundred years ago, when none of the stores and few of the towns we can access today were around. Nowadays, field experiences include some of the same elements but are more varied: vehicle breakdowns miles from anywhere in horribly mosquito-infested terrain or on a warm day when the water supply is running low; being rained in at camp for days at a time with little fuel for the campfire; trucks and other field vehicles sinking into mud in the middle of the night or

Fieldwork: A Little-Changed Art

Facing, 3.1. The wide world of Morrison paleontology: the thrill of victory and the agony of defeat. (A) Night under the stars in southeastern Utah. (B) Zeb Miracle and Kelsie Abrams plastering at an *Allosaurus* site in the Fruita Paleo Area. (C) Waiting for plaster to dry on an eggshell block in the Salt Wash of western Colorado. Vaia Barkas trying to be patient. (D) Five paleontologists to change a tire, southeastern Utah. *Left to right*: Matt Wedel, Jessica Uglesich, Brian Engh (straw hat), and the author (photo by R. Hunt-Foster). (E) J.-P. Cavigelli and crew demonstrate the timeless tripod-and-pulley system for lowering a field jacket into a trailer, Como Bluff, Wyoming. *Photo courtesy of Leslie Elliot. (F) Handiwork of Colorado Plateau biting midges at the Cleveland-Lloyd Quarry; photo courtesy of Joe Peterson. (G) Quarry floods as demonstrated in Montana. Danielle Lytle gets wading duty. Photo courtesy of Cary Woodruff.*

into snow in the middle of the day, getting stuck in deep ruts, sliding over soft ledges, and hanging up on sandstone benches, all a result of accessing remote terrain on unmaintained jeep trails; returning to the car after a day of prospecting, only to find it completely covered in winged harvester ants or simply blown over a cliff; snakes and other animals getting into the field equipment; and dodging lightning storms and their associated hail almost every year.[1] All these complications endured by paleontologists, volunteers, and employees over the decades are a natural part of the work, and perhaps on some level we enjoy the break they give us from the routine.

During 140 years of fieldwork, there has been a lot of planning and effort put into the excavation of vertebrates of the Morrison Formation—and by many people. We owe what we have learned of this Late Jurassic ecosystem and the conclusions we have drawn from the data to them.

First Finds

The early days of fieldwork in the Morrison Formation were times of often year-round work outdoors by men hired by paleontologists to collect for them in teams or alone. Later in the nineteenth century, the work was carried out more by small or large teams led by the paleontologists themselves, collecting for various museums. Most of the best-known dinosaurs of the Morrison Formation were discovered during this period, along with some of the most important quarries.

The first fossil material from rocks later assigned to the Morrison Formation was almost certainly noticed and pondered by Native Americans. Various tribes had legends for how the unusually large bones came to be weathering out where they were, and some of these involved giant animals and monsters of previous worlds; in many cases the recommendation was to leave the material alone and not risk reawakening the spirits of such massive beasts. Although some western fossils, such as trilobites, were seen by some as protective and appear to have been worn as amulets, the large bones were in many cases viewed with caution.[2] And fossils, including dinosaur footprints, were also incorporated into some living structures in the Southwest (Mayor 2005). The scientific study of North American vertebrate fossil material, first carried out by naturalists trained in a European tradition of inquiry that traces its origins back to the science and civilizations of Rome, Greece, and Egypt, began in the mid-1800s with the first organized surveys of the western part of North America.

The first recorded vertebrate fossil collected from rocks attributed to the Morrison Formation was found in 1859. In July of that year, US Army captain John Macomb led an expedition west, mostly along the Old Spanish Trail, to find a good route from Santa Fe into the southern part of Utah and to find the confluence of the Green and Colorado (then Grand) Rivers; he took with him as a scientist John Newberry, a physician and geologist (S. Madsen 2010). On August 17, as the party camped nearby in a desert canyon south of present-day Moab, Utah, Newberry

found bones of a sauropod dinosaur in what was later named the lowest member of the Morrison Formation, the Tidwell Member. Newberry collected bones at the site for a couple of days (part of a day after he found them and then another full day on the trip back toward Santa Fe), and the specimens were returned east with the expedition members when they finished. Eventually the bones were transferred to the paleontologist Joseph Leidy in Philadelphia, who likely identified them but did not describe or name them. That task was done by Edward D. Cope, a former student of Leidy's in Philadelphia. Cope (1877a) gave the few scattered

3.2. Work at J. S. Newberry's *Dystrophaeus* quarry, 2014–2018. (A) Crew working the quarry in 2016. (B) Helicopter lifting field jackets out from the site at the end of the season. (C) Randy Irmis rock-sawing around two larger jackets. (D) Bones as exposed in the quarry, paintbrush for scale.

bones of the sauropod (two forelimb elements, front foot bones, and other pieces) the name *Dystrophaeus viaemalae* in 1877.

And that was pretty much the end of the *Dystrophaeus* story. Or rather, it would have been if not for a naturalist writer from Utah named Fran Barnes, who, at the request of paleontologists Jack McIntosh and Jim Jensen, spent 12 years researching the expedition (luckily, Newberry and Macomb's surveyor, Charles Dimmock, took descriptive notes and measurements along the way; S. Madsen 2010) and, amazingly, was able to relocate the quarry (Barnes 1988, 1989, 2003). This was confirmed in September 1989 by a preliminary dig at the site and comparison with the original material (Gillette 1996a, 1996b).

In August 2014, 155 years to the day after Newberry discovered the material, the Museum of Moab, the Natural History Museum of Utah, Utah Field House, and the Museum of Northern Arizona teamed up to reopen Newberry's quarry and collect more of the skeleton. The teams returned in 2015–2019 (three of the times with helicopter assistance to lift material up and field jackets down) and have so far collected about 50 blocks of bone and rock from the quarry, with plans to return for more (fig. 3.2). This material will likely take several years to prepare out of the rock.

This is an important specimen (Gillette 1996a; McIntosh 1997; Foster et al. 2016b). Although not all workers agree that the Tidwell Member (and thus *Dystrophaeus*) is part of the Morrison Formation as it should be defined, the fact remains that this is the oldest body fossil of a sauropod known from the United States and possibly North America (Gillette 1996b). Until recently, sauropods appeared to be entirely absent from North America before the Late Jurassic and the Morrison Formation. But this specimen shows that some form was present at or just before the beginning of deposition of the Morrison Formation, and tracks from Mexico (Ferrusquía-Villafranca et al. 1996) and from the Entrada Sandstone near Escalante, Utah (Foster et al. 2000), indicate that sauropods were in the area just before *Dystrophaeus*. The bones are the only real way to tell to what other forms the first North American sauropods were related. Sauropods are known from much older rocks in Africa, Asia, and Europe and are quite common in Europe and China during the Middle Jurassic, so what led to their sudden appearance so much later in North America may prove to be a fascinating story. *Dystrophaeus* is the best place to start unraveling the tale.

The next documented Morrison Formation vertebrate fossil find surfaced a few years later, this time in a seemingly unlikely place: the mountains of Colorado. Although the flanks of the ranges of the Rocky Mountains are known to have plenty of Morrison Formation exposure, little occurs in the mountains themselves, and the rocks are often covered by grass, trees, and soil. There is a surprising amount of outcrop of the Morrison Formation at higher altitudes, and outcrops do sometimes yield specimens in this area. In 1869, a Middle Park, Colorado, resident showed F. V. Hayden a partial caudal vertebra of a dinosaur that had been collected locally, probably somewhere in the vicinity of the town of Kremmling, southeast of that minor powerhouse of American Olympic skiing, Steamboat Springs. Hayden was surveying the territories of Colorado and New Mexico for the United States Geological Survey, and when he returned from this trip, the specimen, which had been donated by the Middle Park resident, also ended up with Joseph Leidy in Philadelphia. In 1870, Leidy named the animal *Antrodemus* and recognized it as a carnivorous dinosaur. In later years, it would be determined that the fossil is most likely a caudal vertebra of the famous and abundant Morrison Formation dinosaur *Allosaurus*, named several years later by O. C. Marsh. (In this case, because *Allosaurus* is known from nearly complete skeletons and *Antrodemus* only from a single incomplete vertebra, the name *Allosaurus* is retained even though it was the second name applied to the animal. The *Apatosaurus/Brontosaurus* saga is a similar situation but seems to have created a bit more heat, although there may be a surprising resolution to that, which we will get to later.)

And so things sat in the paleontological world of the Morrison Formation for the next 7 years: two dinosaur finds separated by 10 years, by hundreds of miles, and by the fact that no one then might have guessed that they were of such similar age. These finds were but foreshadowings and preliminary skirmish shots warning of what was to come in 1877. In that year, the Morrison Formation paleontological powder keg in the Rocky Mountains blew big time.

One day in March 1877, railroad worker William H. Reed was returning to the rail station at Como Bluff, Wyoming, with his rifle and a freshly killed pronghorn antelope, descending the north slopes of the bluff, when he noticed several vertebrae and limb bones eroding out of the hillside (Breithaupt 1990, 1998). Recognizing these sauropod bones as those of a very large extinct animal, Reed and the station agent, William E. Carlin, began combing the bluff, and in the next few weeks, they found that it was covered in fossil vertebrate material. Reed and Carlin kept their finds secret for more than four months.

March 1877

Meanwhile, several hundred miles to the south, in a quiet valley known locally as Garden Park, near Cañon City, Colorado, a student from Oberlin College in Ohio found large sauropod bones in the Upper Jurassic rocks (not yet named the Morrison Formation) and wrote to both Edward D. Cope (the former student of Leidy's from Philadelphia) and Othniel C. Marsh (of Yale College) to notify them. The Oberlin student, Oramel Lucas, received a reply only from Cope and was soon at work excavating these bones for the Philadelphia paleontologist. Bones had actually first been found in Garden Park some seven or eight years earlier, and a local newspaper article in the *Cañon City Times* reported the story several months before Lucas made his find, but this was the first active excavation in the area (Monaco 1998).

By late March 1877, dinosaur bones also had turned up west of Denver, Colorado, near the small town of Morrison, found by a local schoolteacher named Arthur Lakes. Lakes had a background in natural history and an interest in Colorado geology, and in March 1877, he went to work measuring the thickness of the rocks near Morrison, Colorado, with Henry Beckwith, a former naval officer in his mid-30s. During their work that day, they ran across several very large fossil bones on the hillside they were climbing. These turned out to be those of a sauropod. Lakes realized the significance of what he and his friend had found (Kohl and McIntosh 1997), because he soon wrote to Marsh to tell the well-known paleontologist of the find. Marsh initially responded promptly, but then Lakes heard nothing for a while. Eventually, however, Lakes was working for Marsh, excavating and shipping the material to Yale.

Cope and Marsh: The Fuel for the Fire

These three major finds in what was to become known as the Morrison Formation (Cross 1894; Eldridge 1896)[3] inspired a frenzy of collecting activity for the next 10 years, a period that produced some of the most important collections of Late Jurassic vertebrates not just in North America but worldwide. The crews employed by Lucas, Lakes, and Reed, on behalf of Cope and Marsh, put in many years of fieldwork, and paleontologists since have benefited greatly from their labor.

In 1877, Edward Drinker Cope was a 37-year-old former student of Joseph Leidy's in Philadelphia. He was primarily a herpetologist, but as an all-around anatomist, he was able to contribute much to vertebrate paleontology as well. He had worked in the field for a number of seasons earlier in the decade, and he now studied out of his home. Othniel Charles Marsh was trained in a broad range of fields also, but thanks to his uncle, he was in charge of vertebrate paleontology at the Peabody Museum at Yale College. He was several years older than Cope but had gotten a later start in his field. Paleontology was actually a second choice for Marsh; he had preferred mineralogy, but at the time he started at Yale, there was no need for such a position.

Cope and Marsh knew each other from a meeting in Europe in the 1860s and corresponded for several years. Their relationship fell apart, however, as a result of several factors: Marsh supposedly buying his way into Cope's New Jersey field sites, Cope restoring a plesiosaur with the skull on the end of the tail instead of the neck and Marsh pointing it out, and the two's field crews interfering with each other in Wyoming in 1872. Although they still wrote each other occasionally, they mostly exchanged insults in print after that, and the feud incited an intense competition between them to outdo each other in terms of finding, naming, and describing paleontological specimens from the western United States.

In 1877, then, Marsh must have been steamed when he learned from Arthur Lakes that in the time since Lakes had first written Marsh and heard nothing, he'd gone ahead and contacted Cope. Cope likewise must have been irked when word probably reached him from Lucas that Marsh had sent his field man, Benjamin Mudge, to Garden Park to check

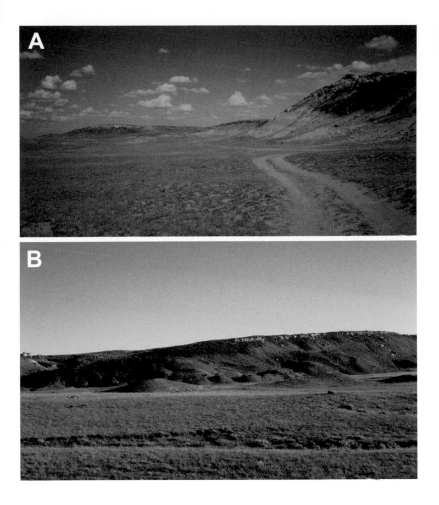

on the situation there. By the time Marsh heard of the Como find from Reed and Carlin in July, secrecy was the name of the game.

Marsh hired Reed and Carlin to collect for him at Como Bluff (fig. 3.3), and although Carlin soon quit, Reed worked for Marsh for years. Marsh also hired numerous helpers over the years for Reed. Reed and Carlin worked their Quarries 1 to 4 at Como Bluff, Wyoming, through 1877 and into the early part of 1878, excavating several *Camarasaurus* specimens, including the type specimen of *Camarasaurus grandis*. In May 1878, they found Quarry 5, which produced the type specimen of the small ornithopod dinosaur *Dryosaurus altus* (Marsh 1878b), along with the first Jurassic pterosaur from North America, *Dermodactylus* (Marsh 1878a). Near this locality during the same month, one of the crew found the first Jurassic mammal in North America, which Marsh named *Dryolestes priscus* (Marsh 1878c).

Cope had, of course, gotten word of the Como Bluff find by late 1877 and soon had hired men working in the area. Reed was writing Marsh

Como Bluff (1877–1889)

3.4. Watercolor by Arthur Lakes showing crews working Reed's Quarry 9 at Como Bluff during the early stages in 1879. Image courtesy of the Yale Peabody Museum.

frequently during this period, describing his efforts to keep Cope's men away. Probably much to Marsh's and Reed's irritation, Carlin was now working for Cope. Carlin was collecting and shipping material to Cope in full view of Reed, who, because Carlin controlled the station house, was stuck boxing his material for shipment to Marsh outside in the cold. Marsh's hired help for Reed came and went. By the spring of 1879, Reed was working mostly alone, trying to work several quarries simultaneously in order to keep the Cope parties out.

Finally, in mid-May 1879, Marsh sent Arthur Lakes up from Morrison, Colorado, to Como Bluff to help Reed. Lakes had worked the Morrison, Colorado, sites for Marsh briefly and now was being transferred. The summer of 1879 turned out to be a busy one for Reed's crews at Como Bluff, and a number of major finds were made during these rare warm months in Wyoming. On June 6, Marsh was visiting his crew and found the Three Trees Quarry (Quarry 7), which contained a well-preserved skeleton of a small ornithopod. In early July 1879, they found Quarry 9, for many years the richest Jurassic mammal quarry in the world (fig. 3.4). The quarry would eventually yield fish, frogs, salamanders, turtles, crocodilians, sphenodontids, fragments of sauropod dinosaurs, theropod dinosaurs, ornithopod dinosaurs, and stegosaurs, plus **multituberculate**, triconodont, symmetrodont, paurodontid, dryolestid, and docodont mammals. Most of these mammals are tiny animals with skulls no more than 5 cm (2 in.) long. Most had, by today's standards, numerous and unusual teeth in their jaws, and they probably fed on insects, worms, small vertebrates, and, in some cases, seeds and other plant material. Quarry 9 is still the most diverse quarry in the Morrison Formation and has also produced more individual specimens than any other (Simpson 1926a; Ostrom and McIntosh 1966; Foster 2003a; Carrano and Velez-Juarbe 2006).

3.5. Skeleton of *Brontosaurus excelsus* (YPM 1980) as reconstructed by O. C. Marsh. This specimen, still mounted at the Peabody Museum at Yale University, was collected by W. H. Reed and his crews mostly during winter 1879–1880 from Quarry 10 at Como Bluff. *From Ostrom and McIntosh (1966).*

At the end of July, the crew found Quarry 10, which yielded one of the best skeletons—and certainly the most famous—ever found at Como Bluff. Members of the crew worked this site for many of the following months, and when the material arrived in New Haven, Marsh named the animal *Brontosaurus excelsus* (Marsh 1879a; fig. 3.5). The specimen, YPM 1980, was eventually mounted at the Yale Peabody Museum and is still there today, a soaring testament to the beautiful structural composition of the sauropod frame. Many sauropod dinosaurs, particularly diplodocids, were constructed like suspension bridges, with tall neural spines over the pelvis supporting long necks and tails, the vertebrae of which were built for maximum strength with the minimum possible weight. This results in vertebrae with large openings (pleurocoels, or pneumatic fossae) and spines and processes supported by extremely thin bony laminae, and these vertebral structures were first best illustrated in the nearly complete skeleton of YPM 1980. Diplodocid vertebrae were known before this, but never from such a good specimen. *Brontosaurus* made the sauropods famous, but YPM 1980 may have been another individual of the genus *Apatosaurus*, which Lakes had found and Marsh had named from Morrison, Colorado. Thus, Reed's most famous sauropod find from Como Bluff may be known as *Apatosaurus excelsus*, the type of the species. More on that later though.

The men took few days off during any of the seasons at Como Bluff, working through all sorts of weather from the constant wind of summer to the lovely ice cold of the Wyoming winter, but they did take off at least major holidays. When the Fourth of July holiday came in 1879, Reed and his assistant, E. G. Ashley, celebrated the dawn by firing their pistols into the air. Ashley at least went outside; Reed simply fired through the roof of his and Lakes's tent.

In August 1879, Ashley found Quarry 12, which yielded *Stegosaurus ungulatus*. This quarry became Lakes's project during the upcoming winter and proved quite challenging. The rock layers were steeply tilted in this area, so the bones were buried very deep, and Lakes had to trench down some 10 m (33 ft) to get all of them. Lakes braved blowing snow, rock cave-ins, and spring water flooding the quarry to get this specimen out. It is still mounted at the Peabody Museum in New Haven. S.

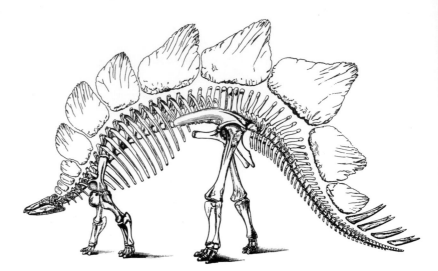

3.6. Marsh's reconstruction of *Stegosaurus ungulatus*, based on a composite of several specimens (but mostly from the skeleton excavated by Arthur Lakes at Quarry 12, Como Bluff). Modern reconstructions would include some changes to this view of *Stegosaurus*: the tail likely was held off the ground; the tail spikes (only four in number, not eight) stuck out to the sides, not up; there were probably more plates; and the plates of *S. ungulatus* were somewhat smaller than shown here. *From Ostrom and McIntosh (1966).*

ungulatus has a femur that is unusually long and slender compared with other known *Stegosaurus* species, such as *S. stenops*, and it was originally believed that the Quarry 12 animal also had eight spikes at the end of the tail instead of the four seen in other species (fig. 3.6). Kenneth Carpenter, then of the Denver Museum of Nature and Science, and Peter Galton, now retired from the University of Bridgeport, have shown that this is not the case and that *S. ungulatus* probably had the typical four-spiked tail (Carpenter and Galton 2001).

During the busy summer of 1879 at Como Bluff, Marsh's crew had some typical camp experiences. Mice made themselves at home in the men's tents, running over all the furnishings and even Lakes's head while he slept, getting into the food, and chewing on cots. Even worse, evening thunderstorms would sometimes inspire thousands of tiger salamanders to make a trek from the nearby lake up to the camp, where they crawled into the tents by the dozens and—owing to their sheer numbers—made more trouble than the mice. On the pleasanter side, Reed once took a break from his normal game-supply duties to fix the men a special dessert of strawberry shortcake (Kohl and McIntosh 1997).

In September 1879, Reed found Quarry 13 on the far eastern end of the bluff. This site produced more dinosaur skeletons than any other single quarry at Como Bluff, being particularly rich in *Stegosaurus* and *Camptosaurus* remains. The site is in a channel sandstone relatively low in the Morrison Formation and yielded the type specimens of the sauropod *Camarasaurus lentus*, the small theropod *Coelurus fragilis*, the stegosaurs *Stegosaurus sulcatus* and *Diracodon laticeps* (=*S. stenops*; Galton 1990), and the ornithopod *Camptosaurus dispar*. The quarry was worked for most of the following decade and ultimately produced parts of at least 43 individuals, including the additional genera *Glyptops* (a turtle), *Amphicotylus* (a crocodilian), *Diplodocus*, and *Dryosaurus*. Quarry 13 produced the first remains of the large ornithopod *Camptosaurus*, and

3.7. Some of the mounts resulting from the Como Bluff years, as seen at the Yale Peabody Museum. The main skeletons are (*left to right*) a mount of juvenile-type *Camarasaurus grandis* material; a mount based on *Camptosaurus* material from Quarry 13; and YPM 1980, *Brontosaurus excelsus*. Behind *Brontosaurus* is the mount of *Stegosaurus ungulatus*. Photo by ReBecca Hunt-Foster. (Courtesy of the Peabody Museum of Natural History, Division of Vertebrate Paleontology, Yale University, peabody.yale.edu).

although this form has since been found at a number of other sites throughout the Morrison Formation, Quarry 13 is one of the most important and productive for these animals. *Camptosaurus* was a bipedal herbivore and weighed about as much as a large horse (~725 kg [1,600 lb]), although some individuals could be even larger. This is close to the size of an average *Allosaurus*, and the other species of bipedal ornithischians in the Morrison (e.g., *Dryosaurus* and *Nanosaurus*) were considerably smaller.

For most of 1880, the crew continued splitting their work between Quarries 9 and 13 because both sites were the most productive for their respective types of vertebrates (microvertebrates and dinosaurs) found in the Como area. By March of that year, Lakes had left to take a job teaching geology at the State School of Mines in Golden, Colorado (now the Colorado School of Mines), and Reed hired Fred Brown to help. In July 1881, Reed's brother, who was out visiting and helping with the work, was killed in a swimming accident in nearby Rock Creek, and after this, Reed's interest in the work at Como began to wane. By the spring of 1883, Reed had finally quit, and for the next few years, Marsh's operations at Como were run by Brown and E. G. Kennedy. These two continued to concentrate on Quarry 9's mammals and microvertebrates and on Quarry 13's dinosaurs. Brown also opened several quarries of his own, mostly on the east end of the bluff, and each of these were small, often single-skeleton sites. By June 1889, Marsh had shut down his operations at Como Bluff. The 12 years of work there had produced the largest collection of Jurassic terrestrial vertebrates in the world up to that time (fig. 3.7), one that is still among the best.

**Garden Park
(1877–1901)**

Around the same time that Reed was finding the first few Como Bluff sites up in Wyoming in March 1877, Lucas found large sauropod bones several miles north of Cañon City, Colorado, in a valley known as Garden Park. These bones came from the slopes of the west side of the valley, near a small, rounded hill of red mudstone later known as the Nipple (fig. 3.8). Lucas wrote both Cope and Marsh and got a response only from Cope, so Lucas and his brother Ira began excavations at two quarries near the Nipple and started shipping the material to Cope in Philadelphia. By August, Cope had named this large sauropod *Camarasaurus supremus* (the supreme chambered lizard; Cope 1877b). These quarries are high in the Morrison Formation, and the specimens on which *C. supremus* is based are very large compared with the average adult size for *Camarasaurus* from lower in the formation. They are so much larger, in fact, that on the basis of scaling of the linear and the volumetric dimensions, *C. supremus* may have weighed more than *Brachiosaurus* and may thus have been the most massive animal known from the Morrison, at some 47 metric tons (103,400 lb).

The Marsh-Felch Quarry is relatively low in the Morrison Formation and lies in a river channel sandstone (fig. 3.8). The bones here came from a level just above some large sedimentary structures in the channel called lateral accretion surfaces; these elongate, sigmoid planes in the sandstone collectively represent the migration of a point bar on the inside curve of a river meander. A number of different dinosaurs were found in this quarry by a crew of Marsh's, including the type specimens of *Allosaurus fragilis* and *Diplodocus longus*. *Allosaurus fragilis*, as we will see later, is the younger of two species of *Allosaurus* in the Morrison Formation, one apparently restricted to the Brushy Basin Member and its equivalents. It is also the more common of the two by far and is the most abundant theropod dinosaur in the formation. *Diplodocus* is one of the three most abundant sauropod dinosaurs in the Morrison, a 24 m (80 ft) long, "slender" sauropod weighing about 12,000 kg (26,400 lb). *Diplodocus longus* was based on a partial tail from the Marsh-Felch Quarry, and numerous specimens (some more complete) were later referred to the species, but *D. carnegii*, found later in Wyoming, is based on better material.[4] After the initial work at the Marsh-Felch Quarry by Mudge and Samuel Williston (another of Marsh's trusted field men), the site was dormant until 1883, when Marshall Felch and his brother Henry began working there for Marsh. The two worked the site until 1888 and collected a number of specimens, most of which were of taxa already known from the formation. They did, however, find a new dinosaur during this period, *Ceratosaurus nasicornis*, a medium-large theropod with a relatively large skull and large, bladelike teeth that are distinctly longer than those of the larger-bodied *Allosaurus*. *Ceratosaurus* is now known from a number of specimens, but it is far less abundant than the dominant *Allosaurus*. This difference in abundance, plus the differences in size and skull/tooth proportions, suggests some kind of separation in feeding strategies between the two carnivores. After the Felch brothers finished working the quarry,

3.8. Localities at Garden Park north of Cañon City, Colorado. (A) Small knoll known as Cope's Nipple, which had several quarries around it. (B) One of Cope's *Camarasaurus supremus* quarries from high in the Morrison near the site in A; bones were excavated from dark-red mudstone just under a sandstone ledge (in dark shadow here). (C) Marsh-Felch Quarry from lower in the Morrison; type locality of *Haplocanthosaurus*, *Diplodocus*, and *Ceratosaurus*. Skeletons were found in an area of lowest piñon and juniper trees, just above thick channel sandstone in the center of the photograph. Diagonal lines in the upper part of the channel sandstone represent migration of sandbars within an ancient river (lateral accretion surfaces, in the language of geology).

the site again went quiet, this time for nearly 12 years. In 1900, William Utterback, under the direction of John B. Hatcher, began excavating the site for the Carnegie Museum of Natural History in Pittsburgh (Monaco 1998). Utterback spent two seasons at Marsh-Felch and found mostly material of common taxa, but he also came across a previously unknown dinosaur. The animal was the relatively small, rare sauropod *Haplocanthosaurus* (Hatcher 1903), a dinosaur that has proved to be restricted to the lower half (or slightly higher) of the Morrison Formation. *Haplocanthosaurus* is now represented by most parts of the skeleton, except, of course, the skull, which is absent from even nearly complete sauropod skeletons with frustrating frequency. Although we don't know what the skull looked like, we do know that this was a moderately robustly built sauropod (not nearly as slender as *Diplodocus*), but it was the smallest of this group in the Morrison and only weighed about 10,000 kg (22,000 lb). By today's standards, of course, that is gigantic for a land animal, but during the Late Jurassic, it was below par for sauropods.

Morrison, Colorado (1877–1879)

After Lakes and his friend Beckwith found their first bones near Morrison, Colorado (fig. 3.9), in late March 1877, they had to figure out how to collect them. It took about a week to get the equipment and manpower to haul the material in. It had snowed heavily the night before, so the bones had to be loaded up and sledded into town. Marsh soon named the material sent by Lakes as the new sauropod species *Titanosaurus montanus*, but he later changed the name to *Atlantosaurus montanus* because it turned out that the name *Titanosaurus* had already been used for another animal. These Morrison bones are of a sauropod and probably belong to what is now known as *Apatosaurus*, but they are too fragmentary to be certain. Lakes and his crew were soon working what they called Quarry 10, just above the town (fig. 3.10), and from here, they would excavate bones in a better state of preservation, including the type specimen of *Apatosaurus ajax*. This is a huge sauropod, and although it is not nearly complete, it was the type and first known good specimen of *Apatosaurus*. The quarry was in a layered, dark mudstone below a sandstone ledge and probably represented a fairly wet environment. Lakes's sketch of the site indicates that the team dug quite far into the hill under the tons of sandstone.

Around the middle of June 1877, while using dynamite to blast sandstone blocks at Quarry 1, the crew happened to uncover the first crocodilian in the Morrison Formation, one Marsh soon named *Diplosaurus felix* (now "*Goniopholis*" *felix*), probably related to but not the same genus as *Goniopholis* from Europe. The fossil consisted of a moderately well-preserved skull, which was laid out in the fresh rock with its pitted bone surface and eye sockets perfectly visible. The men spent the rest of the day chiseling it out of the rock.

The crew worked into the fall of 1878, when Lakes left to work in Marsh's lab in New Haven. Lakes returned in mid-April 1879, and the

3.9. Hogback ridge north of Morrison, Colorado, in the vicinity of several of Arthur Lakes's quarries from 1877 to 1878 (view to the east).

3.10. Watercolor by Arthur Lakes looking northwest at Morrison, Colorado, with out-crops of the Pennsylvanian-age Fountain Formation (now part of Red Rocks Amphitheater) in the background (at the foot of the mountain), c. 1878. The hogback exposure at right is a slope containing Morrison Formation outcrops; Lakes's Morrison Quarry 10 (the local-ity of *Apatosaurus ajax*) is seen above the right (east) end of the town. *Image courtesy of the Yale Peabody Museum.*

crew worked the site until the first week in May before shutting down their operations at Morrison. On May 14, 1879, Lakes left for Como Bluff (Kohl and McIntosh 1997). The Morrison, Colorado, quarries produced some important specimens, including the *"Goniopholis" felix* and *Apatosaurus ajax* fossils, plus *Stegosaurus armatus*, the first evidence of stegosaurs found in the formation (from Quarry 5) (Ostrom and McIntosh 1966). A roadcut along the hogback near Morrison, Colorado, today exposes the Morrison Formation well in the vicinity of Lakes's Quarry 5, and bones are visible at the historic interpretive site. Further north, at the Interstate 70 roadcut, the full section of the Morrison Formation is also exposed.

Henry F. Osborn, as head of the vertebrate paleontology program at the American Museum of Natural History in New York, facilitated the museum's expeditions out to the Rocky Mountain region starting in the late 1890s. Osborn was interested in mammals and in display dinosaurs

The American Museum of Natural History Heads West

3.11. Barnum Brown at the Como Bluff quarry of specimen AMNH 223 (*Diplodocus*) in August 1897. This site is not far from Reed's Quarries 9 and 10. *Courtesy of Department of Library Services, Image #17812 © American Museum of Natural History.*

for the museum. Naturally, one of the first places he had the crews go was Como Bluff, Wyoming. The crew worked for weeks at Quarry 9 removing overburden and digging out the layer from which all the small bones had been collected starting 18 years before. After all the rock had been explored back in the lab, however, very little was found in it—only a few small bones and one new mammal in *Araeodon* (Simpson 1937). It appeared that the years of all-season work by Marsh's men (mostly Reed and Kennedy) had removed all the productive rock. The group had better luck on the dinosaur end of the project. Jacob Wortman found a partial *Apatosaurus* skeleton, and Osborn and Barnum Brown discovered another belonging to *Diplodocus* not far from Reed's Quarries 9 and 10. Both skeletons were collected (fig. 3.11), and the *Diplodocus* proved to be an interesting specimen in that it was one of a handful of these animals with a distinctly slender and tubular femur. Several other individuals have this morphotype, including one from Dinosaur National Monument in Utah (now at the Smithsonian Institution).

The American Museum crew returned in 1898 but this time concentrated on the Medicine Bow Anticline (uplift) several miles north of Como Bluff. The crew hadn't found as much as they'd expected the year before, and they suspected that Marsh's crews had gotten most of the material earlier. In June, they found bones in place at a site on the relatively flat plains north of Como Bluff that they named Bone Cabin Quarry.[5] This site was so named because large fragments of sauropod fossils nearly covered the ground in the area, and some years before, one of the local sheepherders had used the bone fragments to build the foundation for a shelter nearby. The quarry was a reasonably large pit on the flat prairie along the north edge of a shallow draw (fig. 3.12). Bone Cabin became very productive, and the American Museum crews worked there from

3.12. Bone Cabin Quarry, Wyoming. (A) Walter Granger standing among sauropod dinosaur bones in a pit at the quarry in July 1898. (B) Quarry in 1997, with Elk Mountain on the horizon to the southwest. The quarry is a shallow depression occupying most of the foreground. The American Museum of Natural History worked this site for several years around the turn of the 20th century. *Image in A courtesy of Department of Library Services, Image #17838 © American Museum of Natural History.*

1898 through 1905, eventually recovering more than 500 bones (McIntosh 1990c), parts of at least 50 individuals, and a dozen taxa. Most of the dinosaur specimens belong to sauropods and *Stegosaurus*, but there are also crocodilians, turtles, *Allosaurus*, and *Camptosaurus*. Small bits of mammal skulls were found in the collections from Bone Cabin Quarry years later. In 1901, German paleontologist Eberhard Fraas, who was later deeply involved in collecting Late Jurassic dinosaurs from eastern Africa, spent some time at Bone Cabin Quarry as a guest of Osborn's, helping with some of the digging (Maier 2003).

The most important specimen collected from the quarry during these years was the previously unknown small theropod *Ornitholestes hermanni* (Osborn 1903). This slender meat eater was only about 1.5 m

(5 ft) long and had long arms and fingers for grasping prey. In the years since Osborn's description of this new animal, it has been reported only from fragmentary remains from one other site, and thus it remains one of the rarest of the Morrison Formation theropods.

In the first few seasons at Bone Cabin Quarry, the American Museum crew also found a site several miles away called Nine Mile Crossing. This is a site that yielded an *Apatosaurus* skeleton, one that was the first permanently mounted sauropod skeleton in the world and one that is still in the exhibit halls of the American Museum of Natural History today (fig. 6.33).

The Carnegie Museum at Sheep Creek

In December 1898, a New York newspaper published a mistake-riddled article on a large dinosaur from Wyoming, one that Marsh's old collector, Bill Reed, had found.[6] Andrew Carnegie, the Pittsburgh steel tycoon, saw the article and was immediately interested in acquiring it for his new museum. The specimen on which the article was based was still in the ground in the Freezeout Hills of Wyoming, and in truth, it was not a very complete or a well-preserved skeleton. But the museum's director, W. J. Holland, was able to put a crew together to collect it, or any other dinosaur skeleton they could find north of Laramie, Wyoming. The crew spent the month of June in the Freezeout Hills but found little of value. Reed and the crew found that the specimen, which had been blown out of proportion by the newspaper article, was only fragmentary, and Jacob Wortman (the group's leader, whom Holland had hired away from the American Museum) and crewman Arthur Coggeshall were becoming discouraged. They had plenty of company in the area that summer, however, as crews from the University of Kansas, the Field Museum of Natural History in Chicago, and the University of Wyoming were all working in the Morrison Formation nearby. The American Museum of Natural History crews were back at Bone Cabin Quarry and Nine Mile Crossing, and the University of Wyoming's Wilbur Knight was helping lead the Fossil Fields Expedition during July and August throughout the region, visiting some of the quarries (Rea 2001; P. Brinkman 2010).

On July 3, 1899, Wortman, Coggeshall, Reed, and Reed's 17-year-old son were out of the Freezeouts and down in the Sheep Creek area in the flats northeast of Bone Cabin Quarry. They stopped by a site that Reed had noticed the previous spring and hunted around, looking at the surface material and starting to dig in after in situ bone. On July 4, as they began working the quarry, they realized they had a partial skeleton of a sauropod. It later turned out to be two partial skeletons of the new species *Diplodocus carnegii* (CM 84 and CM 94). The crew developed a large pit in excavating the specimens, and during the summer, the American Museum of Natural History crew came by for a visit. The two groups posed for a photograph in the quarry. The photo shows an interesting grouping of workers: the two directors of the paleontology programs at

3.13. Carnegie Museum's Quarry D (or Quarry 3), a shallow depression in rolling hills of prairie at Sheep Creek, Wyoming. This is the type locality of *Diplodocus carnegii*, now mounted at the Carnegie Museum in Pittsburgh. Casts of this specimen are in many museums around the world. Author leaning back into the Wyoming wind to stay upright for scale.

the museums, Holland and Osborn; the Carnegie crewman, Wortman, reunited with his old employer, Osborn; some big names in early 20th-century vertebrate paleontology, W. D. Matthew, Walter Granger, and R. S. Lull; and there with all of them, the man who had been working the region for more than 20 years and who was largely responsible for their being there in the first place, Reed. In 1900, the Carnegie Museum program was taken over by John B. Hatcher, a graduate of Yale who had been a collector for Marsh for a number of years. Hatcher took the Carnegie crew back to Sheep Creek, where they excavated the second individual from the 1899 quarry (known as Quarry D or Quarry 3; fig. 3.13). The Carnegie Museum field men also collected dinosaurs from a number of other quarries in the Sheep Creek area (McIntosh 1981), including one that yielded both an adult *Brontosaurus* (now mounted at the University of Wyoming) and one of the best partial skeletons of a young juvenile *Brontosaurus* known from the formation (now mounted at the Carnegie in Pittsburgh). After the Quarry D *Diplodocus carnegii* material had been removed from the rock, Hatcher described the specimens in a 1901 monograph, still one of the most important ever written on the genus (Hatcher 1901).

Red Fork of the Powder River

After his two seasons working the Marsh-Felch Quarry near Cañon City for the Carnegie Museum, Utterback traveled much farther north for the 1903 field season and began working two sites on the eastern edge of the Bighorn Mountains in north-central Wyoming. He excavated the sites (Quarries A and B) along the Red Fork of the Powder River through 1906 and found a mostly standard assortment of Morrison Formation dinosaurs. However, he uncovered a unique sauropod dinosaur later named *Diplodocus hayi* (Holland 1924), which is now mounted at the Houston Museum of Natural Science. This sauropod was much different from other *Diplodocus* specimens and is clearly a distinct species, and it has recently been renamed as the type specimen of *Galeamopus hayi* (Tschopp et al. 2015).

Island in the Plains: The Black Hills

In 1889, a postmistress in Piedmont, South Dakota, just a few miles north of Rapid City, found dinosaur bones in the Morrison Formation along the eastern slope of the Black Hills. That year, Marsh and his assistant at the time, Hatcher, collected the specimen for the Yale Peabody Museum, and the fossil proved to be that of a new sauropod. Marsh named the animal *Barosaurus* ("heavy lizard") in a paper in the *American Journal of Science* in 1890. Unlike many of the Wyoming finds, the quarry in South Dakota was among grass and a light forest of pine trees on the north slope of a large hill called Piedmont Butte. The quarry, which is still visible among the trees, was actually mostly dug by G. R. Wieland, a Yale collector whom Marsh sent back for the remainder of the specimen in 1898. Wieland worked alone and created a fairly large pit from which he retrieved many important bones, particularly the long cervical vertebrae that are distinctive of *Barosaurus* (Marsh had named the specimen based on caudal vertebrae, which in this genus are less diagnostic than cervicals). Wieland worked well into October that year but was eventually shut down "when cold and dust storms made excavation difficult" (letter to Marsh; see also Wieland 1920). Richard S. Lull, American Museum veteran of the Bone Cabin / Nine Mile Crossing work and later Marsh's successor at Yale, fully described the skeleton (Lull 1919). Good specimens of *Barosaurus* were also found later at Dinosaur National Monument in Utah, at another quarry just a few miles south of the original site in South Dakota in 1980, and in the Bone Cabin Quarry area in Wyoming (Foster 1996a).

Barosaurus is a long, slender animal, as anyone who has seen the mount at the American Museum of Natural History in New York can attest. In the American Museum, the rearing animal's neck soars straight up into the open space of the rotunda. Despite its name, it is not a robust animal, and aside from its long neck, elongate cervical vertebrae, and slightly shorter caudal vertebrae, it looks very much like *Diplodocus*. It is clearly distinct, however, and it is interesting that it is far less abundant in the Morrison Formation than its relative.

Around 1900, crews from the American Museum of Natural History and the Smithsonian Institution (including Lull) collected from several small quarries around Sturgis, South Dakota, north of Piedmont, the former being the quiet town (51 weeks out of the year, anyway) that since the early 1940s has been inundated by Harley-Davidsons every August. These quarries, two worked by the American Museum and one by the Smithsonian, yielded two *Camarasaurus* and a diplodocine (possibly *Diplodocus* or *Barosaurus*) (Foster 1996b).

Elmer S. Riggs was an assistant curator at the Field Columbian Museum in Chicago when he began working the Morrison Formation deposits of Wyoming in the late 1890s. Originally from Indiana, Riggs had been trained at the University of Kansas by Williston. Riggs and his contemporary, Barnum Brown, were both students of Williston's at KU, and by the late 1890s, both were working in paleontology, Riggs in Chicago and Brown in the field for the American Museum of Natural History. Riggs is the paleontologist who later demonstrated, for a time at least, that Marsh's genera *Apatosaurus* and *Brontosaurus* were in fact two names for the same animal, with the former having priority because it had been assigned first (as promised, more on this later).

Riggs had begun making trips out to the Morrison Formation of Wyoming and in 1899 worked in the Freezeout Hills, where he excavated several sites with some success, getting *Diplodocus* material out of what he designated Quarry 3 and *Camarasaurus, Stegosaurus,* and *Allosaurus* out of his Quarry 6. The Freezeout Hills area northwest of Como Bluff is several miles from Sheep Creek, where the Carnegie Museum crew (including Reed, Wortman, and Coggeshall) was also working that summer on the *Diplodocus.* The small Field Museum crew camped in the hills west of the basin where the other museums were working and hunted pronghorn for their meat supply. Despite good material and a number of quarries having been worked, none of the material Riggs got from the Freezeouts was of critical scientific value, nor did it yield any complete display specimens. Still, Riggs planned to return to Wyoming to prospect in 1900 and then to move on to Colorado. By the end of May 1900, however, the Wyoming leg of the expedition had apparently been canceled, and the crew planned to head straight to Colorado, after buying a $40 tent to replace theirs, which was still in the field with the museum's anthropology department.

In the early summer of 1900, Riggs arrived in the Grand Valley of western Colorado to look for dinosaurs in the Morrison Formation. He had made his way there for the summer and early fall with a small field crew from the museum in Chicago. The crew included H. W. Menke, a veteran of the 1897 American Museum expedition and photographer for the Field Museum's 1899 Freezeouts trip, and Victor Barnett, a young man from Indiana that Riggs had hired as a camp cook and assistant. Riggs was enticed to western Colorado by a letter he had received from Dr. S. M. Bradbury, of the relatively new town of Grand Junction, indicating an abundance of dinosaur bone in the area. Riggs had sent letters to a number of towns along the western railroads, inquiring about paleontological finds in the local areas, and Bradbury's letter suggested to Riggs that the area was promising (Armstrong and Perry 1985; P. Brinkman 2010).

Riggs and his crew arrived in the spring and soon afterward began developing a quarry in the Morrison Formation near the mouth of No Thoroughfare Canyon. From this pit (Quarry 12), they collected elements of a *Camarasaurus.* On July 4, 1900, Menke went prospecting in some

3.14. An assistant of Elmer Riggs contemplates a nearly 2 m (6.6 ft) long humerus of *Brachiosaurus altithorax* at Quarry 13 near Grand Junction, Colorado, in 1900. This was the first *Brachiosaurus* ever found. Cliffs in the background are Triassic-Jurassic Wingate Sandstone, now within the Colorado National Monument. *Photographer H. W. Menke, Negative No. CSGEO3893, the Field Museum of Natural History.*

Morrison Formation outcrops to the northwest of the site and found some large bones at the base of the south slope of a hill littered with car-sized blocks of dark, desert-varnished sandstone. Excavation of Riggs Quarry 13 began about three weeks later (Foster et al. 2016a).

The men were working just downslope from several large, dark sandstone blocks, creating a pit that cut down about 1.8 m (6 ft) into the hill, into a soft white sandstone layer otherwise barely visible in the hillside. One of the first bones the crew found was a humerus, and they erected a canvas sunshade over the quarry and worked most of the rest of the summer in view of the looming red cliffs of the Wingate Sandstone to the south and west (fig. 3.14). As they dug into the hill, they also uncovered a series of dorsal vertebrae and several long ribs overlying a femur (fig. 3.15).

On July 26, 1900, after their first day's work at the new site, Riggs wrote to his museum's director, telling of the frustration of nearly a half dozen sites that hadn't worked out and of the moderate success of Quarry 12. Riggs speaks of the first few weeks of the 1900 season as "the most discouraging, if not the most uncomfortable, period that I have known in seven years of collecting." He then happily reports the good news of the Quarry 13 find, which, he recognized immediately, held the largest dinosaur yet known. Interestingly, he also expresses concern about other collecting parties hearing of the Grand Valley and asks the director to keep things quiet until the end of the season.

The crew worked for the next month and a half on Quarry 13, and during this time, Riggs also communicated with the railroads on arrangements for free shipment of the fossil material to Chicago. On September 9, Riggs wrote to the museum that everything was set and that he would "ship tomorrow thirty-seven heavy crates of fossils."[7] The crew wrapped

3.15. Riggs's Quarry 13 in 1900, showing articulated dorsal vertebrae, several long ribs, a femur, and other elements of *Brachiosaurus altithorax* during excavation. *Photographer H. W. Menke, Negative No. CSGEO4028, the Field Museum of Natural History.*

up their work and headed back to Chicago to begin the task of cleaning what they had collected (Armstrong and Perry 1985; P. Brinkman 2010; Foster et al. 2016a).

What Riggs and his crew removed from the site was a partial skeleton of a dinosaur previously unknown from the Morrison Formation, and still one of the biggest. When Riggs published the description of the animal in 1903, he named it *Brachiosaurus altithorax*, the "deep-chest arm-lizard." *Brachiosaurus* has gone on to become, along with *Apatosaurus* and *Diplodocus*, one of the most recognized of the sauropod dinosaurs, appearing in paintings in books for adults and kids alike, as well as in movies. But all that Riggs knew as the bones started coming to light in the fall of 1900 (and indeed, probably during that summer in the quarry) was that he had the very large partial skeleton, in this case minus the neck and head and most of the tail, of an animal that was not *Apatosaurus*, *Diplodocus*, *Barosaurus*, or *Camarasaurus*. One of the most unusual features of the animal was the fact that the humerus, and in fact the whole forelimb, was slightly longer than the femur and hind limb. In all the sauropods found in the Morrison Formation previously, the forelimb and its individual bones were each significantly shorter than in the hind limb. Other differences included details of the pelvis and dorsal and caudal vertebrae.

In 1901, Riggs and crew came back to the Grand Valley and worked again, mostly at Quarry 15, near Fruita, at the base of a steep mudstone slope not far from the Colorado River (fig. 3.16). Here they dug up the back two-thirds of an *Apatosaurus* (or some type of apatosaurine; Tschopp et al. 2015), uncovering mostly vertebrae and also some limb material, which had been found the previous fall. At the time of discovery, the last cervical vertebra and several rib tips were exposed. The hillside

3.16. Riggs's Quarry 15 near the Colorado River in Fruita, Colorado; the quarry "mine" entrance and a monument erected in 1938 are visible at the base of the slope. Riggs and crew excavated an apatosaurine sauropod from this site in 1901 and had to tunnel in to the hillside to retrieve most of the tail.

was so steep and the articulated vertebrae dove so directly into it that, after working with drills and dynamite, a quarry wall 5.5 m (18 ft) high developed. (The anterior portion of the skeleton had most likely eroded away down the hill many years before.) As Riggs wrote in 1903, "These conditions made it necessary to resort to tunneling." Working into the hill and shoring the shaft up with timbers as they went, the group mined 6 m (20 ft) into the ancient floodplain muds and retrieved most of the tail of the sauropod. After getting the plaster jackets out of the quarry, the crew had to take them down a quarter mile to the Colorado River (at that time named the Grand River) to be transported across in a raft they had constructed for the purpose. While working at the site that spring, Riggs nearly had his secret field area discovered when Osborn and Hatcher unwittingly rode through Fruita on the train (P. Brinkman 2010). This

Apatosaurus specimen is still on display at the Field Museum of Natural History in Chicago.

Reed Back at the University of Wyoming

In 1903, Bill Reed was hired by the University of Wyoming as director of the Geological Museum. Reed had been working for the American Museum of Natural History for several years before this, but now he continued collecting Jurassic material north of Laramie and set up a paleontology exhibit in the new Hall of Science at the university. He taught geology and paleontology courses in the department and held his departmental and museum positions until his death in 1915 (Breithaupt 1990). In 1908, Reed collected, from the Morrison near the present-day Alcova Reservoir in central Wyoming, an unusual stegosaur later named *Stegosaurus longispinus* for its long tail spikes. This specimen was recently renamed *Alcovasaurus longispinus* (Galton and Carpenter 2016a). In 2000, some of the other bones that Reed had collected from the Morrison Formation in southeastern Wyoming were sorted and reassembled. Among these bones is an *Apatosaurus* femur that is among the most robust sauropod femurs I've ever measured, thicker in circumference for its length than even the type of *A. louisae* at the Carnegie Museum.

Dinosaur National Monument

Earl Douglass was 40 years old when he joined the staff of the Carnegie Museum in 1902. A native of Minnesota, he studied in South Dakota in college and later at the University of Montana and the Missouri Botanical Gardens in St. Louis. He then went to Princeton on a fellowship and worked under Osborn's old friend W. B. Scott before coming to the Carnegie. He spent his first few seasons in the field collecting fossil mammals in Montana, and it was in 1908, while Douglass was working in the Uinta Basin south of Vernal, Utah, that director Holland came to visit. Holland was there in early September, and in camp early one evening, he and Douglass discussed the south slope of the Uinta Mountains to the north, where some years before, F. V. Hayden had noticed rocks of Jurassic age during one of his surveys through the area. John Wesley Powell had noticed bones in the area in 1871 while passing through, and O. A. Peterson of the American Museum had done the same in 1893. Douglass had devoted very little time earlier in 1908 to prospecting the Morrison in the area, as he was focusing on the Uinta Formation, which is Eocene in age. But he and his assistant had come across a sauropod femur during the summer, and they took Holland to prospect the surrounding area. On their way back, it turns out, they passed right by the uplifted area that would ultimately become the Dinosaur National Monument quarry. Although they didn't find much with Holland, the summer's previous find of the sauropod femur and the looks of the terrain were enough to inspire Douglass to come back to work the Morrison in the area the next year (Douglass 2009).

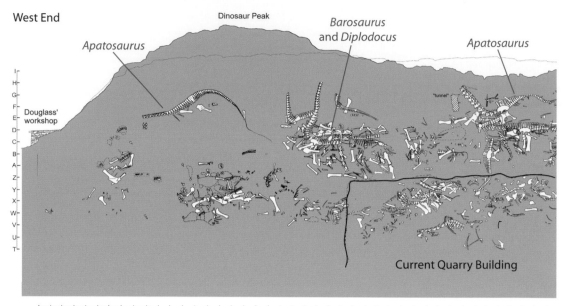

West End

Dinosaur Peak

Apatosaurus

Barosaurus
and *Diplodocus*

Apatosaurus

Douglass'
workshop

"tunnel"

Current Quarry Building

180   160   140   120   100   80   60   40

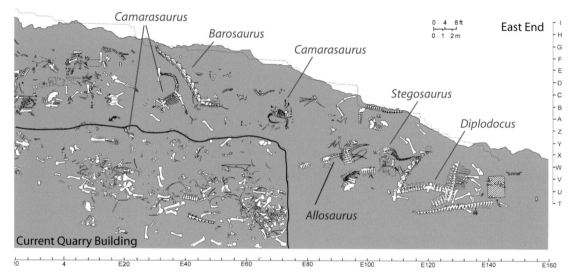

Camarasaurus

Barosaurus

Camarasaurus

0 4 8 ft
0 1 2m

East End

Stegosaurus

Diplodocus

"tunnel"

Allosaurus

Current Quarry Building

0   4   E20   E40   E60   E80   E100   E120   E140   E160

3.17. Map of the Carnegie Quarry at Dinosaur National Monument, Utah, with several skeletons labeled. Note the mixture of articulated skeletons and sections of skeletons along with isolated bones. Notice that the bone layer extended well above the present quarry building and well to the east and west of it. *Labeling by author. Map courtesy of Ken Carpenter.*

Douglass spent the early part of the summer of 1909 prospecting the area, walking miles of steep, rocky terrain and not finding much of anything. He was exploring with the assistance of a local, George Goodrich, and it wasn't until August 17 that the two had any luck. That day, they found eight articulated caudal vertebrae of an *Apatosaurus* laid out in the sandstone of a high ridge a little west of Split Mountain and within sight of the Green River. The sandstone bed dipped to the south at about 67 degrees, so excavation was not easy. Holland came out to see the find sometime afterward, and Douglass eventually sent for his family to move out from Pittsburgh, which they did. He spent most of the rest of his career working what became known as the Carnegie Quarry and living in houses he built in nearby canyons and near the Green River (Colbert 1984; Douglass 2009).

Douglass found two more adult and one juvenile *Apatosaurus* near the original skeleton. He then worked west and found partial skeletons of the sauropod dinosaurs *Diplodocus*, *Barosaurus*, and *Camarasaurus*. After reaching the edge of the sandstone in that direction, Douglass turned east from the original discovery site and, digging deeper, uncovered partial skeletons of the ornithopod dinosaurs *Dryosaurus* and *Camptosaurus*, the theropod dinosaur *Allosaurus*, the sauropods *Camarasaurus* (fig. 3.17) and *Barosaurus*, the plated dinosaur *Stegosaurus*, the turtle *Glyptops*, and a goniopholidid crocodylian (McIntosh 1981). The work took years but was obviously productive.

Removing bones from the sandstone layer required moving rock out of an ever-expanding trench and working down through the hard sandstone. Although dynamite was sometimes used to remove thick sections of overburden, most of the digging was done with hand tools. At times, it looked as if the men were constructing a railroad cut parallel to the bedding but right through the ridge. Mining carts and tracks were used to move rock out of the trench, and the tons of rock hauled out of the cut were dumped to the east and now form the base of the modern parking lot at the quarry. The upturned sandstone bed that was quarried and had bone removed turned out to be huge, growing to 183 m (600 ft) by 24 m (80 ft). The Carnegie Museum worked the quarry from 1909 until 1922, when funding became scarce, and ultimately, 300 metric tons of bones would be removed from the site. Most bones had to be shipped by wagon 81 km (50 mi) south to Dragon, Utah, where they were loaded onto rail cars bound for Grand Junction and eventually Pittsburgh.

In 1923, Charles Gilmore of the United States National Museum at the Smithsonian worked the quarry. He and his crew collected several specimens, including USNM 10865, the *Diplodocus* now on display at the Smithsonian's National Museum of Natural History. Later in 1923, the University of Utah worked the far eastern end of the site (where the parking lot now is) and excavated the *Allosaurus* skull that can now be seen on display at the quarry building at the monument. The quarry was dormant from 1924 to 1952. In 1953, National Park Service excavations began under park paleontologist Ted White.

The quarry produced the biggest collection of dinosaurs from a single locality ever found in the Morrison Formation. Articulated specimens of *Diplodocus*, *Camarasaurus*, *Barosaurus*, and *Allosaurus* were found here, along with individuals of *Camptosaurus*, *Dryosaurus*, *Stegosaurus*, *Torvosaurus*, and *Ceratosaurus*. Crocodilians, turtles, and a sphenodontid were also found. Skeletons mounted in museums today that were collected from the Carnegie Quarry include the *Apatosaurus* and the juvenile *Camarasaurus* at the Carnegie Museum, the *Diplodocus* at the Denver Museum of Nature and Science, the *Barosaurus* at the American Museum of Natural History (a cast was mounted), and the aforementioned

3.18. Carnegie Quarry at Dinosaur National Monument, Utah. (A) Quarry wall within the current building, looking west. (B) Close-up of bones near the western end of the building. (C) Close-up of bones near the eastern end of the building. (D) Anterior caudal vertebra of *Apatosaurus* in anterior view, lens cap for scale. (E) Skull and neck of *Camarasaurus*. (F) Articulated skeleton of a juvenile *Camarasaurus*. (G) Articulated mid-caudal vertebrae of a diplodocine sauropod. (H) Folded leg of *Allosaurus* with (*bottom to top*) femur, tibia, and articulated foot.

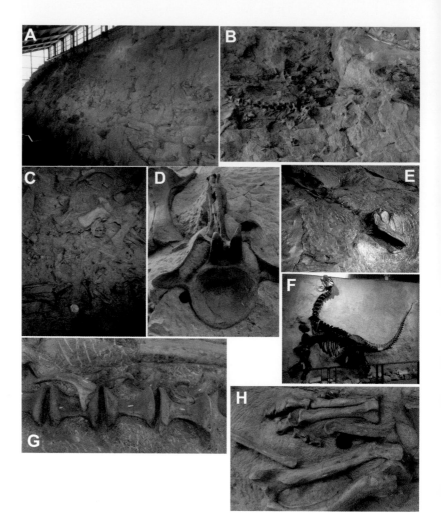

Diplodocus at the Smithsonian's Museum of Natural History. Altogether, the collection would represent more than 120 individuals, and nearly 20 articulated skeletons were found. The type specimen of the turtle *Dinochelys whitei* was found in the quarry, and so were as many as 13 skulls of other taxa, mostly dinosaurs. Tony Fiorillo, now of the Perot Museum of Natural History in Dallas, calculated in a paper from 1994 that the Dinosaur National Monument quarry material was probably deposited over a few years or less, indicating that the animals found there had almost certainly all lived at approximately the same time. It appears that the animals preserved in the quarry likely died noncatastrophically (i.e., individually and not in herds) of malnutrition during severe droughts and accumulated in the deposit over time (Fiorillo 1994; Carpenter 2013). Although for some time it was believed that the skeletons had accumulated on a sand bar on the inside of a bend in a river, in recent decades workers have concluded that the bones were most likely on the bottom of a fluvial channel (Turner and Peterson 1999; Carpenter 2013); there has also been a recent suggestion that the channel that the bones were in was not necessarily a main river channel but a distributary channel to a large crevasse splay (Brezinski and Kollar 2018).

The specimen that Douglass found first in August 1909 (CM 3018) would later be named as the new species *Apatosaurus louisae* and would be mounted (without a skull) at the Carnegie Museum, where it still is today (in a new pose and exhibit hall—and with a skull). *Apatosaurus louisae* (the second part of the name in honor of Andrew Carnegie's wife) is a large and well-preserved specimen. It is also one of the most robust sauropods in the formation, certainly the most stout of its size, weighing as much as 10 metric tons (22,000 lb) more than Marsh's *Brontosaurus excelsus* from Como Bluff (YPM 1980), even though the two are not significantly different in length. In 1915, while work was still going on at the quarry, Woodrow Wilson created Dinosaur National Monument to protect the quarry and some surrounding land, now that the area had been opened for settlement. It wasn't until 1958 that a quarry building was finished, finally protecting the bone layer as one "wall" of the structure (fig. 3.18). Work in the quarry building itself stopped a number of years ago, but excavations continue within the expanded monument boundaries.

More Garden Park

In 1915, a Cañon City, Colorado, resident named Dall DeWeese, who was also an engineer and hunter, found a partial *Diplodocus* skeleton on the east side of Garden Park. This specimen was excavated by the Colorado Museum of Natural History (now the Denver Museum of Nature and Science, where the material is currently stored). Fragmentary remains of turtle and crocodile were also found at this site. This locality was lost for a number of years. By 1997, Pat Monaco and Donna Engard, then of the Dinosaur Depot in Cañon City, had narrowed down the location of this quarry to several acres of the Garden Park Fossil Area, and one day in July of that year, I got to help them wander the hills with old photos, closing in on the old site. After a few hours, we did nail it down, the quarry barely visible on a west-facing gray mudstone slope across the valley from the old Cope and Marsh quarries. The *Diplodocus* from this quarry is one of the best preserved of its kind, so it is unfortunate that only the back portion of the animal was present.

In 1937, local high school teacher Frederick Kessler found and excavated, along with the Denver Museum, a nearly complete *Stegosaurus* skeleton from the Garden Park area (Monaco 1998). This specimen is on display at the Denver Museum of Nature and Science and has been almost continuously since 1938 (fig. 6.44). The quarry also yielded several other vertebrates, including *Allosaurus*, turtle and crocodilian material, and an unidentified ornithopod dinosaur (Foster 2003a).

A Thirsty Lot of Dinosaurs: The Howe Quarry

Several miles north of Shell, Wyoming, 40 km (25 mi) from Montana, the remains of a ranch sit secluded among brush and cottonwood trees with little to the west but badlands and the Bighorn Basin stretching

3.19. View toward the Bighorn Mountains from the Howe Quarry in northern Wyoming. The edge of the pit worked by the American Museum of Natural History in the 1930s is in the foreground; to the right, Brent Breithaupt stands atop the quarry's spoils pile.

out for miles in the direction of Yellowstone National Park. To the east, however, rising abruptly out of the earth, are the curved flatirons of old sandstone forming the serrated western edge of the Bighorn Mountains (fig. 3.19). In this scenic spot, one of the richest and most densely packed bone deposits in the Morrison Formation was worked in the 1930s by a crew from the American Museum of Natural History.

In 1932, residents of the Wyoming town of Greybull showed a bone deposit to Barnum Brown, one of the young paleontologists Osborn had hired for his 1897 Como Bluff expedition (B. Brown 1935). Brown was now nearly 60 years old and had been at the American Museum of Natural History for many years, collecting all over the world and becoming one of the museum's most successful paleontologists. He was working in Lower Cretaceous rocks in Montana that summer of 1932 when, on a trip south into northern Wyoming, he stopped in Greybull, in the shadow of the steep west slopes of the Bighorn Mountains. M. L. Austin, a local resident who often collected rock and bone in the area, told Brown of some bones she knew of north of Shell, several miles closer to the mountains to the east. Brown decided to investigate and soon found himself at the remote ranch of an 82-year-old man named Barker Howe. Howe took Brown up the hill and showed him the bones eroding out. The site looked promising enough that Brown stayed with his assistants Peter Kaisen and D. Harbicht for a week exploring the deposit.

The site on the hillside was in a claystone between sandstone beds and faced east toward the dramatic flatiron sandstones jutting skyward along the west flank of the Bighorn Mountains. Brown decided to return, but there was so much sandstone overburden to be removed that he knew there would be a lot of work involved before they could recover anything. So in 1933, Brown returned alone and hired some help to remove much of the overlying sandstone. By the season of 1934, the dig was ready to go ahead, and it even was now funded by the Sinclair Oil Company, whose mascot of sorts has been a dinosaur for many years.

Brown had a good-sized crew out that summer, usually numbering between 9 and 12. The crew started on June 1 and soon found that the deposit comprised many articulated strings of vertebrae of diplodocid sauropod dinosaurs. The quarry was such a logjam of vertebral columns that they wondered how they would remove everything, particularly because they did not want to actually remove anything until nearly the whole deposit had been uncovered and mapped. The mapping project fell to Roland T. Bird, a motorcycle-riding rookie digger who had, at Brown's invitation, come up from Florida for the excavation. Bird spent two months working on the map and did a good job of it. Only at the end of the summer did the crew begin removing the bones (Bird 1985).

During the summer, the crew had many visitors drive out to the ranch to watch, and as much as some of the men enjoyed giving tours and answering questions, they all worried about getting the project wrapped up before winter set in. Thunderstorms in August created some late night panic as the crew scrambled to lay tarps over the site to protect it from the rain and hail, something they had opted not to do before going to sleep, hoping that they'd luck out and avoid storms. Some visitors complained that the road to the site was too rough. The first snow came on September 19, but it took until November 17 before they were finished with the dig.

Somewhere between 2,400 and 4,000 bones of nearly 25 individual dinosaurs were excavated during the 1934 dig at the Howe Quarry. Most of these were of diplodocid sauropod dinosaurs, some apparently *Barosaurus*. The other elements were of the sauropods *Camarasaurus* and *Apatosaurus* and the bipedal plant eater *Camptosaurus*, and teeth of *Allosaurus*. Some of the more important fossils from the site were skin impressions of sauropods, indicating that the animals had large, rounded bumps along much of the surface of the skin. Such impressions were also found at the site during excavations in the 1990s and showed additionally that some diplodocid sauropods had spines along the back similar to those seen in modern iguanas and tuataras.

The rather dense accumulation at the Howe Quarry also included the bones of a dozen sauropod limbs standing upright and articulated with the feet (Farlow in Bird 1985). This is unusual preservation and, along with the sedimentology, probably indicates that many of the animals at the site were mired in the mud of a waterhole in a river levee or overbank mudflat when they died (Michelis 2004).

Sometime in the mid-1930s, up near the town of Spearfish, South Dakota, in the northern Black Hills, a teenager at the Fuller Ranch came across dinosaur bone weathering out of a grassy hillside below some Lakota Formation sandstone cliffs. Reports of this find eventually made their way to James Bump, then director of the South Dakota School of

Accidental Scholarships: More from the Black Hills

3.20. Fuller's 351 quarry near Spearfish, South Dakota. (A) Excavation led by James Bump (SDSM) underway during the summer of 1935. (B) Same view in 1991, with the quarry still visible. This site yielded *Camarasaurus*, *Allosaurus*, and *Apatosaurus* material (Foster 1996a). *Image in A courtesy of the Museum of Geology, South Dakota School of Mines and Technology.*

Mines and Technology's Museum of Geology in Rapid City. Bump led an expedition to excavate the site during the summer of 1935 (fig. 3.20) and reportedly got the young Fuller a scholarship to attend the School of Mines as a reward for finding what was until quite recently the largest dinosaur quarry in the Morrison Formation of the Black Hills.

The site yielded quite a number of bones. Most seem to belong to an adult *Camarasaurus*, but mixed in with that animal are parts of an *Allosaurus* and a juvenile *Apatosaurus*. The quarry is near the top of the Morrison in the area and is in a whitish sandy siltstone. During a 1991 visit to the site, two theropod teeth were also found in the overlying pebbly sandstone of the Lower Cretaceous Lakota Formation. This was the first major quarry in the Morrison Formation of the Black Hills and the first since the Smithsonian and American Museum work of 1900.

3.21. Holt Quarry near Grand Junction, Colorado. (A) Man believed to be the superintendent of Colorado National Monument among bones of *Stegosaurus* at the site in the 1930s. Note the articulated caudal vertebrae near the man's right hand, the plate on the right side of the photograph, and the tail spikes visible in the center foreground. (B) View from the site in 2005 looking west. The site is in sandstone not far from Riggs Quarry 13 (type locality of *Brachiosaurus*). *Image in A is from the Al Look Collection, 1981.115, Museum of Western Colorado, Loyd Files Research Library.*

Edward Holt was a graduate student at the University of Colorado in Boulder, working on a master's thesis on the Morrison and Summerville Formations in western Colorado, when, in 1937, a collector from the Grand Valley named Ed Hansen showed Holt some bones on the west side of a hill not far from downtown Grand Junction. Holt excavated the site, which was in a whitish sandstone, and found an articulated partial skeleton of *Stegosaurus* (fig. 3.21), along with elements of *Allosaurus* and a sauropod. This quarry is just a few hundred meters northwest of Riggs's Quarry 13, on the same hill where the Field Museum paleontologist had removed the type specimen of *Brachiosaurus* 37 years earlier. Unfortunately, Holt could not finish the dig, and although he buried the site, by 1960, all the material had been stolen; thus, no specimens from this quarry have been preserved. Riggs returned to Grand Junction in 1938 for

Around the Corner at Riggs Hill

the dedication of plaques commemorating his 1900 to 1901 excavations at Quarries 13 and 15, which yielded the *Brachiosaurus* and *Apatosaurus* specimens at the Field Museum in Chicago.

Predator Trap? The Cleveland-Lloyd Quarry

About 40 minutes' drive south of the coal-mining town of Price, Utah, the mudstones of the Brushy Basin Member along the northwest flank of the San Rafael Swell have eroded away to expose one of the most unusual quarries known in the Morrison Formation (fig. 3.22).[8] The Cleveland-Lloyd Quarry has yielded some 10,000 dinosaur bones over the years, and nearly 70% of these belong to one species: the carnivorous dinosaur *Allosaurus fragilis*. In most ecosystems, and thus most deposits, herbivorous species outnumber carnivores by several times. The Cleveland-Lloyd ecosystem of the Late Jurassic was probably no different, but determining why so many more carnivores of one species were eventually preserved here has proved difficult.

This locality has been worked intermittently for the last 90 years or so, being first worked from 1929 to 1931 by Golden York and a crew from the University of Utah. It was later visited by W. L. Stokes and a group from Princeton University from 1939 to 1941. The University of Utah returned for five years in the 1960s under the direction of Grant Stokes, and Jim Madsen of the Utah Division of State History was there for several seasons from 1975 to 1980. Brigham Young University dug at the site from 1987 to 1990 (W. Miller et al. 1996). More recently, the University of Utah returned again in 2001 for the first of several years, and the University of Wisconsin–Oshkosh has worked there most recently.

The bone layer is approximately 50 cm to 1.3 m (1.6 to 4.3 ft) thick and is a light-medium-gray calcareous claystone just below a 50 cm to 1 m (1.6 to 3.3 ft) thick gray limestone. The collections amassed over the years indicate an unusual deposit in which a carnivore dominates the

3.22. Cleveland-Lloyd Quarry in Emery County, Utah. The bone layer is exposed within two buildings. Most of the material already excavated came from the area just to the right of the buildings (much of the area is hidden by the buildings).

3.23. Size range of *Allosaurus* specimens from the Cleveland-Lloyd Quarry, as illustrated by dentaries. Photo by author. Courtesy of the Natural History Museum of Utah, UMNH VP specimens. Scale bar = 10 cm.

herbivores in terms of numbers, as noted above. But the site is actually one of the more diverse among Morrison localities, yielding the carnivorous dinosaurs *Ceratosaurus*, *Torvosaurus*, *Stokesosaurus*, and *Marshosaurus*, in addition to the 46 individuals of *Allosaurus* (fig. 3.23). *Allosaurus fragilis* is one of the best known and most recognizable dinosaurs in the Morrison Formation (J. Madsen 1976a). It was also the dominant carnivore in the Morrison ecosystem, accounting for nearly 75% of the carnivorous dinosaur sample. Herbivorous dinosaur taxa at the quarry include several *Camarasaurus*, one possible *Barosaurus*, *Stegosaurus*, and *Camptosaurus*; the sauropod *Apatosaurus* was only recently identified in the deposit, from a single cervical vertebra (Foster and Peterson 2016). The only nondinosaurian vertebrates from the fauna are a turtle, a possible choristodere, and a crocodilian. The quarry demonstrates that at least five different types of carnivorous dinosaurs overlapped enough in their temporal and geographic distributions to be preserved in the same deposit. How they ended up there, however, remains a mystery.

The real story at the Cleveland-Lloyd Quarry is that mysterious taphonomy (Stokes 1985; Richmond and Morris 1996; Bilbey 1998, 1999; Gates 2005; Hunt et al. 2006; Foster et al. 2016a; J. Peterson et al. 2017). There are, among the *Allosaurus* bones, elements of some of the smallest individuals of that genus known. The bones are mostly disarticulated (out of position and away from the bones they would have been next to in the live skeleton), but the bones are generally not badly broken or weathered. This is a somewhat confusing set of evidence to explain. One of the most commonly presented scenarios is that the predators were lured to the site by prey species mired in mud on a flat or perhaps at the bottom of a shallow pond. As the carcasses rotted, the bones became mixed into the mud and were churned around by subsequent stomping of dinosaurs in the muck. This has been the long-held image of the Cleveland-Lloyd

Quarry, although perhaps no one would argue that the taphonomy of the site has been extensively studied, and thus the traditional interpretation is not strongly defended by anyone. In fact, recent studies have suggested that the animals at the site may have died in a series of droughts instead, which may explain a lot of what is seen at the quarry. Another unusual aspect of the quarry is its apparent high metal-element content in the rock the bones are in (J. Peterson et al. 2017). This high content may be a diagenetic factor or a result of the abundance of carcasses, but it probably was not a cause of any poisoning of the animals, if that occurred.

The high density of carnivores may simply have been because these animals scared off other dinosaurs that might otherwise have tried to access the water at the site. Also, if the animals had been mired in mud rather than physiologically stressed by a drought (which would have made them more susceptible to disease, thirst, and starvation), why do we find no leg elements stuck upright in the deposit like we do at the Howe Quarry, for example? And why are numerous foot bones chewed by carnivores? The drought idea proposed recently by Terry Gates and the idea that carcasses may have been washed into the deposit indeed fit many of the data from this site (Gates 2005; Warnock et al. 2018), although it may be difficult to ever be sure exactly what actually killed these animals (Hunt et al. 2006). Whatever happened to these dinosaurs, the Cleveland-Lloyd Quarry is unique in its preservation of so many *Allosaurus* relative to any other type of animal (Hunt et al. 2006). Other quarries preserve a lot of one or a few types of animals: Howe Quarry with many *Diplodocus* and *Barosaurus*, Mygatt-Moore with many *Apatosaurus*, Reed's Quarry 13 with many *Stegosaurus* and *Camptosaurus*. But Cleveland-Lloyd is unusual in preserving so many carnivores.

Unit 3: The Oklahoma Panhandle

J. W. Stovall was from the border region between Oklahoma and Texas and was a World War I veteran. He had studied with A. S. Romer at the University of Chicago and under R. S. Lull (the veteran of the Como Bluff expeditions of the late 1890s) at Yale. By 1931, Stovall was 40 years old and a geology professor at the University of Oklahoma. That year, a crew working on what is now Highway 64 east of Kenton, Oklahoma, hit a rich bone bed with a grader and notified residents of the town. Kenton is in the extreme northwestern end of the panhandle of Oklahoma, just a few miles from New Mexico and Colorado. The land here is carved by the Cimarron River, which flows east out of New Mexico, but is otherwise fairly flat. When the road crew cutting through the low hillside hit the bones, it was not long before Kenton residents contacted Stovall at the university. Stovall enthusiastically checked into the find and determined that an excavation would be productive. It took several years, however, to round up the funding to carry out the project, and in 1935 he finally got started with a crew comprising students and individuals hired through the Works Progress Administration (Stovall 1938), the Depression-era federal program created to try to keep people employed.

The Kenton field group for the project was known as Unit 3 and included between one and five WPA workers in addition to a Kenton local, Crompton Tate, Stovall's field leader for the project. Unit 3 eventually opened 17 quarries in the Kenton area, although only a handful of these yielded significant material. The site hit by the road crew, Quarry 1, was worked from May 1935 through April 1938 and was in a gray claystone relatively high in the Morrison Formation (fig. 3.24). The bones from the quarry were unusual for the formation in that they were nearly white in color. The field crew had to remove nearly 100 metric tons of overburden to access the bone layer, but they eventually excavated several important specimens, including a large *Apatosaurus* and the type and only known specimen of the large theropod dinosaur *Saurophaganax*. Other dinosaurs found at the site include *Camarasaurus*, *Stegosaurus*, adult and juvenile *Camptosaurus*, and at least one bone of the rare sauropod *Brachiosaurus*, a bone that was not recognized in the Stovall collection until Matt Bonnan and Matt Wedel identified it in 2004.

One of the most important collections from Quarry 1 was that of baby *Camarasaurus* and *Apatosaurus* sauropods (Carpenter and McIntosh 1994). These animals were represented by tiny vertebrae and limb elements that are still among the smallest sauropod bones ever found in the formation. Among these bones are metatarsals of *Camarasaurus* just 5 cm (2 in.) long and a humerus of *Apatosaurus* just 22.5 cm (8.8 in.) long. There seem to be in this collection at least three individuals of each genus.

3.24. Stovall's Quarry 1 near Kenton, Oklahoma; type locality of *Saurophaganax*. The site was first uncovered during local roadwork. The monument featuring a sauropod femur (above the quarry to the left) has since been redone.

Saurophaganax was similar to *Allosaurus* and was closely related (Chure 1995). In fact, the main difference in their skeletal structures is a detail of the vertebrae of the dorsal series. However, *Saurophaganax* weighed between two and three times as much as *Allosaurus* and is known from only one specimen, compared with the more than 100 individuals of *Allosaurus* known. At nearly three metric tons, *Saurophaganax* was a large predator, and it probably fed on large herbivores such as stegosaurs and medium- to small-sized sauropods.

Stovall's crews excavated Quarry 5 from December 1939 to mid-1941. This got to be a large quarry at 73 m (241 ft) wide and with a quarry wall 9 m (30 ft) high. The main elements collected here were of the diplodocid sauropod *Diplodocus*.

Quarry 8 turned out to be one of the most productive quarries in the area. The site produced femurs of one or two types of ornithopod dinosaurs (probably *Dryosaurus*), two species of lungfish, several individuals of two genera of turtles, the southernmost known occurrence of the choristodere *Cteniogenys*, and two ilia of small theropod dinosaurs. Among the large sample from the quarry were two skulls, several limb elements, and numerous vertebrae and teeth of goniopholidid crocodilians (Hunt and Lucas 1987; Hunt and Richmond 2018). In 1964, crocodile expert Charles Mook described one of the skulls as the type specimen of the new species "*Goniopholis*" *stovalli*.

The crew worked Quarries 9 and 10 into early 1942 before funding for the project dried up as the national focus shifted to World War II, affecting most aspects of life in the country.

4.1. Dry Mesa Quarry southwest of Delta, Colorado, in 1997. The huge back wall was cut in 1970s, and the bone layer in the area in the foreground has been worked back slowly since then. The man working near the right-most of the two wheelbarrows (partly hidden by rock, in the bright shirt) is Ed Delfs, who excavated the Cleveland Museum *Haplocanthosaurus* from Garden Park during the mid-1950s.

Renaissance: The Picture Fills In

4

The modern period of work in the Morrison Formation has only accelerated the rate of discovery and learning about the vertebrate fauna. Found during this time have been complete skeletons and new species of salamanders, lizards, and mammals; terrestrial crocodilians; new species of several types of dinosaurs; and new turtles. Despite more than 130 years of fieldwork in the formation, entirely new animals are still being found, demonstrating the incredible and perhaps underappreciated diversity of the Morrison's vertebrate fauna.

Early in 1954, a Louisiana State University geologist discovered bones along Four Mile Creek in the southern part of Garden Park and alerted the Cleveland Museum of Natural History. Ed Delfs and a relatively young crew spent the summers of 1954, 1955, and 1957 excavating the site. They worked under a large sandstone bed and next to a sometimes dangerous creek that flooded at least once (McIntosh and Williams 1988; Monaco 1998). The main skeleton they found in the quarry was that of a large *Haplocanthosaurus*, a genus that had first been found several miles up the road at the Marsh-Felch Quarry. Jack McIntosh and Mike Williams named this skeleton as a new species of *Haplocanthosaurus* (*H. delfsi*) in 1988, in part on the basis of its large size (McIntosh and Williams 1988). Like most specimens of this genus, *H. delfsi* came from reasonably low in the formation. Probably the most important specimens found here during the Cleveland Museum digs were the skull and partial skeleton of a small (probably juvenile) goniopholidid crocodilian, later named as the type specimen of *Eutretauranosuchus* (Mook 1967). This genus has since been found at a couple other sites, but the type is still the best specimen yet described by far.

Cleveland Goes to Garden Park

Paleontology was not new to New Mexico, but as far as work in the Morrison Formation was concerned, it took a while to get going. E. D. Cope had long ago had his collector, David Baldwin, in the territory collecting for him in the Upper Triassic strata, work that led to the discovery and naming of *Coelophysis*, the little dinosaur from the Chinle Group. And Cope himself had worked in the Paleocene of the San Juan Basin of New Mexico in the early 1870s, after tiring of seeing so much of Marsh's men up in Wyoming. In the late 1940s, Edwin H. Colbert and a crew from the American Museum of Natural History excavated a major

Hot Rocks: Early Work in New Mexico

quarry of nearly complete *Coelophysis* skeletons from the Ghost Ranch site in New Mexico. Throughout all this time, however, next to nothing was found in the Morrison Formation, in part because it was not being actively explored anywhere close to the degree to which it had been in areas further north. That all changed in the 1950s, when the post–World War II uranium-mining boom began.

The sandstones of the lower Morrison Formation in the Colorado Plateau region are rich in uranium, and the newly valuable element was mined heavily in northwestern New Mexico, eastern Utah, and western Colorado from the 1950s through the 1970s. Although the activity occurred in a number of formations in the region, the Morrison Formation mines were the most productive, and entirely new towns sprung up as a result of the work (some of these disappeared in the early 1980s, nearly as quickly as they had appeared). Mines in the area of Grants, New Mexico; Moab, Utah; and south of Gateway, Colorado, were particularly numerous—and can still be seen by the dozens along miles of Salt Wash Member outcrop.

One result of the exploration and study of the Morrison Formation for its uranium ore was the identification of a number of sites with dinosaur fossils in the unit in New Mexico. These studies began in the 1950s and were reported by several geologists, but Bill Chenoweth, then a graduate student at the University of New Mexico working on his master's degree, recorded locality data and got identifications of most of the material. We therefore know that within a few years of initiation of work in the Morrison in New Mexico, the known animals from the state had gone from none to a fauna comprising at least *Stegosaurus*, *Allosaurus*, and *Apatosaurus*.

Interestingly, although the Salt Wash Member produced more uranium than the Brushy Basin and may therefore have been more actively explored, it was the latter member that yielded all of the genera of dinosaurs mentioned above. None of the 10 or so localities known from the Morrison Formation of New Mexico through this period was really a major quarry, although major excavations began at new sites by the late 1970s.

| Back to Como Bluff: The American Museum and Yale | Seventy-two years after Osborn's trip, crews from the American Museum of Natural History and the Yale Peabody Museum ran a joint expedition to Quarry 9 at Como Bluff in Wyoming. The groups were led by Tom Rich, Pat Vickers-Rich, and Chuck Schaff and were trying to find more mammal specimens because it had been 30 years since the newest Morrison mammal had been described. In fact, little had been done with Morrison mammals since the end of Marsh's work until George Simpson's work in the 1920s, in which he described several specimens that Marsh had not gotten around to describing (Simpson 1929). A new mammal, *Araeodon*, was among the few that had been collected by the American Museum in 1897, but this remained undescribed for 40 years |

until Simpson named it. By 1967, therefore, little was new in the world of mammals of the Morrison Formation. This lack of activity was the motivation for the AMNH/YPM expedition.

The crews spent the summers of 1968 to 1970 at Como Bluff, mainly in the vicinity of Quarry 9. They found some material at Quarry 9 but also identified four new localities in the area and collected approximately 25 mammal jaws plus a number of isolated teeth. The group also found a new type of dryolestoid mammal named *Comotherium* (Prothcro 1981).

One of the largest and most scenic quarries anywhere in the Morrison Formation sits high on a mesa ridge looking out over the canyons of the Uncompahgre Plateau southwest of Delta, Colorado. The site was found by Eddie and Vivian Jones of Delta, a couple who owned a lumber company and began prospecting as amateurs in the 1940s. James Jensen was a preparator at Harvard University in 1958 when he saw in the Smithsonian Institution in Washington, DC, a sauropod limb bone from western Colorado found and donated by the Joneses. By 1964, Jensen was working at Brigham Young University in Provo, Utah, and, remembering the material he'd seen back east, he decided to contact the Joneses now that he was in the neighborhood. Eddie and Vivian were happy to show Jensen sites they had found in the area, and they kept looking for new ones for several more years. In the fall of 1971, while prospecting on the southern end of a long ridge called Dry Mesa, they found a toe bone of a large carnivorous dinosaur. This phalanx bone turned out to be the first of many elements of a large theropod that would some 8 years later be named *Torvosaurus tanneri*, a meat eater as long as but distinctly heavier than *Allosaurus*. The Joneses showed the bone to Jensen and Ken Stadtman and took the two BYU scientists to the site. Jensen and Stadtman began excavations at the new Dry Mesa Quarry the following summer. The quarry started out in 1972 as a 5.5 m (18 ft) by 91 m (300 ft) excavation, and it only got bigger as the seasons went on (fig. 4.1). Jensen and Stadtman's crews worked the site for seven seasons from 1972 through 1981, after which Jensen retired. Stadtman and crews from BYU have worked the quarry most years since then, although the university recently ceased operations there to catch up on preparation of the previously collected material.

The Dry Mesa Quarry is in a pebbly channel sandstone bed several meters thick in the lower Brushy Basin Member. Some crossbedding is apparent, and the deposit appears to represent a sandy braided stream environment that had highly variable discharge; thus, most of the bones may have been buried during flash floods. It is interesting that at this site, in such a coarse-grained sandstone, small, delicate pterosaur and other microvertebrate bones are preserved among those of some of the largest sauropod dinosaurs known in the formation.

The quarry contains one of the most diverse vertebrate **paleofaunas** in the Morrison, and it has produced several type specimens, including the **pterodactyloid** pterosaur *Mesadactylus*, the large theropod dinosaur

The Dry Mesa Quarry

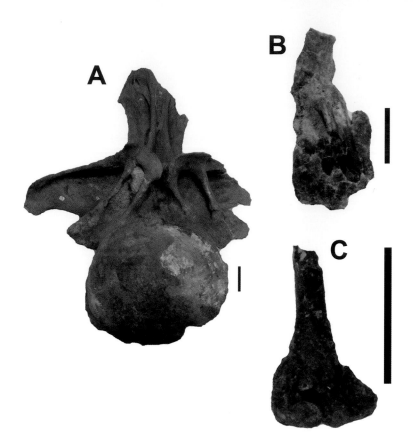

4.2. Selected specimens from the Dry Mesa Quarry, running the range of sizes. (A) Type dorsal of the sauropod *Dystylosaurus* (BYU 4503), possibly *Supersaurus*. (B) Premaxilla of a hatchling or embryo of the sauropod *Camarasaurus* (BYU 8967). (C) Distal half of the humerus of an unidentified mammal (BYU 12111). Scale in A = 10 cm; scales in B and C = 1 cm.

Torvosaurus, and the large diplodocid sauropod dinosaur *Supersaurus*. *Ultrasauros* was originally described as a very large brachiosaurid, but the type specimen (a single dorsal vertebra) has since been shown to be one of the dorsals belonging to the *Supersaurus* partial skeleton in the quarry. In addition to the type specimen of *Supersaurus*, the Dry Mesa Quarry has yielded six individuals of *Diplodocus*, five of *Camarasaurus*, and two each of *Brachiosaurus* and *Apatosaurus* (Curtice and Wilhite 1996). Another new sauropod from the quarry, *Dystylosaurus*, is based on another lone dorsal vertebra and has been described as either from an indeterminate family or a brachiosaurid (fig. 4.2). Brian Curtice believes this specimen also belongs to the *Supersaurus* skeleton. Sauropod guru Jack McIntosh once admitted that he had no idea what *Dystylosaurus* actually is, but he was not yet convinced that the type is part of *Supersaurus*. As one who believes the sauropod fauna of the Morrison Formation has historically been tremendously oversplit, and considering that *Dystylosaurus* is based on a single dorsal vertebra with but one unique character, I tend to regard *Dystylosaurus* as a nomen dubium—that is, a taxon based on material incomplete enough that it is unlikely that future specimens could be confidently assigned to it.

Also found at Dry Mesa were the premaxilla of a hatchling or embryonic *Camarasaurus* and part of a mammal humerus (fig. 4.2), along with fish and amphibian bones among a surprising microvertebrate fauna.

Jim Jensen worked at many sites in the Uncompahgre Plateau area of western Colorado, several of them before getting deep into the Dry Mesa project. One important site he worked was the Dominguez-Jones Quarry south of Cactus Park, which is south of Grand Junction. This site yielded an articulated sauropod skeleton in 1967, one that Jensen named *Cathetosaurus lewisi* in 1988. This specimen, which is nearly complete and is in the Earth Sciences Museum at Brigham Young University, was later suggested to be an old individual of *Camarasaurus,* referred to a new species as *Camarasaurus lewisi.* The specimen shows several areas of extra bone growth on the vertebrae, which likely indicate advanced age of this individual (McIntosh et al. 1996). In addition, Jensen found a *Brachiosaurus* near Potter Creek, Colorado, and a partial skeleton of a small sphenodontian in Cactus Park. Jensen also collected Morrison sauropods from a site at the south end of Cactus Park; this material is interesting in that it contains bones of a number of juveniles of *Apatosaurus.* Kristina Curry-Rogers used some of this material in her studies of growth in sauropods.

Two students at Western State College in Gunnison, Colorado, collected dozens of bones of a sauropod dinosaur from the lower part of the Morrison Formation in 1970. This quarry was along Cabin Creek, about 10 km (6 mi) east of town, and what Jim and Ken Snyder found at their site was an *Apatosaurus* partial skeleton, the only good dinosaur specimen known at the time in the mountainous 241 km (150 mi) stretch between the quarries at Garden Park near Cañon City and the Jensen sites along the Uncompahgre Plateau. The specimen was briefly described by Bruce Bartleson and Jim Jensen in 1988. The Cabin Creek Quarry was at the base of the Morrison Formation, and it appears to be one of the oldest sites in the unit (Bartleson and Jensen 1988). In fact, on the basis of the biostratigraphy of the Morrison and the relative positioning of the formation's quarries proposed by Christine Turner and Fred Peterson, the Cabin Creek *Apatosaurus* would be the oldest member of its genus known so far (Turner and Peterson 1999).

From 2009 to 2013, a fragmentary *Haplocanthosaurus* sauropod was collected from low in the Morrison Formation near the mountain town of Snowmass, Colorado (fig. 4.3), just a little west of Aspen. This specimen had been found by a college student named Mike Gordon on his grandfather's property north of the Elk Mountains; it was donated to the Museums of Western Colorado by the landowners, the Brothers and Bramson families, and although Gordon had set out to find a Cretaceous marine fish (a *Xiphactinus* had been found in the Mancos Shale nearby years earlier), he in fact found one of the rarer dinosaurs in the Morrison Formation. This specimen had the misfortune to be located under a large scrub oak bush, to have been tectonically uplifted, and to have endured millennia of winters near the surface at nearly 2,740 m (9,000 ft) elevation, so it was highly fractured, but as one of few *Haplocanthosaurus*

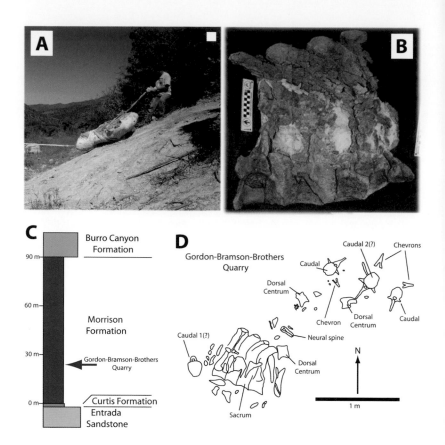

4.3. Sauropod *Haplocan-thosaurus* from the Gordon-Bramson-Brothers Quarry in the lower Morrison Formation at Snowmass, Colorado. (A) Removing a block containing sacrum and dorsals, 2009; Zeb Miracle on brake-rope duty. (B) Sacrum in the right lateral view, showing posteriorly swept neural spines. (C) Stratigraphic position of the site low in the Morrison. (D) Quarry map of the site showing the association of elements. Based in part on Foster and Wedel 2014.

skeletons known, it was important that it was at least identifiable (Foster and Wedel 2014).

The Bighorn Mountains

In 1991, an articulated subadult *Allosaurus* was found on land administered by the Bureau of Land Management north of Shell, Wyoming. This site is in a scenic location along the western flank of the Bighorn Mountains near the Howe Quarry, within view of older upturned rocks that jut upward along the shoulder of the uplift, not unlike the flatirons seen in the Front Range of Colorado or the San Rafael Swell of central Utah. The *Allosaurus* (known as "Big Al") was preserved in the sandstone of an ancient stream channel and is one of the most beat-up dinosaur skeletons known from the Morrison, demonstrating a number of injuries to its ribs, limbs, and pelvis. It also, on further study, proved to be a second specimen of a new species of *Allosaurus* that has been named on the basis of another specimen from low in the Morrison Formation at Dinosaur National Monument. A cast of Big Al is on display at the Geological Museum at the University of Wyoming in Laramie.

Bruce Erickson and crews from the Science Museum of Minnesota began working at the Poison Creek Quarry south of Buffalo, Wyoming, in 1977 and continued through most of the 1980s. The quarry is in a silty claystone just above a channel sandstone near the middle of the formation. The most important specimens to come from this site include a partial *Haplocanthosaurus*, a large *Camptosaurus*, and an unusually long and slender theropod metatarsal. Several other common dinosaurs were found here over the years as well, including *Allosaurus*, *Camarasaurus*, *Apatosaurus*, *Diplodocus*, and *Stegosaurus* (Erickson 2014).

In 1985, Patrick McSherry found a specimen in the Morrison Formation near Buffalo, Wyoming, that was later determined to be the most primitive stegosaur known from the formation. It occurred in the lower 5 m (16.5 ft) of the Morrison in a fine-grained sandstone and is thus one of the oldest stegosaurs in North America. This was, interestingly, the first new species of stegosaur found in the Morrison in a very long time, and it indicates the potential of the lower part of the formation for yielding a rather different fauna from the other levels. The new stegosaur was named *Hesperosaurus* in 2001 and differs from *Stegosaurus* in having a more primitive skull, lower neural arches in the dorsal vertebrae, and smaller, more oval back plates (Carpenter et al. 2001).

Mike Flynn of Sheridan College in northern Wyoming began working several sites around the Poison Creek Quarry shortly after the Science Museum of Minnesota ceased operations there. Two quarries north of but at about the same stratigraphic level as the Poison Creek Quarry have produced a typical Morrison dinosaur fauna, including *Camarasaurus* and *Apatosaurus* (Tucker 2011). Another site in the channel sandstone underlying the other sites has yielded more significant specimens, including a partial *Allosaurus* skeleton and a vertebra of the rare semiaquatic reptile *Cteniogenys*. A much older and smaller relative of the crocodile-like champsosaurs of the Cretaceous and early Cenozoic, *Cteniogenys* was first collected at Quarry 9 at Como Bluff but was not named until 1928 and was not recognized as a member of the champsosaur group until the late 1980s (Evans 1989).

The historic site of so many finds of the American Museum of Natural History from 1898 to 1905 was expanded starting in the 1990s by Western Paleo Labs of Utah. This second series of digs at the quarry proved particularly successful: new specimens and species were recovered, including the ankylosaur *Gargoyleosaurus*; the small, long-legged theropod *Tanycolagreus*; and the scaphognathid pterosaur *Harpactognathus*. Other skeletons found were those of *Stegosaurus*, *Camarasaurus*, and a juvenile *Ceratosaurus*. Also found at the site was the 6 m (20 ft) neck skeleton of a *Barosaurus* diplodocine sauropod.

One site, the East *Camarasaurus* Quarry, yielded the only essentially complete skeleton of a North American Jurassic neosuchian crocodyliform, a goniopholidid specimen consisting of the skull, the limbs, and most of the vertebral column. This quarry is also interesting in that the teeth from the *Camarasaurus* skull had fallen out and were in the process of scattering downstream when they were buried. The river channel sand in which the bones and teeth were encased had flowed to the southeast,

Bone Cabin Revisited

and in the quarry map, one can see a logjam of limb bones on the upstream side of the *Camarasaurus* skeleton and a scattering of bones downstream. And pointing downstream from the skull, up to 2 m (6.6 ft) away and with their heavy ends upstream, are more than 50 teeth of the animal. I've seen few better illustrations of the effects of stream currents on dinosaur skeletons.

"Diplodocid El Dorado": The Howe-Stephens and Howe-Scott Quarries

Kirby Siber and crews from the Sauriermuseum in Aathal, Switzerland, had been working at the historic Howe Quarry in northern Wyoming when they found Big Al, the *Allosaurus*, but that specimen turned out to be on a Bureau of Land Management parcel and was ultimately excavated by the Museum of the Rockies. By 1992, the Siber crews were working a new pit slightly higher stratigraphically at the Howe Quarry, and over the next few seasons, this site yielded a number of partial skeletons. Clearly, the American Museum crew under Barnum Brown had not quite exhausted this area in the 1930s. The Siber group found at least five specimens of diplodocids, along with individuals of *Camarasaurus*, *Stegosaurus*, *Allosaurus*, *Dryosaurus*, and the small neornithischian *Nanosaurus* (fig. 4.4). Several of these taxa were new for the quarry, but the fact that they found so many diplodocines is interesting because Brown and the AMNH crew also excavated many of these sauropods from their pit. Among the material excavated from this new pit are a nearly complete *Camarasaurus* (nicknamed "ET"), the type specimen of the diplodocine *Kaatedocus siberi*, and referred material of another diplodocine, *Galeamopus hayi* (Tschopp and Mateus 2013; Tschopp et al. 2015). The new excavations eventually developed into two distinct quarries just south of the historic Howe Quarry (Howe-Stephens and Howe-Scott).

Perhaps the most important specimen to come out of this new Howe Quarry dig is the articulated and nearly complete skeleton of a baby sauropod, probably less than 2 m (6.5 ft) long (fig. 4.4). In some respects, the skeleton is unlike any adult sauropod from the Morrison Formation: the scapula is unusually large relative to the other elements, and the centra of the dorsal vertebrae are relatively large compared with those of the tail and neck and are much longer relative to their diameter than in any other sauropod I've seen. The limbs and even the feet of this specimen are almost perfectly articulated. It is probably the most amazing dinosaur skeleton yet found in the Morrison Formation. There is, of course, no skull, and neither the neural arches and spines of the vertebrae nor the pelvic bones have been fully prepared, so the identity of the skeleton is still uncertain. This specimen was identified early on as a young juvenile diplodocid but was later reidentified as a basal titanosauriform, possibly a very young *Brachiosaurus* (Schwarz et al. 2007; Carballido et al. 2012).

A Second Heyday at Como Bluff

Little had happened regarding the dinosaurs of Como Bluff since the AMNH crews moved further north after the 1897 season. That changed

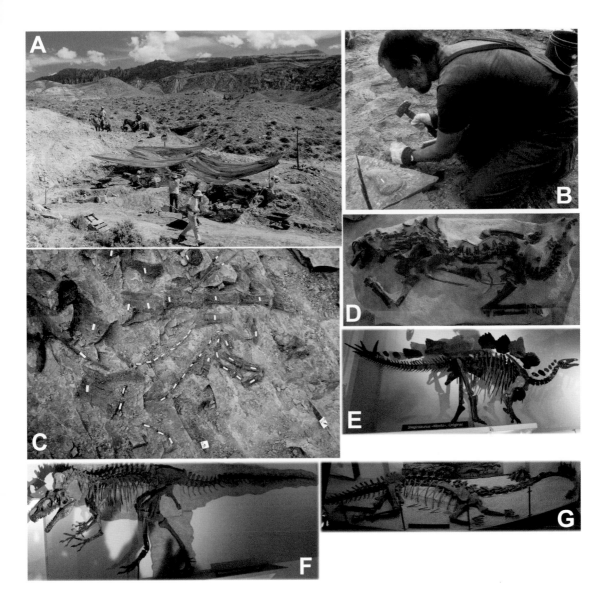

in the 1970s, however, when paleontologist Robert Bakker began working at Como Bluff, particularly on the eastern end, where Bill Reed had had a few quarries. In the last 30 years, Bakker and his crews have revitalized Como as a critical area of exploration in the Morrison Formation and have proved that the area is far from played out. The numerous localities in this area have yielded many specimens and have provided clues to the vertebrate paleoecology of the area during the Late Jurassic. Among the sites are the following:

1. The Bernice Quarry, a locality fairly high in the Morrison and one that produced an *Apatosaurus* as well as pterosaur, turtle, and ?"*Drinker*" material.

2. The Bertha Quarry, a site low in the Morrison in the eastern part of Como Bluff that has produced the type specimen

4.4. The Howe Ranch quarries and material collected from them in recent decades by Sauriermuseum Aathal. (A) Quarry in 1993. (B) Kirby Siber working on part of a *Camarasaurus*. (C) Caudal vertebrae of *Kaatedocus* in the ground. (D) Young brachiosaurid sauropod. (E) *Stegosaurus*. (F) *Allosaurus*. (G) *Diplodocus*. A–C courtesy of Kirby Siber. E–G exhibited at the Sauriermuseum, photos courtesy of Heinrich Mallison; D, cast at the University of Wyoming Geological Museum, photo by author.

of *Brontosaurus* (originally described as *Apatosaurus*, then *Eobrontosaurus*) *yahnahpin* (Filla and Redman 1994; Bakker 1998a) and parts of a *Dryosaurus* and a goniopholidid crocodilian. The site is in a near-channel, overbank mudstone bed, and the main sauropod skeleton seems to be from an animal that became mired in the soft mud.

3. Breakfast Bench, a series of mostly microvertebrate sites in the eastern part of Como Bluff that have produced crocodilians, the large lungfish *Ceratodus robustus*, and the type specimens of the turtles "*Dorsetochelys*" *buzzops* and *Uluops uluops*, the small ornithopod "*Drinker nisti*" (*Nanosaurus*), and the mammals *Foxraptor atrox*, *Morrisonodon brentbaatar*, and *Zofiabaatar pulcher*. The sites are in a series of dark mudstones and appear to be part of a wet floodplain environment near the top of the formation.

4. The Cam Bench Locality, which preserves *Camarasaurus*, *Ceratosaurus*, *Allosaurus*, and *Camptosaurus* material in a dark-gray mudstone above a soil-formed calcium carbonate layer.

5. The Drinker Quarry, high in the Morrison, which produced several adult and juvenile individuals of the small ornithopod "*Drinker nisti*" from a dark-gray mudstone.

6. The Nail Quarry, an important quarry near the middle of the formation, which is in gray smectitic claystone with some limestone. The stratigraphic level of the Nail Quarry is the Talking Rocks Member of the Morrison, and the site has produced a crocodilian and the dinosaurian genera *Allosaurus* (TM 0011), *Apatosaurus*, *Stegosaurus*, *Diplodocus*, and *Camarasaurus*. It has also produced the type specimens of the megalosaurid theropod dinosaur *Edmarka rex*.

7. The Boris Quarry, which produced a partial ankylosaur skull.

8. The Zane Quarry, a locality in the Sheep Creek area that is fairly low in the Morrison and that produced *Camarasaurus*, *Apatosaurus*, *Diplodocus*, *Stegosaurus*, a theropod, and an ankylosaur (Bakker 1986, 1996, 1998a; Bakker et al. 1990; Turner and Peterson 1999).

Some of Robert Bakker's work in the Morrison Formation in recent years has concentrated on studies of shed teeth of dinosaurs and other reptiles from several microvertebrate sites in the Como Bluff area, and these have shown some interesting patterns of apparent dinosaurian feeding behavior that we discuss in later chapters. All carnivorous dinosaurs shed and replaced their teeth continuously throughout their lives, and most teeth were lost during feeding, so these shed teeth can be important indicators as to where carnivorous dinosaurs were feeding and on what prey (Bakker and Bir 2004; Foster 2005a).

Found on private land north of Como Bluff and worked by Chris Weege, Dave Schmude, and Western Paleo Labs, this is stratigraphically one of the lowest quarries in the Morrison Formation and is thus an important site. The quarry is in a sandstone bed and has produced an adult and a juvenile *Allosaurus*, both of which may belong to the new species from low in the formation (*A. jimmadseni*). It has also yielded a primitive stegosaur and elements of an ankylosaur.

Meilyn Quarry

In 1978, ranchers in the northwestern Black Hills of Crook County, Wyoming, found elements of a dinosaur south of Devils Tower National Monument, along Inyan Kara Creek. This site was worked by Phil Bjork, director of the South Dakota School of Mines Museum of Geology at the time. It turned out that the material was a nearly complete forelimb of a juvenile diplodocid sauropod, probably either *Diplodocus* or *Barosaurus*. Unfortunately, no vertebrae were recovered that could have refined the identification. There are some indications that the manus bones of *Diplodocus* were different from this specimen, and it may therefore represent the first good forelimb material of *Barosaurus*, but we can't be certain of that (Foster 1996a).

In 1980, another partial skeleton of a probable *Barosaurus* was found just a few miles south of the type locality and north of Rapid City, South Dakota (Bjork 1983). This specimen consists mostly of dorsal and caudal vertebrae, but it also contains elements of the pelvis and one partial scapula. The Wonderland Quarry site also yielded crocodilian teeth, turtle shell fragments, teeth of *Camarasaurus*, and vertebrae and metatarsals of *Allosaurus*. Then, during preparation of one of the field jackets in 1992, I came across a small theropod ilium right up against a dorsal vertebra of the *Barosaurus*. This small ilium, only about 8 cm (4 in.) long (fig. 6.13), appeared to be that of a juvenile *Stokesosaurus*, a carnivorous dinosaur otherwise known best from the Cleveland-Lloyd Quarry in Utah (Foster and Chure 2000). However, several years later paleontologist Oliver Rauhut noted greater similarities between this Black Hills ilium and a possible tyrannosauroid ilium from the Late Jurassic of Portugal that he had named *Aviatyrannus* than between the former and *Stokesosaurus* (Rauhut 2003).

In 1990, I began a survey for vertebrate fossils in the Morrison Formation of the Black Hills of South Dakota and Wyoming (Foster and Martin 1994; Foster 1996a, 1996b, 2001a). There were at that time three main quarries: Marsh's *Barosaurus* type locality on Piedmont Butte from 1889, Fuller's 351 from 1935, and the Wonderland Quarry, worked in 1980. There were also a handful of other sites that had produced fewer specimens scattered around the Black Hills. My goal was to prospect as much of the Morrison Formation as possible, looking for new sites that had to be out there, and in the process to relocate some of the old sites and check if any new material could be found there.

The Black Hills Revisited

Working the Morrison Formation in the Black Hills is a little different than in other areas. Those who are familiar with the formation on the Colorado Plateau, for example, might recognize it in South Dakota or northeastern Wyoming only because of its distinctive color. Even in the Black Hills, several hundred miles from other outcrops of the formation, the Morrison is largely green, gray, and maroon. Unlike other areas, including the Colorado Plateau, there are few sandstones in the Morrison Formation of the Black Hills, and those that are present are generally thinner, more laterally restricted, and almost always fine grained. The formation itself is also much thinner in the Black Hills; in some places it is as few as 24 m (80 ft) thick. In parts of the Colorado Plateau, for comparison, the Brushy Basin Member alone is commonly 91 m (300 ft) thick. Another difference, and the main problem for finding fossils in the unit in the Black Hills, is that the Morrison is mostly covered with grass and pine trees. This is rarely a problem in the Colorado Plateau, as it can be in other parts of Wyoming, but in the Black Hills, the rocks seem to be particularly obscured by vegetation. Most exposure of the formation comes at roadcuts and in the rare natural exposure in the ring of uplifted Mesozoic rocks that encircles the Black Hills uplift. (In one case, the Morrison Formation was well exposed in the city dump near Sturgis, South Dakota, but it didn't last more than a few weeks because the earthmovers shifted material around and covered this nicely exposed and complete section of the formation.)

At first, the survey was successful in relocating most of the old sites but less so in finding new material of much significance. When I started, the vertebrates known from the Morrison Formation of the Black Hills included the following: *Camarasaurus* from several sites, *Allosaurus* from two localities, *Barosaurus* from Piedmont Butte and Wonderland, *Apatosaurus* from Fuller's 351, some turtle and crocodyliform material from Wonderland (Foster 1996b), and a lungfish from a site near Sturgis (Pinsof 1983). Most of what had been found during the survey up to that point was indeterminate sauropod fragments. In July 1991, however, my assistant, Greg Goeser, and I located a site in the northwestern Black Hills that was more along the lines of what I had been hoping to find: a site with abundant bone and a diversity of vertebrate taxa that would give us a better picture of what Morrison times were like in the Black Hills area. (We were unaware at the time that this site had been noticed as early as the 1970s by paleontologist Peter Houde, now of New Mexico State University.)

The quarry is about 32 km (20 mi) south of Devils Tower National Monument, near the town of Sundance in northeastern Wyoming, and is not far from a marshy, slow-flowing drainage named Little Houston Creek (fig. 4.5). Nearly 25 seasons of fieldwork at the Little Houston Quarry have produced the most diverse vertebrate paleofauna of any site in the Morrison Formation north of Como Bluff (Foster et al. 2020). The quarry contains at least seven individual *Camarasaurus*, elements of

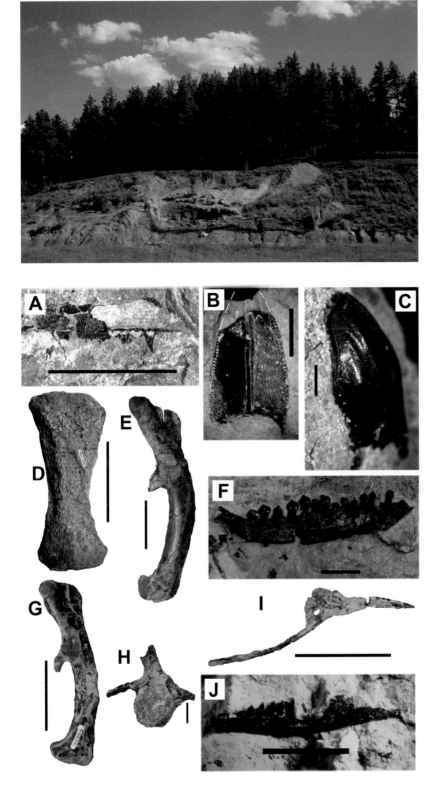

4.5. Little Houston Quarry in the Black Hills of northeastern Wyoming, not far from Devils Tower National Monument. This locality has produced the most diverse vertebrate fauna of any site north of Como Bluff. Note the cover of Black Hills grass and pine trees in this part of Wyoming.

4.6. Specimens from the Little Houston Quarry, Wyoming. (A) Fish jaw, scale bar = 1 cm. (B) Unidentified, unrecurved theropod tooth, scale = 1 cm; very thin and not D shaped in cross section, possibly not from the anterior of the jaw; somewhat similar to teeth in a jaw from the Late Jurassic of Portugal. (C) Tiny *Camarasaurus*-like tooth possibly from a hatchling sauropod, scale bar = 1 mm. (D) Metapodial of a sauropod dinosaur almost completely covered in small circular pits, possibly borings made by dermestid beetles (MWC 5784), scale bar = 10 cm. (E) Left femur of *Nanosaurus* in medial view (SDSM 30490), scale bar = 5 cm. (F) Dentary of *Nanosaurus* (MWC 5822), scale bar = 1 cm. (G) Left femur of *Nanosaurus* in medial view (SDSM 30494), scale bar = 5 cm. (H) Caudal vertebra of an unidentified bipedal neornithischian (SDSM 30496), scale bar = 1 cm. (I) Pubis of an unidentified bipedal neornithischian (SDSM 30503), scale bar = 5 cm. (J) Left dentary of a dryolestid mammal in dorso-labial view, scale bar = 1 cm. All SDSM specimens, except F (MWC 5822); photograph F by R. Peirce.

Apatosaurus, and some bones that could be either *Diplodocus* or *Barosaurus* (Foster and Martin 1994; Foster 1996a); there were also an adult and a young juvenile *Allosaurus* (Foster and Chure 2006; the young juvenile recently assigned to A. *jimmadseni*, Chure and Loewen 2020), several individuals of the neornithischian *Nanosaurus* (Peirce 2006; fig. 4.6), and possible fragments of *Dryosaurus*, *Stegosaurus*, and *Camptosaurus*. Nondinosaurian vertebrates include ray-finned fish, lungfish, frogs, a lizard, sphenodontians, the champsosaur *Cteniogenys*, a goniopholidid crocodyliform, *Theriosuchus*, the turtles *Dinochelys* and *Glyptops*, and the mammals *Docodon* and *Amblotherium*, plus an unidentified multituberculate and ?triconodont (fig. 4.6; Foster 2001a; James Martin and Foster 1998; Foster et al. 2020). There was also at least one very small apparent sauropod tooth (fig. 4.6). Most of the microvertebrates and small ornithopods are disarticulated and occur in several thin but dense layers of bone in siltstone. The quarry matrix consists of interbedded green mudstones and lighter green-gray wavy-laminated siltstones. The bones occur in the siltstones, and the microvertebrates are most densely accumulated in layers of siltstone that contain many small, green clay balls and molds of the shells of freshwater clams. Most of these clamshells seem to have been washed in along with the microvertebrate bones.

The Little Houston Quarry deposit is in the lower middle part of the formation and represents an accumulation in an abandoned channel, which develops when a river changes course upstream and leaves a section of its course stagnant as a pond that slowly fills with mud. In this case, the pond was occasionally reactivated as a flowing part of the river during floods and received the influx of clay balls, clams, and bones, which were quickly buried in the wavy-laminated siltstones. The green mudstones represent periods when the pond was still and was accumulating sediment only as it settled out of the water. Vertebrates are preserved in three levels in the quarry: (1) a basal channel sandstone unit, (2) the main level in an interbedded green claystone and laminated siltstone unit, and (3) an overlying claystone unit. The sandstone probably represents the channel in an active stage; the main bone level would be during a gradual abandonment stage; and the upper deposit, which occurs in claystone above the level of at least one root cast and is lateral to the main part of the quarry, probably represents a stage after infilling of the channel (Foster and Martin 1994; Foster 2001a).

We collected several unusual theropod teeth from the Little Houston Quarry during the early 1990s. Each is relatively flat and compressed, with serrations on each edge, but is unrecurved (fig. 4.6). Former Dinosaur National Monument paleontologist Dan Chure and I have tried to figure out to what theropod these teeth belong but have had no luck. They are unlike any other theropod teeth either of us has seen from the Morrison Formation. Although the teeth remind one slightly of the D-shaped premaxilla teeth of *Allosaurus* and other theropods, these new teeth are nowhere near as thick in cross section, nor are they as recurved as most typical premaxilla teeth. The teeth also are relatively small on average. They may come from a known Morrison theropod that simply has not yet yielded its skull, or it could be a different theropod entirely. We don't yet know. Perhaps someday someone will find such teeth in a skull or jaw somewhere and the mystery will be solved, but for now, this is the only evidence we have for this unusual theropod in the Little Houston Quarry deposit. Oliver Rauhut did describe a small theropod dentary fragment from the Late Jurassic of Portugal that has similar teeth to these, but it is difficult to tell how close the taxa are. The teeth are generally similar to, but do not exactly match, various teeth from the Late Jurassic of Europe and Africa that have been assigned to abelisauroids or carcharodontosaurids (Foster et al. 2020).

Another tooth type from the quarry is different from anything else from the formation, with a nearly circular basal crown cross section and two very strong carinae, the anterior one of which is folded distolingually. The tooth is similar in some ways to those of therizinosaurs, but is likely a variety of anterior tooth from a small neornithischian (Foster et al. 2020).

A screen-washing site just a mile and a half up the road from Little Houston (Mile 175) has produced mammals and salamanders and a wide diversity of taxa but is particularly rich in fish scales, teeth, and tooth patch fragments (Foster and Trujillo 2004; Foster and Heckert 2011).

We also located a site on a private ranch south of Sundance in a rare (for the Black Hills) natural exposure of red claystone. This locality has produced several partial caudal vertebrae of an as yet unidentified diplodocid. I originally identified this specimen as *Barosaurus* (Foster 1996a) but was never comfortable with the idea. It totally lacks pleurocoels and has a deep, narrow ventral excavation on the midcaudal vertebra, similar to the type of *Galeomopus hayi*, although the midcaudal is not as elongate as those of *G. hayi*. I'm still not sure what to make of this specimen.

Another site in the Bearlodge Mountains in the northwestern Black Hills of northeastern Wyoming is in a lacustrine limestone unit at an undetermined level in the Morrison Formation and has produced elements of a sauropod, possibly *Camarasaurus*. This is a rare occurrence of dinosaur bone in limestone within the formation. Few other examples exist, but sauropod bones have also been found in the limestone at the Purgatoire River track site south of La Junta, in southeastern Colorado.

In 1993, Steve Sroka and Russ Jacobson found another locality west of Sundance, Wyoming, which appears to be very close to the top of the formation. This is one of the few quarries in the Morrison Formation to be screen washed, a method of recovering tiny bones and teeth by filtering fine sediments through a screen mesh in water, leaving behind only larger rock fragments and bones. Elements here have also been collected from the surface, and all seem to be coming from green-gray mud in a soft channel sandstone. Vertebrate taxa preserved here include abundant fish, turtles, crocodilians, lungfish, small theropod and ornithopod dinosaurs, the choristodere *Cteniogenys*, and the mammals *Docodon* and *Ctenacodon*. But mostly the site yields fish scales—three varieties and thousands and thousands of fish scales (Foster and Heckert 2011). In many places, the Morrison landscape was not as dry as is commonly pictured.

In 1997, the University of Kansas began working at a site high in the Morrison Formation near Newcastle, Wyoming, in the western Black Hills (Sundell 2005). The site was originally found just before World War II but was then abandoned for more than 50 years. This locality has produced several large *Camarasaurus* and a brachiosaurid, the first known from the Black Hills area. Among other material found at the site are several turtle skeletons, a small hypsilophodontid, and at least one tooth of a probable *Torvosaurus*.

Recently, the Black Hills Institute of Geological Research has excavated a site northwest of Hulett, Wyoming, in the Black Hills that has yielded a skull and skeleton of a large *Camarasaurus*. This quarry is in a light-greenish-gray mudstone, and bone occurs at five levels in the area. Other dinosaurs found at the site include *Stegosaurus*, *Allosaurus*, and a diplodocid sauropod.

Microvertebrate Mother Lode: The Fruita Paleontological Area	California State University Long Beach biology professor George Callison led a small crew to western Colorado during the summer of 1975. The group was exploring Morrison Formation outcrops, and on the recommendations of a local high school teacher and of a CSU Long Beach student with relatives in the area, they stopped at an outcropping of the Brushy Basin Member just across the Colorado River from the town of Fruita (Callison 1987; Kirkland et al. 2005; Foster et al. 2016a). Jim Clark, then an undergraduate student and now a biology professor at George Washington University, noticed the first small bones on a low slope of mudstone. Soon the crew was crawling along the surface, and they noticed that there was quite a bit of very small material there, but in order to get more complete bones, one had to dig in. From then until 1987, Callison led crews excavating two main quarries at what became, in 1977, the Fruita Paleontological Area (FPA). The main quarry is in the middle of the formation near the base of the Brushy Basin Member and is in a near-channel overbank deposit (fig. 4.7); it appears to be about 152 million years old (Foster et al. 2017). The site has produced many specimens of a number of taxa, mostly microvertebrates, including the type of *Fruitachampsa callisoni* (Clark 1985, 2011), a small, gracile, terrestrial shartegosuchid crocodilian; a sphenosuchian crocodilian (*Macelognathus*; Göhlich et al. 2005); and several lizards and mammals, the latter including *Priacodon fruitaensis* (Rasmussen and Callison 1981a). Tom's Place is another major FPA quarry and is at a similar stratigraphic level and in a similar paleoenvironment (fig. 4.7). The quarry is in mudstone laterally adjacent to a channel sandstone bed and has produced numerous microvertebrate remains (Kirkland 2006). This site and the main quarry contain a relatively diverse assemblage of lizard remains, including

4.7. Microvertebrate quarries at the Fruita Paleontological Area, western Colorado. (A) Early work at the Callison (Main) Quarry in 1979, view to the north; Jim Clark swings the pick (*left*). (B) The Callison Quarry in 2017 (foreground) looking west, with the area currently being worked in shadow at left. (C) Tom's Place quarry, right foreground, in 2017, looking south toward the Callison Quarry (middle distance). George Callison and crews worked these sites for approximately a decade and found many microvertebrate specimens in the mudstone. Image in A courtesy of George Callison.

Dorsetisaurus, Saurillodon, and *Paramacellodus* (Evans 1996). In addition to the lizards, the Fruita sites have yielded numerous skulls and skeletons of dryolestid, multituberculate, and triconodont mammals and have produced the type and several referred specimens of the large and robust sphenodontian *Eilenodon* (Rasmussen and Callison 1981b). Most importantly, the FPA has produced the unusual Late Jurassic mammal *Fruitafossor* (Luo and Wible 2005) and the probable snake *Diablophis,* which likely had vestigial limbs (Caldwell et al. 2015). Also named recently from the FPA (but first found in 1976) is the small, adult heterodontosaurid dinosaur *Fruitadens* (Callison and Quimby 1984; Butler et al. 2010, 2012), which weighed only 0.75 kg (1.6 lb). Although Quarry 9 is responsible for more specimens of microvertebrates in the Morrison, no other site in the formation is comparable to the FPA in terms of articulation and completeness of the small specimens. It is simply one of the most important sites in the Morrison.

In the late 1970s, a specimen of *Ceratosaurus* was collected from a sandstone near the middle of the formation in the FPA by Lance Eriksen. The partial skeleton (MWC 1) was designated the holotype of the new species *Ceratosaurus magnicornis* by Jim Madsen and Sam Welles in 2000. The quarry was at the same level as the Callison microvertebrate quarries but in the channel rather than the overbank, and most of the skull was preserved with the skeleton.

Rainbow Park and Dinosaur National Monument

On the northern edge of Dinosaur National Monument is an area that in recent years has yielded a microvertebrate fauna of the same caliber as Quarry 9 and the FPA. Two microvertebrate sites in Rainbow Park (Chure and Engelmann 1989) have produced a diverse assemblage of animals including the type specimen of the small theropod *Koparion douglassi* (Chure 1994) and several partial pipoid frog skeletons (Henrici 1997, 1998a, 1998b), along with other microvertebrate material. Both sites at Rainbow Park (Locality 94 and Locality 96) are high in the formation and occur in gray mudstone-siltstone units (fig. 4.8). Another locality in the area yielded a partial sauropod skeleton that was unusually radioactive, a result of its uranium content.

Site 375 in Dinosaur National Monument (but outside Rainbow Park) is another microvertebrate locality found in recent years. It has produced the type specimen of the new lizard *Schillerosaurus utahensis* as well as undescribed skeletons of small salamanders and some eggshell.

Other sites in Dinosaur National Monument but outside the quarry building include one locality high in the formation that produced an embryonic *Camptosaurus* specimen (Chure et al. 1994), a site with a specimen of the rare Morrison theropod dinosaur *Marshosaurus* (Chure et al. 1997), and another east of the main quarry that produced a nearly complete, articulated *Allosaurus* specimen with skull. The latter quarry is in the Salt Wash Member of the Morrison and is in a gravelly sandstone, and the skeleton is the type specimen of *Allosaurus jimmadseni.*

4.8. Microvertebrate localities in Dinosaur National Monument, Utah. (A) Former National Park Service paleontologists Dan Chure and Scott Madsen at Site 375 in steeply tilted mudstone layers of the Brushy Basin Member of the Morrison Formation in the western part of Dinosaur National Monument. (B) Rainbow Park area in the northern part of Dinosaur National Monument. Upturned beds of the Brushy Basin Member are in the foreground; the Green River flows through the area in the distance. *Photographs by author.*

Around 1979–1980, a teenage girl named India Wood discovered a bone locality in the Brushy Basin Member of the Morrison Formation south of Dinosaur National Monument in northwestern Colorado. She helped Denver Museum of Natural History crews led by Don Lindsey to excavate partial skeletons of *Allosaurus* and *Stegosaurus* from the quarry and then returned on her own in 1985 as part of a college project. The *Allosaurus*

Wolf Creek: More Small Vertebrates

from this site is currently mounted at the Denver museum. The 1985 expedition resulted in finds of a number of microvertebrates, including lizards, frogs, sphenodontians, and a mammal. This locality is quite similar to the Callison quarries at the FPA in being in gray claystones and in containing only a moderate concentration of sometimes articulated material. As one of relatively few microvertebrate localities in the Morrison, the site is important to the understanding of a good part of the Morrison's vertebrate fauna.

Jensen-Jensen Quarry

Jim Jensen seems to have had a knack for finding or working on sites with *Brachiosaurus* specimens, and here he had a site near the town of Jensen in northeastern Utah. The site, relatively low in the Morrison, was excavated by Jensen (the paleontologist, not the town) and is across the Green River southeast of the Dinosaur National Monument quarry. The main specimen here was a fragmentary skeleton of *Brachiosaurus*. The site is in a sandstone unit in the Salt Wash Member and also produced *Camarasaurus* and *Apatosaurus* sauropod material and some limb elements of a small stegosaur.

The Warm Springs Ranch

Thermopolis, Wyoming, is home to some well-formed hot springs just north of the Owl Creek Mountains and west of the southern end of the Bighorn Mountains. This small community became home to of one of the largest and deepest excavations ever dug in the Morrison Formation in 1993. Burkhard Pohl's crews began digging on the Warm Springs Ranch near town and set up the associated Wyoming Dinosaur Center. By having a large backhoe and dump truck running much of each summer, crews here have been able to create the largest open-pit quarry I've ever seen in the formation, and in the excavated hillside lie nearly 40 individual sites. Two of these have been developed into significant quarries yielding many bones of multiple individuals and taxa. Some specimens are articulated, including one *Diplodocus* foot. Most dinosaur material belongs to *Camarasaurus, Apatosaurus, Allosaurus, Stegosaurus,* or *Camptosaurus* (Ikejiri et al. 2006). The size of the overall excavation here and the distribution of bone within it at multiple levels and across the hillside laterally make me wonder if this is what most Morrison hillsides would look like if we could see into them with deep ground-penetrating radar vision. This site is a rare view into the Morrison Formation.

The Earth-Shaker Lizard and a New Mexico Renaissance

In 1978, a specimen later referred to the large sauropod *Camarasaurus supremus* was collected near San Ysidro in New Mexico (Rigby 1982; Johnstone 2017), and this specimen was later suggested to be instead *C. grandis* (Ikejiri 2005). This was the first significant partial skeleton of a dinosaur found in the state. Before this time, most of the known material from New Mexico was rather fragmentary dinosaur pieces found during uranium surveys and mining.

Until the work near San Ysidro, most of what was known of the dinosaurs of New Mexico was based on relatively fragmentary although still identifiable specimens; no quarries with articulated skeletons or multiple taxa had been reported. The San Ysidro *Camarasaurus* (fig. 4.9; NMMNH 21094) had been noticed first in the early 1960s by Will Johnstone and his father, and he returned with his young family 15 years later to try to relocate the bone or two they noticed earlier. It ended up being one of Johnstone's sons who found the site again by accidentally sliding over a bone while playing on the Brushy Basin Member mudstone slopes along a wash (Johnstone 2017). There turned out to be a number of bones at the site, and these were excavated by the Johnstones, many volunteers, and a Bureau of Land Management crew led by Keith Rigby Jr. A number of elements from the site are definitely of a large camarasaur, but the two most complete dorsals are of a size more typical of adult specimens (fig. 4.9), and I wonder if there may be two individuals represented in this quarry, which might in part account for the specimen being identified variously as either *C. supremus* or *C. grandis*.

In 1979, two hikers from Albuquerque, Arthur Loy and Jan Cummings, came across articulated vertebrae of a large sauropod dinosaur in a sandstone in the upper Morrison Formation near San Ysidro, in northern New Mexico. They and two friends, Bill Norlander and Frank Walker, returned to the site later and eventually reported their find to David Gillette, now of the Museum of Northern Arizona but at the time with the New Mexico Museum of Natural History. Gillette and crews developed a large quarry at this site and removed much of the posterior half of the diplodocid skeleton (fig. 4.9), which was determined to have been deposited on a sandbar in an ancient meandering river. The crew tried a number of high-tech search techniques in order to determine where in the nearly 3 m (10 ft) deep sandstone the skeleton was lying (Gillette 1994). The animal was named *Seismosaurus* in 1991 and was estimated to be the longest dinosaur ever found, at 52 m (171 ft). The name was a reference to the idea that this sauropod must have caused microearthquakes as it walked the Jurassic floodplain (Gillette 1991). Whether the skeleton found represented an animal that long has been sometimes hotly debated, but all agree that it was a very large and long sauropod, probably in the range of 28 to 39 m (92 to 129 ft). Some have said *Seismosaurus* wasn't this long, and others have claimed other sauropods of the Morrison were bigger.[1]

Seismosaurus hallorum is similar in most respects to *Diplodocus*, and most of the characters that Dave Gillette used to distinguish the two genera are based on proportional ratios rather than unique aspects of the shapes of the bones. Proportional ratios are not necessarily unimportant, but unique bone characters are less equivocal. Ratios may or may not be important, depending on how much variation between individuals of a species one tends to allow. As one who believes there may have been a significant amount of skeletal variation between individuals within most sauropod species, I believe that it is possible that *Seismosaurus* was a very large, possibly very old, *Diplodocus*. Spencer Lucas and others in the past briefly referred to the specimen collected from this New Mexico quarry as *Diplodocus hallorum* in print, thus officially suggesting a synonymy between the two genus names. A more formal skeletal comparison is in Lucas et al. (2006a). In any case, *Seismosaurus* (*D. hallorum*) demonstrates how large the diplodocid sauropods of the Morrison Formation could become.

After the *Seismosaurus* quarry and San Ysidro *Camarasaurus*, work began in 1989 on the Peterson Quarry, west of Albuquerque. This site had been found in the 1960s by Rodney Peterson, not surprisingly as a result of uranium prospecting. The site is the first major quarry in New Mexico that has produced multiple elements of more than one or two taxa, and it occurs in a channel sandstone in the Brushy Basin Member of the Morrison, downslope from the Jackpile Sandstone Member. The quarry has yielded a diplodocid skull, teeth and other elements of *Camarasaurus*, the first Morrison Formation turtle from New Mexico (*Glyptops*), and a huge allosaurid (possibly *Saurophaganax*; dorsals are not preserved, so it is difficult to say for sure) that seems to be larger than any known *Allosaurus* (fig. 4.9; Heckert et al. 2003a; Williamson and Chure 1996; Lucas et al. 2006b; Lucas and Heckert 2015). The bone preservation at the site is moderate at best, but the true significance of the site lies in the large allosaurid and in the fact that it is the most diverse single quarry in the Morrison Formation known from the state. There must be other sites in New Mexico with as many or even more species and more individuals (and some with other small nondinosaurian vertebrates too), but we just haven't found them yet. Somewhere out there are more Peterson-type quarries (or even larger) in New Mexico. As I alluded to above, we could really use a good nondinosaurian fauna from the Morrison Formation of New Mexico as well.

Most vertebrate fossils in the Morrison Formation in New Mexico have come from the Brushy Basin Member. A pair of *Allosaurus* jaw elements from the Section 19 mine on the southern edge of the San Juan Basin is of interest, then, because these are some of the few identifiable bones from below the Brushy Basin. The bones were found in a channel sandstone in a mine in the Westwater Canyon Member (Salt Wash Member, of Anderson and Lucas) of the Morrison and indicate the potential of some of these lower members for producing well-preserved, identifiable taxa (Heckert et al. 2003b). This is also the level from which the rare *Allosaurus* species *A. jimmadseni* has been collected in other outcrop areas farther north.

Other specimens from the Morrison Formation of New Mexico are generally isolated, scattered around the northern half of the state, and were found in association with uranium mining also. One of the few elements of a small theropod dinosaur was found in Quay County among

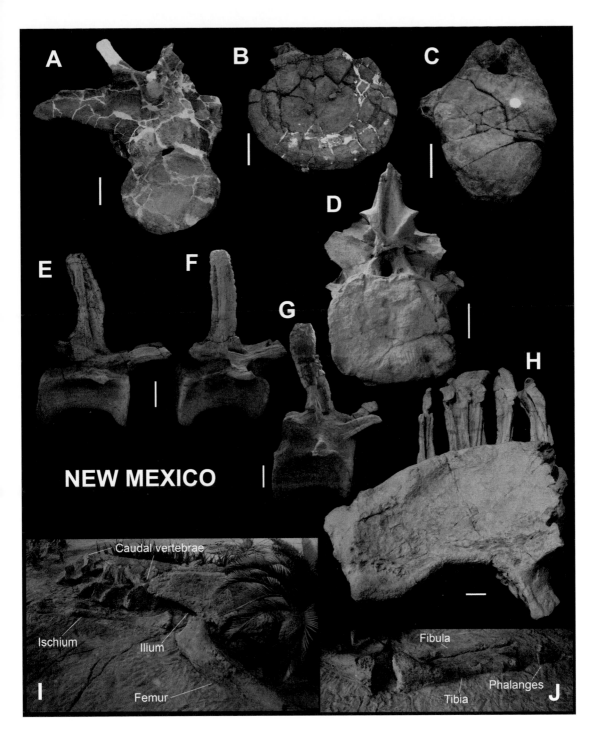

4.9. Dinosaurs from the Morrison Formation of New Mexico. (A–C) San Ysidro *Camarasaurus* (NMMNH 21094). (A) Anterior dorsal in posterior view. (B) Large dorsal centrum. (C) Large anterior caudal centrum. (D–H) *Diplodocus* (*Seismosaurus*) *hallorum* (NMMNH 3690). (D) Caudal vertebra 8 in posterior view. (E) Caudal vertebra 17 in right lateral view. (F) Caudal vertebra 16 in right lateral view. (G) Caudal vertebra 13 in right lateral view. (H) Sacrum and ilium in right lateral view. (I–J) Elements of the tail, pelvis, and hind limb of a large allosaurid from the Peterson Quarry (NMMNH 26083), as on display. This specimen may represent the large allosaurid *Saurophaganax*, but the dorsals (most diagnostic of that taxon) are not preserved. All scale bars = 10 cm. *Photos by author, courtesy of the New Mexico Museum of Natural History and Science.*

fragmentary sauropod remains but has only been able to be identified as a probable coelurosaurid (Hunt and Lucas 2006).

Burning Town: Sites in the Moab Area

The Mill Canyon Dinosaur Trail north of Moab, Utah, consists of a path along a thick, gravelly sandstone in the middle of the Morrison Formation. There are many bones of dinosaurs exposed in the sandstone here probably representing *Camarasaurus*, *Allosaurus*, and *Stegosaurus*, but unfortunately, many of the exposed bones have been vandalized over the years. Crews from Brigham Young University worked in the area several decades ago, but otherwise, this deposit, which probably rivals Dinosaur National Monument as a dense accumulation of bone in a channel sandstone, has never been excavated. The Bureau of Land Management's Moab office has been trying to protect the bones in their natural setting

4.10. Brigham Young University Museum of Paleontology excavation of an apatosaurine (referred to *Brontosaurus parvus*) near Moab, Utah, in 2008. (A) Overview of the quarry during excavation. (B) Dorsal vertebrae in the quarry. *Photos courtesy of Rod Scheetz.*

for some time, and they recently installed new interpretive signs at the site, making this a fun site to visit and see material in place.

Not far from the Mill Canyon trail, in 2008, Brigham Young University crews led by Rod Scheetz excavated a sauropod skeleton that had been found in the Brushy Basin Member years earlier by Moab rock shop owner Lin Ottinger (fig. 4.10). This specimen was identified as *Brontosaurus parvus* by Tschopp et al. (2015) and is one of more complete and least crushed apatosaur specimens I have ever seen. It also includes an impressively large pelvis that is on display at the Museum of Paleontology at BYU.

In 2000, a dinosaur specimen was found in the Morrison Formation in Arches National Park north of Moab, Utah, the first significant one known from that park. Several vertebrae were exposed on the surface of the upper Brushy Basin Member not far from Delicate Arch. This specimen turned out to be probably the stratigraphically highest occurrence of an *Apatosaurus* (or at least an apatosaur) in the Morrison Formation (and thus the geologically youngest; Foster 2005c; Swanson et al. 2005). The specimen was in a green claystone that was dipping 70 degrees from horizontal. Unlike other sauropods from approximately equivalent stratigraphic levels at Cope's quarries at Garden Park, Colorado, this *Apatosaurus* was not unusually large compared with those of its kind from lower in the formation (Foster 2005c). A joint crew from the Museum of Western Colorado and the Utah Museum of Natural History collected part of this specimen in 2003 and found the material closer to the surface to consist of numerous caudal centra and a partial pubis. More may be preserved at the site, but it has not yet been fully excavated.

Another of the recent finds in the Moab area is that of a partial skeleton of *Camarasaurus* from south of town near La Sal Junction. It came from an as yet undetermined level in the Brushy Basin Member of the Morrison Formation. This specimen consists of nine articulated anterior caudal vertebrae, three dorsal vertebrae, much of the sacrum, several fused sacral neural spines, and other elements of the pelvis (Foster 2005c). The specimen (SUSA 515) is in the collections of the Museum of Moab.

In March 1981, J. D. Moore and Pete Mygatt and their wives were out hiking in Rabbit Valley, a remote area adjacent to the Utah border in western Colorado. The group found bone there in a shallow drainage not far from Interstate 70. Nearly 35 field seasons of work later, the site (fig. 4.11) is still producing abundant bones each year, and several thousand have been found already. The most common vertebrates found at this site are the dinosaurs *Apatosaurus* and *Allosaurus* (fig. 4.12; Foster et al. 2007, 2016a, 2018). *Camarasaurus* and *Diplodocus* are rare. The first Jurassic ankylosaurian dinosaur found in North America, *Mymoorapelta* (Kirkland and Carpenter 1994), came from this site, and elements of this heavily armored dinosaur are still found on occasion. *Stegosaurus*, *Camptosaurus*, and most other ornithischians are unknown from the

Jurassic Gladiator Pit: The Mygatt-Moore Quarry

4.11. Mygatt-Moore Quarry in western Colorado, looking northeast. The site has been excavated nearly every year since 1985.

Mygatt-Moore Quarry (except for a jaw fragment of *Nanosaurus* found in 2007 and a single caudal vertebra that may be from a small neornithischian). In addition to the dinosaurs, a few microvertebrates have been uncovered recently, including a sphenodontid (fig. 4.12). No turtles have been found, and only a single vertebra and a single tooth of a crocodilian have been recovered from the site in more than 20 seasons. Small fragments of plant material are abundant in the matrix rock. Gastropods are found in the deposit but mostly in the upper parts of the bone layer and somewhat above it, or immediately below the bone layer. The site also preserved an egg, several skin impressions, and possible coprolites, all of dinosaurs (Chin and Kirkland 1998; Foster and Hunt-Foster 2011). Fish are found in a different lithology about 2 m (6.6 ft) above the top of the bone layer, suggesting that the area later became a permanent standing body of water. This layer can be traced nearly a mile eastward, although the bone layer cannot. Apparently, the later shallow lake that preserved the fish was a more extensive feature than the low, muddy area that preserved the slightly older dinosaur skeletons. The site has been determined to be about 152.18 ± 0.29 million years old (Trujillo et al. 2014); that's an error of about ±0.19% (a fifth of a percent!), far better than we could get even a few decades ago. What's 290,000 years, give or take?

One of the most interesting aspects of the Mygatt-Moore Quarry is the number of shed theropod dinosaur teeth that are found there. Literally hundreds of them are present around the bones in the quarry. And several bones that have been collected have tooth scratch marks. Not all of them are shallow scrapes; some are deep gouges in the bone, indicating feeding action of the theropods (fig. 4.12). Numerous bones at the site exhibit clean, preburial fractures (fig. 4.12).

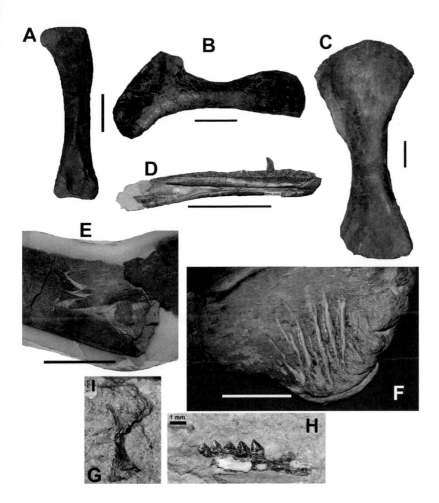

4.12. Fossil vertebrates from the Mygatt-Moore Quarry. (A) Left femur of a juvenile *Apatosaurus* (anterior view), MWC 5439. (B) Left scapula of a juvenile *Apatosaurus* (MWC 1848), lateral view. (C) Right humerus of an adult *Apatosaurus* (MWC 5694), anterior view. (D) Left dentary of a juvenile *Allosaurus* (MWC 5440), lingual view. (E) Fractured sauropod rib with matrix filling the break (MWC MM-1260). (F) Carnivore tooth marks in the distal end of a sauropod ischium (MWC 4011). (G) Unidentified limb element of a possible salamander (MWC 8649). (H) Dentary fragment with teeth of a small sphenodontid, possibly *Opisthias* (MWC 8671). All scale bars = 10 cm, except G and H = 1 mm. *Images in G and H courtesy of Julia McHugh.*

The site was probably a muddy or marshy area with some seasonal standing water, possibly similar to a vernal pool. The plant material indicates abundant vegetation in the area, and the shed teeth and chewed and broken bones indicate significant scavenging of carcasses, but the distinct lack of fish and the paucity of turtle or crocodilian material in the main bone layer suggest that there was little perennial standing water (Foster et al. 2018). The area was probably muddy and surrounded by plants, but the moisture in the soil was likely provided more by a high water table than by surface water. Atmospheric precipitation probably contributed to the dampness and to seasonal ponding.

Not far south of the Mygatt-Moore Quarry in Rabbit Valley and stratigraphically below it in the lower Brushy Basin Member is the Twin Juniper Quarry. This site contained the back end of a partially articulated skeleton of an *Apatosaurus* lying on its right side in a mottled red-and-green paleosol mudstone. Also found with this skeleton was a single, large tooth of a *Ceratosaurus*. The apatosaur skeleton had been only partly buried in the mud and probably spent some time exposed as the well-drained floodplain formed an incipient soil and as plants grew and put down roots in the sediments. The flood of a nearby river later covered the bones with a layer of fine sand, and the specimen was safely entombed for 150 million years, until erosion began to expose it many years ago. The rocks at the site above the quarry form a sequence of alternating thin sandstones and red, mottled mudstones and demonstrate that the cycle of soil formation and flooding went on in this area for many centuries.

Cactus Park

In the early 1990s, Kent Hups, a student at the University of Colorado Denver at the time, collected elements of an articulated ankylosaur similar to *Mymoorapelta* from a fine-grained sandstone in the Cactus Park area of western Colorado, near where Jim Jensen had worked. He was assisted by Adrian Hunt, Martin Lockley, and others, and some of the large blocks were extremely heavy. The specimen (MWC 2610) is important in that it is one of only two associated skeletons of ankylosaurs known from the Morrison and yielded one of the few skulls. Unfortunately, the sandstone in which the bones are preserved is so well cemented with large calcite crystals that preparation is frustratingly slow. Chemical preparation of the specimen, in which the cement is dissolved in a mild acid, is almost no faster or less labor-intensive than mechanical preparation. This is a specimen that will eventually get prepared, but it illustrates the patience that this particular science requires.

Garden Park Renaissance

In the late 1970s, Don Lindsey, a paleontologist at the Denver Museum of Natural History, worked at several sites in the Garden Park area north of Cañon City, Colorado, not far from where Cope and Marsh crews had been years before. Lindsey found one quarry in particular that was moderately rich in bone, yielding fish, frogs, turtles, sphenodontids, crocodiles, and dinosaurs.

This work was followed in the 1990s by DMNH crews led by Ken Carpenter. These groups found a number of quarries, including one of the best egg nest sites in the formation, as well as the Small Quarry (fig. 4.13), a site that preserves one of the most diverse and specimen-rich representations of Morrison vertebrates in the Garden Park area. The Small Quarry has produced fish, turtle, lizard, sphenodontid, crocodile, pterosaur, dinosaur, and mammal material in abundance, including a new species of the mammal *Docodon* (Rougier et al. 2015). The best dinosaur specimen out of the site was a nearly complete *Stegosaurus*, which showed the plates of the animal arranged in two alternating rows.

Uravan: A Concentration of Dryosaurs

In 1973, Peter Galton and Jim Jensen reported on an occurrence of bones of the small, bipedal plant-eating dinosaur *Dryosaurus* at a site in western Colorado. This locality had been found near the uranium-mining town of Uravan in the spring of 1972 by a young future paleontologist named Rodney Scheetz. The site was near the base of the Brushy Basin

4.13. Small Quarry at Garden Park, Colorado, about five years after most of the material had been collected. The site produced an articulated *Stegosaurus*, plus a new species of the mammal *Docodon*, the type material of the pterosaur *Kepodactylus*, an apparent abelisauroid theropod, and other important specimens.

4.14. Cast skeleton of a juvenile *Dryosaurus*, based on material from the Scheetz Uravan locality, as previously mounted at the Museum of Moab. Adults got up to at least twice as long as this 1.2 m (4 ft) composite individual.

Member of the Morrison and was screen washed and surface collected over a number of years, eventually yielding a diverse microvertebrate fauna. Most importantly, about 90% of the approximately 2,500 bones collected from the site belong to embryonic to subadult individuals of the ornithopod dinosaur *Dryosaurus* (fig. 4.14; Scheetz 1991), and there are at least two types of eggshell preserved at the locality also. This is one of the densest concentrations of ornithopod dinosaur material known from anywhere in the Morrison. The associated microvertebrate fauna includes a pterosaur, a turtle, a crocodilian, a number of jaw fragments of two types of sphenodontians, one partial vertebra possibly from the choristodere *Cteniogenys*, and a small jaw fragment of a multituberculate mammal.

Ninemile Hill Microvertebrates

More small vertebrates from the upper part of the Morrison Formation began to surface in 1997, when Kelli Trujillo (at the time a University of Wyoming graduate student) began screen washing several sites at Ninemile Hill, north of Como Bluff, Wyoming (Trujillo 1999). These sites yielded numerous mammals and other small vertebrates, but among the more interesting aspects of the finds were that fish were rather abundant and that a number of vertebrae were found from the small, semiaquatic *Cteniogenys*. These vertebrae showed that on average Morrison *Cteniogenys* were larger than those from mostly Middle Jurassic rocks in the United Kingdom (Foster and Trujillo 2000).

Fox Mesa: New Microvertebrates

Another site found in 1997 has been worked by the Smithsonian Institution near Shell, Wyoming, and the sample is unusual (for many in the Morrison) in that it contains mostly terrestrial vertebrates—and a significant percentage of them are multituberculate mammals. The Fox Mesa site is in the upper part of the Morrison, appears to be in a floodplain deposit, and contains many eggshell fragments as well. Among the vertebrates found are sphenodontids, lizards, theropods, and small ornithopods (mostly teeth). In addition to the multituberculates, there are symmetrodont and possibly dryolestoid mammals. The abundance of multituberculates is interesting, and this site may prove valuable in showing us what the microvertebrate faunas looked like in drier parts of the Morrison floodplain. Most of our microvertebrate sample so far (from sites such as Quarry 9) seems to come from relatively wet environments, so Fox Mesa may provide us with a fresh look at another aspect of the Morrison story (Brett-Surman et al. 2005).

Big Sky Country: The Morrison Formation of Montana

Historically, few expeditions have explored the Morrison outcrops of Montana, but collections were made there in the early part of the twentieth century by crews from the American Museum of Natural History and the Carnegie Museum. The AMNH crews collected parts of the sauropods *Camarasaurus* and *Diplodocus* near the town of Pryor, and the Carnegie Museum worked farther northwest at a site called Wit-tecombe's Ranch and got a relatively diverse though fragmentary fauna including a goniopholidid crocodilian, the lungfish *Ceratodus*, a turtle, a theropod, a possible *Stegosaurus*, and sauropods tentatively identified as *Camarasaurus* and *Diplodocus*.

Most material from Montana has been collected in recent years, however, and it has been excavated by teams from the Museum of the Rockies in Bozeman and from the Academy of Sciences / University of Pennsylvania in Philadelphia. The Strickland Creek, T&J, O'Hair, and Mother's Day sites in Montana have yielded specimens of *Stegosaurus*, *Allosaurus*, *Diplodocus*, and juvenile *Apatosaurus*. The Mother's Day site in particular has been well studied and has yielded a number of appar-ently relatively young *Diplodocus* (~15 individuals) in what appears to be a debris flow deposit (Myers and Storrs 2007; Storrs et al. 2013). One of the specimens from this site is the tiny partial skull of a diplodocid (Woodruff et al. 2015). Recent studies have suggested that some of the diplodocids from this site might have been a dwarfed sauropod species (Waskow 2017), although further work on this idea is still in progress.

The O'Hair site near Livingston has yielded two apatosaurs, a stego-saur, a theropod, two *Diplodocus*, and a jaw of a sphenodontian. MOR 592 from this site is a subadult diplodocine with a very slender ("stove-pipe") femur; it may well be a young *Diplodocus* but has also been hypothesized to be an *Amphicoelias* instead—yet another case of more study needed.

A site in far southern Montana worked by the University of Penn-sylvania has produced the new sauropod dinosaur *Suuwassea emilieae* (Harris and Dodson 2004) and a theropod. *Suuwassea* appears to be a dicraeosaurid diplodocoid sauropod, a type of sauropod previously known more from the Tendaguru and other southern-continent deposits than from the Morrison. It is also relatively small. The fact that coal deposits occur in the Morrison Formation farther north in Montana suggests that the environment was different from most of the region, and this may have been because the plain here was closer to the seaway to the north. If so, perhaps *Suuwassea* was a lowland or coastal animal that preferred habitats slightly different from those of *Camarasaurus*, *Diplodocus*, and *Apatosaurus* (Harris and Dodson 2004).

A site in the Little Snowy Mountains of central Montana has pro-duced the second-northernmost occurrence of dinosaurs in the Morrison Formation, including stegosaurs and, appropriately given its abundance throughout the unit, *Camarasaurus* (Woodruff and Foster 2017). Also re-cently found at a site in this area is what appears to be a haplocanthosaur.

Blue Mesa Beach Brontosaur

A quarry found in the mid-1990s by Tony Fiorillo produced a partial skeleton of *Apatosaurus* from the Morrison Formation along Blue Mesa Reservoir west of Gunnison, Colorado. The site is in Curecanti National Recreation Area right on the shore away from roads and was accessed

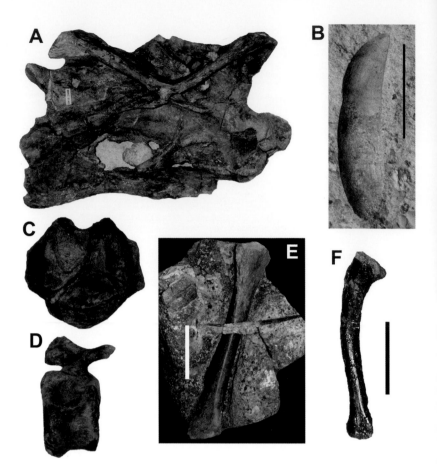

4.15. Vertebrate specimens from Blue Mesa Reservoir's Dino Cove Quarry and other sites in Curecanti National Recreation Area near Gunnison, Colorado. (A) *Apatosaurus* cervical vertebra in right lateral view (MWC 5140), Dino Cove site. (B) Mold of a theropod tooth with root, North Beach site. (C) *Stegosaurus* anterior caudal vertebra in posterior view (MWC 5525), Dino Cove. (D) *Stegosaurus* midcaudal vertebra in right lateral view (MWC 5525), Dino Cove. (E) Tibia of an unidentified bipedal dinosaur, South Beach site. (F) Ulna of a goniopholidid crocodyliform, South Beach site. Scale bars = 10 cm in E, 5 cm in B and F.

by boat during excavation. Eating lunch on a dock at a dinosaur site is a nice change from the usual heat and dust. Among the bones of this *Apatosaurus* specimen were found a theropod tooth and, recently, two vertebrae of *Stegosaurus* (fig. 4.15).

Additional sites, also on the shores of Blue Mesa Reservoir and found by Alison Koch, Forest Frost, and Kelli Trujillo, have been located since the *Apatosaurus* excavation, and several of these have been in conglomeratic sandstones. Animals identified from these sites include *Ceratosaurus*, *Allosaurus*, *Camarasaurus*, *Diplodocus* or *Barosaurus*, a crocodylian, vertebrate burrows, and termites (trace fossils of their nests), among others (fig. 4.15; Koch et al. 2006; Foster et al. 2015). These sites are just off the Colorado Plateau and thus are part of the mountainous province of Colorado, between the plateau and the sites along the Front Range uplift (e.g., Morrison and Garden Park).

San Rafael Swell Area

The Aaron Scott Quarry, on the west side of the San Rafael Swell in Utah, well south of the Cleveland-Lloyd Quarry, was opened relatively recently and has yielded dinosaurs such as *Camarasaurus*, a diplodocine, and *Allosaurus*, plus turtles, goniopholidids, and sphenodontids. These seem to

have come from sediments representing a delta formed by a river flowing into a lake (Bertog 2013; Bertog et al. 2014).

Hanksville Area

The Hanksville-Burpee Quarry has been worked every spring to early summer in recent years and is a huge deposit in a pebbly channel sandstone several miles northwest of Hanksville, Utah, hidden among some of the most barren and beautifully exposed outcrops of the Brushy Basin Member of the Morrison that one is likely ever to see (barren of vegetation, not of fossils). The landscape is so red that it has been used as training ground for potential Mars missions and as the Red Planet during the filming of *John Carter*. Material collected here includes the sauropods *Camarasaurus*, *Apatosaurus*, and *Diplodocus*, plus theropod material (possibly including *Torvosaurus*) and, starting in 2014, elements of an ankylosaur, possibly *Mymoorapelta* (Tremaine et al. 2015; Mathews et al. 2018). The sandstone contains bones along many meters of outcrop, and the quarry (fig. 4.16) consists of numerous individual pits that are usually active concurrently; it is a massive deposit on the order of other typical Morrison sandstone deposits like the Carnegie Quarry at Dinosaur National Monument and the Dry Mesa Quarry—and potentially bigger.

Another site farther west of Hanksville contained a number of sauropods and theropods as well as many large fossil logs in channel sandstones of the Salt Wash Member (fig. 1.31). One of the sauropods included articulated dorsal vertebrae and ribs of the rare *Haplocanthosaurus*. In a study of a number of areas, sedimentologist Sharon McMullen found that, despite its having few dinosaur quarries, the Salt Wash Member in fact contains many bone localities, particularly in the upper half (McMullen 2016). To some degree, I've always felt that some of the lack of development of Salt Wash sites had less to do with their uncommonness in the unit and more to do with their commonly occurring buried by

4.17. Gnatalie Quarry in southeastern Utah. (A) Quarry at the base of thick sandy braided river channel sandstone. (B) Crew at work in the quarry, which consists of several levels.

sandstone blocks, debris, and piñon and juniper trees or at the bases of 10-m-high cliffs of solid sandstone—the Salt Wash is just not an easy unit in which to dig.

Blanding Area

Another site only recently developed, the Gnatalie Quarry is located in the lower Brushy Basin Member southeast of Blanding, Utah (fig. 4.17). The site earned its name early on when the crew from the Natural History Museum of Los Angeles County made the mistake of coming out to the quarry in May and June, during an approximately six-week annual period when southeastern Utah is cursed with billions of cedar gnats (biting midges) that swarm, bite, and generally drive you insane. The quarry is in a coarse, crossbedded sandstone and has produced many bones of a theropod, an ankylosaur, a camarasaur, and a diplodocid (among others) in a concentrated, near log-jam of a deposit (Mocho and Chiappe

2018). The channel appears to be a sandy braided stream that was flowing northeast across the floodplain during Brushy Basin times about 152 million years ago.

In 2016, a new plant locality was found near Blanding by Jim Kirkland and Don DeBlieux of the Utah Geological Survey, and this site preserves abundant taxa such as the ginkgophytes *Czekanowskia* and *Ginkgoites*, ferns such as *Coniopteris*, and conifers like *Brachyphyllum*. Study of this site is just beginning but it also preserves a small fauna in addition to the plants.

River of the Lost Souls of Purgatory

In the 1930s, Roland Bird visited the largest dinosaur track site in the Morrison Formation in southeastern Colorado, south of the town of La Junta, before heading on to the Early Cretaceous track sites of Texas (Bird 1985). The site Bird left along the banks of the Purgatoire River is still there, trackways on an ancient lake shoreline, and it sits in a river valley sunk down into the plains of Colorado southeast of Colorado Springs. But the slopes along the edge of this canyon contain additional outcrops of the Morrison Formation, and here recently several sites have been found high in the formation. Among these, the Last Chance Quarry yielded a significant portion of a large *Apatosaurus*, plus fewer elements of a *Camarasaurus* and a juvenile *Diplodocus* (Schumacher 2008). In addition, there were three dozen shed theropod teeth, mostly around the *Apatosaurus* skeleton, and several sauropod bones with tooth marks, all suggesting scavenging by carnivorous dinosaurs.

Cisco Mammal Quarry

For years the FPA and its microvertebrates in western Colorado appeared to be unique. Although other sites such as Rainbow Park and Wolf Creek produced associated and articulated microvertebrates in a similar taphonomic mode in deposits of gray mudstone with fossils in relatively low abundance, some of the most interesting taxa at the FPA seemed to occur only there. And then Brian Davis of the University of Louisville and Rich Cifelli of the Sam Noble Oklahoma Museum of Natural History and their crews began searching for similar lithologies and deposits in eastern Utah by the near-abandoned town of Cisco near the Colorado River. Surprisingly (but maybe we shouldn't be surprised), they found a layer in the same levels of the Brushy Basin Member that has begun producing some of the same taxa as the FPA. They have so far found a few lizards and the mammals *Fruitafossor*, *Glirodon*, and *Dryolestes*, plus what appears to be a possible *Fruitadens* (heterodontosaurid) and a snake close to *Diablophis*, both previously known only from Fruita (Davis et al. 2018). The site also produced what appears to be a morganucodontid mammal.

Temple Canyon

Low in the Morrison Formation southwest of Cañon City, Colorado, are laminated lacustrine deposits that preserve a unique biota of Morrison taxa, including as yet undescribed fish, phyllopod crustaceans, plants, and insects (Gorman et al. 2008; Smith et al. 2011). This area was developed in the 1990s and early 2000s by crews from the Denver Museum of Nature and Science led mainly by Ken Carpenter, Bryan Small, Jason Pardo, and Mark Gorman. Many of the taxa remain to be worked on.

Jimbo (Douglas) Quarry

Located on private land south of Douglas, Wyoming, this site produced a second partial skeleton of the rare diplodocine *Supersaurus* and a small theropod. This site was worked for years by Bill Wahl, and the specimens are now at the Wyoming Dinosaur Center in Thermopolis; the *Supersaurus* is on display. It also yielded a new genus of small troodontid theropod.

Dana Quarry

Located near Ten Sleep, Wyoming, near the southwestern end of the Bighorn Mountains, the Dana Quarry was developed starting in 2006. It has yielded several well-preserved skeletons suggesting a relatively low stratigraphic position, including the stegosaur *Hesperosaurus* and possibly the theropod *Allosaurus jimmadseni*, plus *Camarasaurus*, a possible brachiosaurid, *Nanosaurus*, *Camptosaurus*, *Ceratosaurus*, *Torvosaurus*, *Coelurus*(?), and *Ornitholestes*(?) among the dinosaurs that have been reported (Galiano and Albersdörfer 2010). There are also diplodocines known from the quarry.

Eggs

Eggs and eggshell fragments were unreported from the Morrison Formation until 1987, when three sites containing eggshell were reported simultaneously (Bray and Hirsch 1998; Carpenter 1999). Those sites were in Garden Park near Cañon City, at the FPA, and near Delta, all in Colorado. In 1986, Grand Junction, Colorado, geologist Bob Young found the site near Delta. This site is in the middle of the Salt Wash Member and is thus the geologically oldest eggshell site in the formation. The eggshell fragments and one egg occur in a reddish mudstone below a sandstone bed and form a dense accumulation. A field jacket of the mudstone approximately 50 cm (1.7 ft) on a side contains thousands of fragments of eggshell, and in many places the shell fragments are clustered. These eggs probably belonged to small theropod or possibly ornithopod dinosaurs, and the site may have been an annual nesting area. In addition to the eggshell, the site has produced a terrestrial crocodylomorph skull and material of turtles, theropods, and other crocodilians.

The first material from Garden Park consisted of a single shell fragment. The fragments from the main quarry at the FPA are not as numerous as at the Young locality, but they are of at least two types. About the same time as the report of these sites, a mostly intact egg was found at the Cleveland-Lloyd Quarry in Utah. This egg has a double layer of eggshell, which suggests the mother dinosaur retained the egg longer than usual as a result of some type of stress or trauma.

By the mid-1990s, crews from the Denver Museum of Natural History had located a site in Garden Park that preserved not only many eggshell fragments but also the intact eggs of a nest. This site is on the west side of Fourmile Creek in Garden Park and produced a nest of at least five subspherical eggs referable to the eggshell genus *Preprismatoolithus*. The eggs are about 10 cm (4 in.) in diameter. In addition, a specimen of *Dryosaurus* was found at the site, and a crocodilian was found nearby in a limestone unit. The site is in the lower middle part of the formation.

Other sites in the Morrison that have been found to contain eggshell are the Uravan Locality, another site in the FPA with juvenile dryosaur and crocodilian bones as well as eggshell (Kirkland 1994); the Mygatt-Moore Quarry; Site 375 at Dinosaur National Monument; and a screen-washing site in the Black Hills of northeastern Wyoming.

In recent years, it has become better understood just how abundant and common tracks of dinosaurs and other creatures are in the rocks of the Mesozoic worldwide. Western North America is no exception. The Jurassic units of the West are full of dinosaur footprints, and although it is far better known for its abundance of dinosaur skeletal remains, the Morrison Formation has in its outcrop area more than 70 known localities of tracks (fig. 4.18) in a number of different rock types, ranging from sandstones and oolitic limestones to mudstone. One can drive into the desert north of Moab, Utah, and see at one site tracks of a limping meat eater and a large sauropod making a right turn. In northeastern Arizona, several trackways of medium-sized theropods lie in sandstone that used to be the banks of a sandy river but is now high on a ridge overlooking the Four Corners area. Crocodile tracks are preserved in the roof of a now-closed uranium mine in the Salt Wash Member of the Morrison Formation on the southeast slope of the La Sal Mountains in eastern Utah. And many footprints of small carnivorous dinosaurs occur on an upturned slab of the Morrison south of Buffalo, Wyoming, a sandstone forced up to a steep angle by the rise of the Bighorn Mountains. Lakeshore limestone deposits preserve hundreds of tracks of sauropods and theropods at two sites in the lower Morrison at Como Bluff, Wyoming, and in the Purgatoire River Canyon in southeastern Colorado.

The first report of dinosaur tracks in the Morrison Formation was one of the last papers published by O. C. Marsh and illustrated several medium to large three-toed tracks from the southern Black Hills of South Dakota. This was in 1899, and it is in some ways surprising that the report came only after the Morrison Formation had already been worked for more than 20 years and from a state with otherwise relatively few Morrison tracks (Marsh 1899). The South Dakota site was near the top of the formation southwest of Hot Springs, South Dakota, and was found by G. R. Wieland, apparently at a time when he was collecting cycad specimens for Marsh from the Lower Cretaceous Lakota Formation. The tracks are

Footprints in the Sands of Time

4.18. Some of the track sites in the Morrison Formation. (A) Purgatoire River locality near the trackways of sauropods attributed to *Parabrontopodus*. (B) Dry Poison Creek near Buffalo, Wyoming, with small tridactyl tracks, Mike and Brian Flynn clearing the surface. (C) Limestone track site with theropods and sauropods low in the Morrison Formation at Como Bluff, Wyoming; Emma Rainforth, Beth Southwell, and Martin Lockley taking measurements.

currently in the Yale Peabody Museum and represent mostly theropod dinosaurs, although one track may be that of an ornithopod.

In the mid-1930s, Roland Bird, the biker-paleontologist who also worked for Barnum Brown excavating Morrison dinosaurs from the Howe Quarry in Wyoming, arrived in southeastern Colorado to investigate what turned out to be the largest known track site in the Morrison Formation, south of the town of La Junta. The site had been found a few years before and was known to local ranchers who worked the plains surrounding the canyon of the Purgatoire River, where the footprints are preserved in limestone along the river. At this site, some 1,300 dinosaur tracks are preserved on several surfaces that represent the shoreline of an ancient lake. The limestone preserves fish, clams, and small, spherical oolites, limestone structures formed by the oscillating action of waves in carbonate environments. Most of the tracks along the old shoreline are those of sauropod dinosaurs, including trackways of five young adult sauropods traveling in a group parallel to each other (Lockley et al. 1986). The other tracks consist of small and large theropods and a few ornithopods. This site in southeastern Colorado was nearly forgotten, however, when better tracks were found shortly afterward in the Lower Cretaceous of Texas and Bird traveled south to investigate those. The Purgatoire River site did contain what were at that time the first identified footprints of sauropod dinosaurs.

Much of the track record of the Morrison Formation has come to light only in the last 30 years, however (Lockley and Hunt 1995; Lockley et al. 1998b; Foster and Lockley 2006; Lockley and Foster 2017). Since 1986, the number of known track sites in the Morrison Formation has more than doubled. Several tracks of medium-sized theropods were found in the Salt Wash Member in northeastern Arizona in the 1990s. Other localities found in recent years have yielded sauropod and theropod tracks at sites near the Colorado River in the Rocky Mountains; sauropod, small theropod, ornithopod, and possible lizard tracks from multiple localities in the northern part of the Uncompahgre uplift in western Colorado; turtle tracks in the Salt Wash Member north of Lake Powell in Utah; and deep tracks of sauropods, small theropods, and small ornithopods in a shoreline limestone near the base of the Morrison Formation at Como Bluff, Wyoming. Two sites near Moab, Utah, preserve sauropod tracks that demonstrate specifics of the environments of preservation. The East Dalton Wells site contains several natural sandstone casts of sauropod tracks that also preserve the casts of mud cracks that formed in the sediment as the muddy floodplain dried after the sauropods had walked through and before the sand was deposited. At the Hidden Canyon Overlook site, the parallel trackways of three sauropods progress westward across what was a sandbar along an east-flowing sandy braided stream. What is particularly unusual about this site is that the tracks are preserved as raised pedestals, remnants of the sediment compacted under the heavy animals' feet and preserved after subsequent floods eroded the sand away from around them.

What we know about the faunas of the Morrison has come through a lot of labor-intensive fieldwork, a little of which we've now seen. In the next two chapters, we take a look at the vertebrate animals of the Morrison Formation in detail.

In with the Old, in with the New: Modern Technologies

Although our field techniques in paleontology have not changed drastically in the past 140 years, and we still use plaster and burlap and hand tools, the lab tools available to us have increased in power dramatically just in the past 25 years. A few tools have been tried to assist in quarrying, from ground-penetrating radar to scintillometers, and these have had varying degrees of success; good old-fashioned back-straining work has never gone away even when we do get help from technology to locate material. You may be able to find a skull with a scintillometer, but you still have to get it out. Rocksaws help, and helicopters make larger specimens easier to extract from remote areas (or even possible to remove at all!), but before any of that, you have to dig around the fossil—no avoiding that. But when it comes to studying material after it has been prepared (or even sometimes before), the world has changed with dizzying speed. J. S. Newberry said that *Dystrophaeus* remained for a future geologist with more time and better tools than he had in 1859; we have no way of knowing if workers like Newberry, Reed, Marsh, or Cope would be jealous of what is available to paleontologists of the 21st century, but it might be a safe bet that they are. CT scanning of bones at large and micro scales, photogrammetry, and radiometric dating of rocks are just a few of the things available to us now that might not have been dreamed of in the 19th century (fig. 4.19). Who knows what future techniques await that we cannot now anticipate?

Facing, 4.19. Some technologies that make paleontologists' work easier. (A) Randy Irmis (*left*) and David Hunt (*right*) CT scanning the ulna of the sauropod dinosaur *Dystrophaeus* at the Smithsonian's National Museum of Natural History, Anthropology Department. (B) MicroCT 3-D image of a sphenodontid skull from Dinosaur National Monument. (C) Laser scanning bones to create 3-D digital models and printed replicas. (D) 3-D digital model of a *Dystrophaeus* ulna made from data obtained in the work showcased in A, showing parts of the bone filled in during preparation (orange). (E) Photogrammetric model of a turning sauropod trackway at Copper Ridge, Utah, showing the depth of the tracks in centimeter-interval lines, from data collection in 2007. (F) Drilling technology existed in the early days of dinosaur paleontology but was rarely if ever employed; here, drilling confirms the presence of dinosaur bone and the productive layer 19 m (62 ft) down and 100 m (328 ft) from the existing edge of the Mygatt-Moore Quarry in Colorado. (G) Some bones fluoresce in UV light; here, bone that blends into the rock in normal lighting (both white) shows up as light purple to the rock's dark purple, allowing easier preparation. (H) An otherwise impossible view; a tooth too delicate to prep out of the rock is CT scanned and reconstructed as a 3-D model, allowing us to see a view that is in the rock still and to learn more about this sauropod. (I) Zircon crystals from Morrison Formation mudstones allow us to get approximate ages of the rocks burying the bones, in some cases to ±190,000 years (precise, when you're talking about 150 million years ago); note scale is 400 microns. (J) Photogrammetric reconstruction of a *Hispanosauropus* trackway from Copper Ridge allows us to see a view that is difficult to make out in the field, data collected in 2007. *Images courtesy of (A) the Smithsonian Institution, James DiLoreto and Randy Irmis; (B) Bhart-Anjan Bhullar; (C) ReBecca Hunt-Foster; (D) the Smithsonian Institution, Randy Irmis; (E), Neffra Matthews (Bureau of Land Management National Operations Center) and Brent Breithaupt (BLM Wyoming State Office); (F–G) author; (H) Randy Irmis; (I) Kelli Trujillo; and (J) Neffra Matthews (Bureau of Land Management National Operations Center) and Brent Breithaupt (BLM Wyoming State Office).*

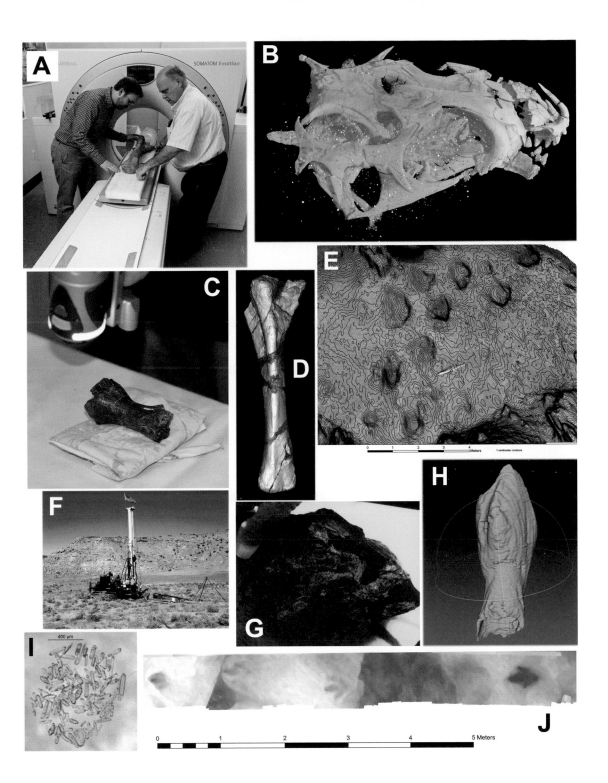

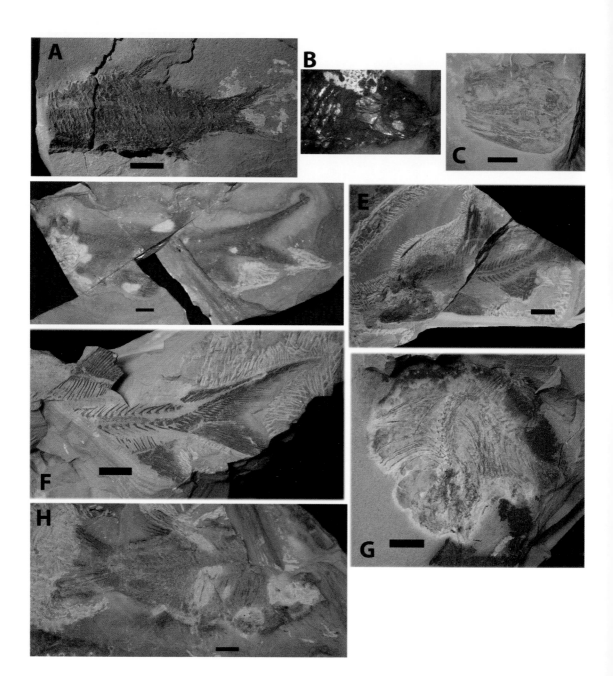

The Morrison Formation has yielded more than 100 known species of fossil vertebrates. The 140-plus years of fieldwork in it has revealed a lot about its animals, but new forms are still being found, and there is undoubtedly much we will learn in coming years. Here and in the following chapter are the beasts we know so far and a little about them.

One of the last things we might expect to find in a rock unit famous for its dinosaurs is something as apparently mundane as a fish, but in fact there are many, mostly fragmentary, fish fossils known from the Morrison Formation. Although the vast Morrison plains were not what we would call junglelike, and although the vertebrates of the time likely experienced a semiarid climate with seasonal rainfall and times of drought, it appears that there still were numerous areas of perennial standing and flowing water and plenty of vegetation. It is in some of these lake and river deposits that we find fish fossils, sometimes in surprisingly large numbers.

Few sites in the Morrison Formation yield fish of significant completeness. The Temple Canyon sites near Cañon City, Colorado, however, have recently yielded lungfish and amiids that preserve much of the skeletons (Pardo et al. 2010; Gorman et al. 2008), and a level just above the Mygatt-Moore Quarry in western Colorado has, for the past 30 years or so, produced a number of small, nearly complete palaeoniscoids, halecostomians, and leptolepids. The Little Houston Quarry in Wyoming and the Dry Mesa Quarry in Colorado have yielded a few indeterminate ray-finned fish. In Wyoming, three screen-washing sites have yielded more fragmentary fish elements than anything else (sometimes up to 89% of the sample) and have produced reasonably diverse microvertebrate samples (Trujillo 1999; Foster and Heckert 2011; King and Heckert 2018). Such occurrences are rare, but good or at least abundant fish specimens can be found.

(In this and the following chapter, taxa followed by an asterisk (*) are ones for which I question their validity as separate taxa in the Morrison Formation. In some cases, they may be based on relatively fragmentary material, and in others they may be **junior synonyms** of taxa named earlier in the scientific literature. These assessments are informal and not rigorous analyses of each taxon, and other paleontologists may agree or disagree with particular cases. I wanted to list most taxa reasonably proposed in the formation but still identify the ones that I see as being based on less sure footing.)

The Forgotten Aquatic Denizens: The Fish

5.1. Fish from the Morrison Formation of Rabbit Valley, Colorado, just 2 m (6.6 ft) above the Mygatt-Moore Quarry. (A) "*Hulettia*" *hawesi*, body minus skull (MWC 5564). (B) "*Hulettia*" *hawesi*, skull and anterior part of body. (C) Skull of a coccolepid, probably *Morrolepis* (MWC 5307). (D) Type specimen of the coccolepid *Morrolepis schaefferi* (MWC 440). (E) Possible *Morrolepis* (MWC 5306). (F) Back portion of *Morrolepis* (MWC 5305). (G) Indeterminate fish (MWC 5941). (H) cf. *Leptolepis* (MWC 3722). All scale bars = 1 cm.

The actinopterygian (ray-finned) fish fauna of the Morrison Formation consists mostly of animals with no close modern relatives. There are at least seven types of ray-finned fish known from the formation, but of these, one, called a pycnodontoid, is known from only one tooth. There is a significant record of fragmentary actinopterygians known from several sites, including Ninemile Hill and two screen-washing sites in the Black Hills, but most of this material is unidentified, being based mainly on at least three very different morphologies of scales found in abundance at the sites (Kirkland 1998; Trujillo 1999; Foster and Heckert 2011). The named and illustrated fish from the Morrison are mainly from other sites.

Morrolepis schaefferi

Morrolepis was first found as a nearly complete skeleton in the "fish layer" in Rabbit Valley, Colorado, near the Mygatt-Moore Quarry, and was named by Jim Kirkland in 1998. It is known from only this site, but since the type specimen was found, two partial skeletons, one nearly complete skeleton, and a nearly complete skull have been found as well (fig. 5.1). On the basis of the specimens found so far, it appears that *Morrolepis* grew to at least 20 cm (8 in.) long and had a tall dorsal fin set posteriorly on the body (Kirkland 1998).

Morrolepis, part of a group of relatively primitive fish called coccolepidids (among the larger group of palaeoniscoids) is different from most modern fish in that it has an asymmetrical tail in which the vertebrae curve up into the longer dorsal lobe. (A group called teleosts, which comprises most modern fish, is more derived, and species may or may not have a symmetrical tail, but the vertebrae stop at the base of the caudal fin and are not incorporated into it like in palaeoniscoids.) *Morrolepis* also would have appeared strange to us because the skull of the fish appears to slant forward so that the large eye sockets were over the front end of the lower jaw. This fish was not particularly large (probably weighing about 113 g [4 oz]), so it likely fed on insects and other small invertebrates in the shallow ponds and lakes it inhabited. It may also have occasionally eaten small fish. The dimensions of the lake in which *Morrolepis* lived in Rabbit Valley have not been determined, but the beds are exposed for nearly 1 km (0.6 mi) to the east, which could be taken as the maximum length or the minimum width of an elongate lake or the approximate size of a nearly circular lake. In either case, on the basis of the rocks in which it occurs, *Morrolepis* lived in a relatively shallow, probably quiet, permanent body of water out on the Morrison floodplain. The mudstones below the lake layer indicate that a low, muddy, and possibly seasonally wet and dry part of the plain gradually filled in with perennial water. The main dinosaur bone layer at the Mygatt-Moore Quarry occurs about 2 to 3 m (6 to 10 ft) below the 1-m-thick (3.3 ft) "fish layer" and contains almost no evidence of aquatic taxa such as fish, crocodilians, or turtles. Above the bone layer, however, crews have found a number of freshwater snail fossils (indicative of quiet standing water), and above that is the layer in which the fish skeletons are preserved. The presence of conchostracans (small aquatic crustaceans) in these levels may suggest that the pond even at this stage was ephemeral. This filling in with water may have been due to a rise in water table or some other local phenomenon but was not necessarily a result of a larger-scale change in climate. Regardless, over time, the area developed into a moderate-sized lake in which *Morrolepis* and other fish lived for at least several thousand years.

There appears to be a second coccolepidid known from material at Temple Canyon, Colorado (fig. 5.2; Gorman et al. 2008).

"Hulettia" hawesi

Hulettia is a genus of Jurassic fish of the group Halecostomi. These are fish more derived than palaeoniscoids but more primitive than the teleosts. *Hulettia americana* is known from Middle Jurassic marine deposits underlying the Morrison Formation in the Rocky Mountain region, but a possible Morrison species was identified by Jim Kirkland in 1998 after it had been found in shallow lake deposits at the Fruita Paleontological Area (FPA) and at Rabbit Valley (where it co-occurs with *Morrolepis*). Both sites are in western Colorado. Jim has since decided that this species is probably not in the genus *Hulettia* after all and may be a new genus, but no one has yet addressed this issue formally.

The skeleton of *"Hulettia" hawesi* is covered in most specimens by its distinctive scale pattern (Kirkland 1998). Unlike *Morrolepis*, whose scales are thin and barely make an impression in the rock in most specimens, *"H." hawesi* fossils are almost all scales, fins, and skull bones, with very little evidence of vertebrae, ribs, or other bones. The scale pattern consists of slightly inclined, nearly vertical rows of scales, articulated by vertical pegs and sockets. The scales along the midline of the body are nearly three times as tall as they are wide. Those on the dorsal and ventral surfaces are more closely equidimensional. There are spines on the scales along the dorsal edge of the fish, anterior to the dorsal fin, that suggest this may be a semionotiform fish unrelated to *Hulettia* (Foster, unpublished data).

Most Morrison Formation specimens of this fish are about 7.6 cm (3 in.) long (fig. 5.1), and when alive, the fish probably weighed just 5 g (0.2 oz). Given its small size, the animal likely fed on insects, other invertebrates, and hatchling fish in the lakes and ponds in which it lived. One of the specimens from the FPA is in a piece of rock with a freshwater snail fossil in it, which suggests that the fish preferred quiet water.

Amioidea indet.

Amioid fish are relatives of the modern bowfin (*Amia calva*), a relatively large predatory fish that lives in rivers of the eastern United States and the Mississippi River valley. Amioids were first found at Quarry 9 at Como Bluff back in the 1800s but were undescribed for many years, and

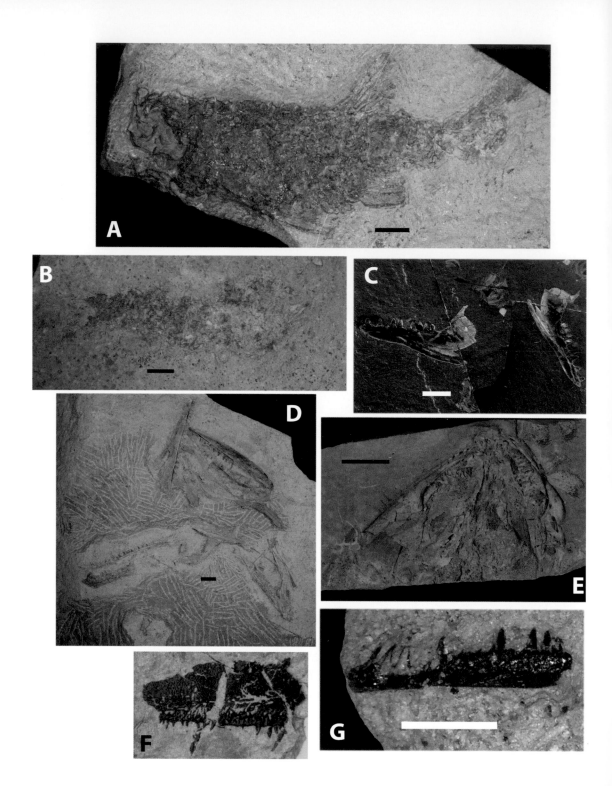

5.2. Fish from the Morrison Formation of Temple Canyon, Colorado, near Cañon City, and other sites. (A) Unidentified actinopterygian, nearly complete (DMNH 58009), Temple Canyon. (B) Complete unidentified pholidophoriform(?) (DMNH 63734), Temple Canyon. (C) Unidentified amioid jaws (DMNH 57642), Temple Canyon. (D) Amioid jaws (DMNH 50674), Temple Canyon. (E) Amioid jaws and partial skull (cf. *Amiopsis*?) (DMNH 54194), Temple Canyon. (F) Unidentified actinopterygian maxilla(?) (SDSM specimen), Little Houston Quarry, Wyoming. (G) Unidentified actinopterygian dentary(?) (BYU 10022), Dry Mesa Quarry, Colorado.

no name has yet been given to the Morrison specimens. Most of the material from Quarry 9 consists of vertebrae, and J. W. Stovall's crews collected many vertebrae in Oklahoma in the 1930s and 1940s. Two nearly complete lower jaws have been found, however, and these are from northeastern Utah and Bone Cabin Quarry in Wyoming. The jaws are similar to *Amia* and other amioids in having two parallel rows of large, sharp, conical teeth along their lengths. Amioids in general have skulls with many connected, sculpted bones and with teeth on several of the bones of the palate. A number of amiid fish have also been found at the Temple Canyon sites in Colorado in recent years (fig. 5.2; Gorman et al. 2008; Pardo et al. 2010), and some of these may be related to freshwater forms known mostly from the Late Jurassic through the Cretaceous of Europe (Grande and Bemis 1998).

The Morrison Formation amioids grew to lengths of at least 51 cm (20 in.) and, at this length, weighed about 2 kg (4.5 lb); some individuals may have been considerably bigger. Modern *Amia* live in slow-flowing rivers and clear ponds with abundant aquatic vegetation; they are predatory and feed almost exclusively on other fish after they pass 10 cm (4 in.) in length. At smaller sizes, their diet also consists of insects, ostracods, and crayfish. Most fossil amioids were piscivorous as well, but the Paleocene genus *Cyclurus* seems to have fed more on invertebrates even as adults (Grande and Bemis 1998). Although lateral teeth of *Cyclurus* are similar to those of other amioids, the teeth of the roof of the mouth are blunter than in other species. The Morrison amioids were likely similar to *Amia* in preferred habitat and feeding, although because of a lack of evidence of the internal teeth, a diet similar to *Cyclurus* cannot be ruled out. What is clear from the two jaws from Utah and Wyoming is that the Morrison Formation amioids were the largest and most dangerous of the predatory ray-finned fish in the rivers of the Morrison basin.

cf. *Leptolepis*

This fish is known from a single nearly complete skeleton from Rabbit Valley, Colorado, and was about 13 cm (5 in.) long (Kirkland 1998). It was deeper bodied than either *Morrolepis* or "*H.*" *hawesi* (fig. 5.1) and is the only teleost that has been found in the formation. As a teleost, cf. *Leptolepis* is more advanced among the Morrison fish. It has, for example, a more modern tail structure than *Morrolepis*. It is found in the same beds in Rabbit Valley as *Morrolepis* and "*H.*" *hawesi* and weighed about 37 g (1.3 oz). Like the other fish with which it lived, cf. *Leptolepis* fed mainly on insects, small invertebrates, and small fish.

Pycnodontoidea

Pycnodontoids are known only from a single tooth from Dinosaur National Monument in Utah. Complete pycnodontoid skeletons from other ages indicate that the fish were deep bodied and laterally compressed, bearing some resemblance to modern butterfly fishes, and their teeth suggest a diet of small invertebrates (Poyato-Ariza 2003).

Pholidophoriformes indet.

One skeleton from Temple Canyon, Colorado, has been identified as a pholidophoriform, but it has not been studied in detail (fig. 5.2; Gorman et al. 2008).

Dipnoi (Lungfish)

POTAMOCERATODUS GUENTHERI

CERATODUS ROBUSTUS

C. FOSSANOVUM

"C." ?FRAZIERI

The fish fauna of the Morrison includes four types of lungfish and five types of ray-finned fish (the former being distinguished from your every-day trout mainly by having fleshy bases to the pectoral and pelvic fins that were a sort of evolutionary precursor to the tetrapod limb; see chap. 2). The lungfish were relatively large, slow-swimming fish with thick fins and four main tooth plates in the skull used for crushing their prey. The tooth plates, two lower and two upper, are large and thick, and they consist of a flattened surface with prominent ridges projecting laterally. Two smaller vomerine teeth are present forward of the tooth plates in the skull. The lungfish fauna of the Morrison comprises the species *Potamoceratodus guentheri* (including *"Ceratodus" felchi*), *Ceratodus fossanovum*, *"C." ?frazieri*, and the very large *Ceratodus robustus* (fig. 5.3; Kirkland 1987, 1998; Pardo et al. 2010). These fish probably closely resembled modern lungfish in size, preferred habitat, and diet.

There are about six species of lungfish alive today, and four of them live in Africa, so the diversity we seem to be seeing in the Late Jurassic of North America is not unusual. *Lepidosiren paradoxa* lives in the Amazon River Basin and the basin of the Paraguay-Parana Rivers of South America, and *Protopterus aethiopicus*, *P. amphibius*, *P. annectens*, and *P. dolloi* live in various parts of Africa. *Lepidosiren* (the South American lungfish) stays mostly in areas with little or no current and feeds on insects, snails, and, as adults, largely on plant material. It grows to just over 1.2 m (4 ft). The four African lungfish species (within *Protopterus*) live in their respective regions of the continent in a variety of habitats, including marginal swamps, floodplains, and quiet areas of rivers and lakes, feeding on clams, snails, fish, frogs, insects, and plant material and seeds. *Protopterus* can grow to lengths of 1 to 2 m (3 to 6.5 ft).

The Australian lungfish, *Neoceratodus forsteri*, can grow to 2 m (6.5 ft) and inhabits the Burnett and Mary River systems of Queensland. At 1 m (3.3 ft) long, individuals of *Neoceratodus* weigh about 20 kg (44 lb).

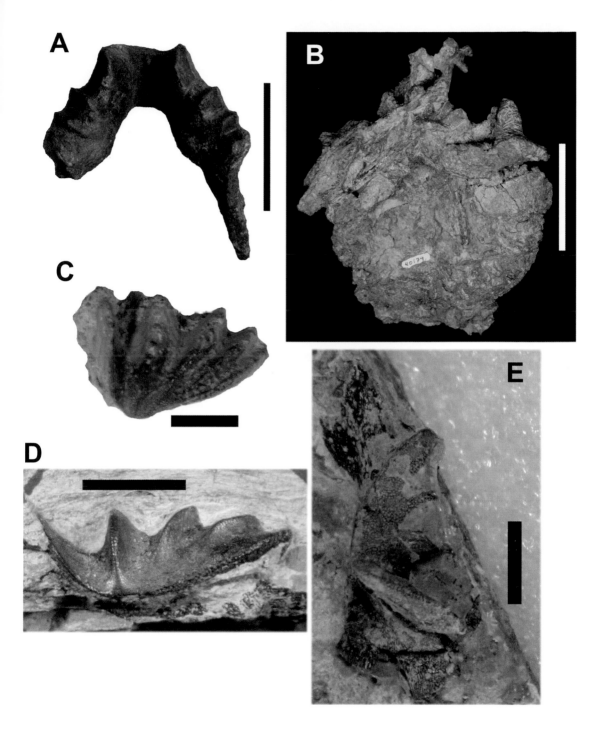

5.3. Lungfish from the Morrison Formation. (A) Lower jaw and tooth plates of *Ceratodus robustus* (MWC 5162), Moffat County, Colorado. (B) Skull with tooth plates of *Potamoceratodus guentheri* (DMNH 40179), Temple Canyon, Colorado. (C) Unidentified lungfish tooth plate (BYU uncataloged), Dry Mesa Quarry, Colorado. (D) Tooth plate of *Ceratodus fossanovum* (MWC specimen), Little Houston Quarry, Wyoming. (E) Left lower jaw and tooth plate of *Ceratodus fossanovum* (MWC specimen), Little Houston Quarry, Wyoming. Scale bars = 5 cm (A and B) and 1 cm (C–E).

This species is probably closest in appearance to the Morrison lungfish and lives in slow-moving or still waters, often deep, in the river systems. *Neoceratodus* feeds on fish, small amphibians, worms, crustaceans, snails, and plants (Froese and Pauly 2005), all of which also would have been available to *Potamoceratodus* and *Ceratodus* in the Late Jurassic. *Neoceratodus* can live up to 70 years. *Neoceratodus* is unlike other modern lungfish, however, in that it cannot estivate, a process of burrowing into the mud and "hibernating" in a cocoon of mucus that is used by the South American and African lungfish. These lungfish can stay in their cocoons during a dry spell, surrounded entirely by dried mud and breathing with their lungs, reemerging and often reproducing when the rains return. Although the Australian lungfish does not estivate, it can breathe with its lungs when necessary in low, stagnant water. It is unclear whether any of the Morrison species of *Potamoceratodus* or *Ceratodus* estivated or were instead similar to the modern *Neoceratodus*, but either way, the two were likely reasonably similar in their size, habitat preference, and diet.

Most specimens of the various species of lungfish in *Potamoceratodus*, "*Ceratodus*," and *C. robustus* from the Morrison indicate animals about 1 to 2 m (3 to 6.5 ft) long and weighing up to about 36 kg (79 lb), although most would have been quite a bit less than this. Lungfish tooth plates have been found at a number of quarries in the Morrison Formation (fig. 5.3), stretching geographically from Oklahoma to Montana and stratigraphically from the Salt Wash or Tidwell Member to the upper Brushy Basin Member and its equivalents.

Hoppers and Squirmers: The Amphibians

Anura (Frogs)

True frogs appeared not that long (geologically speaking) after the first dinosaurs, during the Early Jurassic. Like their modern relatives, ancient frogs had skulls that were large relative to the body and almost semicircular when viewed from above, with large orbital openings; no external tail in adults; a short body with only eight or nine presacral vertebrae; a single sacral vertebra; long, muscular legs with the tibia and fibula fused; and uniquely elongate calcaneum and astragalus bones in the ankle. One of the earliest frog relatives is believed to have been *Triadobatrachus* from the Early Triassic of Madagascar. This small animal appears to be transitional between Paleozoic amphibians and the true frogs that are first found in the Early Jurassic. *Triadobatrachus* had a relatively large, semicircular skull like frogs do, a short tail, and a short neck, but it had a relatively long series of dorsal vertebrae and short back legs. By the time the oldest known frog appeared about 190 million years ago, most characteristics of the modern frog body form were apparent. *Prosalirus* was found in the Early Jurassic Kayenta Formation in Arizona and was able to jump just like modern frogs.

Frogs are known from several sites in the Morrison Formation but are not particularly well represented. The first material came from Reed's Quarry 9 at Como Bluff, Wyoming. Some was named by Marsh

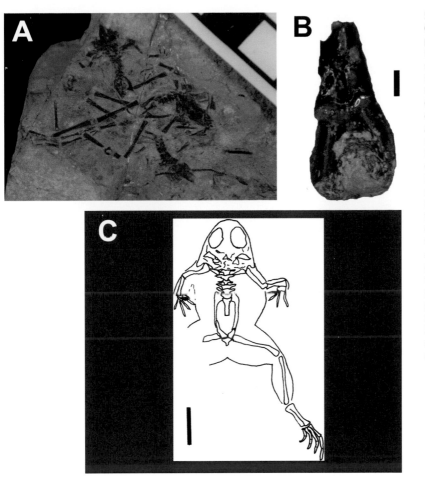

5.4. Late Jurassic frogs. (A) Slab from the northern part of Dinosaur National Monument, Utah, showing numerous bones of several individuals of the Morrison Formation frog *Rhadinosteus* (DINO specimen). Scale is in centimeters. Photograph by Amy Henrici and David Berman, Carnegie Museum of Natural History. (B) Distal half of the humerus of a frog from Little Houston Quarry, Wyoming (SDSM specimen). Scale bar = 1 mm. (C) Partial fossil frog with soft-tissue body outline (*Eodiscoglossus*) from the Late Jurassic of Spain. Although Morrison frogs are generally more fragmentary, this European contemporary (along with *Rhadinosteus*) gives a good idea of what Morrison frogs must have looked like in size and structure. Scale bar = 1 cm. Based on Hecht (1970).

as *Eobatrachus*, and decades later, another piece was named as *Comobatrachus*, but these genera are based on fragmentary material and are now considered nomina dubia (invalid names). The frogs of the Morrison Formation belong to the groups Discoglossidae, Pelobatidae, and Pipoidea, although only two valid species are currently named. The frogs from 150 million years ago look much like those that we know today. The skeletal material we find in the Morrison Formation shows this resemblance, and a nearly complete and articulated frog with a preserved body outline from the Late Jurassic of Spain provides even stronger evidence (fig. 5.4; Hecht 1970). We can infer from this that modern frogs are little changed in form and habit from their Jurassic ancestors, which in turn suggests that most of those ancestors also were semiaquatic insectivores. Modern frogs eat ants, beetles, termites, snails, slugs, crustaceans, and worms, and sometimes small vertebrates if the individual frog is large enough (Mattison 1987). Most frogs capture prey with their tongues. Size and availability of prey are the main factors that determine frog diets—they are not particularly selective eaters. In the water, larger frogs may eat small fish, and on land, frogs have been known to ingest small mammals, birds, snakes, and other frogs. (A zoo-bound African bullfrog once was found to have eaten more than a dozen baby cobras!) Morrison Formation frogs probably fed on prey items similar to those of modern frogs (except the cobras, of course), although few of the Jurassic forms were large enough to have taken many vertebrate prey.

Habitats preferred by Jurassic frogs were probably similar to those of their modern relatives. Although frogs can inhabit cold areas, most prefer warm, moist environments. Few modern species live in large bodies of water like lakes. This may be due to the frogs' vulnerability to predation in open water and direct competition with fish for the insects and other invertebrates that are available there. There are several modern species, however, that live in large lakes in Africa, Asia, and South America. Most modern frogs live around swamps and marshes, foraging in the vegetation in the water and along the shore, and this may have been a preferred environment for frogs of the Morrison Formation as well. Modern species that live along swamp and marsh margins lay their eggs either in the water or hanging from vegetation just above the water's surface. The tadpoles of modern species feed in a variety of ways: some are carnivorous, others are herbivorous, and still others are filter feeders (Mattison 1987). Some adults of modern species live along the banks of rivers and in forests, and these too are environments that must have been inhabited by Jurassic frogs. Like their modern counterparts, the frogs of the Late Jurassic may have vocalized with a similar-sounding croaking. Most fossil frogs from the Morrison Formation have been found in quarries representing pond, river, or abandoned channel environments. Only a handful of sites have yielded frog material, and specimens from Quarry 9 and Dinosaur National Monument reveal the most information, but the sites are geographically widespread, from southern Colorado up to northern Wyoming and from Utah to the eastern edge of Wyoming.

Enneabatrachus hechti

The frog *Enneabatrachus* was found at Quarry 9 at Como Bluff, Wyoming, and is based on an ilium (pelvic bone) only a few millimeters long (Evans and Milner 1993). *Enneabatrachus* is a member of the group of modern frogs known as discoglossids, which may live near water; in small, damp burrows; or under rocks. This species, whose genus name means "nine frog" after the quarry at which it was found, probably weighed only a few grams. A second specimen of *Enneabatrachus* has been described from Dinosaur National Monument. The discoglossids are a family that today is widely distributed geographically.

Pelobatidae indet.

A second type of frog ilium from Quarry 9 is unnamed but probably belongs to the group of frogs known as pelobatids (Evans and Milner 1993; but see Holman 2003 for a view of this specimen as a discoglossid).

This ilium is slightly larger than that of *Enneabatrachus* but is still small. The modern pelobatids are known as spadefoot toads and live in North America, Africa, Europe, and Asia. Some live in arid regions and are specialized to burrow with their hind legs, but other species live in forests and near streams and are not burrowers.

Rhadinosteus parvus

This frog from the Rainbow Park site in Dinosaur National Monument, Utah, was a member of the pipoid group of frogs and more specifically may be a rhinophrynid (Henrici 1998a, 1998b). It was only about 42 mm (1.6 in.) long, and its name means "small slender bone." Pipoids today are largely aquatic species with webbed feet, and they live mostly in and along swamps and lakes. *Rhadinosteus* is known from several partial skeletons found together on several slabs of rock (fig. 5.4).

Caudata (Salamanders)

Salamanders are rare in the Morrison Formation. Most of the known material is incomplete, and only one species name is currently valid (Hecht and Estes 1960; Evans and Milner 1993; Evans et al. 2005). The salamander *Comonecturoides marshi* was named on the basis of a femur from Quarry 9 at Como Bluff, Wyoming, but this species is a nomen dubium because the femur in salamanders is not a distinctive element. There are two distinct, unnamed types of salamander vertebrae at Quarry 9. Other sites with salamander material include the Wolf Creek Quarry, the Small Quarry, the Little Houston Quarry, Jurassic Salad Bar, and Ninemile Hill. At sites outside the quarry building at Dinosaur National Monument in Utah, there are several partial salamander skeletons, only one of which has yet been described (fig. 5.5). One skeleton indicates an animal only a few centimeters long, and the other appears to be a larger, long-bodied animal with short limbs.

Iridotriton hechti

This is the smaller of the two salamanders from Dinosaur National Monument, and it was just recently described and named by Susan Evans et al. (2005). *Iridotriton* appears to be a basal salamandroid, and it is more closely related to North American and European salamanders than to those of Asia (fig. 5.5).

Fossil salamanders from the Middle Jurassic of China show that by this time, the animals were similar in overall form to modern species. The best specimen from the Morrison Formation of Dinosaur National Monument is small, but most fragmentary remains from other sites indicate salamanders similar in size to those of the Middle Jurassic of China—about 15 cm (6 in.) in total length. The salamanders from the Morrison Formation do not appear drastically different from some of

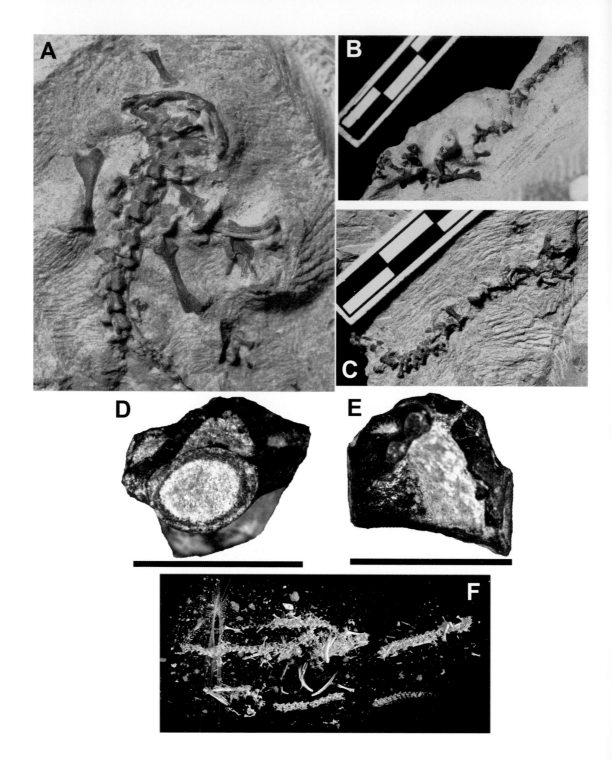

5.5. Salamanders from the Morrison Formation. (A) Skeleton of the salamander *Iridotriton hechti* from the Morrison Formation of Dinosaur National Monument, Utah (DINO 16453). (B–C) Strings of vertebrae and several small limb elements of salamanders from Dinosaur National Monument, Utah. Scales in centimeters. (D) Anterior view of a vertebra of a much larger salamander specimen from the Little Houston Quarry, Wyoming (SDSM specimen). Scale bar = 5 mm. (E) Left lateral view of the same salamander vertebra as in D. Scale bar = 5 mm. (F) MicroCT scan 3-D image of a salamander specimen from Dinosaur National Monument, courtesy of Bhart-Anjan Bhullar.

those of today and may have been quite similar in habits. Modern salamanders live in a variety of environments and exhibit a number of different life history patterns (Hairston 1987; Trenham 2001). Some species metamorphose over several years, but others do so completely within the egg. Some species are entirely aquatic as adults, whereas others are terrestrial as adults, sometimes living in burrows, and return to water only to lay eggs. Field studies and experiments in modern environments suggest that pond-dwelling salamander species occur only in areas with few or no predatory fish, which indicates that the many terrestrial salamanders may have adapted to land partly in order to avoid the larger predators and partly to avoid competition for food with smaller fish. Many salamander species live in temporary ponds, which may have been common on the floodplain of the Morrison Formation, and large predatory fish seem to have been common in some aquatic environments at the time, so at least some of the salamander species of the Morrison Formation were likely terrestrial as adults. The quarries in which Morrison salamanders are preserved generally represent wet environments, including poorly drained floodplains, overbank splays, and abandoned channels. Even terrestrial salamanders may be considered semiaquatic, however, because of their dependence on water for reproduction.

Judging from modern species, salamanders of Morrison times probably ate insects and other small invertebrates while on land and small crustaceans while in aquatic habitats. The Middle Jurassic fossils from China include a larval salamander, with gills, that contains stomach contents of more than a dozen diplostracan ("conchostracan") arthropods (Gao and Shubin 2003). Jurassic specimens from China demonstrate that salamanders of the time already had the ability to regenerate body parts damaged by nonlethal predation attempts (Y. Wang et al. 2016). Some individuals appear to have had some type of disruption of the normal regeneration process and ended up with fused digits, or six or eight digits, for example, which were then fossilized (the animal apparently living with the deformation for some time after the predation and regeneration incident). We can only assume Morrison salamanders were capable of such regeneration too. All Morrison salamander specimens appear to have been from animals the size of small- to medium-sized salamanders of today and likely weighed less than 100 g (4 oz).

Turtles are among the most abundant animals known from the Morrison Formation, with hundreds of skeletons, shells, shell fragments, and limb bones known from all over the outcrop area. In part, this is probably due to the robustness of the turtle skeleton and its high preservation potential, but it also may be because there really were a lot of turtles in the ponds and rivers during the Late Jurassic. Turtles appeared during the Late Triassic around the same time as early dinosaurs and mammals (Joyce et al. 2008), and these shelled reptiles were widespread by the Middle to Late

Mobile Homes: The Turtles

5.6. Turtles from the Morrison Formation. (A–B) Dorsal and ventral views of a partial carapace (dorsal shell) of *Glyptops* (MWC specimen), Bone Cabin Quarry, Wyoming. (C) Dorsal view of the shell of an adult *Glyptops* from Dinosaur National Monument, Utah. (D) Dorsal view of the shell of a juvenile *Glyptops* from Dinosaur National Monument (DINO 992). (E) Partial shell of an adult *Dinochelys* from Rabbit Valley, Colorado. (F) Dorsal view of the shell of an adult *Dinochelys* (DINO 986) from Dinosaur National Monument. (G) Ventral view of a juvenile *Dinochelys* (DINO 993); the arrow indicates the articulate limb skeleton. (H) Dorsal view of the shell of a juvenile *Dinochelys* (DINO 993). (I) Articulated foot of a turtle from Dinosaur National Monument. (J) Hind limb of *Dinochelys* (DINO 986). Note the radiating ridges on the dorsal shells of juveniles in D and H. Scale bars in A, B, G, and I = 1 cm. Scale bars in E and J = 5 cm. Scale bars in C, D, F, and H = 10 cm.

Jurassic (Hay 1908; Joyce 2007). But until 1979, only one valid genus of turtle was known from the Morrison Formation. Since then, three others have been found.

Glyptops ornatus

This appears to be the most common turtle known from the Morrison Formation (Hay 1908). Named by O. C. Marsh, it is likely the same taxon as E. D. Cope's *Compsemys plicatulus* (later renamed *Glyptops*) from high in the Morrison Formation at Garden Park, but Marsh's specimen from Como Bluff has priority due to its being diagnosable (Joyce and Anquetin 2019). A nearly complete skeleton of *Glyptops* was found at the Bone Cabin Quarry in Wyoming, but the species was apparently quite geographically widespread and abundant: it occurs as fragmentary remains at many quarries in the Morrison Formation. *Glyptops* is a type of turtle known as a pleurosternid cryptodire (fig. 5.6), and *G. ornatus* includes the junior synonyms *G. utahensis* and *G. plicatulus* (Gaffney 1984; Joyce and Anquetin 2019). As a cryptodire, or arch-necked, turtle, *Glyptops* retracted its skull by folding the neck back in a vertical plane, as opposed to the side-necked turtles, which fold the neck sideways in a horizontal plane. Other cryptodires include the modern box and pond turtles and the tortoise. *Glyptops* had a relatively long skull, a moderately long tail, and distinct, ridged sculpturing on the surface of the bone that makes up the shell.

Dinochelys whitei

This turtle differs from *Glyptops* in having a nearly smooth rather than sculpted outer shell surface (Gaffney 1979; D. Brinkman et al. 2000). The type specimen is from the main quarry at Dinosaur National Monument (fig. 5.6), but the species has since been identified from other quarries, including several Garden Park sites, the Wolf Creek Quarry, the Little Houston Quarry, the Dry Mesa Quarry, and Quarry 9 at Como Bluff. *Dinochelys* is a pleurosternid turtle similar to both *Glyptops* from the Morrison and several turtles from the Early Cretaceous of Europe, and the partial skeleton and skull of a juvenile *Dinochelys* have been found at the Dry Mesa Quarry. The shell of *Dinochelys* is slightly more elongate anteroposteriorly than that of *Glyptops*.

Uluops uluops

This turtle species is known only from high in the Morrison Formation at Breakfast Bench, Como Bluff, Wyoming (Bakker et al. 1990). The skull is much shorter than the elongate one of *Glyptops*, and the descending process of the pterygoid bone is longer, but the shell sculpture is similar. *Uluops* appears to have been an early baenid cryptodire turtle. The baenids were an exclusively North American group of turtles that were

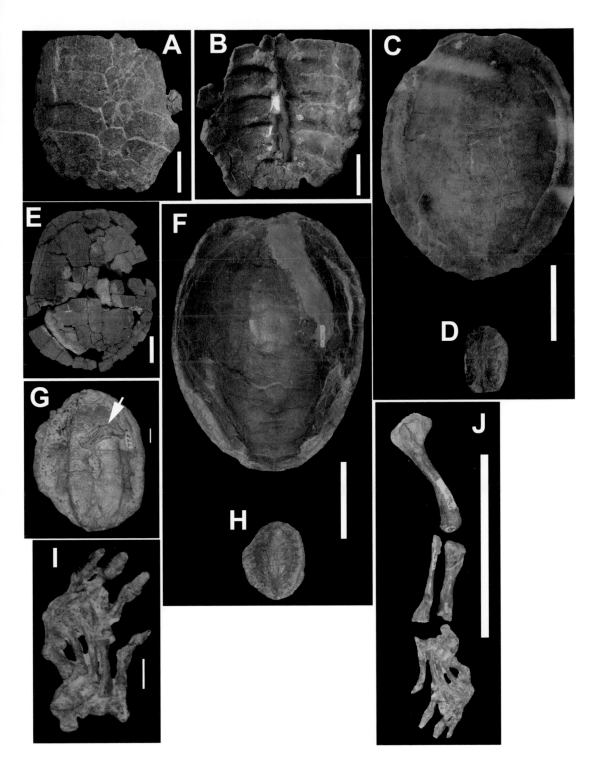

most diverse during the Cretaceous, when they dominated turtle faunas. Unlike today's cryptodire turtles, baenids were unable to fully retract their skulls into their shells, but apparently they were able to partially protect them.

Dorsetochelys occurs in the Lower Cretaceous of Great Britain but has been possibly identified recently at the top of the Morrison Formation as a separate species of the genus. The Morrison species is distinguished by a relatively short skull and a long and wide descending process of the pterygoid bone (Bakker 1998a). The shell sculpturing is similar to that of *Glyptops* and *Uluops*. *"Dorsetochelys" buzzops* appears to be a baenid cryptodire. Some differences in the skull bones suggest, however, that the Morrison form may represent a valid new species but in a separate, as yet unnamed genus different from the European *Dorsetochelys* (Foster and McMullen 2017).

Like their modern counterparts, Late Jurassic freshwater turtles lived in and near ponds, lakes, and rivers. Their remains are commonly found in wet-environment quarries in the Morrison Formation, particularly in areas to the north and east (Foster and McMullen 2017), and there are no dry-land tortoises in the unit. Morrison turtles are estimated to have weighed about 1 to 10 kg (2.2 to 22 lb), with a modal size of about 5 kg (11 lb) (Dodson et al. 1980). Most species presumably were omnivorous and fed on insects, small fish and other vertebrates, and plant material; some modern species vary their diet seasonally or as they grow (Alderton 1988). At least in modern turtles, most eat all of the above items, but individual species often concentrate on one type. Thus, some are largely insectivorous but will also eat mollusks and vertebrates, and others are mostly herbivorous but also feed on fish and invertebrates. Among those modern species that are insectivorous to a large degree, population densities are lower in permanent bodies of water, where there is greater competition from insectivorous fish, and this may have been the case during Morrison times as well. Some modern species are more insectivorous during the summer and eat plants during the winter.

Ancestry of the Tuatara: The Sphenodontians

Living on islands off the coast of modern-day New Zealand is a rare and endangered reptile known as the tuatara. Tuataras are lizard-like in appearance, and the two are close relatives, but tuataras are not true lizards. They grow up to 61 cm (2 ft) in length and weigh about 1 kg (2.2 lb). Tuataras are most active at night and feed on insects (particularly beetles), worms, snails, slugs, spiders, lizards, young birds, and eggs. They can burrow and often live in areas of bushes and low forest, and they may live for up to 120 years (Robb 1986). The tuatara is the last living member of an ancient group of reptiles known as the **Sphenodontia**. This name comes from the scientific name of the tuatara, *Sphenodon*. During the Mesozoic, sphenodontians were both much more diverse and more geographically widespread than today, and many had unique and varied tooth forms and varied individual ecologies (M. Jones 2009; Rauhut et al. 2012a); in fact, in some faunas, they were more abundant than their true-lizard relatives.

The first sphenodontian from the Morrison Formation was found at Como Bluff, Wyoming, by O. C. Marsh's crews but was not recognized until several decades later.

Opisthias rarus

Opisthias was named by Charles Gilmore on the basis of a specimen from Como Bluff's Quarry 9 but has since been identified from numerous quarries and is by far the most common sphenodontian species from the Morrison Formation (fig. 5.7; Gilmore 1910; Foster 2003a). Specimens are known from most areas of the formation and from quarries as varied as the Carnegie Quarry and Rainbow Park sites at Dinosaur National Monument, the FPA, the Wolf Creek Quarry, the Uravan Locality, and the newly discovered Kings View Quarry. The genus is also known from the Purbeck Formation in England (Evans 1992). *Opisthias* is known mostly from lower jaw specimens, but in these elements it is generally similar to the modern *Sphenodon*, though the former's teeth are a little broader, and it was probably similar in size, appearance, diet, and general habits

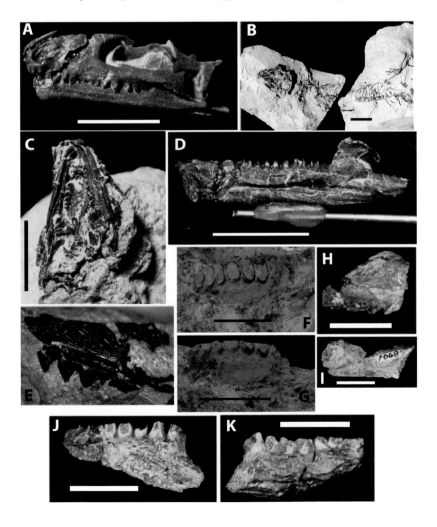

5.7. Sphenodontians of the Morrison Formation. (A) Left lateral view of the skull of *Opisthias rarus* (DINO 16454) from Dinosaur National Monument, Utah. (B) Nearly complete articulated skeleton of an unidentified sphenodontian from the Fruita Paleontological Area. (C) Palatal view of a skull with lower jaws, *Opisthias rarus*, DINO uncataloged. (D) Internal view of a lower right jaw, *Opisthias rarus* (DINO 16391). (E) Left palatine of *Opisthias*? (SDSM specimen), Little Houston Quarry, Wyoming. (F–G) Occlusal and labial(?) views of the lower jaw of *Eilenodon robustus* (BYU 11460), Dry Mesa Quarry, Colorado. (H–K) *Eilenodon robustus* (DMNH 10685) from Garden Park, Colorado. (H) Right maxilla. (I) Left palatine. (J) Left dentary. (K) Right dentary. All scale bars = 1 cm. Image in B courtesy of Jim Kirkland.

to that genus. How similar terrestrial Mesozoic sphenodontians such as *Opisthias* were in habits to modern tuataras is difficult to say, however, because the modern forms may not be typical, marginalized as they are to cool-temperate parts of New Zealand. The Morrison forms probably would not have become heat stressed at relatively low temperatures like *Sphenodon* does; but then, the Morrison floodplain must have been much warmer overall than the environment of the modern animals. Recent studies of *Opisthias* have suggested that it may be a stem eilenodontine (M. Jones et al. 2014).

Theretairus antiquus

On the basis of a single dentary from Como Bluff, and named by mammal expert and prominent vertebrate paleontologist G. G. Simpson, this sphenodontian's name means "ancient wild beast companion" because of its having been found among the many Jurassic mammal specimens of Quarry 9 (Simpson 1926b). It has been considered by some workers to be a juvenile of *Opisthias* and thus a junior synonym of that genus, although others do not think that this synonymy is yet solidly established. Preliminary analysis of new material from northern Wyoming suggests *Theretairus* may well represent a distinct lineage (Demar and Carrano 2018). It is the rarest sphenodontian in the Morrison Formation, known from just two or three specimens. It is, at least in the dentary, fairly different from *Opisthias*, with a large, conical tooth set well posterior to the front of the jaw and alternating large and small teeth posterior to this. Why *Theretairus* is so rare is hard to say, but it may simply have been a small and environmentally restricted species during the Late Jurassic and thus less likely to be preserved and found. It was, like *Opisthias* and *Sphenodon*, probably a carnivore feeding on various insects and small vertebrates, but we can't be sure how it might have specialized to separate its diet from that of *Opisthias*. The few specimens of *Theretairus* indicate an animal somewhat smaller than *Opisthias*, and thus this species probably weighed much less than 1 kg (2.2 lb).

Eilenodon robustus

Eilenodon was first found at the FPA in western Colorado (Rasmussen and Callison 1981b). It is a relatively large and robust sphenodontian with a thick, deep lower jaw and dentary teeth that are wide, robust, and worn on both the top and outer edge, with thicker enamel than other sphenodontians (fig. 5.7; M. Jones et al. 2016). The thickness of the enamel in the teeth appears to have increased from the tip down toward the base, something that may have helped create sharp enamel wear surfaces as the teeth wore down (M. Jones et al. 2018). Fragments of the upper jaws also indicate thick, heavy bones and teeth unlike anything seen in modern sphenodontians. Four new specimens and localities have been identified in recent years in Colorado and Utah (Foster 2003b). *Eilenodon* is a member of a group of large sphenodontians including *Toxolophosaurus* from the Early Cretaceous of Montana and *Priosphenodon* from the Late Cretaceous of Argentina. Considering its broad and robust teeth, *Eilenodon*, along with its relatives, was largely herbivorous and possibly omnivorous, feeding on plant matter (possibly including horsetails), seeds, insects, and small vertebrates. *Eilenodon* was in all probability a terrestrial animal that lived on the ground surface in and near low vegetation or in shallow burrows. It is difficult to estimate the mass of an animal known from so little of the skeleton, but on the basis of comparison with the modern *Sphenodon*, most individuals would have weighed close to 1 kg (2.2 lb).

There are four types of lizards found in the rocks of the Morrison Formation, three scincomorphans and one anguimorphan (Prothero and Estes 1980; Evans 1996; Evans and Chure 1998a, 1998b, 1999). Lizards today range from 0.5 g to 250 kg in weight (a fraction of an ounce up to 550 lb), but all of the Morrison lizards were on the smaller end of this range and probably weighed only a few grams (less than an ounce; fig. 5.8). Today,

Squamata I: The Lizard Kings

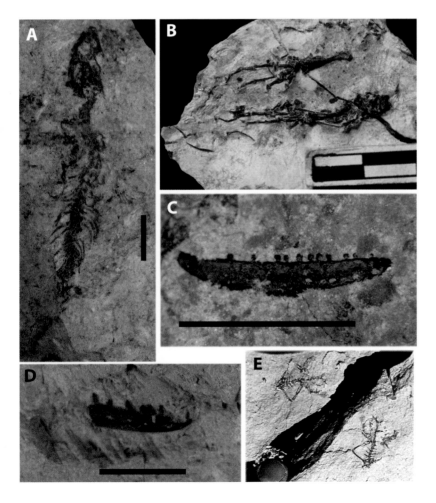

5.8. Lizards from the Morrison Formation. (A) Nearly complete, articulated but unidentified lizard (UMNH VP 13829) from near Hanksville, Utah. Photo by author, courtesy of the Natural History Museum of Utah. (B) Articulated lizard legs from Dinosaur National Monument (DINO 13861). (C) Right dentary and teeth of a small scincomorph (paramacellodid?) lizard from the Black Hills, Wyoming (SDSM uncataloged). (D) Partial right dentary and teeth of the anguimorph lizard *Dorsetisaurus* (DMNH 18371) from the Wolf Creek Quarry in northwestern Colorado. (E) Part and counterpart hind limbs and tail of a small lizard from the Fruita Paleontological Area, Colorado. Image in E courtesy of George Callison. Scale bars in A and D = 1 cm; scale in B in cm; scale bar in C = 5 mm; fingertip for scale in E.

there are many types of scincomorphs and anguimorphs of a variety of habits (Pianka and Vitt 2003), but these two sister clades are together more derived than the geckos or the most primitive lizards, the iguanas.

Scincomorphs and anguimorphs today are generally insectivorous, eating all types of insects, spiders, scorpions, worms, and invertebrate larvae (Mattison 1989; Cooper 1994). Many modern species have diets consisting of a large percentage of termites, although some specialize on mollusks and some switch to eating seeds or other plant matter when insect populations are low. Other species may eat mostly small vertebrates. Anguimorphs of the Morrison Formation like *Dorsetisaurus* had sharp teeth and may have fed on some tiny vertebrates, but this lizard genus was fairly small, and the supply of ingestible small vertebrates may have been limited. Thus, even the Morrison anguimorphs may have fed mostly on insects and other invertebrates.

Just as lizards today have morphological features that allow them to lose part of their tail and still regrow it, lizards as far back as the Eocene have been fossilized with direct osteological evidence of having lost and regenerated their tails. And tail vertebrae with structures suggesting this ability have been found in the Jurassic (ElShafie 2017). So despite not having any fossilized regenerated tails, Jurassic lizards, including those of the Morrison, probably had the ability to escape from predators the same way their relatives do today.

The Morrison lizard fauna of paramacellodids and anguimorphs has been characterized informally as a "Purbeckian-type" squamate assemblage along with the faunas from the Purbeck and Guimarota; elements missing from the Morrison lizard fauna include gekkonomorphs and iguanians known from the Late Jurassic to Early Cretaceous of the more eastern parts of western Europe and India, respectively (Nydam et al. 2016).

Schillerosaurus utahensis

Schillerosaurus is a small lizard known from a partial skeleton from Dinosaur National Monument and was named by Susan Evans and Dan Chure (1999), although the genus name had to be changed recently (from *Schilleria* to *Schillerosaurus*) because the original had already been used for a subgenus of modern arachnid (Nydam et al. 2013). Other than being a scincomorph, its relationships to other lizards are unknown, and it is the only lizard genus known so far that is endemic to the Morrison Formation.

Saurillodon sp.

Saurillodon is a scincomorph lizard known from the Morrison Formation at the FPA in western Colorado. It had been found previously in the Middle Jurassic of England and Scotland and in the Upper Jurassic of Portugal.

Paramacellodus sp.

This scincomorph lizard was identified in new collections from Quarry 9 by Don Prothero and Richard Estes (1980), and it is now known as well from the FPA and the Rainbow Park sites at Dinosaur National Monument. *Paramacellodus* is a long-ranging genus, having been found in the Middle Jurassic of Scotland, the Upper Jurassic of Portugal and North America, and up into the Lower Cretaceous of Spain, Morocco, and England—a span of nearly 60 million years. It is a small lizard with short, relatively blunt teeth. Paramacellodids generally appear to have been nearly global in their distribution by the Late Jurassic (Nydam et al. 2016)

Dorsetisaurus sp.

Dorsetisaurus is an anguimorph lizard and was also identified from the Quarry 9 collections by Prothero and Estes (1980). It has long, sharp, and slightly laterally compressed teeth and occurs at a number of other Morrison Formation sites. It was originally named on the basis of finds in the Upper Jurassic of England and occurs in rocks of similar age in Portugal as well. Anguimorph lizards appear to have been restricted to the western part of Laurasia during the Late Jurassic (Nydam et al. 2016).

Diablophis gilmorei

This species was named *Parviraptor* as an anguimorph lizard from the Morrison Formation by Susan Evans. Its type material is from the FPA, and it has recently been identified as a new genus of snake (*Diablophis*—the

Squamata II: Why'd It Have to Be Snakes?

2 mm

5.9. Partial lower jaw of the snake *Diablophis gilmorei* in lingual (*top*) and labial (*bottom*) views (OMNH 78560, housed at the Sam Noble Oklahoma Museum of Natural History). Note the recurved conical teeth. From the Cisco Mammal Quarry, Grand County, Utah. Photos courtesy of Randy Nydam, Brian Davis, and Michael Caldwell.

"devil snake"; Caldwell et al. 2015).[1] *Diablophis* has now also shown up at the Cisco Mammal Quarry in Grand County, Utah (fig. 5.9).

There is not complete agreement on this material's representing a snake (it is rather contentious in fact), but if it does, it suggests that the small mammals and lizards of Morrison times were under predatory risk not only from small carnivorous dinosaurs and terrestrial crocodilians but also from snakes that may well have retained vestigial limbs. The possibly related snake genus *Parviraptor* (*P. estesi*) also occurs in the Upper Jurassic through Lower Cretaceous of England, and the snake *Eophis* is known from the Middle Jurassic of England; in the Upper Jurassic of Portugal is the snake *Portugalophis* (Caldwell et al. 2015). This Morrison snake species *Diablophis gilmorei* is currently known from upper and lower jaw material and a handful of vertebrae.

When George Callison first started working at the Fruita Paleontological Area in 1975 (it was not yet called that), he was specifically looking for microvertebrates and those that might shed light on the origin of snakes. He reported that they had found a snake in the fauna some time later (Callison 1987), but it took another 28 years for this to be backed up by others. As it turned out, he may have been right!

Early Champsosaurs: The Choristodera

The **Choristodera** (or "champsosaurs") were a group of reptiles with several common taxa that existed during the Late Cretaceous and Paleocene, about 80 to 90 million years after the time of the Morrison Formation. These animals, such as *Champsosaurus*, were similar in size, form, and habit to modern crocodilians but were only distantly related. Crocodilians are, along with birds, dinosaurs, and pterosaurs, part of a group of reptiles known as archosaurs. The choristoderes are more closely related to this group than any other but do not appear to be true archosaurs. It is possible that they originated in the latest Triassic (Storrs and Gower 1993), so *Cteniogenys*, from the Middle to Late Jurassic, would not be the oldest choristodere.

Cteniogenys antiquus

Several small vertebrate specimens, which had been collected by O. C. Marsh's crews working at Quarry 9 at Como Bluff, sat undescribed for many years after work there had ceased. Among these was a small jaw that paleontologist Charles Gilmore named *Cteniogenys* in 1928. Gilmore (1928) admitted that he wasn't completely sure what *Cteniogenys* was; he tentatively identified it as a lizard in his description but added that it might instead be a frog. It wasn't until the late 1980s that the relationships of the genus were figured out by paleontologist Susan Evans, after more material that could be referred to *Cteniogenys* was found in Europe (Evans 1989, 1990, 1991). These elements showed that *Cteniogenys* was in fact a small and much older relative of the champsosaurs (Choristodera)

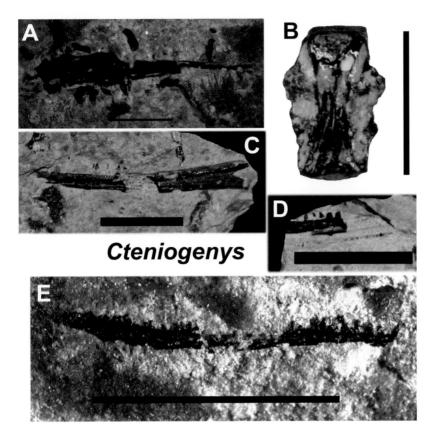

Cteniogenys

5.10. Elements of the choristodere reptile *Cteniogenys antiquus* from the Morrison Formation. (A) Left maxilla in labial view. (B) Vertebra in dorsal view. (C) Right dentary in lingual view. (D) Labial view of the dentary fragment missing from the center of the image in C. (E) Lower jaw of a small individual. A, B, and E are SDSM specimens, scale bars = 5 mm; C and D are an MWC specimen, scale bars = 1 cm. All from the Little Houston Quarry, Wyoming.

of the Late Cretaceous and Paleocene. During those epochs, the semi-aquatic champsosaurs lived much like crocodiles do, in and near waterways. During the Late Jurassic, however, early champsosaurs in the form of *Cteniogenys* were much smaller, although they seem to have been semiaquatic like their later relatives (Matsumoto and Evans 2010). For a long time, the genus was known in the Morrison Formation only from Quarry 9, but in recent years, specimens have been identified at Dinosaur National Monument in Utah; Ninemile Hill, the Sheridan College quarries, and the Little Houston Quarry in Wyoming; the Wonderland Quarry in South Dakota; Stovall's Quarry 8 in Oklahoma; and possibly the Uravan Locality in Colorado (fig. 5.10). *Cteniogenys* has also been found now in Middle and Late Jurassic rocks in the United Kingdom and Portugal, and the vertebrae of *Cteniogenys* from the United Kingdom sites seem to be, on average, somewhat smaller than those from the Morrison Formation (Chure and Evans 1998; Foster and Trujillo 2000).

Cteniogenys grew to approximately 25 to 50 cm (1 to 2 ft) in total length and probably weighed little more than 500 g (17 oz). The skull was long and slender with a tapered snout, and the lower jaws (fig. 5.10) contained many bluntly conical teeth with small, vertical ridges lining the inside surfaces of each crown near the tip. The vertebrae we find are small, spool shaped, and relatively long for their diameter. The contact surface between the centrum and the neural arch is relatively wide and sometimes quite long compared with the length of the vertebra, and the neural arches are almost always detached from the centrum when found. The floor of the neural canal of each vertebra also has a distinctive hourglass shape with a long central ridge. Most specimens from the Morrison Formation consist of teeth, jaw fragments, or vertebrae, so little else of the skeleton is known. Most of what we know about the skull is based on the European material.

Cteniogenys lived semiaquatically in and near rivers and ponds and probably fed on insects and small fish, although we can't be sure of the animals' diets because they are so small and different from any potential modern analogs. The teeth suggest that the animals were predators, and their small overall size indicates that they may best have hunted small surface insects and fish of the Morrison waterways. *Cteniogenys* is a rare animal in the Morrison Formation, with just under 60 specimens known; that accounts for about 2% of the more than 2,800 specimens of vertebrates of all taxa from the formation. Certainly, a large part of this rarity is due to *Cteniogenys* being such a small animal, because bones of small vertebrates are less likely to be preserved and to survive intact to be found by paleontologists than are bones of large dinosaurs, for example. Ecologically, small animals should be much more common than large ones, but among a group of a certain body size, some animals will be less common than others, and among the smallest Morrison Formation vertebrates, *Cteniogenys* may have been one of these rare species. *Cteniogenys* is mainly known from quarry sites that preserve freshwater ponds or rivers, or at least wet floodplains. At Ninemile Hill, Quarry 9, the Little Houston Quarry, and the adjacent Mile 175 site, *Cteniogenys* remains are in pond or abandoned channel deposits and are associated with abundant fish remains. At Como Bluff and Stovall's Quarry 8, *Cteniogenys* remains are associated with abundant teeth of crocodiles (Bakker and Bir 2004); at the Sheridan College Quarry, the genus was found in sandy deposits of a stream channel; and at the Wonderland Quarry in South Dakota, *Cteniogenys* was found in a fine white siltstone with turtle, crocodilian, and freshwater snail fossils.

Of the 10 or 11 localities in the Morrison Formation that are now known to contain *Cteniogenys* fossils, most are in the northern part of the formation's distribution. One site is in South Dakota, seven (possibly more) are in Wyoming, and one each occur in Oklahoma and northern Utah. There is a half centrum that may belong to this taxon at the Uravan site in western Colorado. The Oklahoma and Colorado specimens were found only in the last few years in museum collections where they had gone unnoticed. Before these finds, *Cteniogenys* was entirely restricted to areas of Morrison outcrop from Dinosaur National Monument north. It now appears that the genus was widely distributed but that it and its preferred environments were much more common to the north, where the Morrison Formation floodplain may well have been wetter than in the Colorado Plateau region (Foster and McMullen 2017).

Tropical regions today commonly have one or more species of crocodilian in their rivers and ponds (and sometimes in marine environments close to shore). Even for those of us who don't live in areas currently within the range of modern crocodiles and alligators, the animals are familiar elements of the tropical fauna. Many more of us, however, live within the range of ancient crocodilians, which, during Morrison times, ranged into Wyoming, South Dakota, and Montana. The striking thing about some Morrison crocodilian species is how similar they were in overall outward appearance to today's *Crocodylus* and *Alligator*. Crocodiles (in the form of crocodylomorphs) have been around, little changed, since shortly after the first dinosaurs appeared on Earth, and for most species, the habitat and ecology of the animals are similar to those of modern forms. A common Morrison Formation genus, *Amphicotylus*, would have reminded us of the crocodilians we know (fig. 5.11), spending much of its time in water and feeding mostly on aquatic animals and those terrestrial ones that ventured too close to the shore. There were also, however, crocodilians during the Jurassic that neither looked nor lived like those of modern times. These were small, cursorial animals that lived entirely on dry land.

Familiar (and Less Familiar) Carnivores: The Crocodylomorphs

5.11. A Morrison Formation goniopholidid crocodyliform, as displayed at the North American Museum of Ancient Life, Lehi, Utah. Cast based on material from the Bone Cabin Quarry, Wyoming.

The diverse fauna of crocodilians from the Morrison Formation, then, includes several species of "traditional" semiaquatic forms and at least four (possibly six) small terrestrial animals.

As defined systematically, Morrison crocs lie outside what are correctly called **Crocodylia** (comprising only modern crocodiles and alligators and their common ancestors). Even the more advanced crocs of the Morrison Formation are more primitive than the ancestors of modern crocodiles and alligators. Thus, I use the term *crocodilian* throughout the book in a very general sense. There are many crocodile-like animals early in the evolution of the group (**Crocodylomorpha**), and the transitions were so blurred that defining exactly what qualifies as a true crocodilian can be difficult without some precise rules. As currently defined, then, the 11 proposed crocodilian species of the Morrison Formation are more accurately described as indeterminate crocodylomorphs (2 species), sphenosuchian crocodylomorph (1 species), atoposaurid crocodylomorph (1 species), shartegosuchid crocodylomorphs (2 species), and neosuchian crocodyliforms (5 species). Each of these, in order, is progressively closer to modern crocodiles, but all are outside Crocodylia itself, so technically no Morrison crocs are crocodiles, systematically speaking. However, I use the general term *crocodilian* to refer to them, if for no other reason than to keep the terminology shorter for our purposes.

Semiaquatic Crocodilians

AMPHICOTYLUS

Goniopholis (a neosuchian) was named in the 1800s by Sir Richard Owen, who also coined the term *dinosaur*. The first specimen was from the Lower Cretaceous of England (Salisbury and Naish 2011; Salisbury et al. 1999), and the family has also been identified in the Upper Jurassic–Lower Cretaceous of Europe (England, Portugal, Germany, Spain, and Belgium), the Lower Cretaceous of South America, and purportedly the Upper Jurassic–Upper Cretaceous of North America (De Andrade et al. 2011; Jeremy Martin et al. 2016). In overall body form, *Goniopholis* (and goniopholidids generally) was similar to modern crocodilians, and the skeleton resembled those of today as well, except that the vertebrae were not **procoelous** and the dermal armor or **osteoderms** along the top of the back were rectangular with a prominent projection on one end. Most later crocodilians have a ball-and-socket articulation between the vertebrae in which the anterior surface of the vertebra is deeply concave and the posterior surface strongly convex. In *Goniopholis* and some of its relatives, however, this wasn't the case, and both anterior and posterior surfaces are only slightly concave. Recent studies of dermal armor in archosaurs suggest that the ornamented surface of crocodylian osteoderms and the vascularization of these bones may have evolved in part to assist sun-basking thermoregulation in these animals (Clarac et al. 2017, 2018).

The first goniopholidid specimens in the Morrison Formation showed up at Morrison, Colorado, and eventually five species were named. All were semiaquatic carnivores that resembled today's crocodiles and alligators in their ecologies. Most Morrison species recently have been referred back to Cope's *Amphicotylus* instead because it appears that Late Jurassic goniopholidids of North America differ somewhat from the European *Goniopholis* and may form their own **clade** (E. Allen 2010, 2012; Pritchard et al. 2013), or they may be closer to different respective taxa (De Andrade et al. 2011).

Amphicotylus probably fed mainly on fish and small reptiles, dinosaurs, and mammals that had the misfortune to get too close to the water at the wrong time. Like extant crocodilians, some may have ingested snails and shellfish also.

AMPHICOTYLUS LUCASII

This specimen was the type of *Amphicotylus lucasii* named by Cope in 1878. It had been found at Garden Park near Cañon City, Colorado, and was later referred to the European genus *Goniopholis* as a new species before recently being redesignated as *Amphicotylus*. The type specimen consists of vertebrae, ribs, and dermal scutes, but a referred skull was uncovered later (fig. 5.12). The skull belongs to an individual that was about 3.04 m (10 ft) long and approximately 132 kg (291 lb).

Amphicotylus lucasii may be more closely related to the European goniopholidids *Nannosuchus* and *Goniopholis* than it is to its Morrison contemporary *Eutretauranosuchus* (De Andrade et al. 2011; Puértolas-Pascual et al. 2015).

AMPHICOTYLUS GILMOREI*

On the basis of a skull found in the Morrison of Wyoming, in the Freeze-out Hills north of Como Bluff, this species was named by W. J. Holland in 1905, originally as *Goniopholis gilmorei*. If built like an alligator, it would have been about 2.9 m (9.5 ft) long and 108 kg (238 lb) in weight.

AMPHICOTYLUS STOVALLI*

This is another species named from a nearly complete skull, one found in the Morrison quarries near Kenton, Oklahoma. The skull suggests an overall body length of 3.10 m (10.2 ft) and a mass of 143 kg (315 lb). As with *A. gilmorei*, it was originally named as a species of *Goniopholis*.

"GONIOPHOLIS" FELIX

O. C. Marsh named this first species to be found in the Morrison Formation from a skull found at Lakes's Quarry 1 at Morrison, Colorado, in 1877. This was likely a subadult individual, as the total body length of the animal was only about 1.95 m (6.4 ft) and the weight about 26 kg (57 lb). It is unclear how it relates to other goniopholidids of the Morrison Formation. (Length and mass estimates are based on scaling comparisons to modern alligators and are from Farlow, unpublished data, and Farlow et al. 2005.)

5.12. Goniopholidid crocodyliforms of the Morrison Formation. (A) Dorsal view of the skull of *Amphicotylus lucasii* from the Morrison Formation at Garden Park in Colorado (AMNH 5782). (B) Skull of the type specimen of *Eutretauranosuchus delfsi* in palatal view, from the lower Morrison Formation at Garden Park, Colorado (Cleveland Museum of Natural History specimen, CMNH 8028). (C) Right lower jaw of *Eutretauranosuchus* (CMNH 8028); the jaw is broken and twisted midlength so that the front half is in occlusal view and the posterior half is in lateral (labial) view. (D) Left lower jaw of *Eutretauranosuchus* (CMNH 8028) in lateral (labial) view. (E) Dorsal view of a *Eutretauranosuchus* referred specimen (BYU Museum of Paleontology specimen, BYU 17628) from the Dry Mesa Quarry, Colorado. (F) Right lateral view of the same skull as in E. All scale bars = 10 cm.

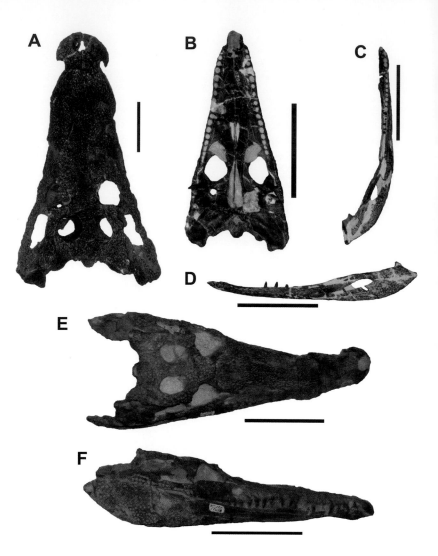

There are subtle differences among these species of *Amphicotylus* (and "*G.*" *felix*), but most of them relate to degrees of variation in, for example, the size and shape of the external nares (nostril openings), the development of the notch in the snout just posterior to the nose, the shapes of the openings in the skull behind the orbits, and the character of the pitted texture of the skull. By far most of the goniopholidid material from the formation is fragmentary teeth, scutes, or vertebrae not assigned to any species, and almost all the good skull and partial skeleton specimens have their own names, which are based on the above minor variations. Only A. *lucasii* has a character apparently unique among the four species: a shallow indentation on either side of the base of the snout. This feature may be significant as a specific character, but without more specimens exhibiting the same condition, it is for now more of an anomaly.[2]

It is possible that there was more than one species of *Amphicotylus* in the Morrison, but most of the specimens representing the four species listed above do not vary sufficiently to be convincing evidence of this.

None contains features that demonstrate unequivocally that it came from a biologically separate species (such as an individual bone that is clearly different in shape and not just proportion); the differences are just too subtle. More likely, the variations in "*G.*" *felix,* *A. lucasii,* *A. gilmorei,* and *A. stovalli* are due to individual variation within one species, and all might eventually be referred to *A. lucasii* as the lone Morrison representative of *Amphicotylus* (Foster 2006).

Determining whether there were truly one or two species of *Amphicotylus* would require more and better material, but I doubt that there were four, and I suspect that this group has been oversplit—that is, that more species have been named than really existed. This can happen early in the understanding of a group of animals, when they aren't well known and minor proportional differences in the little good material that exists are taken as significant. As more specimens are found, these differences of degree often fade into irrelevance as the real characters that differentiate the species come out. At this point, *Amphicotylus* is known from too many teeth, scutes, and fragments and too few skulls and partial skeletons for there to be overwhelming evidence of multiple species, and the differences among the type specimens are unconvincing.

EUTRETAURANOSUCHUS DELFSI

In the 1950s, during excavations at their Garden Park quarry, crews from the Cleveland Museum of Natural History removed a skull and partial skeleton of a subadult goniopholidid from the lower Morrison Formation (fig. 5.12). In a 1967 publication, Charles Mook gave this specimen the name *Eutretauranosuchus* and noted several unique features of the skull (Mook 1967). Among these is a secondary opening in the palate anterior to the main opening (i.e., anterior to where the main nasal air passage enters the back of the mouth). This anterior opening is not present in the type specimen of *Goniopholis* from England (or perhaps in *Amphicotylus*) and so would seem to set *Eutretauranosuchus* apart as a unique Morrison goniopholidid. This region of the skull is not well preserved in *A. lucasii* or "*G.*" *felix,* so we can't compare the character in all Morrison goniopholidids. The lower jaw of *Eutretauranosuchus* is relatively slender, and the vertebrae are similar to *Goniopholis* in being slightly concave on both anterior and posterior surfaces. Some of the vertebrae in the type specimen have unfused neural arches, suggesting that the individual was not fully grown.

Eutretauranosuchus was also later identified from the collections of Quarry 9 at Como Bluff, the Bone Cabin Quarry, Dinosaur National Monument, and the Dry Mesa Quarry (fig. 5.12; Foster 2003a; D. K. Smith et al. 2010; Pritchard et al. 2013). *Eutretauranosuchus* may be more closely related to the Asian goniopholidid *Sunosuchus* than to *Amphicotylus* (which in turn may be more closely related to *Goniopholis*), although this is not entirely clear (De Andrade et al. 2011; Pritchard et al. 2013). Such ties between Morrison taxa and both Asia and Europe are not unexpected, as there are plenty of ties between the Morrison and European Late Jurassic faunas, and among crocodylomorphs the shartegosuchids are related to Asian taxa too (see below).

Like other goniopholidids, *Eutretauranosuchus* was a semiaquatic carnivore, spending most of its time in rivers and ponds (or sunning on their shores) and feeding on fish and small to medium-sized terrestrial vertebrates. The type specimen from Garden Park represents an individual that was about 1.77 m (5.8 ft) long and weighed approximately 18 kg (40 lb), but older individuals got considerably larger, as shown by a specimen found at the Dry Mesa Quarry.

Goniopholidids as a group are relatively derived among well-known Morrison Formation crocodilians, being one of the sister taxa to the group including modern crocodiles and alligators (Crocodylia; Steel 1973; Clark 1994). Among the other Morrison crocodilians, most are more primitive than this group.

THERIOSUCHUS MORRISONENSIS

One left mandible from the Little Houston Quarry in northeastern Wyoming is different from other neosuchian crocodyliforms of the formation (fig. 5.13). Years ago I described this specimen rather cautiously as a juvenile goniopholidid (which would have demonstrated strong allometric growth in the relative elongation of the lower jaw; Foster 2006), but in the years since I changed my mind and now believe this is something previously unrecognized in the Morrison Formation.

This specimen (MWC 5625) is about 14 cm (5.6 in.) long, has a short tooth row, and is dorsoventrally deep compared with goniopholidids. Closer examination showed that the teeth (which are missing) were probably of at least three types along the tooth row: small with round roots, very large with round roots, and laterally compressed. This size and morphology showed that this was actually not a goniopholidid but a new species of the European Middle Jurassic to Early Cretaceous atoposaurid crocodylian *Theriosuchus* (Foster 2018).

Theriosuchus was first named from the Lower Cretaceous Purbeck Formation of England (Owen 1879; Salisbury 2002) and has since had members of its genus or close relatives reported from Asia, Portugal, Spain, Romania, and possibly a few other places, usually in the Late Jurassic to Early Cretaceous (Jeremy Martin et al. 2010; M. Young et al. 2016; also possibly ranging into the Late Cretaceous). *Theriosuchus* and its relatives were somewhat small neosuchians with short and triangular skulls in top view, with large eyes and short, narrow snouts (Salisbury 2002; Jeremy Martin et al. 2014).

MWC 5625 is relatively large among these forms and differs from the atoposaurid *Knoetschkesuchus* (from the Late Jurassic of Portugal) in lacking an external opening in the mandible and in having a mandible less elongate relative to its maximum depth (fig. 5.13; Schwarz and Salisbury 2005; Schwarz et al. 2017). It differs from the atoposaurids *Atoposaurus*,

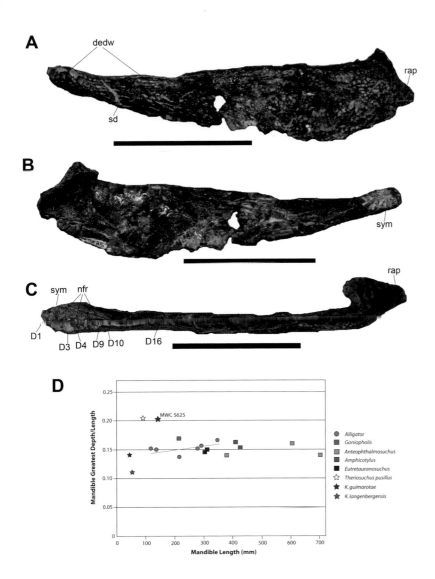

A dedw, rap, sd

B sym

C sym, nfr, rap, D1, D3, D4, D9, D10, D16

D

Mandible Greatest Depth:l/Length vs Mandible Length (mm)

MWC 5625

Legend:
- ● Alligator
- ■ Goniopholis
- ■ Anteophthalmosuchus
- ■ Amphicotylus
- ■ Eutretauranosuchus
- ☆ Theriosuchus pusillus
- ★ K. guimarotae
- ★ K. langenbergensis

5.13. Atoposaurid crocodyliform *Theriosuchus morrisonensis*, from the Little Houston Quarry, Wyoming (MWC 5625). Left mandible (lower jaw) in (A) Lateral (labial) view, (B) medial (lingual) view, and (C) occlusal view. Note the enlarged dentary alveoli D2 and D3; the small, circular D4 through D9; and the slightly crushed but large and elongate D10 through D16. (D) Comparison of the mandibular depth:length ratio as a function of lower jaw length, showing a greater ratio in *Theriosuchus pusillus* and MWC 5625 (stars) than in *Knoetschkesuchus* (stars), *Alligator* (circles), and European and North American goniopholidids (squares). The green line indicates the trend in *Alligator*. The length of the *T. pusillus* mandible is estimated due to the missing tip in BMNH 48328. The data in D are updated from Foster (2006) with additions from references cited in the text. Scale bars in A–C = 5 cm. Abbreviations in A–C: D1, dentary alveolus 1; dedw, "waves" in dorsal edge of dentary; nfr, nutrient foramina row; rap, retroarticular process; sd, external sculpturing of dentary; sym, symphysis.

Alligatorium, and *Alligatorellus* in lacking homodont dentition and from *Alligatorium* and *Alligatorellus* specifically in lacking the external mandibular opening and a smooth external surface of the mandible (Tennant et al. 2016; Tennant and Mannion 2014). MWC 5625 is most similar overall to a referred lower jaw of *Theriosuchus pusillus* (BMNH 48328) illustrated by Salisbury (2002) and Foster (2018). Although the anterior tip of that specimen is missing, an estimation of the full jaw length suggests that the depth:length ratio of the mandible is very similar to that of MWC 5625 (fig. 5.13).

This is the first evidence of the atoposaurid *Theriosuchus* in the Late Jurassic of North America and a new crocodylian form for the Morrison Formation (Foster 2018). This discovery thus strengthens faunal ties once again between the Morrison and the Late Jurassic–Early Cretaceous of Europe. The new occurrence also increases the diversity of crocodylomorphs in the Morrison Formation, with now at least two goniopholidids, two shartegosuchids, two hallopodid sphenosuchians, a species of *Theriosuchus*, and a possible protosuchian. Different atoposaurids appear to have been terrestrial or semiaquatic, depending on the species (Tennant and Mannion 2014), and with little of the postcranial skeleton known, the specific ecology of *Theriosuchus* is not clear. The association of *T. morrisonensis* with fish, amphibians, and abundant turtles, and its paleoenvironmental setting in an abandoned channel pond deposit (Foster 2001a), suggest this Morrison species may have been semiaquatic.

Terrestrial Crocodylomorphs

FRUITACHAMPSA CALLISONI

Fruitachampsa was one of the house cats of the Morrison Formation. It was a small, long-limbed carnivore with a short, almost triangular skull and sharp, canine-like teeth. It was also apparently fully terrestrial, unlike its Morrison neosuchian crocodilian contemporaries. Whereas all crocodilians we know today are essentially semiaquatic and have relatively short limbs, some ancient crocodilians, including *Fruitachampsa*, were long-legged and slender carnivores built for speed (fig. 5.14). Most specimens of *Fruitachampsa* indicate animals slightly less than 1 m (3.3 ft) in total length, and most probably weighed less than 5 kg (11 lb). The humerus and other known limb bones are among the most relatively slender of any crocodilian and are far more so than in semiaquatic forms. The skull is much shorter in front of the eyes than in most other Morrison crocodilians, and the orbits and fenestrae are much larger (relative to overall skull size). Unlike most other Morrison crocodilians, the palate is covered with grooves and ridges to nearly the same degree as the external bones of the skull, the teeth are slightly laterally compressed, and there is a pair of caniniform teeth near the anterior end of each dentary (fig. 5.15). In addition, *Fruitachampsa* differs from other Morrison crocodilians in having procoelous vertebrae (Clark 1985, 2011).

This terrestrial crocodilian has been found at several sites in the lower Brushy Basin Member in the FPA in western Colorado and almost nowhere else. It is interesting that *Fruitachampsa callisoni* has not been confirmed from any other microvertebrate collection, and yet it is not rare at the Fruita localities; there are nearly a dozen individuals represented in the material from those sites.

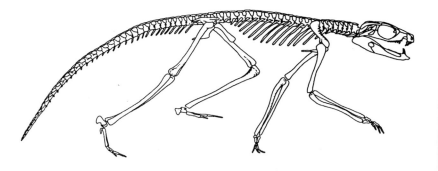

5.14. Skeletal reconstruction of the mesoeucrocodylian *Fruitachampsa callisoni* from the Morrison Formation at the Fruita Paleontological Area, Colorado. Total length ~1 m (3.3 ft). *Drawing by Rick Adleman based on Clark (1985).*

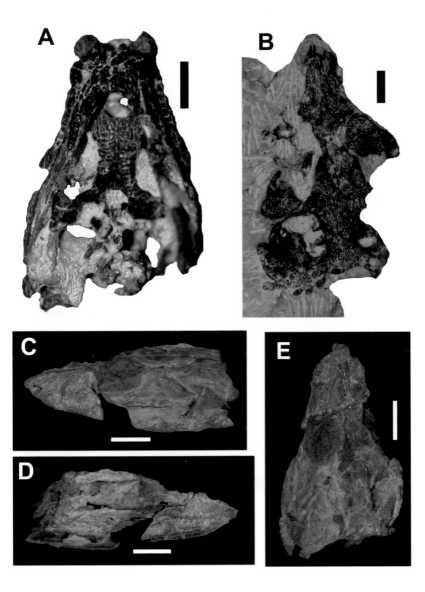

5.15. Shartegosuchid crocodylomorphs of the Morrison Formation. (A) Skull of *Fruitachampsa callisoni*, Brushy Basin Member, Fruita Paleo Area, in palatal view, showing sculpturing of bone on the roof of the mouth. (B) Skull of a second specimen of *Fruitachampsa*, Fruita Paleo Area, in dorsal view. (C–E) New shartegosuchid from the Salt Wash Member near Delta, Colorado; skull in (C) left lateral, (D) right lateral, and (E) dorsal views. All scale bars = 1 cm. A and B are LACM specimens; C–E is an MWC specimen.

Fruitachampsa was a relatively fast terrestrial carnivore that, on the basis of its size, probably fed on small lizards, amphibians, and mammals and possibly on dinosaur juveniles and eggs (Clark 2011; Kirkland 1994, 2006). There are at least two sites where terrestrial crocodilian remains have been found among baby ornithopod dinosaur remains or eggshells. One is a site in the FPA, where the remains are among various young *Dryosaurus*, and the other is the Salt Wash site near Delta, Colorado. These sites may indicate that terrestrial crocodilians were targeting very young dinosaurs or the eggs of the species.

Fruitachampsa was described by Jim Clark in his 1985 master's thesis at the University of California, but a name is not official until it appears in a more widely available format such as a scientific journal or memoir, so the name was formally published in 2011 (Clark 2011). Compared with *Amphicotylus* and *Eutretauranosuchus*, *Fruitachampsa* is more primitive

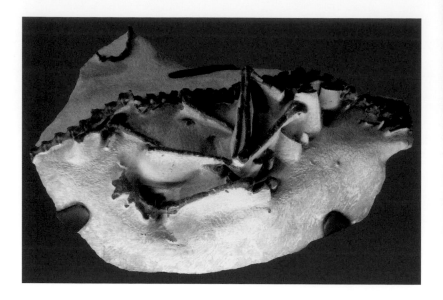

5.16. Small crocodylomorph skeleton (LACM 154921) from the Fruita Paleo Area, Colorado, as displayed at the Natural History Museum of Los Angeles County. This specimen may be a well-preserved individual of *Fruitachampsa*.

and thus more distantly related to modern crocodilians than those two genera. It is a member of a group of small terrestrial crocs known as the Shartegosuchidae, which comprises mainly Asian taxa from the Late Jurassic.

An articulated partial skeleton from the FPA (fig. 5.16) was recently very briefly described as a new sphenosuchian with elongate carpals (Arcucci et al. 2013). This specimen now appears to be a more complete postcranial skeleton of *Fruitachampsa* (Chiappe, pers. comm., 2017) than was previously known, and it indicates that the form had unusually long wrist bones.

NEW SHARTEGOSUCHID

A small shartegosuchid skull was found in 1987 in material collected from the Young Egg Site in the Salt Wash Member of the Morrison (fig. 5.15). It is probably several million years older than *Fruitachampsa*, occurring as it does below the Brushy Basin Member in the middle of the Salt Wash Member, and differs from it in having a very lightly sculptured palate and a groove down the midline of the roof of the mouth. Otherwise, it is rather similar to *Fruitachampsa* from the Brushy Basin Member of the Morrison and to *Shartegosuchus* and *Nominosuchus* from Asia. It appears to have dispersed to North America separately and earlier than did the lineage that led to *Fruitachampsa* (Foster and Clark, in prep.). This indicates that the Morrison Formation had faunal similarities not just to Europe and Africa during the Late Jurassic but also to Asia.

MACELOGNATHUS VAGANS

The type specimen of *Macelognathus* was first found at Quarry 9 at Como Bluff, Wyoming, and the type is based on a pair of anterior fragments of dentaries from one individual. These were found in 1880 by William

Macelognathus

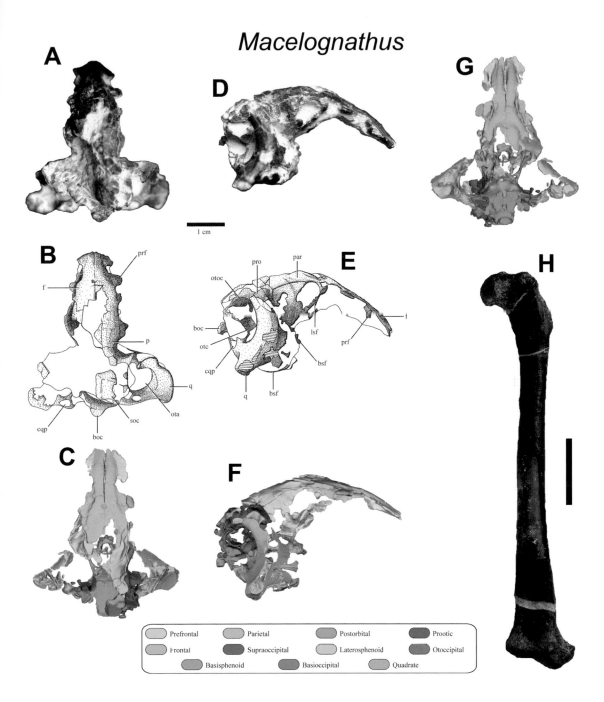

5.17. Sphenosuchian crocodylomorph *Macelognathus vagans* from the Fruita Paleontological Area, Colorado. (A–G) Partial skull. (A–C) Dorsal view. (D–F) Right lateral view. A and D are photos; B and E are drawings labeling bones; C and F are CT scan reconstructions of the skull with bones color-coded as indicated below. (G) CT scan palatal view of the partial skull. (H) Left femur in anterior view. Both scale bars = 1 cm. *Images in A–G courtesy of Jim Clark.*

Reed and briefly described by Marsh in 1884. The systematic association of the genus bounced around for a while, being identified initially as a member of the Dinosauria and even as being related to turtles before John Ostrom finally suggested that, most likely, *Macelognathus* was a crocodilian (Ostrom 1971). Recently, more complete material referable to *Macelognathus* was found in the LACM collections from the FPA in western Colorado (fig. 5.17), and these specimens suggested that the genus was a sphenosuchian crocodylomorph (Göhlich et al. 2005). The most recent study of a braincase among the FPA material (Leardi et al. 2017) indicates that *Macelognathus* is the sister taxon to *Hallopus* (in the family Hallopodidae) as part of a clade that is the sister taxon to Crocodyliformes (i.e., slightly more derived than spheonosuchians). What makes this species apparently unique among Morrison Formation crocodilians is that the dentaries contain no teeth anteriorly and together form a flattened, spatula-like anterior end to the lower jaws. There are tooth sockets behind this structure, but what the "jaw spatula" was for is difficult to say.

HALLOPUS VICTOR

Hallopus was named by Marsh as a species of the neornithischian *Nanosaurus*. It was found at Garden Park in the 1800s and is instead another cursorial, terrestrial crocodilian (fig. 5.18). There has been some uncertainty about from exactly what level the specimen was collected, but it is likely from high in the Morrison Formation (Ague et al. 1995). It has long, slender limb bones (Walker 1970), and it is the sister taxon to *Macelognathus* within the Hallopodidae (Leardi et al. 2017). In fact, due to the lack of skull material with the known specimen of *Hallopus*, it is possible that it and *Macelognathus* may prove to represent the same taxon (Leardi et al. 2017).

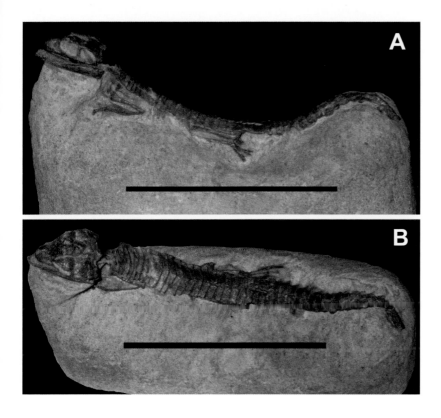

5.19. Crocodylomorph *Hoplosuchus kayi* from the Morrison Formation at Dinosaur National Monument, Utah, in (A) left lateral view and (B) dorsal view. Note the folded limbs and dermal armor. Scale bar = 10 cm. (CM 11361; photo by author; courtesy of the Carnegie Institute, Carnegie Museum of Natural History.)

HOPLOSUCHUS KAYI

This tiny crocodilian was found, the story goes, while blasting with dynamite through a sandstone bed during road construction at Dinosaur National Monument in Utah (fig. 5.19). It came from a level somewhat below the Carnegie Quarry and its building that you see today. The skeleton is only about 20 cm (9 in.) long, and this individual, although possibly quite young, probably weighed about 26 g (0.9 oz). *Hoplosuchus* may be the most primitive crocodilian known from the formation, and it appears to be related to *Protosuchus* (Buscalioni 2017). It appears to have an **antorbital fenestra** (an indentation in the skull anterior to the eyes that is lost in more derived crocodilians, including all others from the Morrison). *Hoplosuchus* also has relatively long limbs (although not as long as in *Fruitachampsa*), a strongly inclined quadrate, and a heavy covering of rectangular osteoderms along most of the body. The osteoderms do not appear to be as strongly pitted or sculptured as in other Morrison crocodilians and are in two rows of tightly overlapping plates from the neck to the pelvis. The osteoderms are smaller on the tail of the animal but seem to completely enclose it. *Hoplosuchus* was named in 1926 by Charles Gilmore as an aetosaur but was later identified as a primitive crocodilian. It is similar to one of the most primitive crocodilians known, *Protosuchus* from the Lower Jurassic of the western United States. Certainly, in retaining the antorbital fenestra, *Hoplosuchus* is the

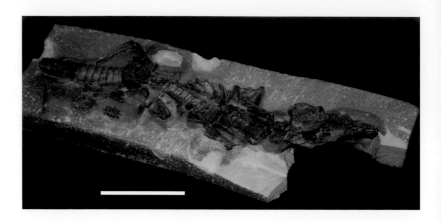

5.20. Crocodylomorph skeleton from the Morrison Formation of northeastern Arizona (UCMP 34634). Scale bar = 10 cm. Note the triangular skull, articulated limbs, and near-complete covering of osteoderm plates along the tail. Although it has not yet been identified and although there is some question as to whether it indeed was collected from the Morrison Formation, this specimen does demonstrate some similarities to the type specimen of *Hoplosuchus*.

most primitive of the Morrison crocodilians. It is difficult to say how large *Hoplosuchus* might have gotten, if indeed the type and only known specimen is very young—perhaps about 1 m (3.3 ft), maybe more. Or perhaps the type specimen is average adult size, and it rarely grew bigger—we don't know for sure.

An unidentified specimen possibly from the Morrison Formation of northeastern Arizona may indicate that *Hoplosuchus* indeed grew larger (fig. 5.20). This partial articulated skeleton was collected by Charles Camp of the University of California; it appears to be from the Morrison Formation, though there are conflicting data about the locality (P. Holroyd, pers. comm., 2016). The end of the tail is missing, but the full skeleton would have been about 50 cm (20 in.) long, and in general appearance, it is similar to *Hoplosuchus*. The plate-like osteoderms are in a similar pattern to the type, with the tail almost completely enclosed; the limbs are moderately long and folded near the body; and the skull is relatively short and triangular, apparently with a short preorbital area. Unfortunately, the skull is not well preserved, so we cannot identify this specimen as *Hoplosuchus* (or any other crocodilian, for that matter) with

5.21. *Hoplosuchus* life restoration. *Artwork by William Berry (© National Park Service, courtesy of Dan Chure).*

any confidence, but it is exciting to see that something at least reminiscent of *Hoplosuchus* has been collected.

What did *Hoplosuchus* eat? We can't say for certain because the crowns of the teeth are not well preserved, and in many cases, only the roots are intact. It was likely carnivorous, feeding on insects and small vertebrates or on larger vertebrates, depending on how large the individual was (fig. 5.21), but we can't rule out its being omnivorous because not all crocodilians have conical teeth meant just for snapping up fish and unsuspecting terrestrial vertebrates. Some rare crocodilians had teeth of unusual shape that may have been effective on plant material as well. These teeth are almost like those of primitive mammals. So we're not sure what *Hoplosuchus* was eating, although on the basis of the relatively long limbs, we can say that it was probably terrestrial.

The presence of semiaquatic neosuchian crocodilians in the Morrison Formation is important in that it provides us some indication of the environmental conditions of the time. By analogy with modern crocodilians, and in particular the alligator, we can infer that Morrison times, even in Wyoming and Montana, were relatively warm, with temperatures rarely getting below 3°C (37°F). The modern species *Alligator mississippiensis* has body temperature range limits of 3°–38°C (37°–100°F). These limits roughly approximate the temperatures of the Morrison floodplain because, as far as we know, Jurassic crocodilians were what is popularly known as cold-blooded. They are not exact ranges, however, because we don't know that Morrison crocodilians had the same range limits as alligators and because there needs to be a little room on either end of the range for the animals' abilities to last a certain amount of time when the environment is at the hot or cold limits, mainly achieved by behavioral regulation of their temperatures. Although alligators are ectothermic (subject to outside conditions that control their body temperature; Cassiliano 1997), they can regulate their own temperature to some degree through sunning when it is cool and staying in the shade or water when it is warm. Alligators can survive colder temperatures longer if they have access to water. Thus, the Morrison floodplain may have gotten colder than 3°C (37°F) occasionally during the cooler part of the year (just as it occasionally freezes in the alligator's habitat in the southeastern United States today) but only for short periods of time, and the average annual temperature was much higher than today's in the Rocky Mountain region.

Soaring Overhead: The Pterosaurs

The winged reptiles known as pterosaurs are well known to the public in the form of the Cretaceous-age *Pteranodon*, but they are quite well preserved in the Upper Jurassic Solnhofen deposits of Germany as well. Pterosaurs have also been found in the Upper Jurassic of western Cuba and, of course, a lot of areas around the world. As a group, pterosaurs dominated the skies for about 160 million years, during the Mesozoic era. Most of these animals were much smaller than *Pteranodon*, but some

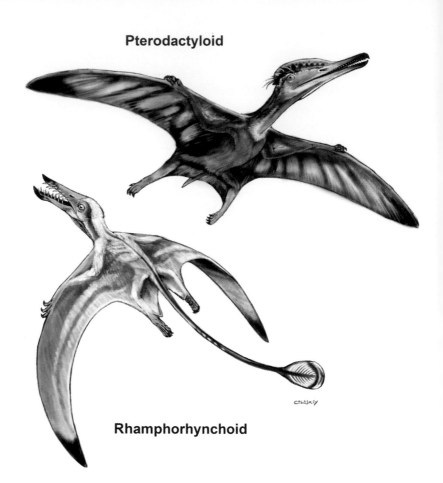

Rhamphorhynchoid

5.22. Two general types of pterosaurs known from the Morrison Formation. These reconstructions are based on more complete forms from other regions. The pterodactyloid is based on *Pterodactylus*; this gives an idea of the possible general appearance of such Morrison forms as *Kepodactylus* and *Mesadactylus*. The rhamphorhynchoid is based on *Rhamphorhynchus*. Artwork by Matt Celeskey.

grew to have wingspans of about 2.75 m (9 ft) or more. For a long time, pterosaurs were represented in the Morrison Formation by a single fragment of a wing bone. Then, in the 1970s and 1980s, they began showing up in quarries and museum collections a little more frequently, although they are still one of the rarest vertebrates known from the formation. They are even rarer than mammals, in part because pterosaur bones are so exquisitely thin walled. In order to lighten the skeleton as much as possible to facilitate flight, pterosaurs evolved bone that is hollow inside, with walls just 1 mm (0.04 in.) or so thick. With such delicate bones, pterosaurs are, not surprisingly, infrequently preserved in the Morrison Formation because the bones would have been easily destroyed before being buried. Pterosaurs are rare in many terrestrial deposits, but their bones and tracks are abundant in nearshore and marine rocks. Pterosaurs may have been more common near the ocean than inland, but there probably were plenty that fed on freshwater fish of the Morrison's rivers and lakes. Their bones are so delicate that they likely just are not preserved frequently.

Pterosaur species generally belong to one of two types: the pterodactyloids, which have short tails, and the "**rhamphorhynchoids**," which have long tails, sometimes with a diamond-shaped bit of skin at the distal end (fig. 5.22). "Rhamphorhynchoids" are more of a grade of less derived pterosaurs than an actual group. It is hard to imagine predators getting much meat out of a pterosaur: the animals needed to stay light, and what muscle mass they had was concentrated in the body. However, a recent find from the Cretaceous of South America demonstrates that theropod dinosaurs occasionally bit pterosaurs (Buffetaut et al. 2004). We have always suspected that pterosaurs laid eggs like other reptiles, and the recent find of a tiny pterosaur embryo inside an egg from rocks in China has confirmed this (Wang X. and Zhou 2004).

Pterosaur trackways from the Tidwell and Salt Wash Members of the Morrison Formation indicate that pterosaurs spent at least some of their time walking in sandy environments of the Morrison. The pterosaurs were, however, quite obviously flying animals, as indicated by the muscle attachments of their sternum and wing bones. Morrison pterosaurs probably fed on fish and perhaps insects. The stomach contents of some pterosaurs from Europe demonstrate clearly that these animals fed mostly on fish. When feeding on fish, pterosaurs of the Morrison Formation may have skimmed over the surface of a lake or river and plucked the prey from the water, or they may have plunged from the air into the water, grabbed a fish, and taken off again, similar to the method of some modern seabirds, if their bones were strong enough. The Morrison pterosaurs may well have scavenged dinosaur carcasses too.

The hollowness of the bones and the fact that they are mostly made up of wings result in pterosaurs being extremely light for their size. Most from the Morrison Formation were probably less than 2 kg (4.4 lb). Pterosaur bone histology and specimens preserved with a hairlike covering on the bone indicate that these flying reptiles were warm-blooded. Most pterosaurs are found along the ancient coasts, and it is believed that they lived in trees or on cliffs in these areas and fed in the ocean. Morrison Formation pterosaurs probably were inland species that lived in trees and specialized in feeding on fish in the rivers and lakes, or those that we find may be rare individuals that ventured inland from coastal habitats and happened to be preserved in the sediments.

Recent research indicates that some pterosaurs had enlarged lobes in their brains for processing large amounts of sensory information from the skin and muscle membranes of their wings, which suggests that they were capable flyers (Witmer et al. 2003), but on land, they had an unusual gait as a result of the long, folded wings and short legs. Whether moving on flat ground or climbing cliffs or trees, pterosaurs had to move with their elongate wing fingers folded up past their "elbows." Although such apparently awkward stances have been fuel for past speculation on why some types of animals weren't adapted to their own modes of life, trackway evidence and the fact that pterosaurs were around for some 160 million years indicate that they did just fine.

Of seven proposed taxa, there appear to be at least two or three pterodactyloid species (monospecific genera *Mesadactylus*, *Kepodactylus*, and possibly *Utahdactylus*) and one "rhamphorhynchoid" species (*Harpactognathus*) in the Morrison Formation that are valid (King et al. 2006).

Two named forms may or may not be nomina dubia (*Dermodactylus* and *Comodactylus*), and another appears to be a clear nomen dubium (*Laopteryx*).

(Wingspan estimates below are based on Wellnhofer [1991].)

Dermodactylus montanus*

This species of Morrison pterosaur is based on half of a fourth metacarpal (part of the hand and wing) and was found by Marsh's crews at Quarry 5 at Como Bluff (Marsh 1878a). This was the first evidence of Jurassic pterosaurs in North America. It was a small pterodactyloid pterosaur with a wingspan of approximately 1 m (3.3 ft) and a weight of approximately 1.5 kg (3.3 lb). *Dermodactylus* was for nearly 100 years the only named pterosaur in the Morrison Formation, but it has been recently delegated to a nomen dubium because of its fragmentary nature and the fact that other material cannot be confidently referred to it.

Comodactylus ostromi

Comodactylus was collected by Marsh's crews at Quarry 9 at Como Bluff in the late 1800s but was not identified in the collections at the Yale Peabody Museum until Peter Galton named it in 1981. This "rhamphorhynchoid" species was based on a single complete metacarpal and was one of the larger Morrison pterosaurs (Galton 1981a), as pterosaur expert Peter Wellnhofer has estimated the wingspan at about 2.5 m (8.2 ft; Wellnhofer 1991); it may have weighed about 8.6 kg (18.9 lb). *Comodactylus* has been viewed by some as a nomen dubium because of its being based on a single wing bone.

Mesadactylus ornithosphyos

This moderately small pterodactyloid pterosaur was named in 1989 from material collected at the Dry Mesa Quarry southwest of Delta, Colorado (fig. 5.23; Jensen and Padian 1989). The material had first started showing up in this scenic and large quarry starting in the 1970s, among large sauropod and other dinosaur bones in a coarse sandstone deposit. Although none of the material is articulated and not all bones of the skeleton have been found, it was believed to be the most completely known of Morrison pterosaurs. It may, however, be a chimera of a couple different pterodactyloid pterosaurs, including a possible ctenochasmatoid (Witton 2013) or possibly rhamphorhynchoid, dsungaripteroid, and ornithocheiroid groups (Sprague and McLain 2018). If more than one pterosaur is involved in the type material, the individual forms may still be new. Elements of a right wing of a pterodactyloid appeared in the Kings View Quarry near Fruita, Colorado, in October 2002, and these have been referred to *Mesadactylus* (fig. 5.23; King et al. 2006), although if only the

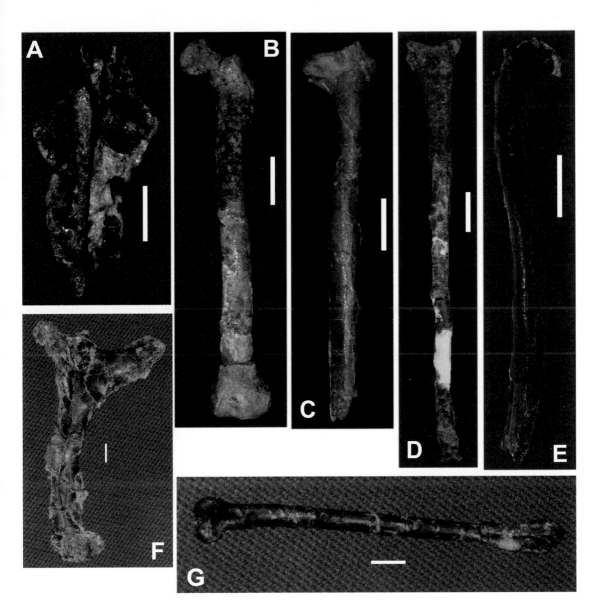

5.23. Pterosaurs of the Morrison Formation. (A) Type synsacrum of *Mesadactylus* (BYU 2024), from the Dry Mesa Quarry, Colorado. (B) Left femur of *Mesadactylus* (BYU 17214). (C) Distal half of the left radius of *Mesadactylus* (BYU 9493). (D) Right radius of *Mesadactylus*(?) from the Kings View Quarry, Colorado. (E) Small limb element of an unidentified pterosaur or maniraptoran theropod (BYU 2023). (F) Humerus of *Kepodactylus*, from the Small Quarry, Garden Park, Colorado. (G) Wing element of *Kepodactylus*. All scale bars = 1 cm.

type synsacrum of *Mesadactylus* can be assigned to the taxon, the wing material from Fruita is now unidentified.

As a pterodactyloid pterosaur (or several), *Mesadactylus* probably had a short tail, a small body, and an elongate skull. Like all pterodactyloids, the wing was supported by elongated forearm bones and those of the hand and fourth digit. Pterosaurs flew around on wings largely formed by the equivalents of human ring fingers that were several times the length of their torsos. The metacarpal of the fourth digit forms the structure of the wing in the hand (the other hand bones only support three small fingers). A pulley-like distal end to the fourth metacarpal allows folding of the wing back up the forearm, while a hook-like process on the proximal end of the first wing phalanx of the fourth digit prevents hyperextension of the wing.

Mesadactylus (even if based only on a synsacrum) had a wingspan of about 1.8 m (6 ft), and it likely flew near bodies of water (probably rivers, based on the deposits we have found it in so far), plucking fish from just under the surface. Like its smaller German contemporary, *Pterodactylus*, its skull probably had many small, conical teeth for catching the prey, and it may have scavenged terrestrial carrion as well. *Mesadactylus* probably weighed about 1.5 kg (3.3 lb).

Kepodactylus insperatus

Kepodactylus was a large pterodactyloid pterosaur with a wingspan of about 2.5 m (8 ft) and a weight of approximately 1.5 kg (3.3 lb). This species of pterodacyloid was named from a humerus and several other elements from the Small Quarry at Garden Park, Colorado (fig. 5.23; Harris and Carpenter 1996). It was found with other small vertebrate material and a *Stegosaurus*. *Kepodactylus* was thought to be related to several Cretaceous pterosaurs from Asia called dsungaripteroids (Wellnhofer 1991; Padian 1998), including the genera *Phobetor* and *Dsungaripterus*. These pterosaur genera have long, crested skulls with a reduced number of conical teeth, pincerlike tips to the mouth, and small, bony extensions off the back of the skull. Recent analyses, however, have suggested that *Kepodactylus* may have been a less derived pterodactyloid, among the ctenochasmatids with genera such as *Ctenochasma* and *Gnathosaurus* (Andres et al. 2014). The taxa from Asia had wingspans of about 1.5 m (4.9 ft) to 3 m (9.8 ft), and *Kepodactylus* may have had a wingspan of about 2.5 m (8.2 ft) and a mass of about 8.6 kg (18.9 lb).

Harpactognathus gentryii

Recent work at the Bone Cabin Quarry has unearthed two types of "rhamphorhynchoid" pterosaur, one of which was named *Harpactognathus* on the basis of the anterior part of a skull. This species is related to the Late Jurassic *Scaphognathus* from the Solnhofen deposits of Germany. Both *Harpactognathus* and *Scaphognathus* are characterized by having sharp, conical teeth widely spaced throughout the jaws, and *Harpactognathus* is unique in having a relatively wide snout with a thin, low crest along the top margin (Carpenter et al. 2003). The skull was moderately long and tapered toward the tip of the mouth, and the end of the long tail had a vertically oriented and diamond-shaped expansion of skin. The wingspan of *Harpactognathus* has been estimated at about 2.5 m (8 ft), and it likely weighed approximately 6 kg (13.2 lb).

Utahdactylus kateae

Utahdactylus was named by Czerkas and Mickelson (2002) on the basis of an apparent partial but unprepared skeleton, at the time thought to be of a "rhamphorhyncoid" pterosaur, from the middle Tidwell Member of the Morrison Formation north of Moab, Utah. This is one of few vertebrate specimens described from the Tidwell Member, but Chris Bennett suggested that the specimen (at the time) was identifiable only as an indeterminate **archosaur** and that there was no clear evidence that it was a pterosaur (Bennett 2007). Czerkas and Ford (2018), however, described newly prepared material from the same specimen, including a cervical vertebra, pectoral girdle, synsacrum, and partial lower jaws, that indicate *Utahdactylus* may have been a ctenochasmatid pterodactyloid pterosaur.

Laopteryx priscus*

Marsh described *Laopteryx priscus*, a small, fragmentary skull from Quarry 9, as that of a Jurassic bird. John Ostrom, also of Yale but more than 100 years later, determined that there was no reason to identify this specimen as a bird and that it was more likely an indeterminate pterosaur.

Fragmentary and indeterminate pterosaur material is also known from several other quarries but is generally rare in the Morrison.

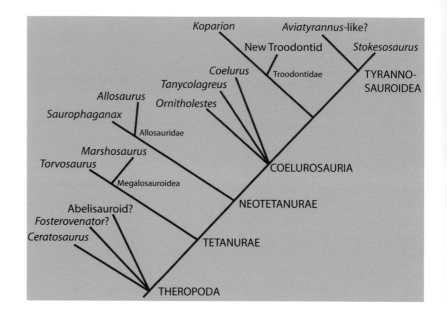

6.1. Relationships of the theropod dinosaurs of the Morrison Formation. Question marks indicate less complete taxa whose validity and positioning are uncertain. Based on Carrano and Sampson (2008) and Carrano et al. (2012).

Gargantuan to Minuscule: The Morrison Menagerie, Part II

The dinosaurian carnivore fauna of the Morrison Formation includes at least 11 genera and probably more (fig. 6.1), although of the nearly 200 individual specimens collected so far, nearly 75% belong to the genus *Allosaurus*. Some of the theropods were bigger than *Allosaurus*, some were much smaller, and some had horns, but the diversity of the fauna demonstrates that the carnivores likely divided their diets and feeding strategies in order to avoid ecological overlap as much as possible. Possibly not all the theropods known from Morrison Formation quarries lived in the same geographic areas and during the same times, but the fact that six of the genera are found at the Cleveland-Lloyd Quarry in Utah demonstrates that at least those did live together in the same area and at the same time. Thus, the theropods must have fed in different ways on different prey species, although it is difficult for us to determine exactly what these habits were. We can, however, infer some habits from the fossils that we have and from their distributions.

Theropod dinosaurs were small to large bipedal animals that walked and ran with erect hind limbs held close to the body's centerline under the hips, and they most often progressed in a digitigrade stance (i.e., walking on the toes with the actual ankle joint above the ground). On the basis of trackway evidence that we see in the Morrison and other formations, theropods most often walked and only rarely trotted or ran. This does not necessarily mean that they were slow or incapable of running, just that like most of us, they walked a much greater proportion of the time. Although carnivorous dinosaurs are probably most closely related to modern birds and are often compared with them, most members of the two groups walked quite differently. In *Allosaurus*, for example, the femur was normally held vertically under the pelvis and swung anteroposteriorly during walking or running. In birds that run, such as the ostrich, the femur is held with the knee forward and not vertically, and the main rotation of the limb is anteroposteriorly from the knee joint down. The cause of this difference may relate to the fact that the bird center of gravity is more forward than that of large carnivorous dinosaurs. The long tail and overall body mass distribution of *Allosaurus* results in a center of gravity that is much closer to the pelvis than in birds and thus allows the dinosaur to maintain a nearly vertical femur (Farlow et al. 2000; T. Jones et al. 2000).

Whether *Allosaurus* and other carnivorous dinosaurs used their walking and running abilities to hunt has been a source of speculation since

Beast Feet: The Theropod Dinosaurs

the first theropods were unearthed. Most juvenile theropods (including *Allosaurus*) were significantly more gracile in the limbs than were adults of the same species, and they thus likely hunted smaller and more agile prey than their older counterparts (Foster and Chure 2006). Tooth marks gouged into sauropod bones found in numerous quarries in the Morrison Formation indicate that large carnivorous dinosaurs certainly fed on adult sauropods, but it is unknown whether the theropods actually killed the prey animal in any of these cases. It seems unlikely that an *Allosaurus* or another large theropod dinosaur of the Morrison Formation could have attacked and brought down a healthy adult sauropod, mainly because of the size difference between the two. A full-size *Apatosaurus* weighed somewhere in the range of 24 to 34 times as much as an adult *Allosaurus*, and few predators today routinely attack prey animals more than 10 times heavier than themselves. And there seems to be a limit to how big a carnivore can get and still be able to attack larger prey, as it begins to reach its physical limits of strength and mobility (Farlow 1976; Farlow and Pianka 2003). It is much more likely that adult theropods attacked juvenile sauropods and other smaller dinosaurs and fed on adults only as carrion.

Some of the other types of dinosaurs on which theropods fed include stegosaurs, ornithopods, and the rare ankylosaurs. Large theropods may have sometimes hunted in small groups or packs, and these might have formed as associations of otherwise solitary individuals as conditions dictated. If individuals' midsized prey species became less abundant and large theropods were forced to feed on larger, perhaps subadult, sauropods, they may have formed temporary groups to try to bring down these herbivores, but the groups would have disbanded as soon as smaller prey species returned. This is only one situation in which lone large theropods may have temporarily hunted as a group. There are many other potential scenarios, but it is difficult to determine with certainty whether group hunting in a permanent social structure was a species characteristic for any large Morrison theropods. Perhaps some hunted in packs most of the time, or perhaps none did, but I doubt that all of them had such a social structure.

It has also been suggested that one way theropods may have dealt with prey so many times larger than themselves was to take meat from the animals without actually killing them. This idea, proposed by Greg Paul and inspired by behavior in some modern marine mammals, indicates that two or three good-sized bites off a large sauropod might have been enough to feed an adult *Allosaurus* for a day—an injury that would be survivable for a 34-ton sauropod (Paul 1998). Tom Holtz has termed this possible mode of attack in theropods "flesh grazing."[1]

Another possibility is that some large theropods switched diets entirely away from other dinosaurs when prey species became rare. Robert Bakker has suggested that *Allosaurus* from the Como Bluff region switched to aquatic vertebrates for food for part of the year as dinosaurian prey left the area during migrations. That theropods would have eaten any vertebrate available is logical, and if indeed dinosaurian herbivores became scarce for part of the year, there is no reason allosaurs wouldn't have begun frequenting bodies of water to pursue animals that had not left the area. Bakker has data from shed carnivore teeth that seem to indicate that this was the case (Bakker and Bir 2004). Specifically what aquatic animals allosaurs would have targeted and how they attacked them is not known, however.

At times, theropods may well have fed on each other. As described by paleontologists Dan Chure, Tony Fiorillo, and Aase Jacobsen, a partial skeleton of *Allosaurus*, collected in 1902 by Bill Reed at Quarry R in Wyoming, contains tooth marks on the end of the pubis bone that seem to match the large theropod *Ceratosaurus* or *Torvosaurus* and demonstrate that the carnivorous dinosaurs occasionally did consume other carnivores in the Morrison paleoecosystem (Chure et al. 1998b; Farlow and Holtz 2002). On the basis of the location of the tooth marks, the bite was more likely made during feeding on the carcass than during an attack, but clearly the carnivores went after each other when meat was available. One can imagine that carcasses were also routinely the sites of scuffles between predators fighting over the newly available food. And as has been shown recently with the large theropod *Majungasaurus* from the Cretaceous of Madagascar, theropods were sometimes cannibals (Rogers et al. 2003). Obviously, it sometimes would be counterproductive for a species to hunt its own kind, but feeding on carrion of their own species would have made for an easy meal.

Tooth marks gouged into the bones of herbivorous dinosaurs are more common in the Morrison than those made into carnivore bones (Chure et al. 1998b). Most marks consist of single scratches, but some consist of deep, parallel grooves made by numerous teeth at once, and few show evidence of more than a few bites. The bones do not appear to have been chewed intentionally by the animal as an attempt to consume any part of the skeleton; rather, the marks seem to have been incidental contacts made during removal of the muscle around the bones. It is interesting that many of the chewed bones in collections are those of the pelvis: *Camarasaurus* and *Mymoorapelta* ilia and the ischium of an *Apatosaurus* are among the best examples. The significant mass of muscle around the back legs and hips of most dinosaurs probably led carnivorous dinosaurs to concentrate on this area. At some sites in the Morrison Formation, shed theropod teeth are concentrated around the pelvis of herbivorous dinosaurs.

Small theropods, of which there were a number of genera in the Morrison, probably fed somewhat differently than large theropods like *Allosaurus*. *Ornitholestes*, *Tanycolagreus*, *Coelurus*, and *Koparion* were quite different physically and ecologically from their large cousins. These carnivores were of much lighter body weight and probably did not attack even juvenile sauropods or other dinosaurs except for the smallest. Most of their diets were composed of small vertebrates such as mammals, lizards, frogs, salamanders, and possibly aquatic animals such as fish. Some taxa may even have been partly insectivorous.

The theropod dinosaurs included the largest meat-eating terrestrial vertebrates of all time, and the Late Jurassic was unusual in having a greater diversity of large theropods than other geologic ages. A number of formations have much larger carnivorous dinosaurs than does the Morrison Formation (including the Late Cretaceous–age Hell Creek Formation, with its star *Tyrannosaurus*), but no other time had as many types of theropods that weighed so many kilograms. As paleontologist Don Henderson has noted, the large size of some Morrison theropods may have helped them minimize their prey choice overlap by allowing each to feed on a wider range of sizes of prey (Henderson 1998).

The fact that many theropods were so much larger in mass than most large mammalian carnivores of today has been of interest to paleoecologists for some time. In a 1993 study, paleobiologist Jim Farlow noted that these large carnivores are squeezed, in a sense, by competing needs: first, to maintain high enough population sizes (enough individuals) to prevent the extinction of the species as a result of a disaster or other chance event; and second, to keep population densities low enough to prevent the species from overexploiting its resources (Farlow 1993). Juveniles and adults of the same species may have fed on different prey items in order to minimize overlap and competition, and carnivorous dinosaurs may have simply had lower calorie requirements than would a similar-sized mammalian carnivore. Another possible manner in which theropods got away with their larger body size and avoided the problems of low population sizes was if their prey species were actually more densely distributed than would be expected for similar-sized mammalian herbivores. This may in fact have been the case in the Morrison Formation, which, as I noted above, has an unusual diversity of large theropod predators. Recent experiments in biology suggest that in some ecosystems, a greater diversity of predator species actually results in an increase in population density of the prey species (Finke and Denno 2004).

Farlow also developed equations to predict the density of theropod dinosaur populations. According to calculations based on this work, we might expect there to be approximately two to six individuals of large theropods such as *Allosaurus*, *Torvosaurus*, and *Saurophaganax* for every 100 km^2 (38.4 mi^2) of the Morrison floodplain. There might be 12 individuals of *Ceratosaurus* in the same area, 28 of a medium-sized theropod such as *Marshosaurus*, and nearly 500 of small theropods such as *Ornitholestes*. These numbers are based on comparisons with modern ectothermic reptiles, however, and if theropod dinosaurs had higher metabolisms (as seems likely, at least for smaller species), the densities would be lower. It is also important to remember that these numbers are based solely on the average adult weight of the animals and do not take into account the actual numbers of different species found in the Morrison Formation. Among large theropods of the Morrison, for example, despite the predicted densities and on the basis of the fossil record, *Allosaurus* was much more common than other large forms and must have had higher population densities than the others (although it is unknown whether the density for *Allosaurus* was higher than predicted by the equation or if perhaps the densities of other taxa were correspondingly lower).

Ceratosaurus

Ceratosaurus was named by O. C. Marsh on the basis of a partial skeleton (USNM 4735) from the Marsh-Felch Quarry at Garden Park, Colorado, and was described in detail by early 20th-century paleontologist Charles Gilmore (1920). Several partial skeletons of other specimens of this genus have been found at the Fruita Paleo Area in Colorado and at the Cleveland-Lloyd Quarry in Utah. *Ceratosaurus* is generally regarded as part of the primitive theropod group Ceratosauroidea, which is placed between more primitive taxa, such as *Coelophysis* from the Late Triassic Chinle Group and *Dilophosaurus* from the Early Jurassic Kayenta Formation, and more derived theropods like allosaurs and coelurosaurs. *Ceratosaurus* is thus a Late Jurassic holdover of a relatively primitive theropod grade in the Morrison, and most of its contemporaries (*Allosaurus*, *Torvosaurus*, *Ornitholestes*, etc.) were more evolutionarily derived. *Ceratosaurus*'s closest relative among the Ceratosauria may have been *Genyodectes*, an Early Cretaceous ceratosaurid from South America (Rauhut and Carrano 2016); *Ceratosaurus*'s Late Jurassic contemporary *Elaphrosaurus* from Tanzania is a more distantly related noasaurid ceratosauroid, and the Cretaceous-age abelisaurids (a sister taxon to the ceratosaurids) are almost exclusively from southern continents (Delcourt 2018).

Ceratosaurus is best distinguished by the horn on its snout formed by the paired nasal bones (figs. 6.2 and 6.3). Like *Allosaurus*, it also had small horns above and just in front of each of the eyes (part of the lacrimal bone). A number of theropod dinosaurs of all geologic ages had horns, crests, and ridges on the skulls, and *Ceratosaurus* is among these. The horn on *Ceratosaurus* was likely used as a display or species-recognition device, rather than for defense or attack. Most of the bone crests and ridges on theropod skulls were fairly thin and were probably not strong enough to function in combat; *Dilophosaurus*, for example, had paired thin crests along its skull that may well have broken if butted up against the skull of a conspecific. The horn of *Ceratosaurus* and the crests and ridges of other theropods may have been brightly colored as part of their display or recognition function.

Another distinctive character of *Ceratosaurus* is the fact that although on average its vertebrae and limb bones were smaller than those of *Allosaurus*, its teeth were quite a bit larger. *Ceratosaurus* teeth are long, recurved, and laterally compressed and are different from those of *Allosaurus*. The forelimbs of *Ceratosaurus* were also more slender than those of *Allosaurus*, and *Ceratosaurus* had a shorter hand that retained four metacarpals (Gilmore 1920; J. Madsen and Welles 2000). The ratio of the combined length of the three standard phalanges of digit 2 to the length of the radius in the two theropods is 1.55 in *Allosaurus* and 0.67 in *Ceratosaurus*, indicating that *Allosaurus* had a relatively short, stout radius

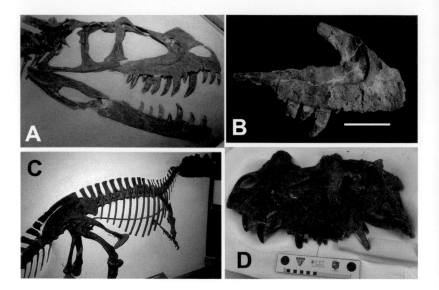

6.2. Morrison Formation theropod dinosaur *Ceratosaurus nasicornis*. (A) Skull of a juvenile specimen from Wyoming as displayed at the North American Museum of Ancient Life. (B) Maxilla of the MWC 1 skull and skeleton from the Fruita Paleo Area in Colorado. (C) Old mount of the type skeleton from the Marsh-Felch Quarry (USNM 4735). (D) Skull of the type specimen. Scale bar in B = 10 cm; scale bar in D in cm. Photos A–C and D by author and R. Hunt-Foster, respectively; C and D courtesy of the Smithsonian Institution.

and a larger gripping hand than did *Ceratosaurus*. This may indicate that the two were utilizing slightly different types of prey, and it may also result from the fact that *Ceratosaurus* was part of a lineage that ultimately reduced the relative size of the manus (forefoot) to a significant degree (Carrano and Choiniere 2016).

Ceratosaurus had other distinctive characters, including having only three teeth in the premaxilla, dual pits on the sides of the cervical vertebrae, unusual pits under the anterior articulations (prezygapophyses) of the dorsal vertebrae, fused metatarsals (foot bones) in some specimens, and four digits on the hands. It was a relatively large theropod at approximately 6 m (19.7 ft) long; estimates of its mass range from 524 kg (1,155 lb) on the basis of models and calculations made by dinosaur artist and researcher Greg Paul to 275 kg (606 lb) and 452 kg (996 lb) for the FPA and Cleveland-Lloyd specimens, respectively, on the bases of measurements of the femur and an equation proposed by J. F. Anderson et al. (1985).

The large teeth and clawed hands indicate that *Ceratosaurus* was a carnivore and almost certainly concentrated on the meat of vertebrates.

6.3. Restoration of *Ceratosaurus*. Total length ~6 m (19.7 ft). *Drawing by Thomas Adams.*

Other dinosaurs, particularly juvenile sauropods and other species as big as or smaller than *Ceratosaurus*, would have been favorite prey items. The skull of *Ceratosaurus* appears to have been slightly flexible to accommodate the stresses of feeding. Holtz has suggested that *Ceratosaurus* and its relatives rarely used their hands in feeding and instead relied mainly on their bite, and this seems reasonable considering the small hands and gracile forearms in *Ceratosaurus* compared with other large theropods such as *Allosaurus*. Bakker has studied shed teeth of various theropods, however, and on the basis of these interesting distributions, he has suggested that ceratosaurs of the Morrison concentrated their feeding efforts on large lungfish (Bakker and Bir 2004). That theropods would have occasionally, perhaps seasonally, targeted aquatic species as a main food source is quite possible, but no doubt Morrison ceratosaurs were equipped to take on other dinosaurs, and did.

In addition to the Marsh-Felch, FPA, and Cleveland-Lloyd Quarries, *Ceratosaurus* specimens have been found at Dinosaur National Monument in Utah, both Quarry 9 and Nail Quarry at Como Bluff in Wyoming, the Mygatt-Moore Quarry and Dry Mesa Quarry in Colorado, the Bone Cabin Quarry in Wyoming, and the Agate Basin Quarry in Utah. The Bone Cabin specimen is particularly interesting in that the partial skeleton is a juvenile (fig. 6.2). *Ceratosaurus* specimens (or ceratosaurids) have also been identified from Late Jurassic deposits in Portugal, from Tanzania in east Africa, and from the same age in Switzerland.

CERATOSAURUS NASICORNIS

Ceratosaurus nasicornis is based on a partial skeleton with a nearly complete, articulated skull found at the Marsh-Felch Quarry after the initial late 1870s excavation (Marsh 1884). The specimen is now at the Smithsonian Institution (fig. 6.2) and is the holotype of the genus *Ceratosaurus* (Gilmore 1920). Few specimens found since have been initially referred to this species, the good skeletons found subsequently having been designated as types of additional species.

CERATOSAURUS MAGNICORNIS*

The new species *Ceratosaurus magnicornis* was based on a single specimen (MWC 1) from a channel sandstone at the FPA in Colorado and was named by paleontologists Jim Madsen and Sam Welles in 2000. The specimen was collected by Lance Eriksen in the mid-1970s and consists of most of the skull, several cervical and dorsal vertebrae, and several elements of the hind limbs. This specimen is slightly larger than *C. nasicornis*, and although the nasal horn core is a bit longer and notably lower in *C. magnicornis*, few large differences exist between the two. Most of the differences between the two species are proportional or size related. Although the diagnostic characters listed by J. Madsen and Welles (2000) serve to distinguish the two specimens (USNM 4735 and MWC 1) and thus the two species, Matt Carrano and Scott Sampson (2008) synonymized *C. magnicornis* with *C. nasicornis*. *Ceratosaurus magnicornis* may be from a stratigraphic level slightly higher in the Morrison than is *C. nasicornis*, but we can't be certain because correlating between the Colorado Plateau region and the Front Range can be difficult, and the difference in time may not be significant in any case. Both specimens were found in channel sandstones, but other fossils of this genus are known from other depositional environments as well.

CERATOSAURUS DENTISULCATUS*

This species was based on a partial but disarticulated and somewhat scattered skeleton (UMNH 5278) from the Cleveland-Lloyd Quarry in Utah. It is the largest of *Ceratosaurus* skeletons yet known from the Morrison Formation. It was distinguished on the basis of proportional and size differences and was synonymized with *C. nasicornis* by Carrano and Sampson (2008).

Fosterovenator churei

This new species of small theropod was described and named from a type tibia and referred fibula from Reed's Quarry 12 at Como Bluff (Dalman 2014a). The tibia appears to belong to a small ceratosauroid, possibly an abelisaurid; the tibia does have an unusual structure in proximal view, but the material may be too limited to say much more for certain at this point. The type tibia is somewhat similar to a tibia from Garden Park that has been referred to *Elaphrosaurus* (Chure 2001), although the latter was later noted to be perhaps closer to indeterminate abelisauroid tibiae from Tendaguru in Tanzania (Rauhut 2005). *Fosterovenator* and the Garden Park tibia could then represent the first Morrison abelisauroid ceratosaurs, though more complete material is needed to support this (Dalman 2014a).

Based on the material available, this species is very possibly a nomen dubium; but the tibia is unusual in proximal view, and it is the only name we have for possible abelisauroid material from the Morrison at the moment—and it now appears that none of the Morrison material truly fits *Elaphrosaurus*. (Additionally, I admit that I have an obvious personal motivation not to add an asterisk to this one—but I really am trying to be objective.)

Elaphrosaurus sp.*

Elaphrosaurus bambergi was found in what is now Tanzania by German crews working at Quarry "dd" there in the early 20th century. The specimen was named in 1920 and mounted in 1926 in Berlin (Janensch 1920; Maier 2003). An unidentified species of this genus was identified from some isolated bones at a couple of spots in the Morrison Formation. Peter Galton first identified *Elaphrosaurus* sp. in the Morrison on the basis of a humerus found previously at the Marsh-Felch Quarry near Cañon City, Colorado (Galton 1982b). Another specimen, a proximal tibia from the Small Quarry not far from the Marsh-Felch, collected in the 1990s, was referred to the genus also (Chure 2001).

Elaphrosaurus was a medium-sized and slender theropod, with a maximum total length of about 6 m (19 ft) and a weight of approximately 210 kg (463 lb). The skull and hands are missing from known specimens, but the animal probably had a long, low skull with many small, serrated teeth. The fact that the tibia is significantly longer than the femur suggests it was a fast runner, unlike large individuals of *Allosaurus*, in which the femur is just longer than the tibia. *Elaphrosaurus* probably fed on small dinosaurs, mammals, lizards, and other small vertebrates of its time.

The slender nature of the hind limbs (particularly the metatarsals) of *Elaphrosaurus* led to early hypotheses that it might be related to the ornithomimid theropods of the Cretaceous period (*Ornithomimus* and *Gallimimus*, for example) (Gauthier 1986), but more recent work has indicated that it is more likely an abelisauroid ceratosaur and in a larger group including its contemporary *Ceratosaurus* (Holtz 1994, 1998).

The material from the Morrison Formation that had been referred to *Elaphrosaurus* has since been reclassified as indeterminate possible abelisauroid material (Rauhut and Carrano 2016), so it appears that North America does not share this genus (*Elaphrosaurus*) with Africa after all. But what the Morrison's possible abelisauroid is remains a mystery.

Torvosaurus tanneri

Named by Galton and Jim Jensen from the diverse Dry Mesa Quarry in western Colorado (Galton and Jensen 1979), this species of megalosaurid theropod dinosaur has been described in detail by Brooks Britt of Brigham Young University (Britt 1991). *Torvosaurus* was a large and very robust theropod and was the first "megalosaurid" found in the Morrison Formation (fig. 6.4). It is among a group of theropods including *Spinosaurus*, *Megalosaurus*, and *Afrovenator* that are closer to allosauroids and coelurosaurs than to the more primitive Ceratosauria (Holtz et al. 2004).

Torvosaurus was a much larger and heavier carnivorous dinosaur than was the more common *Allosaurus*. Among the elements of *Torvosaurus*

6.4. Restoration of *Torvosaurus*. Total length ~11 m (36 ft). *Drawing by Thomas Adams; coloring by Matt Celeskey.*

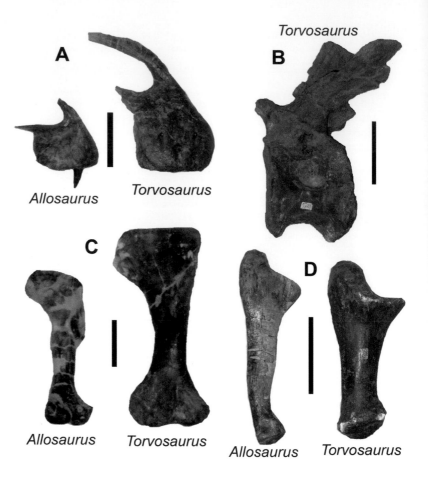

6.5. Some elements of *Torvosaurus* and comparison with an average adult *Allosaurus*. (A) Right premaxillae. (B) Vertebra of *Torvosaurus*. (C) Left humeri. (D) Ulnae. All scale bars = 10 cm. Note the much greater size of the humerus and premaxilla of *Torvosaurus* and its similar length but much greater robustness in the ulna. *Torvosaurus* specimens from the Dry Mesa Quarry are BYU 2002 (humerus and ulna) and BYU 4882 (premaxilla).

from the Dry Mesa Quarry stored at the Museum of Paleontology at Brigham Young University is a premaxilla, which demonstrates the size of the teeth of this carnivore in its tooth sockets. The tooth sockets are huge—nothing special compared with a *Tyrannosaurus*, perhaps, but among Morrison theropods and relative to *Allosaurus*, the size of a *Torvosaurus* tooth is remarkable. *Torvosaurus* probably weighed about 1,950 kg (4,299 lb), and its large teeth were laterally compressed. The humerus alone was noticeably larger and more robust than that of *Allosaurus*, and the ulna was shorter but thicker than that of *Allosaurus* (fig. 6.5). *Torvosaurus* likely fed on medium to relatively large dinosaurs of several types but is generally rare. It has been recently identified from Late Jurassic deposits in Portugal too.

The discovery of a very well-preserved megalosauroid theropod likely just basal to *Torvosaurus* and *Megalosaurus*, a form from the Late Jurassic–age Rögling Formation of Germany named *Sciurumimus* (the "squirrel-mimic," so called because of its bushy tail), has shown that even relatively primitive theropods well removed from dromaeosaurs and other near-birds had filamentous integumentary structures (i.e., "protofeathers"). Most analyses place this megalosauroid (*Sciurumimus*) basally within Megalosauridae and, in some cases, that whole family basal within Tetanurae (Rauhut et al. 2012b); the significance of this is that the position of *Sciurumimus* within **Theropoda** speaks to the possibility that Morrison taxa between it and known feathered theropods such as dromaeosaurs, namely *Torvosaurus* and *Allosaurus*, may have had protofeathers as well, at least at some stage of their development. So we do not know that *Torvosaurus* had fuzzy protofeathers, but it appears possible that it could have.

*Edmarka rex**

This megalosaurid species was described by Bakker and others in 1992 on the basis of several elements from the Nail Quarry in the eastern Como Bluff region (Bakker et al. 1992). The specimen has only subtle differences from *Torvosaurus* in the type jugal (a skull bone below and behind the eye). *Edmarka* is also similar in size and weight to the referred material of *Torvosaurus*, and *Edmarka* may in fact be simply an individual variant of *Torvosaurus*.

Marshosaurus bicentesimus

Marshosaurus was named by Madsen on the basis of fragmentary type pelvic material (ilium; see fig. 6.12) and referred skull elements from the Cleveland-Lloyd Quarry in Utah (J. Madsen 1976b). It could not be assigned to any known family then, but it has recently been suggested to have been a basal megalosauroid of the family Piatnitzkysauridae, a group with ties to South America (Carrano et al. 2012). A moderate-sized theropod at approximately 250 kg (551 lb), *Marshosaurus* was probably a predator of medium-sized and small dinosaurs. Only a few specimens

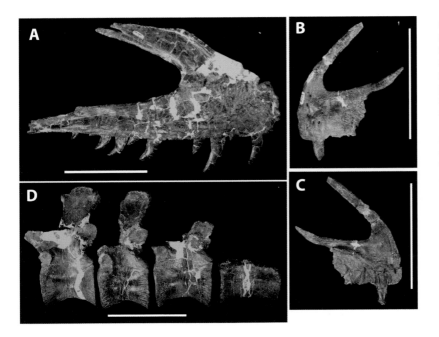

6.6. Elements of a piatnitzkysaurid theropod from near Dinosaur National Monument, Colorado (DMNH 3718). (A) Right maxilla. (B) Left premaxilla in lateral view. (C) Left premaxilla in medial view. (D) Four caudal vertebrae in left lateral view. All scale bars = 10 cm. This specimen is close to but slightly different from *Marshosaurus* (Sertich 2017).

are known from outside the type locality, and much of the skeleton is still unknown. The teeth of the premaxilla of *Marshosaurus* may be laterally compressed, if a possible referred specimen is correctly identified.

There may be a second, new piatnitzkysaurid from a site in northwestern Colorado (Sertich 2017); this specimen seems to differ significantly in its skull elements from those referred to *Marshosaurus* (fig. 6.6).

Allosaurus fragilis

Allosaurus (Marsh 1877; J. Madsen 1976a) is by far the most common theropod genus found in the Morrison Formation (Foster and Chure 1998, 2000), and it is also the best known osteologically (J. Madsen 1976a). The species *Allosaurus fragilis* was briefly described by Marsh from Garden Park, Colorado, and includes *A. lucaris* and the genera *Antrodemus, Laelaps, Labrosaurus, Creosaurus* (Williston 1901), and in part *Hypsirophus* as junior synonyms. The main osteological description of *Allosaurus fragilis* was done by Madsen in 1976. Paul (1988a) considered *A. atrox* as a species distinct from *A. fragilis*, and Sebastian Dalman (2014b) named the new species *A. lucasi* based on isolated and somewhat fragmentary material from southwestern Colorado. However, paleontologist David Smith, in his studies of variation in most *Allosaurus* material, has seen no evidence for more than one species (*A. fragilis*; D. K. Smith 1998). And there appears to be a fair amount of individual variation within that species at the Cleveland-Lloyd Quarry (Carpenter 2010). Most recent analyses suggest that there are two species in the Morrison, *A. fragilis* and a very recently named one (see below; Loewen 2004).

Allosaurus is more derived (systematically advanced) than the ceratosaurs and megalosaurs but is more primitive than the coelurosaurs. It appears to be a basal member of the Allosauroidea, and within that larger grouping *Allosaurus* is the sister taxon to the Carcharodontosauridae, a family of large, mostly Cretaceous allosauroids including *Acrocanthosaurus, Carcharodontosaurus,* and *Giganotosaurus* (Brusatte and Sereno 2008).

Allosaurus was a large and powerful predator (fig. 6.7). Its limb structure suggests that it could move well, but the fact that its femur was longer

6.7. Restoration of *Allosaurus,* the most abundant carnivorous dinosaur in the Morrison Formation. Total length ~7.5 m (24.6 ft). *Drawing by Thomas Adams.*

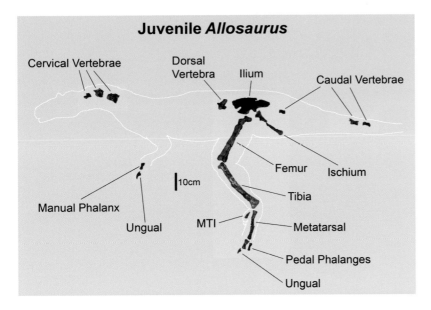

Juvenile *Allosaurus*

Cervical Vertebrae

Dorsal Vertebra

Ilium

Caudal Vertebrae

Manual Phalanx

Ungual

10cm

MTI

Femur

Ischium

Tibia

Metatarsal

Pedal Phalanges

Ungual

6.8. One of very few associated skeletons of young juvenile *Allosaurus* known from the Morrison Formation (SDSM 30510). This partial skeleton from the Black Hills in northeastern Wyoming shows significant differences from the proportions seen in adults: the femur is longer relative to the ilium; the tibia is longer than the femur; the metatarsal is slender and longer relative to the femur; and the hind-limb total length is ~33% longer relative to the ilium. These differences suggest that young allosaurs were better runners than adults of their species. Scale bar = 10 cm.

than the tibia indicates that it was not really built to be a fleet pursuer, chasing other large dinosaurs all over the Jurassic plain. Rather, it saved its muscular hind legs for charging and overpowering smaller prey, and it used its great bulk for subduing that prey or for running off smaller predators and scavengers from existing kills. As young juveniles, however, *Allosaurus* had long legs (fig. 6.8) and were better adapted to running than as adults (Foster and Chure 2006). They may well have spent much of their hunting time chasing after small vertebrates.

Allosaurus was a large carnivorous dinosaur but certainly not the largest that existed in the area during the Late Jurassic. Still, as predators today go, it was huge, and on top of that, it was armed with teeth and claws unlike anything around now. An average adult could be up to 8.5 m (30 ft) long, and the largest individuals represented at the Cleveland-Lloyd Quarry are nearly 12 m (40 ft) long. Published estimates of adult weights range from 1,000 to 4,000 kg (2,205 to 8,818 lb; P. Christiansen and Fariña 2004; Campione et al. 2014). An estimate of 1,000 kg (2,205 lb) is probably reasonable for large adults of the genus, although I have measured a number of average sized femurs of *Allosaurus* and have come up with estimates more in the 700 kg (1,543 lb) vicinity. Even at that size, *Allosaurus* was still a large meat eater.

Allosaurus likely fed on a wide variety of other dinosaurs, including young sauropods and stegosaurs and a variety of smaller ornithopods. In terms of diversity, there may have been as many species of sauropod for *Allosaurus* to choose from as there were ornithopods, but this depends on how one splits the named species of each prey group. Certainly, however, in terms of numbers of individuals, there were more ornithopods to grab than anything else—and I mean grab. The teeth of *Allosaurus* are not all that big or robust, and the skull is nowhere near as heavily built as that of a *Tyrannosaurus*, so as terrible as the bite of an *Allosaurus*

was, it wasn't as strong as it might have been. Significant weapons in the allosaur arsenal were the giant meat hooks of claws it had on its hands (fig. 6.9). These bones as we find them today are approximately 10 to 20 cm (3.9 to 7.9 in.) long, although in life they would have had a keratin sheath that would have extended their length by several centimeters and sharpened the tip even more. *Allosaurus* had hands that were longer and a forearm (radius and ulna) that was relatively shorter than those of more primitive theropods like *Ceratosaurus*; thus, *Allosaurus* would have had a better and stronger grasping ability. For an adult *Allosaurus* of about 1,000 kg (2,205 lb) attacking an adult *Camptosaurus* ornithopod nearly its size at around 800 kg (1,763 lb), it might be hard to do enough damage with an initial injuring bite to subdue the prey without having to fight it. Injuries in allosaur skeletons do indicate that they led a rough life, but we do not know for certain that these injuries occurred while hunting. It would seem that an allosaur's best bet would be to subdue the prey dinosaur with a strong grip with both forelimbs and several bites pressing down on the prey's throat until it quit struggling, similar to the mode of attack of many modern cats. Claws on the hind feet, although not as sharp as those of the hands, may have helped with this attack strategy. In this way, the *Allosaurus* avoids having to dodge kicks and swings of the tail in trying to deliver one huge, damaging bite, and it is able to use the moderate strength of its jaws to crush the trachea of the prey while only then needing its teeth to feed. In this scenario, the claws of the hand (and to a lesser degree the feet) were critical to the allosaurs' hunting style. Some recent analyses indicate that while feeding, *Allosaurus* used head motions similar to modern avian raptors, pulling meat off a carcass with upward and backward movements (Snively et al. 2013).

Ken Carpenter and Matt Smith compared the forelimbs of several theropod dinosaurs and found that those of *Allosaurus* and *Tyrannosaurus* were particularly strong and well suited to holding on to prey animals while delivering bites with the mouth (Carpenter and Smith 2001). For the same amount of muscle force, *Allosaurus* may have been able to hold a 36% greater load in its arms than the average human. Previous studies of *Allosaurus* and other theropod hands also have found that among large Morrison theropods, the allosaurs were best adapted for grabbing and grappling prey (Hartman 1996). Holtz (1995, 2003) has suggested that the allosaur skull may not have been strong enough to withstand staying clamped on a struggling prey item, and so allosaurs may have held on with the forelimbs while using what I'd call multiple bite-and-release strikes with the mouth. Either way, the forelimbs must have been important to the predation techniques of *Allosaurus*.

Other hunting styles have been proposed for *Allosaurus*, however, and I may be in the minority (or alone!) in believing that a hold with the hands and a bite-and-hold strategy with the jaws was the primary method of subduing prey in *Allosaurus*. More flashy strategies have been proposed, but the hands are often ignored as part of the equation. Still, these alternatives suggest some interesting ways in which larger prey may have been taken. Bakker has pointed out that the musculature is stronger in the allosaur neck than in the jaws and that the skull functioned somewhat differently than in more primitive theropods. He also offers an explanation for the puzzlingly unintimidating teeth seen in *Allosaurus*, compared with those of *Ceratosaurus* and *Torvosaurus*, for example. Bakker suggests that the allosaurs had the ability (in addition to feeding on smaller prey in a more conventional style) to open their mouths to an extra-wide gape and slash long gashes in the hide of large dinosaurs, using their teeth effectively as individual, giant serrations on a full-jaw cutting edge (Bakker 1998b). This would appear to contradict the idea that adult sauropods were mostly safe from theropod predation. If Bakker's idea is true of allosaurs, this feeding strategy still may have worked only on relatively young sauropods. A three-quarters-grown, subadult *Apatosaurus louisae* (about 17 m [56 ft] long) would still have outweighed an adult *Allosaurus* more than 17 to 1. Scaling specimen CM 3018 down in size gives a mass of about 15,315 kg (33,694 lb), compared with a large *Allosaurus* of just under 1,000 kg (2,205 lb). However, a half-grown, juvenile *Apatosaurus louisae* (about

11 m [36 ft] long) would scale to approximately 3,828 kg (8,423 lb), and although it was still four times as massive as large Morrison theropods, it would have been within the conceivable range of either a tooth-slashing or grasping-and-neck-biting *Allosaurus*. Either way, even young sauropods were large but were likely targets of adult allosaur attacks.

Bakker's slashing-teeth proposal helps explain some of the oddities of the *Allosaurus* skull and does not preclude regular feeding on smaller prey (as mentioned above and also in his paper), and it thus suggests that *Allosaurus* was a versatile feeder. Bakker has also found evidence from shed teeth that suggests that when larger prey were scarce, allosaurs fed near water sources and on aquatic taxa. If true, this also would indicate a tremendous diversity to the prey *Allosaurus* would utilize—a generalist predator with a skull adapted to allow it to target large prey that would normally have been outside its abilities. Models I've worked on that simulate prey-to-predator energy flow and relative abundances of predator and prey taxa suggest that in order to have been as abundant as it was relative to other theropods, *Allosaurus* would have had to have been a generalist, feeding on a greater variety of prey species than other carnivorous dinosaurs (see chap. 7).

In an interesting study that relates to the tooth-slashing hypothesis, Emily Rayfield from the University of Cambridge and others found an unexpected contrast in that the *Allosaurus* skull was overbuilt for the bite force that it is calculated to have had on the basis of the apparent size of the jaw-closing muscles (Rayfield et al. 2001). They suggest that individuals of *Allosaurus* brought their skulls down on prey like a hatchet (more or less) and that they used this method to go after both small and large dinosaurs. The contrast between the strength of the skull and the power of the bite is interesting in that it seems to fit Bakker's slashing model. The skull of *Allosaurus* seems to have been able to withstand many times more force directed upward against the upper jaw than could be generated by the jaw-closing musculature. Overbuilding is not uncommon, as many mammalian skulls are apparently stronger than necessary, but *Allosaurus* is more so. It would be valuable to see if the strength built into the allosaur skull also applies if the force is directed downward from the top of the snout rather than up from the jaw. Maybe the additional skull strength, beyond what is built into other vertebrates, is unrelated to bite force or feeding. Possibly some of the extra strength protects against downward forces on the skull resulting from contact with other objects. Could this, along with the lacrimal horns and nasal ridges seen in this and other theropods, be related to some form of intraspecific "jousting" the animals engaged in? It's possible, but we need to see if the skull strength Rayfield and her colleagues found works in both directions (up and down) and to compare results with other large Morrison theropods (which have not yet been studied similarly).

The relatively weak bite of *Allosaurus* seems unexpected. Although the bite, as calculated by Rayfield et al. (2001) for the anterior teeth, was equivalent only to the range of modern wolves, mountain lions, and

Allosaurus

6.10. *Allosaurus* specimens from several sites in the Morrison. (A) Partial skeleton AMNH 666 from Como Bluff, here as mounted at the American Museum feeding on a sauropod tail. (B) Skull of the *Allosaurus jimmadseni* type specimen from Dinosaur National Monument, left lateral (*top*) and internal (*bottom*) views. (C) Partial skeleton LACM 46030 from Utah as mounted at the Natural History Museum of Los Angeles County.

leopards, the force they found for the back part of the jaw, where there is more leverage, was closer to that of a lion—not as strong as you might expect from such a large animal, but strong nonetheless, and conceivably enough to easily do in an ornithopod dinosaur. These calculations were done with computer vector simulations from computed tomographic scans of the skull of Big Al (MOR 693) from northern Wyoming, a specimen that may belong to the new and more lightly built species of *Allosaurus*. It would be interesting to see what calculations based on an *Allosaurus fragilis* skull would reveal.

Although I question the likelihood of *Allosaurus* successfully and habitually hunting healthy and fully adult sauropods, shed allosaur teeth show that they often fed on large adult sauropod carcasses. Teeth are commonly found around the skeletons of sauropods, particularly near the pelvis, and tooth marks have also been seen on sauropod bones from a number of sites. These tooth marks are often found on pelvic bones. The theropods seem to have enjoyed the main muscle mass around the huge hind legs and hips of the large plant eaters. So, although they probably rarely killed an adult sauropod, and although they were likely as often hunters as scavengers, *Allosaurus* would gladly help themselves to a sauropod carcass any time they found one.

Since the first *Allosaurus* were found in the Morrison Formation in Colorado and Wyoming, they have been found in numerous quarries all over the Rocky Mountains (fig. 6.10). *Allosaurus* is the most common and best-known theropod in the formation, accounting for 70% to 75% of all Morrison Formation theropod specimens (Foster and Chure 2006). The other 10 genera comprise the remaining 25% to 30%. *Allosaurus fragilis* has been identified from the Late Jurassic of Portugal as well, and this is the first confirmation of a dinosaur species occurring on two different continents. The presence suggests a land connection between Europe and North America sometime during the Late Jurassic, just as faunal similarities suggest a connection between North America and Africa (perhaps through South America) at another time during the Late Jurassic.

Allosaurus also appear to have led relatively active lives, susceptible to injuries such as bone breaks. Several complete specimens are known to show a number of healed bone breaks, and often on the same side of the animal (suggesting a possible single-injury event). Big Al has broken limb elements and a rib, among other pathologies, and Big Al 2 has injured neck vertebrae, scapula, and ribs (Hanna 2002; Foth et al. 2015).

Allosaurus jimmadseni

A new specimen from low in the formation at Dinosaur National Monument (DINO 11541) represents a newly named species of *Allosaurus* (Chure 2000; Chure and Loewen 2020). This specimen is one of the most complete and best articulated of all known *Allosaurus* specimens and has a number of differences from A. *fragilis*, including an extremely thin coracoid. Also, nearly complete skeletons that may be referred to this species have been excavated from the Big Al and Big Al 2 quarries near Shell and the Meilyn Quarry north of Como Bluff, both in Wyoming. The Meilyn Quarry specimen is a juvenile. The very young juvenile from the Little Houston Quarry (Foster and Chure 2006) also has been referred to this species (fig. 6.8). This species appears to occur stratigraphically lower than A. *fragilis* in the Salt Wash Member and possibly equivalent levels (Loewen 2004).

Saurophaganax maximus

This species was found high in the Morrison Formation at Stovall's Quarry 1 in Oklahoma and was originally named *Saurophagus* (Ray 1941). Dinosaur National Monument paleontologist Dan Chure altered the genus name in 1995 because the original name apparently had been preoccupied by another animal and because some of the original material was not distinctive. The new name, *Saurophaganax*, means "king of the reptile eaters" because this is the largest carnivore known from the Morrison Formation, reaching lengths nearly equal to that of *Tyrannosaurus* from some 85 million years later. *Saurophaganax* was a large theropod and has been estimated to have weighed approximately 2,700 kg (5,952 lb) and to have been close to 12.5 m (42 ft) long (Chure 1995). The new type specimen is a neural arch from one of the dorsal vertebrae that displays a unique horizontal projection from each side of the base of the neural spine. In other aspects of the skeleton, *Saurophaganax* is similar to its possible relative *Allosaurus*. Only two individuals of the genus have been found, and both came from Stovall's Quarry 1, so it seems to have been rare. It may have lived only in late Morrison times, but either way, it was undoubtedly a predator of relatively large vertebrates.

*Epanterias amplexus**

Named by E. D. Cope in 1878 from his quarries high in the Morrison Formation at Garden Park, Colorado, this large species has been considered valid and has been identified at another locality in the northern Front Range of Colorado by Bakker and others, although Chure has noted that *Epanterias* is skeletally indistinguishable from *Allosaurus*. The type cervical neural arch of *Epanterias amplexus* (AMNH 5767) is approximately 20% larger than that of a moderately large *Allosaurus fragilis* (AMNH 666) from the Bone Cabin Quarry but is morphologically almost indistinguishable (fig. 6.11). For now, I suspect that *Epanterias* represents a large individual of *Allosaurus*.

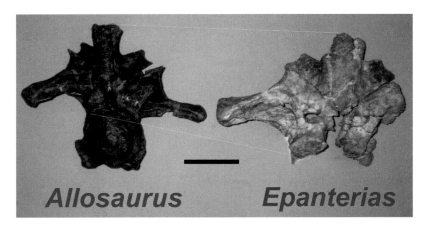

Allosaurus   **Epanterias**

6.11. Comparison of the anterior dorsal neural arch of *Epanterias amplexus* (AMNH 5767, *right*) with a complete anterior dorsal vertebra of adult *Allosaurus* (AMNH 666, *left*). *Epanterias* is 20% larger but is essentially indistinguishable. Scale bar = 10 cm.

Koparion douglassi

This recently described theropod, whose genus name translates roughly as "scalpel," is based on a single small tooth from the DNM-94 quarry at Rainbow Park in the upper Brushy Basin Member in the northern part of Dinosaur National Monument, Utah (Chure 1994). *Koparion* was referred to the Troodontidae (although some recent analyses have questioned this), a group of small, carnivorous (or possibly omnivorous) theropods mostly known from Cretaceous rocks. Nearly all troodontids were small, highly cursorial predators with relatively large numbers of teeth for the size of their jaws. Just recently, a possible troodontid skeleton was found in the Morrison in Wyoming, although it is unclear how it relates to *Koparion*. Considering the small size of *Koparion* (and some other Morrison theropods, for that matter), it is possible that its diet was not entirely composed of vertebrate meat. Theropod specialist Tom Holtz and others have suggested that some troodontids may have been omnivorous (Holtz et al. 1994). Even some modern, small carnivoran mammals almost never eat meat. The aardwolf of southern and eastern Africa, for example, has relatively small teeth and specializes in eating snouted harvester termites. If a modern, dog-sized hyaenid mammal such as the aardwolf can be effectively insectivorous, we need to consider the possibility of some unexpected diets for some of the small theropod predators of the Late Jurassic.

Hesperornithoides miessleri

The partial skeleton of this very small theropod (~0.89 m [3 ft] in length) with relatively long legs was found in the Jimbo Quarry near Douglas, Wyoming, slightly above the *Supersaurus* skeleton. It was found in an arrangement suggesting it may have been buried in a resting position, and it had strongly recurved teeth quite distinct from those of *Koparion* (Hartman et al. 2017). *Hesperornithoides* has proved to be a relatively basal troodontid, most closely related to several forms from Asia (Hartman et al. 2019).

Stokesosaurus clevelandi

Stokesosaurus was described by Madsen on the basis of a specimen from the Cleveland-Lloyd Quarry in Utah. The ilium of this species has a distinct vertical ridge rising from the acetabular hood to the dorsal edge of the iliac blade (fig. 6.12), and a second ilium, some vertebrae, and a premaxilla, also from Cleveland-Lloyd, were referred to the genus (J. Madsen 1974; Chure and Madsen 1998). *Stokesosaurus* has been referred provisionally to the Tyrannosauroidea, making it among the oldest members of the group that led to and includes the most famous carnivore of all time. Few specimens of *Stokesosaurus clevelandi* have been found, however, and little of the rest of the skeleton is known, although a separate species of *Stokesosaurus* from the Kimmeridge Clay of England may give us some indication (Benson 2008). The fact that so little of the skeleton is known makes inferring its habits and relationships difficult.

On the basis of its status as a theropod dinosaur and the premaxilla found in Utah, we can infer that *Stokesosaurus* was a bipedal terrestrial carnivore. It has been estimated that *Stokesosaurus* weighed approximately 80 to 85 kg (176 to 187 lb), but it is almost impossible to estimate

Stokesosaurus type

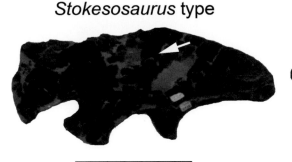

Stokesosaurus referred

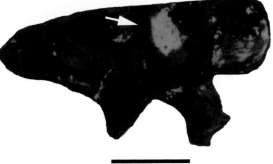

Marshosaurus

Allosaurus

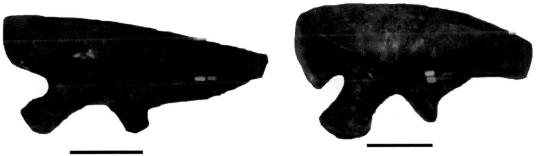

the weight of an animal with no limb material and so little else of the skeleton. Most likely *Stokesosaurus* had a mass between 70 kg and 100 kg (154 to 220 lb), and therefore it might have hunted prey from as much as 200 kg (441 lb) down to small vertebrates such as mammals and lizards. Potential prey dinosaurs would have been ornithopods and small stegosaurs and sauropods. In addition to the mammals and lizards, there were many other small vertebrates available at the time.

Aviatyrannus-like Form

A very small, partial ilium of a theropod from the Morrison Formation of the Black Hills in South Dakota may represent a second tyrannosauroid in the unit (fig. 6.13). This specimen represents an individual approximately just 60 cm (24 in.) high at the hips and is one of the smallest theropods known from the Morrison. It was found among diplodocine sauropod material during preparation of field jackets collected in 1980, and it was originally described as a young *Stokesosaurus* (Foster and Chure 2000), but recognition of differences between it and that taxon, plus similarities to the new *Aviatyrannus* from the Late Jurassic of Portugal, led to its being removed from the former and identified as possibly an *Aviatyrannus*-like form from the Morrison (Rauhut 2003, 2011). On reading Oliver Rauhut's papers, I had to admit that this ilium demonstrates more in common with

6.12. Ilia of the theropod *Stokesosaurus* from the Cleveland-Lloyd Quarry compared with other theropod taxa from the same locality. *Top* is the type *Stokesosaurus* (UMNH VP 7434), *left*, and referred (UMNH VP6383), *right*, ilia; arrows indicate the vertical bar on each. *Marshosaurus* is UMNH VP 6372; *Allosaurus* is UMNH VP 5410. All are left ilia except the referred *Stokesosaurus* (*upper right*). All photos by author. Courtesy of the Natural History Museum of Utah. Scale bars = 10 cm.

6.13. Partial right ilium of the *Aviatyrannus*-like form from the Wonderland Quarry north of Rapid City, South Dakota. (A) Lateral view. (B) Medial view. (C) Ventral view. Scale bar = 5 cm. Note the vertical bar in the iliac blade over the **acetabulum** in A.

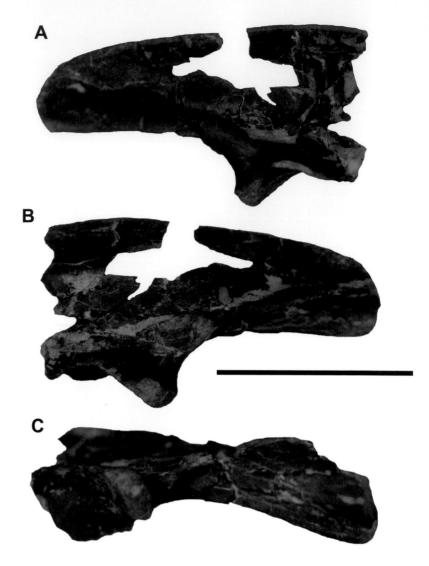

that of *Aviatyrannus* than that of *Stokesosaurus*. So this may represent yet another previously unrecognized small theropod taxon in the formation.

Ornitholestes hermanni

Ornitholestes was named by H. F. Osborn in 1903 on the basis of a partial skeleton with skull (AMNH 619) excavated from the Bone Cabin Quarry in Wyoming (Osborn 1903; Ostrom 1980; Carpenter et al. 2005b; Senter 2006). Also from Bone Cabin was the manus of what was thought to be a second individual but which has recently been identified as belonging to *Tanycolagreus*. *Ornitholestes* was a small, lightly built predator, approximately 2 m (6.5 ft) long and weighing about 12.6 kg (28 lb; fig. 6.14). The skull is one of the lightest built of Morrison theropods and is 14.5 cm (5.7 in.) long with nearly conical teeth, the longest and straightest of which are in the anterior maxilla–posterior premaxilla region. *Ornitholestes* is

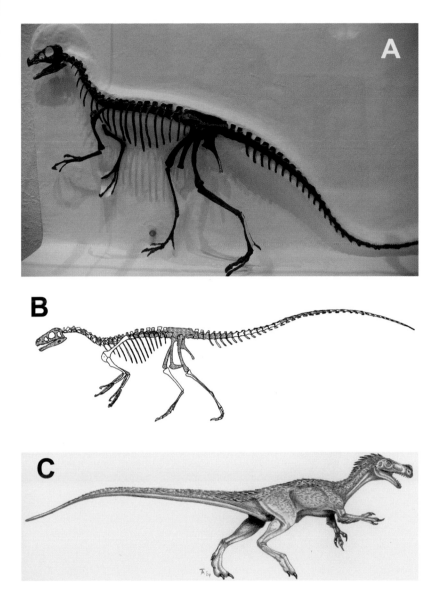

6.14. Small theropod *Ornitholestes hermanni.* Total length ~2 m (6.6 ft). (A) Mount of a skeleton at the American Museum of Natural History (AMNH 619). (B) Skeletal restoration by Ken Carpenter, preserved elements in gray. (C) Life restoration by Thomas Adams. *Photo in A courtesy of R. Hunt-Foster; B courtesy of K. Carpenter.*

a primitive coelurosaur, which means that it is one of the less advanced members of a theropod group including the tyrannosaurids, ornithomimids, oviraptorosaurs, and dromaeosaurids. Near the base of this group are several forms including *Scipionyx* (from Italy) and *Ornitholestes* and its Morrison contemporary *Coelurus.* Because *Ornitholestes* was small, light, and agile, it probably hunted small vertebrate prey like mammals, lizards, amphibians, and small dinosaurs and did not overlap in diet with the larger theropods. However, it may have competed with very young *Allosaurus,* which were small, long legged, and light. No confirmed specimens of *Ornitholestes* have been found at other quarries, but some elements possibly belonging to it were recovered from the Dry Mesa Quarry.

Recent finds in China of small theropod dinosaurs with feathers have led to questions of when feathers appeared and for what purpose. These discoveries have also inspired some to take a second look at the first bird, *Archaeopteryx*. Some have suggested that feathers arose as insulation in small theropods that were sometimes **arboreal** and that these feathers evolved into a type adapted for flying; others have suggested that feathers appeared as insulation but then were used to cover eggs in the nest while brooding, and then were further used later in flight (Hopp and Orsen 2004; Chatterjee and Templin 2004; Wellnhofer 2004). In any case, many paleontologists think that the Chinese feathered theropods were probably not the only ones with such a covering, just the only ones so far found with feathers preserved. Some paleontologists suspected that the first theropods with some form of feather covering may have been basal coelurosaurs, but now, with even megalosauroids seeming to have some type of covering like this, it appears likely that *Ornitholestes* may well have been a feathered theropod of the Morrison floodplain. These feathers, if present, would not have been as developed as those of birds like those found in the Lower Cretaceous in China but rather would have covered the body except the legs in a short coat, perhaps with longer feathers lining the top of the skull or neck and the back edges of the forearms.

Coelurus fragilis

This small theropod was named by Marsh from Reed's Quarry 13 at Como Bluff. Marsh also named a second species, *C. agilis*, but this has been determined to be the same as *C. fragilis* (Marsh 1879b). For some time, there was uncertainty as to whether *Coelurus* and *Ornitholestes* might in fact be the same animal (Ostrom 1980). This may have been partly because neither had had a full, well-illustrated description published. In the 1980 study, paleontologist John Ostrom determined that this genus was indeed distinct from *Ornitholestes*. A small, long-legged, and highly cursorial theropod, *Coelurus* had been assigned to the Maniraptora until recently. It is now classified with *Ornitholestes* as a basal member of the Coelurosauria, a larger group also including tyrannosaurs, dromaeosaurs, and ornithomimids.

Coelurus was approximately 20 kg (44 lb) in weight and 2.3 m (7.5 ft) long and was probably a fast runner (fig. 6.15). It had long cervical vertebrae, hollow dorsal and cervical vertebrae, unusually long metatarsals, and a relatively longer leg in general than other Morrison theropods. Compared with *Ornitholestes*, *Coelurus* had a longer and more slender neck and body, slightly shorter arms, and longer hind limbs (Carpenter et al. 2005b). Like *Ornitholestes*, it probably fed mostly on small vertebrate prey. It seems to have been somewhat more abundant than *Ornitholestes* and was probably contemporaneous with it. Although the two taxa are not known from the same quarry, *Coelurus* ranges from Quarry 13 low in the Morrison to Quarry 9, which is relatively high. Although the stratigraphic level of the Bone Cabin Quarry is still debated, it is probably between the levels of Quarries 13 and 9, and thus *Ornitholestes* would overlap with *Coelurus* temporally.

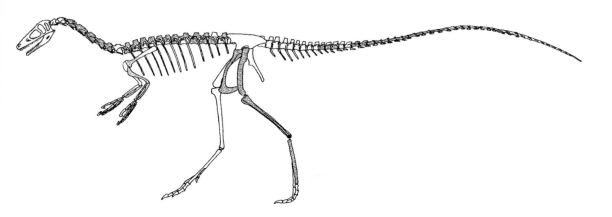

6.15. Skeletal reconstruction of *Coelurus*, preserved elements in gray. Courtesy of Ken Carpenter.

How would these two small theropods minimize competition for the same prey items? *Coelurus* may have targeted slightly larger species because it is itself a little larger than *Ornitholestes*, but the skull is unknown in *Coelurus*, so a direct comparison is so far impossible. The skull of *Ornitholestes* has a large orbital opening and therefore may have held a large eye. The opening is much larger for the size of the skull than is seen in tyrannosaurs, allosaurids, or dromaeosaurs and is nearly the same relative size as in ornithomimids. If *Ornitholestes* did have the relatively largest eye of the Morrison theropods, it may have been adapting to maximize low-light vision. Given that some of the small vertebrate prey, such as mammals, may have been most active during the dawn, twilight, or nighttime hours, perhaps *Ornitholestes* was minimizing competition with its slightly larger contemporary by targeting specifically these **nocturnal** or **crepuscular** small species and leaving the daytime-active species to *Coelurus*. Accomplishing what is called **niche partitioning** by having different times of activity during the day is not uncommon in modern ecosystems; there are other forms of partitioning the resource, but this particular one is similar to "taking shifts." It is not something that is done consciously by the species (as far as we know) but is simply a result of the availability of resources in any given environment and each species finding a way to get by.

Of course, in the case of our Morrison ecosystem, we are really speculating in our comparison because the skull is unknown in *Coelurus* and some other Morrison theropods. *Coelurus* may well have had large orbits also, in which case it might have been as well adapted for nighttime vision. For that matter, how might *Coelurus* and *Ornitholestes* have tried to minimize competition for the same prey as very young *Allosaurus*, which would have been (for a time) the same body size as the two coelurosaurs and which definitely overlapped with them in time? We really don't know. Perhaps, as Bakker has suggested in his work at Como Bluff, small allosaurs really weren't out chasing small vertebrates on their own like *Coelurus*; maybe they were being fed parts of medium to large dinosaur carcasses by adults. In either case, *Coelurus* was an important member of the theropod dinosaur community of the Morrison

Formation—it is in fact one of the most common theropods in the fauna after the dominant *Allosaurus*.

Tanycolagreus topwilsoni

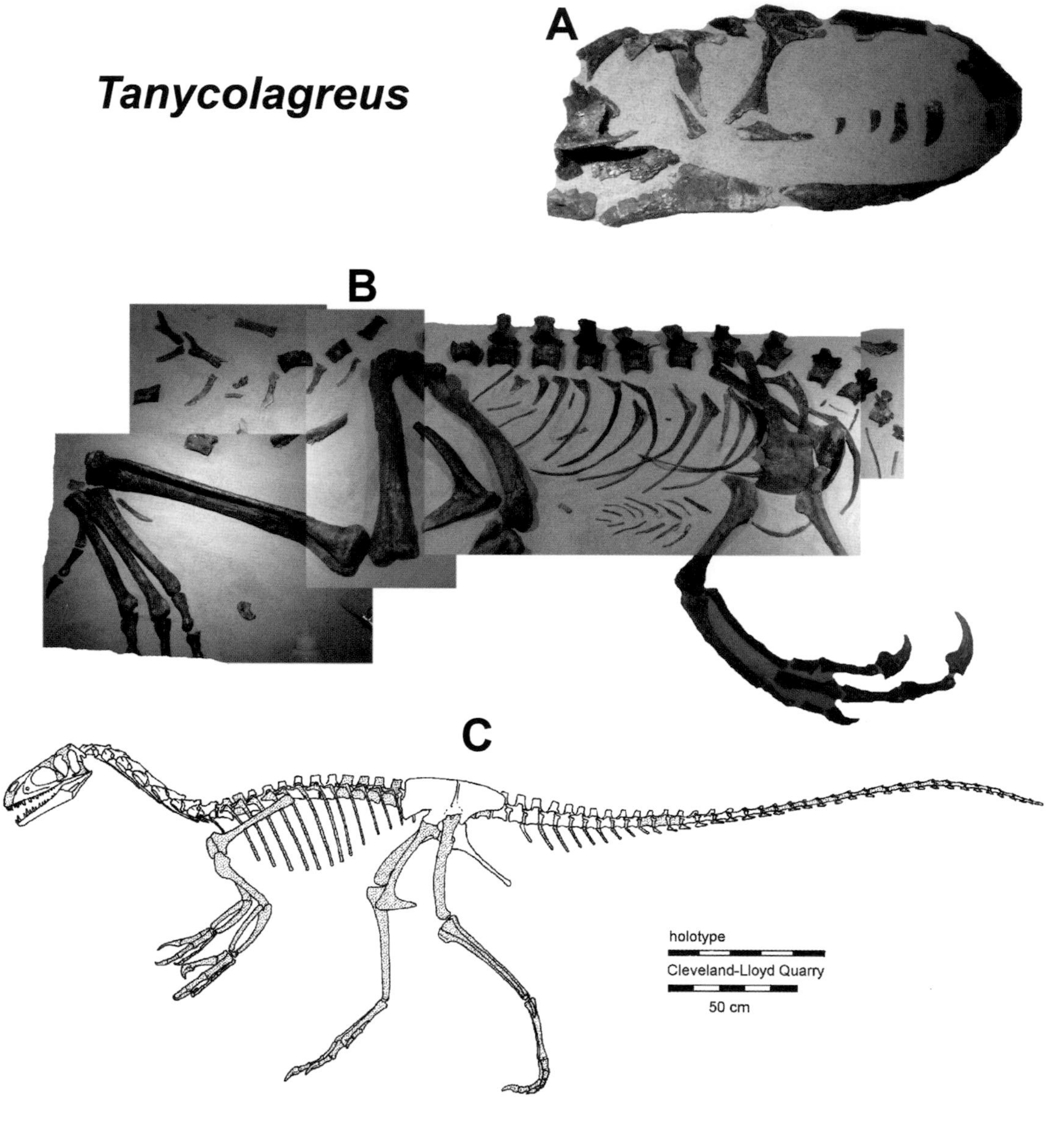

6.16. Small theropod *Tanycolagreus*. (A) Skull elements. (B) Postcrania. (C) Skeletal reconstruction by Ken Carpenter, preserved elements in gray. A and B as displayed at the North American Museum of Ancient Life.

This small theropod was collected from near the old Bone Cabin Quarry in Wyoming in 1995 (Carpenter et al. 2005a). It would have been about 3.4 m (11 ft) long and a little more than about 1 m (3.3 ft) high at the hip and weighed approximately 36 kg (79 lb; fig. 6.16). The skull is partly preserved and was apparently that of a typical small carnivorous dinosaur; *Tanycolagreus* therefore probably fed on small dinosaurs and other small vertebrates of the Morrison community. Referred specimens

A

Tanycolagreus

B

C

holotype

Cleveland-Lloyd Quarry

50 cm

have been identified from the nearby old Bone Cabin Quarry and from the Cleveland-Lloyd Quarry in Utah. *Tanycolagreus* had relatively long forelimbs (similar to *Ornitholestes*) and had a tibia longer than the femur (similar to *Coelurus*), which is an adaptation suggesting good running capabilities. The hind limbs are not as elongate as in *Coelurus*, however. Because no ilium of *Tanycolagreus* has been found and there is no other overlap of elements between it and *Stokesosaurus*, it remains possible that *Tanycolagreus* in fact represents a partial skeleton of *Stokesosaurus*.

Indeterminate Theropods

Among the isolated bones and teeth of theropods from the Morrison are several elongate metatarsals, including specimens from the Small Quarry in Garden Park, the Little Houston Quarry in the Black Hills, and the Poison Creek Quarry in north-central Wyoming (fig. 6.17). These specimens may indicate unrecognized occurrences of small theropod taxa such as *Coelurus* and *Tanycolagreus*. A proximal tibia fragment from the Small Quarry may represent an abelisauroid from the formation (fig. 6.17).

The tooth type represented by the two flattened, unrecurved theropod teeth from the Little Houston Quarry (illustrated in fig. 4.6B) is also hard to identify, although as noted in chapter 4, there are some

6.17. Elongate limb elements of indeterminate small theropods from the Morrison Formation. (A) Left metatarsal IV (DMNH 4250) from the Small Quarry, Garden Park. (B) Right metatarsal II from the Little Houston Quarry, Black Hills (SDSM uncataloged; field number JRF 95215). (C) Metatarsal from the Poison Creek Quarry, Wyoming (SMM specimen). (D) Proximal half of the right tibia (DMNH 36284) of a possible abelisauroid from the Small Quarry. All scale bars = 10 cm.

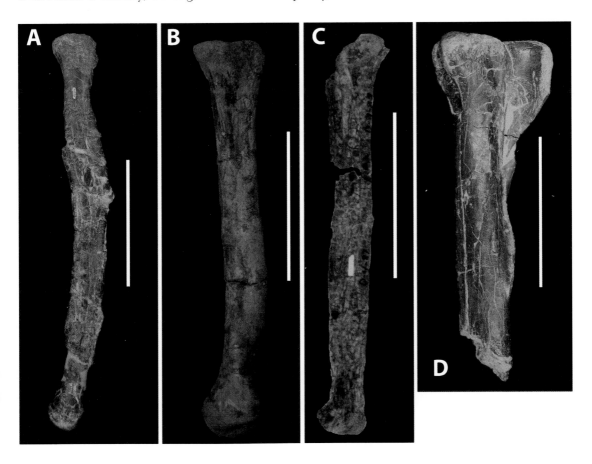

similarities to teeth in a small jaw described from the Late Jurassic of Portugal. The teeth are also somewhat similar to some specimens of abelisaurids and carcharodontosaurids but not all of them. There are at least two teeth of this form known from the Little Houston Quarry (Foster et al. 2020), but otherwise the teeth are a bit of a mystery.

Thunder Feet: The Sauropod Dinosaurs

The sauropod dinosaurs of the Morrison Formation are largely what gave the rock unit its fame. The Morrison has a well-deserved reputation for containing some of the best-preserved evidence of the time of giants. The sauropods are in some sense what it's all about in the formation—the Morrison is almost synonymous with large sauropods.

So many specimens of *Camarasaurus* have been found that we are closer to having a good picture of its species definitions, individual and populational variation, ontogenetic change, paleobiogeographic and biostratigraphic patterns, and sexual dimorphism than for any other sauropod. So what good is another *Camarasaurus* specimen? Plenty, if we really want to understand the animal. Ironically, paleontologists often lament finding "yet another" of a common type of dinosaur, but although another specimen of a "well"-known animal won't necessarily help much in figuring out the relationships of a clade, it will help understand that species just a little bit more—and we can't forget that the sample sizes of most dinosaur species are pathetically small. (Still, having been a few times on the business end of rolling a heavy field jacket containing but a single sauropod bone, I experience the same reluctance to collect all but the most significant sauropod specimens.) So I don't think the sauropods have outworn their utility; rather, the animals that follow below are the real heart of the Morrison vertebrate fauna.

Sauropods are biological marvels. No other land animal has ever achieved their size or majesty, and so it should not be surprising to us if some aspects of their biology and ecology prove to be unusual compared with what we know today. The early view of sauropods was that their teeth were too weak to eat any but the softest plants and that they were too large to support their own weight on land and thus spent most of their time in ponds and swamps eating aquatic vegetation. The water in this view was also seen as a refuge from predators. Not all paleontologists were convinced of this, however, and a few thought that the massiveness of the limbs was easily enough to support the bulk of the animals' weight. Elmer Riggs was one of those favoring a terrestrial interpretation. The view today has come around to Riggs's vision, largely thanks to two relatively recent reanalyses of sauropod structure by Bakker in 1971 and Walter Coombs in 1975. These authors noted aspects of the structure of the skull, feet, limbs, and tail that suggested a terrestrial lifestyle more than an aquatic one (Bakker 1971; Coombs 1975). As far back as 1904, Riggs had also noted that the structure of the limbs and feet were more similar to those of terrestrial animals, and that the pneumaticity of the vertebrae seen in sauropods was something he had been unable to find in any modern aquatic or semiaquatic animal. In addition, sauropod tails were not specialized for aquatic propulsion (as had been proposed long ago), sauropod skeletons are sometimes found in well-drained floodplain muds and in moderately developed paleosols, and trackways of sauropods show that they walked on wet sediments (such as floodplain muds) that later dried and cracked under a likely blazing sun. So most evidence suggests that sauropods were more than capable at living out of the water.

It was assumed early in paleontological studies that sauropods and other dinosaurs laid eggs, but there was no direct proof of this. It had even been suggested that sauropods in fact gave birth to live young. Recent finds, however, have confirmed that sauropods laid eggs about 15 cm (6 in.) in diameter and in clutches of sometimes more than two dozen eggs. Sauropods appear to have dug shallow, round nests into soft sediment and laid the eggs, but they did not bury them in the depression (Chiappe et al. 2001, 2004). The only embryonic sauropod remains from the Morrison Formation were found at the Dry Mesa Quarry in western Colorado (fig. 4.2; Britt and Naylor 1994) Juvenile and baby sauropod specimens occur at a number of localities in the Morrison Formation (Foster 2005b), but the largest sample comes from Stovall's Quarry 1 in the Oklahoma panhandle (Carpenter and McIntosh 1994). Here, a number of limb, pelvic, and vertebral bones are preserved that represent animals with bodies little larger than a big dog. The smallest bones in the sample from Quarry 1 are from sauropods about 17% the length of an adult. For *Apatosaurus*, at 21 m (70 ft) long, this indicates an overall length of the baby dinosaur of 3.6 m (12 ft), of which approximately 1 m (3.3 ft) would be the body without the neck and tail. Baby sauropods just slightly larger than this have been found at several other quarries in the Morrison Formation, including Dinosaur National Monument, Quarry E at Sheep Creek, and the Mygatt-Moore Quarry.

Comparisons of the bones of juvenile and adult sauropods show an unusual pattern in that the limb bones grow nearly isometrically (i.e., the proportions of the individual bones are nearly the same in young juveniles and in adults). This has been found for sauropod limb elements by several researchers (Foster 1995; Curtice et al. 1997; Wilhite and Curtice 1998; Bonnan 2004, 2007). The femur of a one-fifth-size *Apatosaurus*, for example, looks like a scaled-down version of an adult femur and is not noticeably more robust or slender. In a number of animals, from horses to theropod dinosaurs, the hind limbs of juveniles are remarkably longer and more slender relative to body size than those of adults (see Foster and Chure [2006] for a long-legged juvenile *Allosaurus*); in young dogs, the paws are relatively larger than in adults. The skulls of young sauropods appear to have been relatively larger than in their older relatives, so the isometric growth seen in individual limb bones and in some of the rest of the body in sauropods was unexpected. Studies of bone histology in sauropods suggest that sauropods were similar to reptiles in that they continued to grow significantly after reaching sexual maturity, but they were like modern large mammals in that growth was determinate (Sander 1999;

Curry 1999). Thus, sauropods were not directly comparable to either modern reptiles or mammals. Young sauropods grew at a high rate right after hatching and slowed only a little as juveniles. After reaching sexual maturity, they continued to grow until late adulthood, when their growth essentially stopped (Dunham et al. 1989). Regardless of the ultimate size different species attained as adults, they seem to have matured by about the same age (less than 20 years), with larger species simply retaining their adult growth period longer in their lives than smaller species. It is still unknown how long sauropods lived, but it may have been a long time.

Sauropods had some unusual features in their skeletons. Although all were big, with solid and massive limb bones, the vertebrae of most were highly pneumatic. Cervical, dorsal, and in some cases anterior caudal vertebrae contained huge openings in the sides and often multiple internal chambers in what appears to be solid bone from the outside. Cervical vertebrae in many sauropods look like I-beams in cross section as a result of the deep lateral chambers formed on each side. Only the center wall of bone remains. The ball-like anterior ends are smooth and solid on the outside but honeycombed with small chambers on the inside. The structures of pneumaticity throughout much of the axial skeleton probably housed a system of diverticula and pulmonary air sacs similar to but less developed than those of modern birds (Britt 1997; Reid 1997; Wedel 2003a, 2003b, 2005, 2009). In birds, the lungs have several directly connected air sacs in the body cavity, and some of the sacs have diverticula that branch up into and along the vertebral column, causing pneumatic structures in the vertebrae. This system of diverticula and air sacs and the resulting reduction in bone mass in the skeleton—if present in sauropods, as we suspect—would have lightened the overall mass of the animal, assisted breathing through such a long neck and overall breathing efficiency, and improved thermoregulation through heat evaporation in the sacs. The presence of this system of air sacs in a not-quite-avian form in sauropods hints that indeed the sauropod metabolism was not low and "reptilian," but neither was it perhaps at the high level of birds and mammals. The thermoregulatory function of the air-sac system may well have served to help moderate temperature resulting from either high metabolism or other factors in a moderately **bradymetabolic**, herbivorous animal of very large size.

The reinterpretation of sauropod habits in the past 40 years to a more terrestrial lifestyle has also opened a debate as to the physiology of the animals. The view that at least some dinosaurs may have been warm-blooded to the same degree as modern mammals and birds leads one naturally to wonder where the sauropod behemoths fit on the metabolic spectrum. The early assumption was that sauropods, as plodding, water-loving reptilian creatures, must have been as cold-blooded and sluggish as modern reptiles. This assumption has changed, and they are now viewed as being more active and perhaps even nearly warm-blooded. As we have already seen, their growth rates are nearly mammalian in character (though clearly less), and the patterns of growth fit neither reptiles nor mammals exactly (Werner and Griebeler 2014), so the truth, not surprisingly, may be a bit more complicated. (Also note that Myhrvold [2016] argued that there was no relationship between growth rate and metabolic rate.) Many workers view sauropods as having been neither cold-blooded and sluggish nor warm-blooded and active to the same degree as mammals but somewhere in the middle. Before exploring this idea, though, there are some aspects of these two end-member conditions that need to be discussed.

Having a slow metabolism (bradymetabolic) often goes with being **poikilothermic** (having a body temperature that varies), and a higher metabolism (**tachymetabolic**) is associated with being **homeothermic** (maintaining a steady temperature). What are commonly called "warm-blooded" animals are generally endothermic, meaning their fast metabolism produces enough heat for them to maintain a higher body temperature. The "cold-blooded" animals are mostly ectothermic, meaning that they have a slow metabolism that does not regulate the body temperature. Rather, body temperature varies depending on the outside environment—they are bradymetabolic and poikilothermic. There are trade-offs to each, as endothermic animals, with high metabolism, may be able to sustain higher activity levels for longer periods (and in colder environments), but they must eat much more than ectothermic animals. And neither category is as strictly defined as outlined above. For example, although most endothermic animals today (mammals and birds) maintain a constant, higher temperature than the lower, variable temperature of reptiles, there are several exceptional species: some small birds lower their body temperatures at night, and the naked mole rat (a mammal) is poikilothermic.

The reason I outlined the four elements of endothermy and ectothermy (poikilothermy, homeothermy, bradymetabolism, and tachymetabolism) is that each defines part of two variables within what are usually called cold- or warm-bloodedness: metabolic rate and temperature variability. Although modern endotherms are generally tachymetabolic and homeothermic, it is entirely possible for a bradymetabolic animal to be homeothermic, and some modern animals in fact are. Were sauropods this way?

Ancient animals can be difficult to fit neatly into the modern categories of endothermic and ectothermic. There has been plenty of debate as to the standard metabolic rate of the dinosaurs (Bakker 1980, 1986; Spotila 1980; Coe et al. 1987; Farlow 1987, 1990; Colbert 1993; Reid 1997, 2012; Seymour and Lillywhite 2000), and although some dinosaurs (especially small, feathered theropods) may well have been tachymetabolic and homeothermic, I doubt all dinosaurs were, and few dinosaurs probably fit the model of endothermy the way modern birds and mammals do. Although tachymetabolic small theropods would benefit from higher activity levels and the ability to operate in lower temperatures that would hinder them if they were ectothermic, tachymetabolic sauropods would only be saddled with significantly increased food-intake requirements and heat-loss problems. Jan Weaver, in a study of *Brachiosaurus*, calculated that intake requirements for that animal were not much greater than those of modern large mammals unless *Brachiosaurus* were endothermic, in which case the daily needs tripled on average (Weaver 1983). As paleontologist Peter Dodson wrote, sauropods have "more to lose . . . than to gain" from being tachymetabolic (Dodson 1990). Sauropods could well have been homeothermic despite being bradymetabolic; they may have had a relatively slow metabolism but maintained a consistent (and probably high) body temperature. The body temperature may not have varied simply as a result of the immense mass of the animal. Because the ratio of surface area to mass goes down as an animal gets bigger, a larger individual will gain and lose body heat at a much slower rate than a smaller individual of the same species. At adult body weights of 10,500 to 47,000 kg (23,148 to 103,617 lb), sauropods of the Morrison Formation likely had plenty of mass to maintain what has been called **inertial homeothermy**—having a body so large that on cool nights and warm, sunny days, the animal's core was simply impervious to significant temperature change. A large, ectothermic dinosaur of this type could have benefited from some of the same effects of endothermy but with a savings of not having to devote nearly as much energy expenditure to heat production, freeing itself to put that energy toward growth.

Another possible contribution to constant body temperature (one that might have increased that temperature overall) is the huge amount of plant material that would have undergone digestion in a sauropod stomach at any given time. As anyone who has felt the warmth coming from a bag of recently cut lawn clippings can attest, the slow decomposition of browse material would have provided a source of heat for the sauropod's body core—what has been called **fermentative endothermy**.

Evidence from bone histology, as we have seen, shows that sauropods grew quickly, but there are also combinations of characters of the bone structure that suggest sauropods (and indeed other dinosaurs) were unlike any modern vertebrate. Fibrolamellar bone, which is formed without interruptions in deposition, is common in large and fast-growing birds and mammals; lamellar-zonal bone is often characterized by "growth rings" (lines of arrested growth, LAGs) formed by periodic and temporary slowing or cessation of growth due to lack of resources and is typical of modern reptiles. Both types of bone are common in dinosaurs, and both types can be found in different bones of the same individual. So dinosaurian physiology is not a simple matter of one modern analog or another. It appears more and more likely that the dinosaurs were unique. The bone histology indicates that sauropods and other dinosaurs grew quickly and continuously for part of their lives but that this growth could be slowed when and where necessary in the skeleton. This also implies that the circulatory and respiratory systems of the sauropods were as efficient as those of modern mammals, and most dinosaurs probably also had fully four-chambered hearts.

Although the growth patterns and other aspects of sauropod paleobiology seem to indicate a physiology unlike that of modern reptiles, and although their body temperatures may have been relatively high and constant, their metabolism was probably between a reptilian low level and the tachymetabolism of modern birds and mammals. They may have been what R. E. H. Reid has called "subendothermic super-reptiles." To paraphrase his assessment, physiologically, dinosaurs were more like mammals and birds than modern reptiles are but more like reptiles than birds and mammals are (Reid 1997, 2012; Werner and Griebeler 2014).

Estimates of speeds of sauropod dinosaurs based on trackways consistently show that sauropods walked most often at speeds of about 4.8 km (3 mi) per hour (Lockley 1991; Lockley and Hunt 1995). Almost no trackways show speeds significantly faster or slower than this, so sauropods seem to have moved around at what for us would be a good walk. On the basis of comparative leg length, the sauropods seem to have taken it easy most of the time. Walking speed estimates for *Brachiosaurus*, made on the basis of limb proportions, are similar at 4.2 km (2.6 mi) per hour, and upper speed limits for sauropods calculated in the same study were 16.9 km (10.5 mi) per hour (Christian et al. 1999). Bone bed assemblages and trackways of sauropods suggest that many species were age segregated with individual herds sometimes composed of adults to subadults or of juveniles (Myers and Fiorillo 2009).

What did sauropods eat? And how did they feed? Most of the plants available to sauropods were ferns, tree ferns, seed ferns, horsetails, cycadophytes, ginkgoes, and various conifers. All Morrison sauropods were herbivorous and would have fed on most of these plant types to varying degrees. Given the different plants' nutrition, growth rates, digestibility, and abundances, sauropods probably concentrated on araucarians, horsetails, cheirolepidiaceaens, ginkgoes, and then other conifers (Gee 2011); ferns and cycads probably were less frequent food sources. Studies of skull-jaw strength, microwear on sauropod teeth, and overall tooth wear of sauropods indicate that different species used slightly different feeding strategies and probably concentrated on different plant types (what is called **resource partitioning**), but we are still not sure what plants individual species focused on (Fiorillo 1998b; P. Christiansen 2000; Button et al. 2014). Riggs first proposed that diplodocid sauropods may have been adapted for feeding in treetops (Riggs 1904), standing on their hind legs and tail. Bakker has been the main proponent of this view in the last 40 years (Bakker 1978, 1986), though others agree too (e.g., Hallett and Wedel 2016), but as we will see in a later section, some recent work suggests instead that diplodocids were designed for feeding low to the ground, a view that I tend to lean more toward.

Although many polished-looking stones can be found on the eroded surfaces of Morrison Formation outcrops, and although many of these are often assumed to be gastroliths of sauropods, there are very few occurrences of gastroliths and sauropod skeletons in direct association. They do occur together rarely, so at least some sauropods had these gastroliths, but the vast majority of isolated stones popularly called gastroliths are probably current-deposited cobbles from the lower parts of overlying

6.18. Relationships of eleven sauropod dinosaurs of the Morrison Formation. Based primarily on Wilson (2002), Whitlock (2011b), and Upchurch et al. (2004a). Dashed lines indicate uncertain placements. *Haplocanthosaurus* has most frequently been placed as a basal diplodocoid but has also been found recently (and in previous analyses) to be a basal macronarian (Royo-Torres et al. 2017, in supplementary information) or even a non-neosauropod (Harris 2006). *Dystrophaeus* is only known to be outside Diplodocoidea (Foster et al. 2016b), and its position is the most uncertain of the Morrison sauropods.

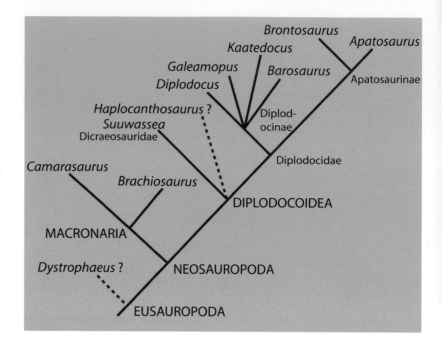

formations. I've seen such stringers of cobbles in mudstones in lower levels of the Lakota Formation in the Black Hills and of the Cedar Mountain Formation in Utah and have never seen such occurrences in the Morrison Formation. Considering this, the direct evidence for an avian-style gastric mill in sauropods is weak (Wings 2015).

The sauropods of the Morrison Formation were traditionally classified in several families: Diplodocidae, Camarasauridae, Brachiosauridae, and Cetiosauridae. This classification worked until relatively recently, but it no longer reflects the phylogeny of the animals (fig. 6.18; see also Upchurch 1998; Wilson and Sereno 1998; Wilson 2002). *Camarasaurus* and *Brachiosaurus* still fit in their respective families (though these have become somewhat tenuous too), but the newly named *Suuwassea* is the first Morrison sauropod that is a "diplodocoid" but outside the family Diplodocidae (probably a dicraeosaurid). All members of Diplodocidae are diplodocoids, but not all diplodocoids (e.g., *Suuwassea*) are members of the Diplodocidae.[2] In addition, most old members of the Cetiosauridae (including the Morrison's *Haplocanthosaurus*) are no longer considered part of that group but have now been found to be more closely related to other species.

Sauropods first appear in the Late Triassic and are rare through the Early Jurassic (Buffetaut et al. 2000); by Middle Jurassic times, they have diversified in faunas known from the United Kingdom and from China, but they reached their peak during the Late Jurassic. Figure 6.18 shows the relationships of sauropods of the Morrison Formation as currently understood. I discuss the Morrison Formation species in an order that roughly reflects that cladogram.

Mass estimates of sauropods were based on circumference measurements of their humerus and femur (J. Anderson et al. 1985; Campione et al. 2014) and then reduced by 10% to account for extensive vertebral pneumaticity (Wedel 2005).

Dystrophaeus viaemalae

This sauropod was found by J. S. Newberry in 1859 about an hour's modern-day drive south of what is now Moab, Utah. The specimen he collected included a fragment of scapula, a complete ulna, half a radius, and three metacarpals (Gillette 1996a; McIntosh 1997); for years, this sauropod was classified as part of an indeterminate family or possibly a diplodocid (von Huene 1904). Rediscovery of the original site by Fran Barnes and additional collecting by him and then–state paleontologist Dave Gillette in 1989 began the process of revealing more about this geologically very old specimen. Found in the lower Tidwell Member of the Morrison Formation (~158 mya), Dystrophaeus is by several million years the oldest well-preserved body fossil of a sauropod from the formation (Gillette 1996b; Foster et al. 2016b). Its relationships to other Morrison sauropods are thus important. Excavations since 2014 have turned up additional material of this animal, including tail and back vertebrae and teeth, the latter of which indicate that Dystrophaeus is not a diplodocoid, but exactly what it is is still not clear (Foster et al. 2016b). It is possibly a macronarian or is the first nonneosauropod from the formation (see fig. 9.1).

Suuwassea emilieae

One of the more recently discovered, described, and named sauropods in the Morrison Formation, Suuwassea (pronounced Soo-OO-wass-ee-a) is from the southern part of Montana (Harris and Dodson 2004) and is part of an apparently unusual northern fauna in the formation, one that may have been better adapted to the wetter conditions that seem to have prevailed up there. The type was a relatively small, subadult sauropod about 15 m (49 ft) in length and was peculiar in having a second small hole in the top of its head, sometimes called a parietal foramen. This foramen behind and between the eyes is unknown in any other Morrison sauropod, and although the animal is similar to Apatosaurus in aspects of its postcranial skeleton, it is in fact less closely related to that genus than are Diplodocus and Barosaurus. Suuwassea is part of the diplodocoid group, which includes Diplodocus, Apatosaurus, Barosaurus, Dicraeosaurus (the African contemporary of the Morrison diplodocids), and a number of others (Harris 2006). And although the relationships of Suuwassea within this group are unclear, it appears to be a basal dicraeosaurid (Whitlock 2011b; Hedrick et al. 2014), a group otherwise known only from South America and Africa, although there has been a recent suggestion that a long-neglected specimen described by Marsh in 1889 may be a previously unrecognized, second dicraeosaurid taxon (Whitlock and Wilson 2018).

Diplodocus

DIPLODOCUS LONGUS

This original type species of the genus *Diplodocus* was found by Marsh's crews at the Marsh-Felch Quarry near Cañon City, Colorado. The type specimen now consists of only a few tail vertebrae (fig. 6.19) and a few other elements (best illustrated by Tschopp et al. 2018c), although in the field the vertebrae were part of an articulated series (McIntosh and Carpenter 1998). Nearly complete skeletons of this genus have been found since then, including two individuals from one pit forming the type of *D. carnegii* from Sheep Creek, Wyoming (CM 84 and 94), and several referred specimens of *D. longus* from Marsh-Felch and from Dinosaur National Monument.

Diplodocus is a well-known genus with unique aspects of its skeleton that help clearly distinguish it from most other sauropods of the Morrison Formation in most parts of its anatomy. *Diplodocus* is characterized by a long, lightly built, narrow skull with external nares (the nasal openings in the skull) on top between the eyes; slender, pencillike teeth restricted to the fronts of the upper and lower jaws; moderately long cervical vertebrae with neural spines V-shaped in anterior or posterior view; cervical ribs about the same length as the centrum; anterior dorsal vertebrae with short, V-shaped neural spines and posterior dorsals with I-beam-shaped and tall neural spines; tall sacral neural spines; anterior caudals with deep pleurocoels and ventral grooves on the centra, winglike caudal ribs, and tall neural spines; elongate midcaudal vertebrae with pleurocoels and ventral grooves; chevrons elongated both anteriorly and posteriorly; and relatively slender pelvic and limb elements.

6.19. Type caudal vertebrae of the sauropod *Diplodocus longus* (YPM 1920) from the Marsh-Felch Quarry, Colorado. Scale bar = 20 cm. *Photo by author. (Courtesy of the Peabody Museum of Natural History, Division of Vertebrate Paleontology, Yale University, peabody.yale.edu.)*

Diplodocus was about 28 m (92 ft) long and probably weighed about 12 metric tons (fig. 6.20). My calculations, which are based on the Smithsonian's specimen USNM 10865, indicate a mass of 11,391 kg (25,060 lb). Groups of *Diplodocus* likely moved around the open Morrison floodplain in small herds, feeding on low-growing plants such as ferns, horsetails, and cycadophytes and occasional midlevel plants such as tree ferns, seed ferns, ginkgoes, and small araucarians. Comparisons of skull muzzle shape and tooth wear indicate that *Diplodocus* and *Apatosaurus* were likely lower-level **browsers** whereas *Camarasaurus* and *Brachiosaurus* were higher-level browsers (Fiorillo 1998b; Whitlock 2011c). Matt Bonnan's work on sauropod limbs suggests that *Diplodocus* may have had a longer walking stride than its contemporaries *Apatosaurus* and *Camarasaurus* (Bonnan 2004). *Diplodocus* would have been able to raise its head several meters above the ground but probably kept it most of the time within about 3 m (10 ft) of the surface, feeding on all the vegetation right in front of it.

Paleontologist Mike Parrish and computer programmer Kent Stevens have modeled the neck vertebrae of a number of different sauropods and found that the necks of some of these animals could not flex as much in the upward direction as we had supposed. It appears that *Diplodocus* could raise its head only to 4.3 m (14 ft) above the ground and in neutral pose would have been just 80 cm (31 in.) off the surface (Stevens and Parrish 1999, 2005a, 2005b). This suggests that most feeding done by *Diplodocus* would have been close to the ground on low-growing horsetails, ferns, and other plants. This seems to agree with the results of sauropod tooth-wear studies (e.g., Fiorillo 1998b). However, comparisons with modern animals and their necks suggest that sauropod necks may have in fact been carried maximally extended (i.e., up) rather than in neutral pose (Taylor et al. 2009; Taylor and Wedel 2013a), although this is also not entirely agreed on (Stevens 2013); the habitual orientation of the necks in sauropods is something still quite contested. Perhaps the neutral or habitual pose is less important than the range of motion, as it relates to feeding. Possibly, sauropods carried their necks a bit higher than within a few feet or meters of the ground but not likely up vertically like a swan. Regardless, we can agree that *Diplodocus* and its relatives were

6.20. Restoration of the sauropod *Diplodocus*. Dermal spines are indicated by diplodocid specimens at the Howe Quarry in Wyoming. Total length ~28 m (91.9 ft). *Drawing by Thomas Adams.*

feeding differently from each other and especially from camarasaurs and brachiosaurs.

Diplodocine sauropods (diplodocids closer to *Diplodocus* and *Baro-saurus*, for example, than to *Apatosaurus*; Taylor and Naish 2005) were probably low-browsing herbivores feeding relatively close to the ground, although they certainly could reach reasonably high when they wanted to; they were seemingly among a rather diverse **guild** of low-browsing dinosaurs in the Morrison Formation (Rees et al. 2004). Studies of biomechanical stress in the skull of *Diplodocus* suggest that it was sufficiently built for biting and stripping branches and other vegetation (M. Young et al. 2012).

It has been suggested that diplodocids were specifically adapted for rearing up on their hind legs to feed on treetops, but I see no reason for *Diplodocus* to have wasted its energy by habitually walking from stand to stand of trees, rearing up, and feeding. Such a level (~20 m [65 ft] up) would have had a lower density, overall biomass, and diversity of plant material than the first few meters above the ground. *Brachiosaurus* was much more elegantly designed to exploit the high feeding niche and was much less common, which argues against rich browse up high. It also could simply stroll around feeding, rather than having to continuously raise itself up and down like a drawbridge to reach its food. Two of the characteristics of the diplodocid skeleton that have been proposed as adaptations for rearing up during feeding are the tall neural spines over the sacrum and the relatively short front limbs, but these are likely adaptations for suspension support of an elongate neck and tail and a primitive condition inherited from short-forelimbed ancestors, respectively. It isn't that that diplodocids never would have reared up (they would have occasionally), but I think it much more likely that they would have saved their energy and fed mostly on the low-growing vegetation all around them.

Given that the teeth of *Diplodocus* were small and restricted to the front of the jaws (fig. 6.21), it seems that the animals would have stripped off plant material and swallowed it without chewing. We find isolated teeth of diplodocids in the rocks of the Morrison Formation, and they are often worn heavily, the blunt tip of the natural tooth shape now nearly flat. So whether the teeth were being worn down by contact with other teeth while biting off mouthfuls of food or whether the wear is strictly from stripping off tough, gritty food, this wear does not result from actual chewing of the food in the mouth. Food was stripped and swallowed by sauropods, and most breakdown of the material occurred in the stomach.

Footprints of diplodocid sauropods indicate that these animals often walked in the soft, wet mud of the floodplains not far from rivers, as well as along the sandbars of rivers and the shorelines of lakes. Sauropods also have been found as fossils in soils that indicate better-drained parts of the floodplain, and these large reptiles also may have moved in groups around the dry parts of the landscape.

Skulls of diplodocids have recently been studied in more detail, with indications emerging that some of the bones of the snout may have been

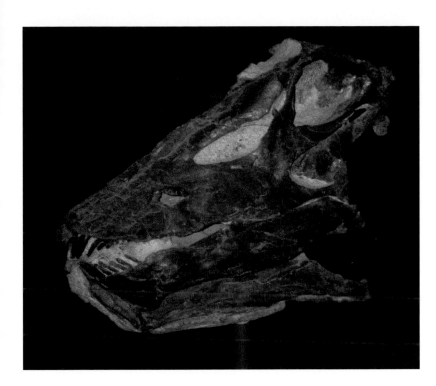

6.21. Left lateral view of the skull of *Diplodocus*, showing thin, pencillike teeth restricted to the anterior part of the jaws (USNM 2673). This specimen was recently referred to *Galeamopus pabsti* by Tschopp and Mateus (2017). *Courtesy of the Smithsonian Institution. Photograph by author.*

loosely joined so that they could slide along the joint contacts slightly, which would have added minor flexibility to the snout, something that may have come in handy during feeding (Tschopp et al. 2018a). Based on skull specimens from Wyoming, it has also been proposed that at least some diplodocines had possibly cartilage-based projections over the eyes ("sun visors," you might say) and beaks bracing the tooth row (Tschopp et al. 2018b). If diplodocines had beaks just outside the row of their pencillike teeth, this might help explain why in several known cases, rows of diplocodine teeth (and nothing else) are preserved in life position in sandstone, for example. A recently described skull of a very young diplodocine indicates that juveniles of these taxa had shorter muzzles, relatively larger eyes, and teeth not restricted to the front of the jaws (Woodruff et al. 2018). Somewhat surprising about these teeth farther back in the jaw, away from the front of the mouth, is that they appear to have been less pencil shaped (and thus more like those of nondiplodocid sauropods) than the teeth in the front and than those of adults. However, from what we see in modern embryos, a young individual temporarily demonstrating traits more like those of ancestors than those of adults of their own species is not exactly shocking either.

DIPLODOCUS CARNEGII

The sauropod species that the Carnegie Museum crews excavated from Sheep Creek, Wyoming, in 1899 is one of the best known of the early *Diplodocus* specimens (fig. 6.22; Hatcher 1901; McIntosh 1990b), much better than the type of *D. longus* (although several referred specimens

6.23. Skeleton of *Diplodocus* (possibly *D. hallorum*) (USNM 10865) as mounted previously at the National Museum of Natural History (Smithsonian) in Washington, DC.

of *D. longus* are reasonably well preserved). In fact, it was recently proposed that *D. carnegii* be designated the new type specimen of the genus *Diplodocus* (Tschopp and Mateus 2016; Tschopp et al. 2018c), although this has not been universally supported (Carpenter 2017). Most of the differences between *D. carnegii* and *D. longus* are minor and likely due to individual variation, and thus the validity *D. carnegii* as a name is possibly questionable, despite its being a better specimen. There may well be just one well-known species of *Diplodocus* represented in the Morrison Formation. In his examination of sauropod species differences, however, Jack McIntosh (1990b) retained *D. carnegii* because it was based on such good material (Hatcher 1901). The morphotypes represented by USNM 10865 (fig. 6.23) and AMNH 223, with the "stovepipe" femurs, may be the same single species also, although they have most recently been tentatively assigned to *Diplodocus hallorum* (Tschopp et al. 2015).

DIPLODOCUS HALLORUM*

This species is based on *Seismosaurus* from northwestern New Mexico (fig. 4.9). The partial skeleton consists of several dorsal and caudal vertebrae and most of the pelvis, and it was found in a channel sandstone (Gillette 1991, 1994). Originally described as a new genus and species, it was later placed in *Diplodocus* (Lucas et al. 2006a). The differences with other specimens of *Diplodocus* are rather subtle: neural spines of the caudal vertebrae are taller and more vertical (although this feature depends on the relative position of the vertebrae within the tail series and may vary anyway), the pubis is slightly more massive, and the pleurocoels in the caudal vertebrae are less well developed. The hooklike distal end of the ischium turned out to be the tip of a neural spine taphonomically abutted to the end of the former bone (Lucas et al. 2006a). So there is little separating this specimen from *Diplodocus* and its best-known specimens. Still, Emanuel Tschopp et al. (2015) retain *D. hallorum* and even refer a few other specimens to it (such as DMNH 1494; fig. 6.24).

In habits and diet, *D. hallorum* was likely similar to other diplodocines, although its daily food requirements would have been significantly

6.24. Skeleton of *Diplodocus* (DMNH 1494) as mounted at the Denver Museum of Nature and Science. Traditionally identified as *D. longus*, this was recently tentatively referred to *D. hallorum* by Tschopp et al. (2015). (Six-year-old demonstrating how not to read labels, for scale.)

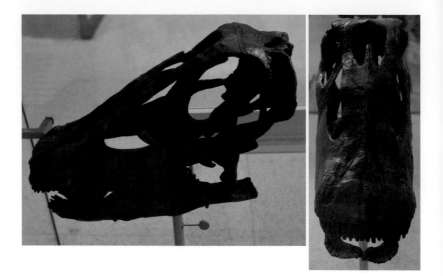

6.25. Skull of the sauropod *Galeamopus* sp. (AMNH 969) as referred by Tschopp et al. (2015). Previously identified as *Diplodocus*. *Photos courtesy of R. Hunt-Foster.*

greater. The type specimen of *Diplodocus hallorum* may have been an individual as much as 33–36 m (108–118 ft) long (Herne and Lucas 2006) and 38,250 kg (84,150 lb) in weight.

Galeamopus

GALEAMOPUS HAYI

This species was based on a single specimen (CM 662; now HMNS 175) from the southeastern edge of the Bighorn Mountains in Wyoming. The partial skull and skeleton was initially named *Diplodocus hayi* and was described by W. Holland (1924), and it is now mounted at the Museum of Natural Science in Houston, Texas (McIntosh 1981). This species is similar in most respects to *Diplodocus* but has distinct differences, particularly in the tail and in the type and referred skull material from the Bighorns and from the Bone Cabin Quarry north of Como Bluff (fig. 6.25; Tschopp and Mateus 2017). It is characterized by having relatively robust forelimbs with a relatively long ulna, plus midcaudal vertebrae with shallow pleurocoels, long and low neural spines, and an unusually deep and narrow groove on the ventral surface of the centrum (B. Curtice, pers. comm., 1998). The species was placed in the new genus *Galeamopus* ("want helmet") by Tschopp et al. (2015) based on characters of the type skeleton and the addition of the referred skull.

GALEAMOPUS PABSTI

Named in 2017, *Galeamopus pabsti* is based on SMA 0011, a partial skeleton from the Howe-Scott Quarry, a site geographically near and slightly stratigraphically above the historic Howe Quarry of Barnum Brown's day. This species is distinguished by details of the skull, cervical vertebrae, limbs, and pelvis, but a reduction in length of the centrum from the last cervical vertebra to the first dorsal (Tschopp and Mateus 2017) is among

Facing, 6.26. Elements of the type specimen of the sauropod *Barosaurus lentus* (YPM 429) from the Morrison Formation at Piedmont Butte, South Dakota. (A) Dorsal vertebrae, Jack McIntosh's hand and tie for scale. (B) Anterior caudal vertebrae. (C) Series of six midcaudal vertebrae in left lateral view. (D) Ventral view of a midcaudal vertebra showing the markedly shallow ventral excavation of the centrum. (E) Partial pubis. Scale bars = 20 cm. *Photos by author. (Courtesy of the Peabody Museum of Natural History, Division of Vertebrate Paleontology, Yale University, peabody.yale.edu.)*

the more significant features. The Smithsonian skull USNM 2673 from the Marsh-Felch Quarry (fig. 6.21) was referred to this species as well (Tschopp and Mateus 2017).

Barosaurus lentus

Marsh named *Barosaurus* on the basis of YPM 429, which was found north of Rapid City, South Dakota, in 1889 (fig. 6.26; Marsh 1890). This species is similar to *Diplodocus*, and in many parts of the skeleton, the two are indistinguishable, but there are distinct differences, particularly in the neck and the number of dorsal vertebrae (McIntosh 2005). Additional partial skeletons of *B. lentus* have been found at the Dinosaur

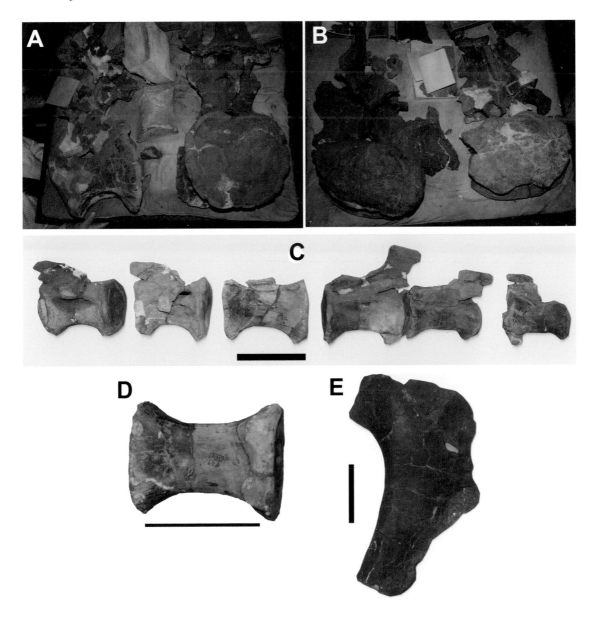

National Monument quarry in Utah and possibly at the Howe Quarry in Wyoming, the Cleveland-Lloyd in Utah, and the Wonderland Quarry several kilometers south of the type locality in South Dakota. A second species named by Marsh as *B. affinis* from the type locality is an indeterminate, smaller diplodocid, possibly also *B. lentus*. An additional species, *B. africanus*, was reported from the Tendaguru deposits in Tanzania, but this referral has not been fully established. A juvenile sauropod found at Dinosaur National Monument's Carnegie Quarry was recently identified as a young *Barosaurus* (Melstrom et al. 2016).

Barosaurus was about 24 m (80 ft) long and weighed about 10,761 kg (23,675 lb) (an estimate made on the basis of my measurements of AMNH

6341, from Dinosaur National Monument). It lived on the Morrison floodplain and fed on low-growing vegetation (ground level up to about 4.6 m [15 ft]), probably including ferns, horsetails, and occasional cycadophytes. Distinguishing *Barosaurus* from *Diplodocus* can be difficult because the two are similar skeletally (and thus presumably phylogenetically). In general, *Barosaurus* had a longer neck with individual vertebrae nearly 50% longer than in *Diplodocus* (fig. 6.27). The trunk and tail were shorter in *Barosaurus*. The dorsal vertebrae of the two diplodocids are nearly identical, but *Barosaurus* may have had only 9 dorsals, whereas *Diplodocus* had 10. The caudal vertebrae are similar, except that those of *Barosaurus* are slightly shorter for the same diameter and have ventral excavations that are generally shallower, the tops of the anterior caudal neural spines do not have a notch as in *Diplodocus*, and the pleurocoels do not occur as far back in the caudal vertebral series as in *Diplodocus*. Other comparisons between the two include the fact that the ischium shaft is slightly thicker in *Barosaurus* and that the forelimb of *Barosaurus* is longer and somewhat more slender. The shoulder region thus may have been slightly higher in *Barosaurus*. Diplodocid specimens from the Howe Quarry in Wyoming (some probably *Barosaurus* or *Diplodocus*) preserve skin impressions that provide evidence that these dinosaurs had a median row of dermal spines along the top of the tail and perhaps the back and neck (Czerkas 1992). The spines do not appear to be composed of bone.

Kaatedocus siberi

This new diplodocine is based on a neck and a disarticulated and associated skull from the Howe Quarry in Wyoming (fig. 6.28; Tschopp and Mateus 2013). The skull elements were disarticulated from the end of the neck and from each other but were nearby. The vertebrae are similar to those of *Diplodocus*, although they have a few differences of varying significance; but the clearly diplodocine skull has a unique posterior edge to the external nares (a U-shaped notch between the frontal bones), unlike any other diplodocid skull out of the Morrison Formation. This taxon shows that the diversity of diplodocids in the Morrison Formation may well have been a bit higher than we expected. In fact, recent experiments on the nutritional value of plants grown under simulated, Mesozoic-like elevated CO_2 conditions suggest that sauropods may have been able to get more energy from the plants of the time than we thought and that this would have allowed greater population densities than previously calculated (Gill et al. 2018); and greater population densities may have made it easier (assuming some niche partitioning) for the Morrison Formation landscape to have accommodated a higher diversity of large herbivores.

Amphicoelias altus*

Amphicoelias is known from two dorsals, a femur, and a few other fragments (fig. 6.29) and is similar to the "stovepipe femur" morphotype of

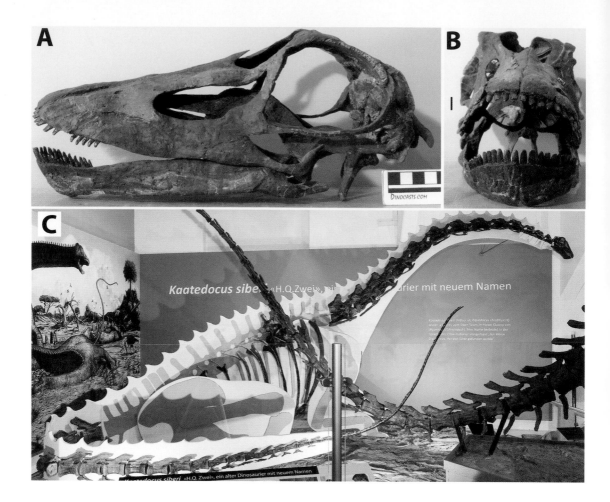

Text within image C: **Kaatedocus sibe**... «H.Q.Zwei»... urier mit neuem Namen

Text at bottom of image C: *Kaatedocus siberi* «H.Q. Zwei», ein alter Dinosaurier mit neuem Namen

Text in image A: DINOCASTS.COM

6.28. Material of the sauropod *Kaatedocus siberi* (SMA 0004) from Howe Ranch, Wyoming. (A) Skull in left lateral view, scale in cm. (B) Skull in anterior view, scale bar = 1 cm. (C) Neck and tail as displayed at the Sauriermuseum Aathal. *Photos in A and B courtesy of Emanuel Tschopp; C courtesy of Kirby Siber.*

Diplodocus. It was named by Cope in 1877 from specimens at Garden Park, Colorado. Some sauropod workers think that *Amphicoelias* is a distinct and valid genus, but I have to disagree and believe instead that it is probably simply a large *Diplodocus*—with so few elements known from *Amphicoelias*, however, it may be impossible to say which of these views is correct.[3] I don't think there is enough material to maintain a solid, separate genus name, and I also don't believe that there is any character in the material we have to contradict its being *Diplodocus*, so in the absence of further evidence, I'd tentatively identify it as "?*Diplodocus*." For now, however, I leave the name *Amphicoelias* assigned just to that one specimen in the American Museum of Natural History in New York; I remain unconvinced by referrals of other material, which often were based almost entirely on the form of the femur (which is of a type seen in several *Diplodocus* specimens). If someone proved, though, that *Amphicoelias* and *Diplodocus* were the same thing, because of the priority rule, we'd be faced with switching all of our museum *Diplodocus* labels to *Amphicoelias*. Luckily, I don't think this is likely to happen. The late sauropod guru (and professional nuclear physicist) Jack McIntosh and I agreed that *Amphicoelias* is a nomen dubium, but Jack believed that

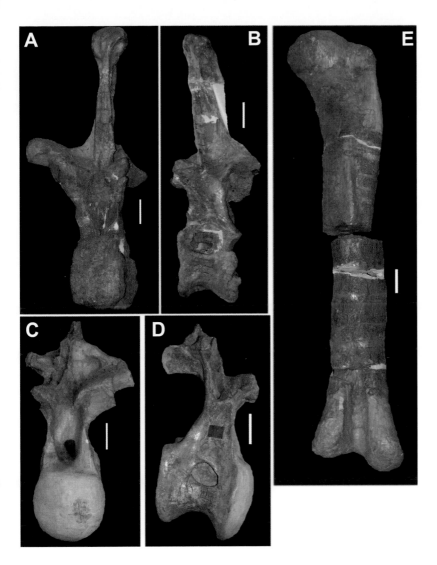

6.29. Type material of the sauropod *Amphicoelias altus*. (A) Posterior dorsal vertebra in anterior view. (B) Same in left lateral view. (C) Dorsal vertebra in anterior view. (D) Same in right lateral view. (E) Femur. Scale bars = 10 cm. *Photos courtesy of Cary Woodruff.*

whatever *Amphicoelias* is, it is not *Diplodocus* because the former has straight, vertical, anteroposteriorly deep neural spines on one of the dorsal vertebrae (unlike those of *Diplodocus*, which are inclined forward and are narrow). Here, Jack and I simply had a difference of opinion. I tend to assume that the differences between the dorsal neural spines of *Diplodocus* and *Amphicoelias* are due to individual variation until we have more specimens of the latter, all demonstrating the same (and preferably additional) differences from the former.

Maraapunisaurus fragillimus

A separate species of *Amphicoelias*, "*A.*" *fragillimus*, was named by Cope on the basis of a diplodocoid neural arch and spine that is fully twice the size of the same element in *Apatosaurus* or *Amphicoelias altus*. McIntosh's reevaluation of "*A.*" *fragillimus* produced estimates that the animal may have been 49 m (162 ft) long, and a study by Carpenter (2006) suggested it could have been an astounding 58 m (190 ft) long! The femur alone may have been nearly 3 m (9.8 ft) long. That's far larger than any other Morrison dinosaur. Unfortunately, the material on which Cope based this animal is now lost, so we cannot be sure of its true size. It is possible, with the type material missing, that Cope's published size measurements were a misprint of some kind (Woodruff and Foster 2014), but the facts that Cope never corrected the publication and that his rival Marsh never questioned it both suggest that, even if the measurements were off, they likely were close and that this one bone was of the largest individual dinosaur ever found in the Morrison Formation.

However, a recent reanalysis of "A." *fragillimus* by Carpenter (2018) found, quite unexpectedly, characters of Cope's original material that indicated that this specimen represented the first Jurassic rebbachisaurid diplodocoid sauropod and the first in North America. The specimen was renamed *Maraapunisaurus fragillimus*, a new genus name that means "huge lizard" in the native Ute language (with some Greek thrown in). This same study also found that *Maraapunisaurus* was probably about 32 m (105 ft) long and about 8 m (26 ft) high at the hips, smaller than originally estimated but still deserving of its name "huge." Having a rebbachisaurid present in North America in the Late Jurassic suggests that the group may have originated here and dispersed through Europe and Africa to South America (Carpenter 2018).

*Dyslocosaurus polyonychius**

This specimen was collected by F. B. Loomis in the Lance Creek area of Wyoming and consists of a partial foot and lower hind limb. Without better locality data, it is difficult to be certain of the specimen's age, but given that it is a sauropod and that there are at least limited Morrison Formation outcrops in that part of Wyoming, it is probably from the Upper Jurassic. The genus and species were designated by McIntosh et al. (1992), who suggested that it was a diplodocid distinguished by having at least four and maybe five claws on the foot (instead of the usual number of three for diplodocids), although this latter interpretation has been questioned recently. Tschopp et al. (2015) indicated that this specimen may be a dicraeosaurid, but I treat it as an indeterminate diplodocoid here.

Supersaurus vivianae

The Dry Mesa Quarry has yielded a number of large sauropods, but this is the best known. *Supersaurus* was a large diplodocid sauropod (fig. 6.30; Jensen 1985), possibly related to *Barosaurus*; it may have been close to 38 m (125 ft) long and 36,180 kg (79,596 lb) in weight. It had possibly the longest neck of all Morrison Formation dinosaurs, at an estimated 15 m (49 ft) in length (Taylor and Wedel 2013b)! Except for its large size, little in the skeleton that Jim Jensen described from Dry Mesa at the

Facing, 6.30. Type and referred specimens of the sauropod *Supersaurus*. (A) Cast of the type scapula-coracoid from the Dry Mesa Quarry. (B) Type specimen cervical vertebra (BYU Museum of Paleontology specimen, BYU 9024) in left lateral view. (C–G) Parts of a referred specimen from the Jimbo Quarry near Douglas, Wyoming. (C) Cervical vertebra. (D) Cervical vertebra. (E) Dorsal vertebra. (F) Fibula. (G) Caudal vertebrae and chevrons. C–G are as displayed at the Wyoming Dinosaur Center, Thermopolis, a 501c(3) nonprofit organization.

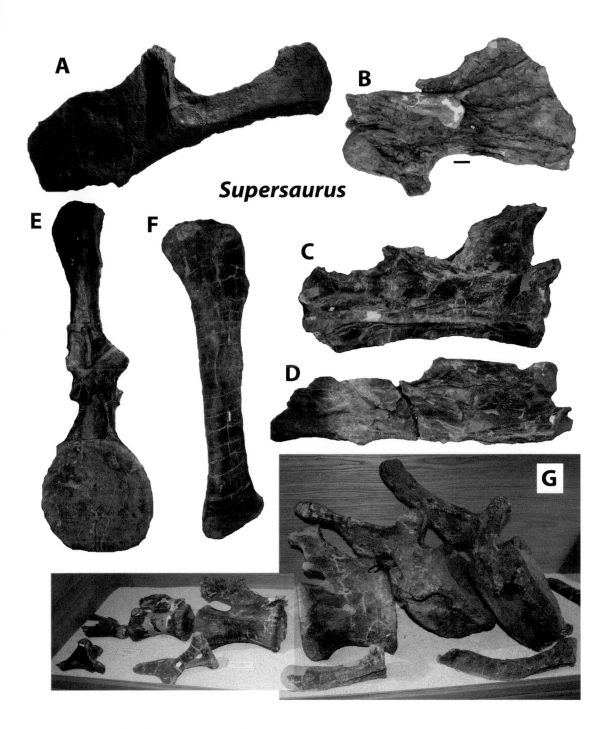

Supersaurus

time distinguished it from *Barosaurus*; the main characters Jensen used to demonstrate the uniqueness of *Supersaurus* were based on the size of the scapula-coracoid (shoulder blade; fig. 6.30), a pair of bones that is in fact quite similar to those of *Diplodocus* and *Barosaurus*. It seemed possible that this skeleton was simply a large individual of *Barosaurus*. The midcaudals that Jensen himself referred to *Supersaurus* are much like those of *Diplodocus* and *Barosaurus*, but there is another series from Dry

Mesa that are suspected to belong to *Supersaurus* instead, and these are very similar to those of *Apatosaurus* and *Brontosaurus*.

Enter Jimbo. This partial skeleton was found associated in a quarry near Douglas, Wyoming (in an apparent fire-induced debris flow; Lovelace 2006), and was eventually identified as second specimen of *Supersaurus* that helped fill in some of the picture (fig. 6.30; Lovelace et al. 2007).[4] Like the type specimen, this *Supersaurus* from near Douglas (now on display at the Wyoming Dinosaur Center in Thermopolis) is huge, with long, wide ribs nearly 2.7 m (9 ft) long and very large vertebrae. Interestingly, the cervical vertebrae of the neck are very long (somewhat like *Barosaurus*), and the giant caudals look not entirely unlike those of apatosaurs. Perhaps not surprisingly then, *Supersaurus* has been variously identified as either an apatosaurine (Lovelace et al. 2007) or a diplodocine (Tschopp et al. 2015).

Apatosaurus/Brontosaurus

Apatosaurines appear to be an endemic group of diplodocid sauropods that are unique to North America (Foster and Peterson 2016). They have a long and complex history of discovery in the formation that started when the first elements found were named *Apatosaurus ajax* by Marsh in 1877 on the basis of a relatively young specimen from the lower dark-mudstone level of Lakes's Quarry 10 at Morrison, Colorado (Marsh 1877). This specimen consists of a three-vertebra sacrum, caudal vertebrae, one cervical vertebra, a tibia, a coracoid, and several other elements of a large sauropod, but it is not quite as complete as later specimens from Como Bluff and Dinosaur National Monument. Several other species named by Marsh from the same area are invalid or otherwise referable to *A. ajax*, and so this is the primary example of *Apatosaurus*. In 1879, Marsh named *Brontosaurus excelsus* on the basis of a nearly complete specimen from Reed's Quarry 10 at Como Bluff (Marsh 1879a). (Yes, that's a separate Quarry 10 from the one in Colorado.) This is still one of the best sauropod skeletons known from the Morrison and the one still mounted at the Yale Peabody Museum (fig. 3.7). *Brontosaurus*, because of its more complete nature, went on to become much better known to paleontologists and nonpaleontologists than *Apatosaurus*, but in 1903 Riggs determined that the two animals were the same and formally, if briefly, synonymized the names. Because Marsh named *Apatosaurus* first, that name has priority, meaning it is the formally recognized name. But in 2015, Emanuel Tschopp, Octavio Mateus, and Roger Benson restudied the Morrison diplodocids and proposed that in fact both *Apatosaurus* and *Brontosaurus* are valid genera with a few species assigned to each. In this classification, *Apatosaurus* includes *A. ajax* and *A. louisae*, and there are three species in *Brontosaurus*, *B. excelsus*, *B. parvus*, and *B. yahnahpin* (Tschopp et al. 2015). *Apatosaurus* and *Brontosaurus* have also been argued to have been separate genera based on anatomical differences between two skulls tentatively assigned to each (Bakker and Mossbrucker 2016).

I have to admit to preferring the name *Brontosaurus* for most of the specimens we know from the Morrison, but that's not how priority rules work; we will have to see if the "resurrection" of *Brontosaurus* holds, but the Morrison may prove to have had at least two genera of apatosaurines. I tend to place genus (and often species) designations a little lower on the tree than many of my colleagues do, allowing for a certain degree of individual variation in the skeletons, so what Tschopp and his coauthors have as the subfamily Apatosaurinae, including *Apatosaurus* and *Brontosaurus* and five species between them, is equivalent to what I would see as the genus *Apatosaurus* (possibly with fewer than five species).[5] The truth is that for most species of apatosaurs we have very few specimens, so we have no real handle on what the variation was within the populations; this small sample size and our tendency to treat potentially continuous variation in skeletons as representing discrete characters mean we may be doing nothing more, in a worst-case scenario, than documenting variation within one species (Foster 2015a). But we don't know that for sure. So, until or unless someone publishes a formal opposing view, arguing with data to back them up, that *Apatosaurus* and *Brontosaurus* really are the same genus and that Riggs was right more than 110 years ago, the classification of Tschopp et al. (2015) is the most up-to-date one we have, and so it is the one I use here. (And at 297 pages, it is thorough!).

For years, several museum mounts of *Brontosaurus* or *Apatosaurus* had what we all would recognize as a *Camarasaurus* skull on the end of the neck. The reason for this goes back to when Marsh reconstructed the skeleton of *B. excelsus* in a drawing. No skull was found with that specimen from Quarry 10 at Como Bluff, so in the drawing Marsh used the partial skull of a *Camarasaurus* from Quarry 13, some 6.4 km (4 mi) east along the bluff. At this time, the whiplash tail shared by *Diplodocus*, *Brontosaurus*, and *Apatosaurus* was unknown in either of the latter; the closer similarity in foot structure between *Diplodocus* and apatosaurs was not realized because no complete feet of an apatosaur had yet been found; and the caudal vertebrae of apatosaurs that were known were those that are in fact somewhat similar to *Camarasaurus* and more so than to *Diplodocus* or *Barosaurus*—so Marsh apparently thought *Brontosaurus* was related to *Camarasaurus* and restored *B. excelsus* that way. And so this became the standard view. But later paleontologists began to suspect that apatosaurs were not as close to *Camarasaurus* as Marsh had thought, and by the time the Carnegie Museum collected two good skeletons of *Apatosaurus* from what was to become Dinosaur National Monument, it was clear that in fact apatosaurs had a *Diplodocus*-like whiplash tail and similar "double-beam" chevrons. The Carnegie crews also found a skull (similar to that of *Diplodocus* but wider) just 4 m (13 ft) from the end of their first *Apatosaurus* skeleton, and the skull fit the neck perfectly. So when William Holland described *A. louisae* in 1915, he suggested that *Apatosaurus* really possessed a wide, *Diplodocus*-like skull. H. F. Osborn of the American Museum, however, said (more or less) that Holland was crazy; Holland backed off, at least publicly, and thus the Carnegie

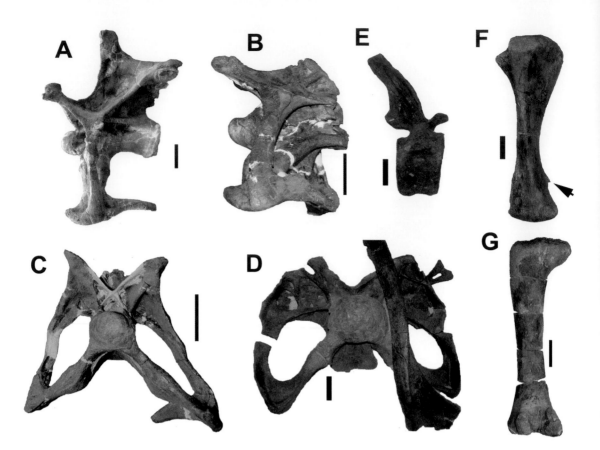

6.31. Elements of apatosaurs from western Colorado and eastern Utah. (A) Cervical vertebra 7 of *Brontosaurus parvus* (BYU 18531) from the Mill Canyon Quarry north of Moab, Utah, in left lateral view. (B–G) *Apatosaurus* from the Mygatt-Moore Quarry in western Colorado. (B) Anterior cervical vertebra in left lateral view (MWC 3829), showing short cervical ribs. (C) MWC 3829 cervical vertebra in anterior view. (D) Posterior cervical vertebra, dorsal rib, and chevron, as preserved (MWC specimen). (E) Right lateral view of an anterior caudal vertebra (MWC 2847). (F) Ulna showing a pathological spur of bone near the distal end (arrow) (MWC 5044). (G) Large femur measuring 1.876 m (6.15 ft) in length. Scale bars in A–F = 10 cm; scale bar in G = 25 cm.

Museum's *Apatosaurus louisae* stood in the dinosaur hall in Pittsburgh for 20 years with no head at all. It wasn't until Holland died that museum staff later put a skull on the skeleton—a *Camarasaurus* skull! Finally, in 1978, following various lines of evidence, the Carnegie Museum's David Berman and sauropod expert Jack McIntosh showed that indeed Holland had been right all along and that the skull of *Apatosaurus* was like that of *Diplodocus*, just a bit wider, and was nothing like that of *Camarasaurus* (Berman and McIntosh 1978). Eventually, the museum mounts of *Apatosaurus* and *Brontosaurus* were changed over to the correct skull, including the Carnegie's specimen (certainly much to the relief of Holland's ghost) and even that in the American Museum.

Apatosaurs were very large animals, whether you are talking about the original *Apatosaurus ajax* or Como's *Brontosaurus excelsus* or (especially) Dinosaur National Monument's *Apatosaurus louisae*. By measuring the circumference of the humerus and femur in different sauropod skeletons, we can estimate the weights of the animals. Measurements of YPM 1980 indicate that *Brontosaurus excelsus* weighed approximately 21,822 kg (48,008 lb), but measurements of CM 3018 show that the robust morphotype of apatosaur, *A. louisae*, weighed about 30,631 kg (67,389 lb)! The whole animal was often about 21 m (69 ft) long as an adult, and some were up to 25 m (82 ft). Apatosaurs are some of the most common

dinosaurs in the Morrison Formation and have been reported from every state in which the formation is exposed; it is the only Morrison dinosaur taxon, as represented by bone, reported from Arizona (Curtice 1999). The limb bones and vertebrae of apatosaurs are relatively robust, and the cervical ribs are short (fig. 6.31); in fact, it has been hypothesized that the robust neck bones of apatosaurs may have been in part strengthening for giraffe-like, intraspecific neck-butting competition (M. Wedel and M. Taylor, pers. comm., 2016). Dodson et al. (1980) noted that apatosaurs may be found with others of their genus less often than some other sauropods, suggesting that they may have been more solitary animals. In this case, *Apatosaurus*'s greater bulk than the similarly long *Diplodocus* could have been a result of selection for individual size protection against predators, as opposed to herd protection. Another possibility is that the bulky bodies of apatosaurs were in part an adaptation for knocking down vegetation such as trees during feeding; apatosaurs are particularly abundant at the Mygatt Moore Quarry, which has evidence (both palynological and macrofossil) of abundant and diverse conifers, so it is possible that these animals preferred wooded habitats. If so, they may have been colored for such settings, especially as juveniles (fig. 6.32).

Recent finds at the Mygatt-Moore Quarry in Colorado and restudy of older finds in Wyoming indicate that apatosaurs had a greater number of replacement teeth stored in their jaws at any given time than did diplodocines (McHugh 2018; J. Peterson et al. 2018), suggesting that apatosaurs may have been eating tougher vegetation requiring more frequent tooth replacement. Microwear studies of diplodocid teeth (e.g., Fiorillo 1998b) have not yet identified significant variation in the pits and scratches of the wear surfaces within the family (just between them and camarasaurs), but such differentiation may prove to be there with more study. Or perhaps the apatosaurs' need for more teeth per socket relates to more frequent tooth loss while pulling at tree branches, if these animals did in fact frequent the woods. Actual stripping of the leafy vegetation from those branches may not have resulted in significantly different microwear (on the tooth wear facets) from that of diplodocines. We don't yet know what differences there were in diet between diplodocines and apatosaurs, but the apparent differences in muzzle width and now replacement tooth

6.32. Restoration of the sauropod *Apatosaurus*. Total adult length ~22 m (72 ft). Color patterns in most dinosaurs are purely speculative; here the pattern is for a young individual, assuming camouflage for wooded areas of the Morrison floodplain. *Drawing by Thomas Adams; coloring by Matt Celeskey.*

number suggest that apatosaurs were specializing on something that was, in some way, tougher to deal with.

Some juvenile apatosaur specimens have been found in a number of quarries, including Dinosaur National Monument, the Mygatt-Moore Quarry, Stovall's Quarry 1, and the Bone Cabin Quarry (Foster 2005b). One of the best specimens, however, was collected from the Sheep Creek area in Wyoming and described as *Elosaurus parvus* (O. Peterson and Gilmore 1902). This very small individual in fact represents a baby *Brontosaurus* and is one of the few partial baby sauropods ever found in the Morrison. It has since been identified as a separate species (*B. parvus*) by Paul Upchurch et al. (2004b) and then by Tschopp et al. (2015).

Studies of bone histology, as we have seen, have shown some interesting growth patterns for dinosaurs. Like some other dinosaurs, apatosaurs seem to have grown relatively quickly to near-adult sizes and then slowed considerably in old age. The study by Kristina Curry-Rogers of the Science Museum of Minnesota, which was based on juvenile material from Cactus Park in Colorado and adult material from Sheep Creek in Wyoming, suggests that young apatosaurs grew to 34% of adult size in about 5 years and 56% of adult size in approximately 10 years (Curry 1999). These age estimates, however, have recently been suggested to be underestimates; growth may have slowed down somewhat later than after 8 to 10 years and continued at a low rate for most of the animals' lives (Chinsamy-Turan 2005).

Vertebral pneumaticity is not as developed in apatosaurs as it is in diplodocines such as *Diplodocus* and *Barosaurus*; although the presacral vertebrae are well pneumatized in apatosaurs as in other diplodocids, the tail vertebrae generally lack the pneumatic fossae regularly present in the diplodocines. But the fact that some apatosaur specimens have pneumatic fossae on some caudals (and sometimes on one side but not the other) suggests that the air-sac system still was present in the tail even in apatosaurs (Wedel and Taylor 2013) and that characters based on the presence or absence of pneumatic fossae or in what positions they occur may be suspect.

APATOSAURUS AJAX

The genotype of *Apatosaurus* was the specimen found at Lakes's Quarry 10 at Morrison, Colorado, in 1877. A few specimens have been referred to this species since; two of the more complete were specimens from near Thermopolis, Wyoming (Upchurch et al. 2004b), and from north of Como Bluff (fig. 6.33).

APATOSAURUS LOUISAE

This species is based on CM 3018 from Dinosaur National Monument, now mounted at the Carnegie Museum in Pittsburgh (fig. 6.34). This was the first skeleton found by Earl Douglass when he found the Carnegie Quarry in 1909 (Holland 1915), and it was described in a monograph a

6.33. Skeleton of the sauropod *Apatosaurus ajax* (AMNH 460) from Wyoming, as mounted at the American Museum. *Photo courtesy of R. Hunt-Foster.*

6.34. Skeleton of *Apatosaurus louisae* (CM 3018) from Dinosaur National Monument, as mounted at the Carnegie Museum (seemingly conversing with *Diplodocus*, upper left). *Photo by author. (Courtesy of the Carnegie Institute, Carnegie Museum of Natural History.)*

few decades later (Gilmore 1936). As mentioned above, it is a fairly large and robust specimen. It is named after Andrew Carnegie's wife, Louise.

BRONTOSAURUS EXCELSUS

Marsh's best apatosaur skeleton, *B. excelsus*, came from Reed's Quarry 10 at Como Bluff (fig. 3.7). One of the first sauropods ever mounted, YPM 1980 is still one of the most beautiful displays of a diplodocid anywhere (humble opinion). Maybe that's personal bias because of the history behind it or because it towers over you, but I never tire of staring at that

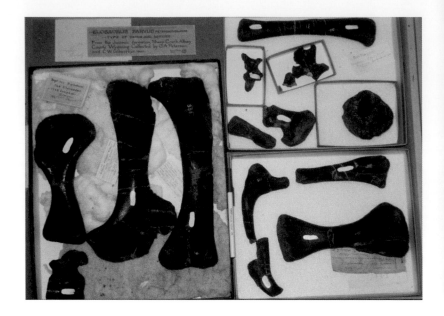

6.35. Elements of the type material of the sauropod *Brontosaurus parvus* (CM 566) from Sheep Creek, Wyoming, at the Carnegie Museum. Pen is approximately 15 cm long. *Photo by author. (Courtesy of the Carnegie Institute, Carnegie Museum of Natural History.)*

specimen. (In fairness, I could stare at CM 3018 nearly as long.) This type specimen for the genus *Brontosaurus* has historically been the example of what an apatosaur is.

BRONTOSAURUS PARVUS

When *Elosaurus parvus* was named from the juvenile material from Sheep Creek, Wyoming, in 1902, perhaps no one would have predicted that it would eventually be viewed, in turn, as a junior synonym of *B. excelsus*, then *A. excelsus*, then *A. parvus*, and eventually *B. parvus*. The main specimens of this species of seemingly everyone's favorite sauropod genus include the type juvenile from Sheep Creek (CM 566; fig. 6.35) and an adult specimen mounted at the University of Wyoming (UW 15556; fig. 6.36), also from Sheep Creek (Upchurch et al. 2004b; Tschopp et al. 2015).

BRONTOSAURUS YAHNAHPIN

Down low in the Morrison Formation, from a site on the east end of Como Bluff, apatosaurs are represented by an older and slightly more primitive species named *B. yahnahpin*. This was originally named as a species of *Apatosaurus* (Filla and Redman 1994). Bakker recently set this species in its own genus, called *Eobrontosaurus*, as *E. yahnahpin* (Bakker 1998a), but its referral to a species within *Brontosaurus* came in 2015 (Tschopp et al. 2015). By either genus name, it appears to be a more primitive species of apatosaur, distinguished from other apatosaurs in having a slightly more expanded distal end of the scapula, a more oval-shaped coracoid, and slightly longer cervical ribs.

Brontosaurus yahnahpin was found in what was during the Late Jurassic a muddy overbank deposit in which the animal appears to have

6.36. Referred skeleton of *Brontosaurus parvus* (UW 15556) from Sheep Creek, Wyoming, at the Geological Museum (University of Wyoming).

become mired. It was found as a single disarticulated but associated skeleton with a couple of bones of an ornithopod and a crocodilian. An apatosaur specimen from Cabin Creek near Gunnison, Colorado, appears to be from nearly the same stratigraphic level as *B. yahnahpin* (if such long-distance correlations are accurate; we don't know for sure yet), but it is unclear whether the Cabin Creek specimen can be referred to the species because, of the characteristic bones, only the scapula is preserved, and its distal end is only moderately expanded. (A degree of distal expansion does occur in individual specimens of apatosaurs from higher up in the formation.) The more common apatosaur species (*A. ajax*, *B.*

excelsus, and *A. louisae*) were found much higher stratigraphically than *A. yahnahpin*; all three were in a roughly similar level of the mid-upper Brushy Basin Member or equivalents. The apatosaur specimen from the top of the Brushy Basin Member at Arches National Park in Utah may be the stratigraphically highest of the genus in the formation, apparently at the level of Cope's quarries at Garden Park (Foster 2005c). Unlike the *Camarasaurus supremus* specimens from those sites, however, the Arches apatosaur is no larger than other members of its genus from lower in the formation. Unfortunately, not enough of this specimen is yet known to be able to determine to what species it should be assigned.

Haplocanthosaurus priscus

Haplocanthosaurus priscus was first found at the Marsh-Felch Quarry in Garden Park (fig. 6.37) and is one of the smallest sauropod dinosaurs in the Morrison Formation (Hatcher 1903). A second species, *H. utterbacki*, was named from the same site and is included in *H. priscus*.

6.37. Haplocanthosaurids. (A–G) Relatively small sauropod from the lower Morrison Formation at Garden Park, Colorado, *Haplocanthosaurus priscus* (CM 572). (A and B) Dorsal vertebrae demonstrating the characteristic high neural arches and dorsolaterally swept transverse processes. (C) Dorsal vertebra in lateral view. (D) Caudal vertebra in anterior view. (E) Same in posterior view. (F) Caudal in right lateral view. (G) Sacrum and ilia in left lateral view. (H) Unidentified haplocanthosaurid from the Salt Wash Member west of Dinosaur National Monument (FHPR 1106), two dorsals in left lateral view. Scale bars = 10 cm. *All photos by author. A–G courtesy of the Carnegie Institute, Carnegie Museum of Natural History; H courtesy of the Utah Field House of the Natural History Museum.*

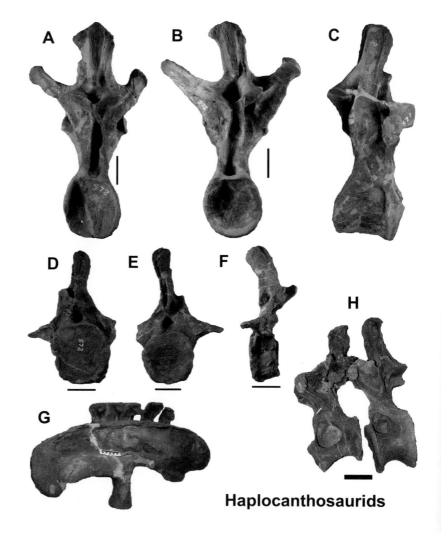

Haplocanthosaurids

Haplocanthosaurus has since been found at several other quarries in the formation, including a couple of sites at the foot of the Bighorn Mountains of Wyoming, near Snowmass, Colorado, and possibly Montana (Erickson 2014; Foster and Wedel 2014), although generally it is rare. One of the best haplocanthosaur specimens ever found was collected from just west of Dinosaur National Monument near Vernal, Utah, in 1999 (fig. 6.38; Bilbey et al. 2000). Haplocanthosaurs appear to be restricted to the middle to lower Morrison Formation. No skull is known for *Haplocanthosaurus*, but the skeleton is characterized by dorsal vertebrae with relatively tall neural arches and unsplit neural spines, relatively small centra, and transverse processes angled dorsolaterally; sacral vertebrae with short neural spines; and simple caudal vertebrae with short neural spines and large chevron facets. In most recent analyses, *Haplocanthosaurus* appears to be a diplodocoid—or it may be the sister taxon to *Camarasaurus* + Titanosauriformes (i.e., a basal macronarian), or it may be a

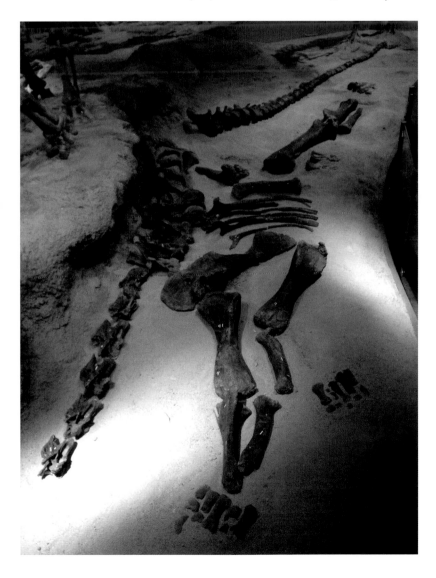

6.38. One of the most complete haplocanthosaurids from the Morrison Formation (FHPR 1106) as displayed at the Utah Field House of Natural History State Park Museum, Vernal, Utah. This specimen has not yet been identified.

nonneosauropod—no one seems to be able to agree (Wilson 2002; Harris 2006; Whitlock 2011b). The vertebral pneumaticity in *Haplocanthosaurus* is less well developed than it is in diplodocids or camarasaurids (Wedel 2003a, 2003b, 2009; Foster and Wedel 2014). *Haplocanthosaurus* (likely based on *H. priscus*) has been estimated to have weighed about 6,300 kg (13,860 lb), although one haplocanthosaur specimen I measured with nearly the same length femur as the type of *H. priscus* came out at about 10,500 kg (23,148 lb).

*Haplocanthosaurus delfsi**

This noticeably larger species of *Haplocanthosaurus* is based on a partial skeleton that was found low in the Morrison Formation at the Cleveland Museum of Natural History quarry south of Garden Park, Colorado. It was collected in the 1950s and is now mounted at the museum in Ohio. *Haplocanthosaurus delfsi* is distinguished by having bones 35% to 50% larger than the equivalent bones of *H. priscus*; also, the femur is more robust, as is the pubis. Whether these differences amount to much of anything biologically I think remains to be proved. The skeleton is 21 m (70 ft) long, and these dimensions indicate an animal weighing as much as 18,900 kg (41,580 lb), or three times as much as *H. priscus* (McIntosh and Williams 1988). Another haplocanthosaur specimen of about this size was recently uncovered in Montana.

Haplocanthosaurus* may be part of a distinct fauna from the lower half of the Morrison, characterized by species of dinosaurs possibly ancestral to the more common ones in the upper part of the formation. These lower Morrison species seem to include the theropod *Allosaurus jimmadseni*, the sauropod *Brontosaurus yahnahpin*, possibly the stegosaur *Hesperosaurus mjosi*, and *Haplocanthosaurus*. More specimens are needed from the Salt Wash Member and equivalent lower levels of the formation, however, before we can be confident of this pattern. At this point, it is intriguing, but more data are needed.

Camarasaurus

Camarasaurus supremus* was first excavated by E. D. Cope's men working near Cañon City, Colorado, and is the type species of the genus *Camarasaurus* (Cope 1877b; Osborn and Mook 1921). *Camarasaurus* itself includes the later-named junior synonyms *Morosaurus* and *Uintasaurus*—and possibly *Cathetosaurus*. Although Cope's species *C. supremus* was the first named, it has in fact proved to be one of the least abundant of *Camarasaurus* species, with *C. lentus* and *C. grandis* accounting for far more of the referred specimens (Ikejiri 2005).

Camarasaurus* is the most common dinosaur found in the Morrison Formation (fig. 3.18F), among sauropods or any other group (Foster 2003a). It was a moderately large to very large and bulky sauropod with stout limbs, a relatively short neck and tail, and a large skull (fig. 6.39)

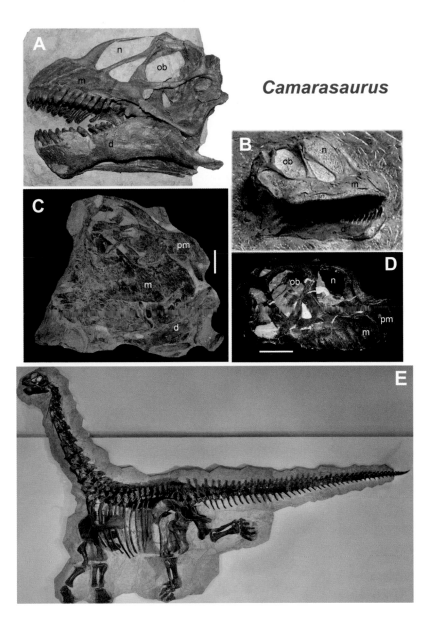

Camarasaurus

6.39. Examples of the common Morrison Formation macronarian sauropod *Camarasaurus*. (A) Slightly mediolaterally flattened skull, SMA 0002, from the Howe Ranch sites. (B) Only very slightly mediolaterally flattened skull from the Carnegie Quarry wall at Dinosaur National Monument. (C) Very mediolaterally flattened skull, SDSM 114501, from the Little Houston Quarry. (D) Somewhat dorsoventrally crushed and distorted skull, GPDM 220, from the Little Snowy Mountains, Montana. (E) Nearly complete skeleton, SMA 0002, from Howe Ranch, Wyoming. Scale bars = 10 cm. d, dentary; m, maxilla; n, external naris; ob, orbit; pm, premaxilla. *Photo in A courtesy of Cary Woodruff; E courtesy of Kirby Siber. For another full* Camarasaurus *skeleton, see fig. 2.3.*

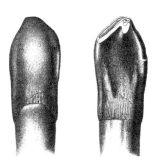

with many thick, spoon-shaped teeth set in robust upper and lower jaws (fig. 6.40). Some of its distinguishing characters are the strong skull, which is more robust and round than that of *Brachiosaurus*; long ribs of the cervical vertebrae; 12 cervical and 12 dorsal vertebrae; U-shaped split neural spines of the posterior cervical and anterior dorsal vertebrae; wide, fan-shaped neural spines of the posterior dorsal and sacral vertebrae; short spines on the sacral vertebrae; scapulae with expanded distal ends; a robust pubis with a long contact for the ischium; and an ischium with a narrow, curved shaft and a long contact for the pubis.

The neck and teeth of *Camarasaurus* suggest that this animal fed at low to moderate heights above the ground surface on a variety of vegetation. My high-feeding head-height estimate for *Camarasaurus*, which

6.40. *Camarasaurus* tooth in external view (*left*) and internal view (*right*). *Camarasaurus* teeth are robust and spoon shaped for feeding on tough vegetation; note the wear facet on the tip in the internal view. The smoother lower area is the root; the crenulations of enamel on the tooth itself are typical of many sauropod teeth. *From Ostrom and McIntosh (1966).*

is based on skeletal reconstructions, is about 5 m (16 ft), although the animal probably fed most often in a range down to about 2 m (6 ft), near its neutral neck position. The relatively large teeth of this sauropod may have been useful for the tougher vegetation of the Morrison basin. The bite forces calculated for *Camarasaurus* seem to indicate that it was eating relatively tough material (Button et al. 2016).

One indication of possible high browsing by *Camarasaurus* is the difference between microwear patterns of adult and juvenile specimens of this genus and those of adult *Diplodocus*. Adult *Camarasaurus* show rougher microwear than juvenile *Camarasaurus* and adult *Diplodocus* (Fiorillo 1998b), which indicates that juvenile *Camarasaurus* possibly had dietary overlap with *Diplodocus* until reaching adult size, at which time they began feeding at higher levels in the canopy. Teeth of *Brachiosaurus* and *Camarasaurus* indicate different chewing styles: *Brachiosaurus* possessed a more slicing tooth action, whereas the thick teeth of *Camarasaurus* contacted directly and thus resulted in some degree of oral processing of food (Calvo 1994). It has been suggested that the finer microwear of *Diplodocus* was a result of its supposedly feeding consistently at higher levels away from silica grains in the leaves, but then the similarity between *Diplodocus* microwear and that of obviously low-feeding juvenile *Camarasaurus* would appear contradictory. Perhaps the shared fine microwear of adult *Diplodocus* and young *Camarasaurus* indicates feeding on soft vegetation near the ground surface, and the coarser wear in teeth of adult *Camarasaurus* is a result of a diet of much tougher plant material such as conifers.

Camarasaurus in general and its individual species are widespread within the formation. Several species can co-occur in one area, and the ranges of species seem to overlap—there is no obvious pattern to the occurrences yet (Woodruff and Foster 2017). Several *Camarasaurus* have been found in well-drained floodplain paleosols in the Como Bluff area in Wyoming (Connely 2002).

CAMARASAURUS SUPREMUS

The large species *C. supremus* has been found mostly high in the formation and is some 50% larger in most linear dimensions than are the three smaller *Camarasaurus* species. Too little of *C. supremus* is known to measure the bones that we need to estimate the length and weight, but if we scale up from the smaller *C. lentus* (from which *C. supremus* is otherwise nearly indistinguishable except perhaps in the caudal neural spines; Ikejiri 2005), we get an animal nearly 23 m (75 ft) long and weighing approximately 42,300 kg (93,060 lb), the heaviest dinosaur known from the Morrison Formation. It is important to remember that most specimens of *Camarasaurus* in the Morrison Formation are not from animals this big; it is just the few specimens of *C. supremus* from the upper part of the formation that indicate such large sauropods, and compared with typical *Camarasaurus* bones, those of *C. supremus* are very big.

This species was named by Marsh and is, as mentioned above, nearly indistinguishable in its osteology from *C. supremus*. It includes the species *Camarasaurus annae* and *Uintasaurus douglassi* as junior synonyms. *Camarasaurus lentus* is the species to which most of the *Camarasaurus* material from Dinosaur National Monument has been referred, and the skeletons from the Morrison Formation indicate an animal up to about 15 m (50 ft) long. A nearly complete and articulated, one-third-grown, juvenile *C. lentus* was found at the Dinosaur National Monument quarry and described by paleontologist Charles Gilmore in 1925 (fig. 3.18F). This wall-mounted specimen is one of the most famous skeletons of a sauropod in the world and is about 5 m (16 ft) long (Gilmore 1925).

CAMARASAURUS GRANDIS

Marsh named *C. grandis* in 1877 from Reed's Quarry 1 at Como Bluff (Marsh 1877). Included in this species as junior synonyms are *Morosaurus impar* and *Morosaurus robustus*, both taken from the same quarry (fig. 3.7). The skull of *C. grandis* is nearly indistinguishable from that of *C. lentus* (J. Madsen et al. 1995), and the animals overall are nearly the same length and weight, but *C. grandis* is characterized by having the neurocentral sutures of the dorsal vertebrae elevated well above the level of the neural canal (in *C. lentus* the suture is down near the base of the neural canal; McIntosh 1990a, 1990b). Unfortunately, this difference can be seen only in juveniles. *Camarasaurus grandis* was, like *C. lentus*, about 15 m (50 ft) long. Specimens of average adults of both species vary somewhat in size and demonstrate a range of estimated masses. My measurements have ranged from 8,389 kg (18,456 lb) for a smaller individual (YPM 1901) to 16,572 kg (36,458 lb) for one on the large end of the spectrum (SDSM 351). Most were probably about 12,600 kg (27,720 lb).

CAMARASAURUS LEWISI

This species was described by Jensen as *Cathetosaurus lewisi* in 1988 but designated a new species of *Camarasaurus* eight years later. It is a two-thirds-complete skeleton (BYU 9047) from the Dominguez-Jones Quarry in western Colorado and likely represents an old individual because there is much ossification of the ligaments and tendons around the pelvis (Jensen 1988). One of the main features that distinguish *C. lewisi* from other species of *Camarasaurus* is that the split in the neural spines of the dorsal vertebrae continues back nearly to the sacrum, whereas in other species, that split ends around the middle part of the dorsal column (McIntosh et al. 1996). There also was a rotation of the ilia in relation to the sacral vertebrae, but it is unclear whether this was unique—the condition is not really clear in other species of *Camarasaurus*. Several other specimens have since been tentatively referred to *C. lewisi*, including a very well-preserved small individual from the Howe-Stephens Quarry in northern Wyoming (Mateus and Tschopp 2013; Tschopp et al. 2014) with a patch of possible soft tissue partially covering part of the lower jaw (Wiersma and Sander 2017). *Camarasaurus lewisi* was probably similar in length and weight to *C. grandis* and *C. lentus*, though some specimens are slightly smaller.

Brachiosaurus altithorax

All dinosaur-addicted children know two sauropods: *Brontosaurus* (or *Apatosaurus*, depending on the species now), and the less controversial *Brachiosaurus*. Although the latter, one of the largest among Morrison sauropods, has had a more staid paleontological resurrection than its diplodocid cousin, it is in some ways the more spectacular of the two because it breaks some sauropod conventions in opting not to become ludicrously long like *Barosaurus* and some of the Chinese sauropods, nor to just become big all around like *Supersaurus vivianae* and *Camarasaurus supremus*. No, *Brachiosaurus* is interesting because it instead went up. The front legs became longer than those in back, the shoulders raised up and the chest got deep, and the neck vertebrae were long.

When Riggs excavated the first *Brachiosaurus* (fig. 6.41) from his Quarry 13 near downtown Grand Junction, Colorado, in 1900, most of the Big Six of Morrison sauropods were already known: *Camarasaurus*, *Apatosaurus* (and *Brontosaurus*), *Diplodocus*, and *Barosaurus* were old news, and the last, *Haplocanthosaurus*, was to be published in a monograph the same year Riggs got his *Brachiosaurus* paper out, 1903. So when Riggs realized he had something new in the Field Museum's new specimen (Riggs 1903, 1904), he must have been excited, for among the six main sauropod genera, specimens of *Camarasaurus*, *Apatosaurus* and *Brontosaurus*, and *Diplodocus* far outnumber the other three. Riggs had found an animal that was to prove rare in the Morrison fauna but geographically wide ranging, with relatives turning up in ensuing years in places as far away as Africa and Europe. In the Morrison, confirmed *Brachiosaurus* have been found at only about seven sites (including in the Black Hills and Oklahoma; Bonnan and Wedel 2004), and since its original appearance at Quarry 13, significant partial skeletons have been found at only a few of those sites. Represented at these sites are only about nine known individuals, so compared to *Camarasaurus*, for example (with more than 200 individuals), *Brachiosaurus* is quite rare. Jim Jensen of Brigham Young University in Utah managed to connect with people who had found three of the *Brachiosaurus* sites, all within a couple hours' drive of the type locality: Jensen-Jensen, Potter Creek, and Dry Mesa (Jensen 1987). A recently described specimen of *Brachiosaurus* from the Black Hills includes the largest hind foot of a sauropod ever reported (Maltese et al. 2018), an indication of just how large *B. altithorax* was relative to most of its Morrison Formation cousins. And a specimen discovered by paleoartist Brian Engh in Utah in 2019, consisting of both humeri and several rib fragments (FHPR 17108), may be the geologically oldest of

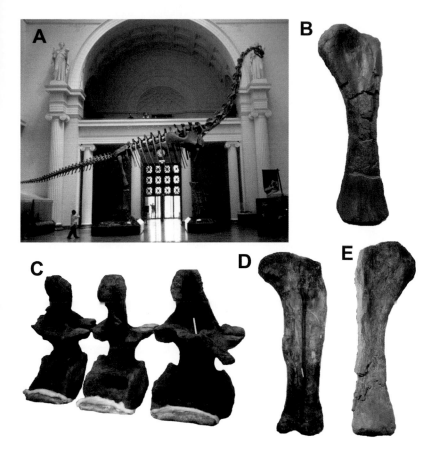

6.41. Sauropod *Brachiosaurus altithorax* from the Morrison Formation. (A) Mounted cast skeleton as it once appeared at the Field Museum of Natural History in Chicago. (B) Cast of the left humerus of a specimen from the Potter Creek Quarry, Colorado. (C–E) Type specimen FMNH 25107 from Riggs Quarry 13, western Colorado. (C) Three dorsal vertebrae; note the length of the centra. The pen at the base of the neural spine of the vertebra on the right is ~15 cm long (6 in.). (D) Posterior view of the right femur. The pen is 15 cm. (E) Anterior view of a cast of the right humerus. *All photos by author; A, C, and D courtesy of the Field Museum of Natural History; B and E courtesy of the Museums of Western Colorado.*

the genus in the Morrison and includes what may prove to be the best preserved of the mere four humeri now known from the animal (Potter Creek and Riggs's Quarry 13 type being the only other two with humeri).

A decade after Riggs's Colorado discovery, German expeditions at Tendaguru in eastern Africa found even more of a closely related brachiosaur than Riggs had, turning up the elongate cervical vertebrae that had not been preserved in the Grand Junction specimen. They also found skull material for what they named *Brachiosaurus brancai*; skull bones were totally unknown for the North American *Brachiosaurus* until the 1990s, when Ken Carpenter of the Denver Museum of Natural History prepared some material from the early days of the Marsh-Felch Quarry near Cañon City, Colorado, and uncovered both dentaries, a left and right maxilla, and several other skull elements. The North American brachiosaur skull bones were slightly different from those from Tanzania, and along with some subtle differences in the postcranial skeleton, the bones confirmed the species separation of the North American *Brachiosaurus* and *B. brancai*. Some of the cranial differences of the Morrison *Brachiosaurus* relative to the brachiosaur from Tanzania are a longer muzzle in the Morrison form with a more sloping snout (less angular than *B. brancai*) and a greater number of maxillary teeth (fig. 6.42; Paul 1988b; Carpenter and Tidwell 1998). It was later determined that the Tanzanian

6.42. Comparison of the likely appearances of the heads of *Brachiosaurus* from the Morrison Formation and the closely related *Giraffatitan* from the Tendaguru Beds of Tanzania. The Morrison reconstruction is based on a recently described specimen found at the Marsh-Felch Quarry. *Artwork by Matt Celeskey. Based on data in Carpenter and Tidwell (1998).*

Brachiosaurus sp.
Morrison

Giraffatitan brancai
Tendaguru

material represented a brachiosaur probably of a separate genus from *Brachiosaurus* from the Morrison, so the former became *Giraffatitan brancai* (Taylor 2009).

 Brachiosaurus, the Morrison original, was about 18 m (59 ft) long and weighed as much as 39,506 kg (86,914 lb) (on the basis of my measurements of FMNH P25107). Other estimates of the mass of *Brachiosaurus* have run a fairly wide range, from 10 to 40 metric tons up to 87 metric tons. It should be pointed out, however, that some of these latter estimates are based at least in part on some of the *Giraffatitan* material; a recent mass estimate specifically for that Tanzanian brachiosaur of approximately 38,000 kg (83,600 lb; Gunga et al. 2008), along with comparison of their respective limb bones, suggests that the Morrison and Tendaguru taxa were of approximately similar (and very large) size. Despite children's-book reconstructions of *Brachiosaurus* in deep lake waters, feeding on soft aquatic plants and functioning like a gargantuan reptilian snorkel, it walked the solid (although sometimes muddy) ground of the Morrison floodplain, feeding on high vegetation (~5 m [16 ft] and up) above the other herbivores of the time. It likely spent time browsing on ginkgoes, conifers, tree ferns, and taller cycadophytes and may have eaten about 200 to 400 kg (441 to 882 lb) of plant material each day (~0.5% to 1.0% of its body weight). *Brachiosaurus* may also have had a resting heart rate of just 15 beats per minute, although the heart could have weighed as much as 386 kg (851 lb), so the amount of blood pumped could still be substantial. The total blood volume for *Brachiosaurus* could have been close to 3,659 L (967 gal), and it would have had blood pressure higher than any known modern animal. The pressure would have been required in order to transport blood from the heart up to the brain, a distance of nearly 8 m (26 ft; Gunga et al. 1999; Weaver 1983). It is plausible, even probable, that *Brachiosaurus* occasionally fed below 3 to 5 m (10 to 16 ft) above the ground, as giraffes today can reach up to 5 m (16 ft) into trees but also feed below 2 m (6.5 ft) (Dagg and Foster 1976). Even if the neck of *Brachiosaurus* were not held at a high, nearly vertical angle, the head would have been well above the ground. The high shoulder region and long humerus, as well as the long neck, indicate that *Brachiosaurus* was adapted for feeding on high foliage, and estimates based on scaled reconstructions by Greg Paul (McIntosh et al. 1997), adjusted for a less vertical neck position (Stevens and Parrish 1996, 1997, 1999), indicate that neutral neck position for *Brachiosaurus* may still have been some 9.4 m (31 ft) above the ground.

The distinctive osteology of *Brachiosaurus* includes a broad and robust muzzle with thick maxilla, premaxilla, and dentary bones. The teeth are thick and spoon shaped. The unique shape of the *Brachiosaurus* skull in lateral view is in part due to the nasal process of the premaxilla rising posterodorsally from a posteriorly positioned point of the premaxilla (although this is less extreme in the Morrison *Brachiosaurus* than in the African *Giraffatitan*). The cervical vertebrae are long, and the dorsal series is unique among Morrison sauropods in that the neural spines are highest in the anterior dorsal vertebrae while those of the posterior dorsals and the sacrals are unusually short. The caudal vertebrae have relatively short neural spines as well, and the tail as a whole is not as long as in most other Morrison sauropods. The humerus is long (fig. 6.41)—just longer than the femur—and the metacarpals are also long.

The sauropod *Ultrasauros* was originally described as a brachiosaurid from the Dry Mesa Quarry in western Colorado. The type specimen of *Ultrasauros* now appears to be a dorsal belonging to the type skeleton of *Supersaurus* from the same quarry (Curtice et al. 1996), and the referred scapulae of *Ultrasauros* are indistinguishable from *Brachiosaurus*. So we now have only one brachiosaurid at Dry Mesa (*Brachiosaurus*), and *Ultrasauros* (*Ultrasaurus*) as a distinct, valid dinosaur name no longer exists (Curtice et al. 1996).[6]

*Dystylosaurus edwini**

Jensen described this species (whose generic name means "two-beam lizard") on the basis of a single large dorsal vertebra from the Dry Mesa Quarry (fig. 4.2; Jensen 1985). The main distinguishing character of the species was that the dorsal had a unique dual-strut support of its prezygapophyses (the processes that articulate with the arch of the next vertebra in front). The vertebra indicates a large sauropod, probably pushing 40 metric tons (88,184 lb), but other than this, there is little else we can say about this animal because so little of the skeleton is known (McIntosh 1990a). Jensen was not even sure to what sauropod family it belonged. McIntosh was not sure what *Dystylosaurus* is either, although a number of years ago he had it listed as a possible brachiosaurid. Brian Curtice and Ken Stadtman have suggested recently that it may in fact be part of the type skeleton of *Supersaurus* (Curtice and Stadtman 2001).

Roof-Lizards: The Stegosaur Dinosaurs

Switching now to the ornithischians (Sereno 1986), stegosaurs are some of the most visually interesting and unique dinosaurs. Any animal that can be up to 7.5 m (25 ft) long and weigh up to 5,000 kg (11,000 lb) and yet still be bristling with long, sharp spikes and huge, fan-shaped plates has got to be on anyone's list of favorites. *Tyrannosaurus* may be the star dinosaur of all time, but in truth it has only one appeal: that it's a big walking head with huge teeth. As impressive as tyrannosaur skeletons are to all of us, I've always found the unusual ornamentation of a stegosaur skeleton more interesting. Take *Kentrosaurus*, from the Late Jurassic of Tanzania. This is a dinosaur with a front half lined with dermal plates similar to but slightly smaller than those of *Stegosaurus*, but the entire back half of the animal, from over the hips back down the tail, is arrayed with double rows of huge spines that make it look like a dinosaurian *Hallucigenia*.[7] There are even long spikes over the shoulders of *Kentrosaurus*. It makes the four tail spikes of *Stegosaurus* look not quite so intimidating. *Stegosaurus*, however, had by far the largest plates of any stegosaur, some of them pushing 1 m (3.3 ft) in height. This line of plates along the back of a *Stegosaurus* must have been impressive (figs. 6.43 and 6.45).

The plates of *Stegosaurus* have been a source of controversy regarding this dinosaur for decades. There are about 17 plates on *Stegosaurus stenops*, with smaller oval ones over the neck, then increasing in size to over the hips and base of tail, where they are large and almost diamond shaped, and decreasing in size again progressing out the tail. The plates are thin, often grooved, and have thick, rugose bases that were embedded in thick skin (fig. 6.45). They projected up and slightly laterally, and their bases were just off center from the tops of the neural spines of the vertebrae and not directly above them. Early workers debated the arrangement of the plates, the main questions being whether the plates were in one or two rows and, if two, whether the plates were paired or alternating. Gilmore's view of two alternating rows was the predominant one for some time, until Stephen Czerkas proposed an arrangement in which the alternating rows gave way to a single row from the middorsal vertebrae back to the last plate on the tail. Recent *Stegosaurus* specimens from Jensen, Utah, and Cañon City (Garden Park), Colorado, have indicated, however, that Gilmore's original two-alternating-rows reconstruction is probably correct (fig. 6.44; Gilmore 1914; Czerkas 1987; Carpenter 1998a).

The plates most likely functioned in several roles, including species recognition, display, defense, and thermoregulation (Farlow et al. 1976; Buffrénil et al. 1986; Carpenter 1998a; Main et al. 2005). The sizes of the plates appear to have differed between the coexistent Morrison species *S. stenops* and *S. ungulatus*, and with possible color differentiation, the plates may have aided in the identification of conspecifics. Given the appearance of larger size that the plates afforded a *Stegosaurus*, and if the plates were colored, they may have functioned as a display to attract mates, like enlarged feathers on some birds. A similar effect could have been useful to try to ward off predators or conspecific rivals. When viewed from the side, the apparently larger *Stegosaurus* would look much like a cat arching its back. (The plates themselves would be of little value as defensive armor, however.) The plates seem to have had well-vascularized skin (or keratin?) covering them, so the animals may in fact have been able to change the color tone of the plates by flushing the skin with increased blood flow. This blood flow in the skin may also have served to help the animal warm up and cool down quickly. This does not necessarily imply that *Stegosaurus* was cold-blooded, like a lizard sunning or shading itself to regulate temperature. An animal with a higher metabolism would still have use for a way to dump excess heat, just as elephants use their ears to cool down today.

The tail spikes of stegosaurs were clearly defensive weapons (one of Arthur Lakes's Como Bluff field assistants called the spikes "devils tails"). A sheath of hornlike material probably covered the spikes in life, but these sheaths likely did not significantly lengthen the effective size of the spike. Despite early reconstructions of the skeleton, life restoration paintings, and models that show the tail spikes projecting up and out from the tail, the spikes must have projected almost straight laterally from the tail, with only a slight posterior and dorsal tilt. They simply are more effective weapons in such an arrangement, and the old-school orientation of pointing up and only slightly laterally (still surprisingly pervasive in models and mounts) must have been a product of Marsh's original reconstruction of the animal with the tail dragging on the ground.

Some of the *Stegosaurus* tail spikes in collections around the country (roughly 10% of those surveyed in one study) show evidence of injury and healing, suggesting that they were broken in use. Some of the injured specimens indicate that the animals lived with osteomyelitis in the tail spikes (McWhinney et al. 2001). At least one *Allosaurus* caudal vertebra was found with an apparent puncture wound to the left caudal rib, most likely due to a strike by a stegosaur tail spike (Carpenter et al. 2005c). Bakker (2017) highlighted a pelvic bone of an allosaur that appeared to have been punctured by a stegosaur spike and suggested that stegosaurs struck by rotating the tail near the distal end and almost clustering the tail spikes before striking (in this case) upward at the predator. This would suggest far more flexibility and dexterity with the tail than suspected in the traditional view of sideways lashing that has been envisioned for what the comic *The Far Side* once termed the stegosaur "thagomizer" (Larson 2003).[8]

6.43. Skeletons of *Stegosaurus* from the Morrison Formation. (A) Nearly complete specimen (NHMUK PV R36730) from the Red Canyon Ranch Quarry, Wyoming. (B) Partial specimen (MWC 81) from the Bollan Quarry in Rabbit Valley, western Colorado. *Image in A from Maidment et al. (2015) © The Natural History Museum and courtesy of Susannah Maidment.*

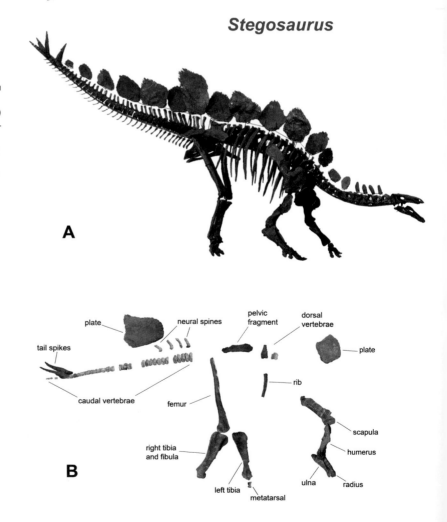

Stegosaurus

6.44. Close-up of a *Stego-saurus* skeleton from Garden Park, Colorado, at the Denver Museum of Nature and Science, showing the double row of alternating plates along the back and dermal ossicles forming an armor layer protecting the underside of the throat.

Besides the plates and tail spikes, stegosaurs are characterized by generally long, low skulls with short, leaf-shaped teeth; dorsal vertebrae with tall neural arches; anterior caudal vertebrae with dorsoventrally elongate caudal ribs and tall neural spines; robust limb bones; and forelimbs that are much shorter than hind limbs (fig. 6.43). There were also many small dermal ossicles embedded in the skin of the throat region (fig. 6.44) and possibly dermal ossicles around the hips and hind legs as well. The recent discovery of integument impressions on stegosaur remains from northern Wyoming indicates that the skin was covered with polygonal, tuberculate scales 2–7 mm (0.08–0.27 in.) in diameter with some larger scales near the dorsal surface that are up to 20 mm (0.79 in.) in diameter; the plates also appear to have had a presumably keratinous cover that was smooth other than fine, low ridges running dorsoventrally (N. Christiansen and Tschopp 2010).

Stegosaurus

At up to 7.6 m (25 ft) long, *Stegosaurus* was the largest nonsauropod herbivore in Morrison times (fig. 6.45). Adults seem to have consistently reached masses of 5 metric tons or more; measurements of partial skeletons DMNH 1483 and MWC 81 give mass estimates of 5,284 kg (11,649 lb) and 4,997 kg (11,016 lb), respectively. *Stegosaurus* was terrestrial and

6.45. Restoration of the armored dinosaur *Stegosaurus*. Total length ~7.6 m (24.9 ft). *Drawing by Thomas Adams.*

fed on low-growing plants probably within 1 m (3.3 ft) of the ground (Galton 1997); these plants most likely consisted of ferns, cycadophytes, and low conifers.

Good specimens of *Stegosaurus* are known from sites including Dinosaur National Monument, Quarries 12 and 13 at Como Bluff, the Marsh-Felch Quarry, the Utah Field House Quarry, the Bollan Quarry, the Small Quarry, the Red Canyon Ranch Quarry, and the Kessler Quarry. Bones of young juvenile *Stegosaurus* from Dinosaur National Monument and Quarry 13 at Como Bluff (from animals about 1.5 to 2.6 m [5 to 8.5 ft] long) have shown that there were some proportional differences in the skeletons of juveniles and adults in this genus (Galton 1982a). Particularly the scapula and femur were more slender in the juveniles, and although tail spikes were present, it is not known whether very young *Stegosaurus* had dermal plates.

STEGOSAURUS ARMATUS*

The first *Stegosaurus* specimen in the Morrison Formation was that of *S. armatus* (YPM 1850), which was found in a whitish sandstone at Lakes's Quarry 5 at Morrison, Colorado, in 1877. The specimen was described briefly by Marsh, although it was then and is still not yet completely prepared. One block of sandstone from the quarry has five articulated midcaudal vertebrae (actually just the centra) and one chevron. An anterior caudal from YPM 1850 indicates that this is one of the biggest stegosaur individuals ever found in the Morrison. Unfortunately, because of the preservation of the material, the fact that only some of the bones have been prepared, and the incomplete nature of the specimen, there seems to be no way to tell at this point what later-named species may in fact belong to *Stegosaurus armatus*. Probably some do, but YPM 1850 is not yet in good enough shape to be able to refer any other material to it.

STEGOSAURUS UNGULATUS*

The next *Stegosaurus* species was named in 1879. Arthur Lakes didn't toil away in the trench of Como Bluff's Quarry 12 for nothing, because what came out of the site was a better *Stegosaurus* skeleton than he'd found in Colorado. *Stegosaurus ungulatus* had a long femur, radius, and ulna and relatively smaller plates along its neck, back, and tail than did S. *stenops*. Marsh originally thought that the animal had eight spikes on the tail instead of the typical four (fig. 3.6), but Carpenter and Galton showed that Marsh most likely was mistaken and that it had only four spikes. This individual may be the same species as S. *armatus*, but there is too little of S. *armatus* available to be sure, and the limb proportions of S. *ungulatus* may or may not be of any significance (Galton 1990; Carpenter and Galton 2001).

STEGOSAURUS STENOPS

This is the best-known species of *Stegosaurus*, with nearly complete specimens from several quarries in the Morrison Formation (Marsh 1881, 1887b; Maidment et al. 2015), and in fact, due to the incompleteness of S. *armatus*, S. *stenops* has been proposed as a new type species for the genus *Stegosaurus* (Galton 2010, 2011; this proposal was later adopted, ICZN 2013). The femur is less elongate than in S. *ungulatus*, the plates are larger, and the sacral vertebrae are ventrally keeled. There are two pairs of tail spikes. The best "road-kill" specimens of S. *stenops* come from the Marsh-Felch and Small Quarries at Garden Park near Cañon City, Colorado, and the Red Canyon Ranch Quarry near Shell, Wyoming (fig. 6.43; Gilmore 1914; Maidment et al. 2015). *Stegosaurus stenops* is most closely related to S. *homheni* from Asia, *Miragaia* from Europe, and *Hesperosaurus* from Wyoming.

Alcovasaurus longispinus

During his time at the University of Wyoming, Reed collected a specimen of a stegosaur from central Wyoming near the present-day Alcova Reservoir (Gilmore 1914). This animal was generally stegosaur in most of its skeleton, but the tail spikes were much longer than in any other known member of the family in the Morrison Formation (90% of the length of the femur; fig. 6.46). The partial skeleton consisted mostly of caudals, the pelvis, a femur, posterior dorsals, and a cervical vertebra. Gilmore named the new species *Stegosaurus longispinus* in 1914, and it was placed in the new genus *Alcovasaurus* by Galton and Carpenter (2016a). According to those authors, in addition to the very long tail spikes, A. *longispinus* differs from other Morrison stegosaurs in having six pairs of sacral ribs, short distal caudal centra, and transverse processes on all caudals. Raven and Maidment (2017) found *Alcovasaurus* to be a relatively primitive thyreophoran outside stegosaurs but understandably were skeptical of this result and suspected that the position was due to a large amount of missing data from the skeleton; their analysis did seem to show that it was a separate genus, however. *Alcovasaurus* is almost certainly a stegosaur and almost certainly a third Morrison genus, in addition to *Hesperosaurus*

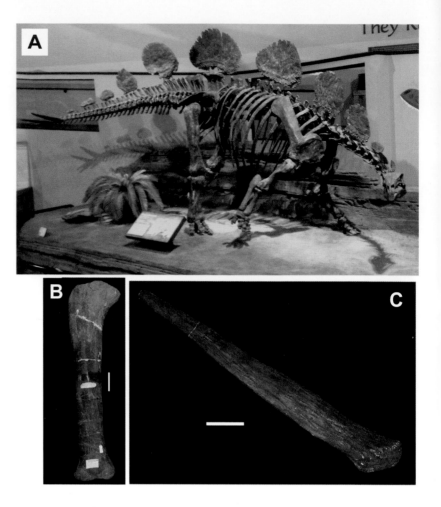

6.46. Other stegosaurs from the Morrison Formation. (A) Cast of *Hesperosaurus mjosi* from near Buffalo, Wyoming, mounted at the North American Museum of Ancient Life. (B and C) Remaining elements of *Alcovasaurus longispinus*. (B) Femur (UW 20503). (C) Cast of the elongate tail spike that gives the species its name. Scale bars = 10 cm.

and *Stegosaurus*, but its exact position in the stegosaur tree was unclear. Recently, Costa and Mateus (2019) noted characters suggesting that *Alcovasaurus longispinus* was actually a dacentrurine stegosaur close to *Miragaia* and possibly a species of *Miragaia* itself. Dacentrurine stegosaurs are an otherwise European group, and *Alcovasaurus* (or *M. longispinus*) might be the first North American representative.

Hesperosaurus mjosi

Arthur Lakes found the first remains of *Stegosaurus* in the Morrison Formation in 1877, and for 124 years, that genus was the only stegosaur reported from the rock unit. Yet a possibly non-*Stegosaurus* stegosaur was found in northern Wyoming, near Buffalo. What may have made the difference was that the specimen was from the lower 5 m (16.4 ft) of the formation and thus may be as many as several million years older than most faunas from the Morrison Formation. This new, primitive stegosaur partial skeleton was named *Hesperosaurus mjosi* in 2001 (fig. 6.46). It is distinguished from *Stegosaurus stenops* by having a larger antorbital fenestra; relatively larger teeth; three more cervical vertebrae; dorsal vertebrae with much shorter neural arches; and at least 10 smaller, oval plates along the dorsal midline of the body (Carpenter et al. 2001). *Hesperosaurus* also had four tail spikes and was approximately the same size as or slightly smaller than *Stegosaurus*. Although this species was once considered a possible species of *Stegosaurus* (Maidment et al. 2007, 2015; N. Christiansen and Tschopp 2010) recent analyses appear to confirm its status as a separate genus, and it and *Miragaia* (from Europe) seem to form a clade that is the sister taxon to *Stegosaurus* (Raven and Maidment 2017). Several additional specimens from northern Wyoming and Montana (Maidment et al. 2018) have recently been formally or tentatively referred to the species also.

Hesperosaurus may have been similar to *Stegosaurus* as a low-level terrestrial herbivore. Given its position low in the formation, it may have been an ecologically equivalent predecessor to that genus, but if the two were contemporaries (as is still possible), they should have had some geographic separation, contrasting habitat preferences, or dietary differences to lessen direct ecological overlap of their lifestyles. At this point, it is impossible to say what those might have been or if instead the two stegosaurs were restricted to different times of Morrison history or different geographic areas (Maidment et al. 2018). We'll never know for certain that they were restricted; only finding the two together can prove that they were contemporaries, but you can't prove that they weren't. The point at which we say they were most likely temporally or geographically separated would be after many more *Hesperosaurus* are found and only low in the formation or to the north where no *Stegosaurus* have been found, but one specimen of either turning up in the other's geographic or stratigraphic territory would shoot down the whole idea and demonstrate that the two coexisted, at least for a time or in an area.

Ankylosaurs were unknown from the Jurassic of North America for nearly 120 years of active collecting in the Morrison Formation. Somehow paleontologists had managed to find plenty of *Camarasaurus* and *Allosaurus* and other types of dinosaurs, but apparently not one piece of ankylosaur material surfaced in that time. Admittedly, ankylosaurs were quite rare during the Late Jurassic, but it is surprising that a distinctive type of armored dinosaur that weighed about as much as a full-grown horse could have eluded us for so long.

That all changed quickly, however. Starting in about 1990, bones that were clearly those of an ankylosaur began appearing at the Mygatt-Moore Quarry in western Colorado (fig. 6.47). Soon after, these dinosaurs popped up at the Hups Quarry and Dry Mesa Quarry, both on the Uncompahgre Plateau in western Colorado, and at the Bone Cabin Quarry and another site near Sheep Creek in Wyoming. Within about three years, ankylosaurs were known from about a half-dozen quarries in the

Jurassic Knights: The Ankylosaur Dinosaurs

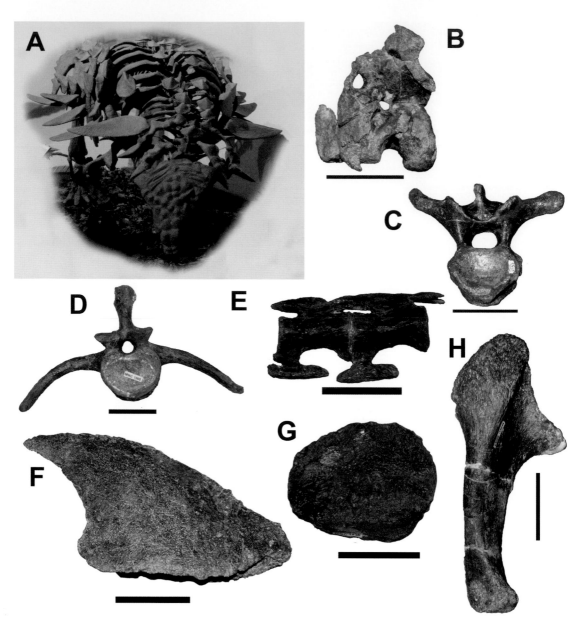

6.47. Elements of the polacanthid ankylosaur *Mymoorapelta maysi*. (A) Cast skeleton at the Museums of Western Colorado. (B) Braincase (MWC 5435) in cross-sectional left medial view. (C) Vertebra in anterior view. (D) Anterior caudal vertebra (MWC 1806) in anterior view. (E) Distal caudals in lateral view. (F) Lateral spine. (G) Dorsal dermal plate. (H) Ulna (MWC 5643). Scale bars = 5 cm.

Morrison Formation. And now more have shown up at the Hanksville-Burpee and Gnatalie Quarries in Utah, as well.

Ankylosaurs were heavily armored, squat animals with osteoderms embedded in the skin along most of the back and neck and with spikelike plates along the sides from the neck all the way down the tail. In addition, the osteoderms fused into a large, bony shield over the pelvis. Some forms had stegosaur-style spikes on the shoulders, and later Cretaceous forms had large clubs on the end of the tails (*Ankylosaurus* being the prime example of this). There are three families of ankylosaurs: the nodosaurids, polacanthids, and ankylosaurids, and both ankylosaurs known so far from the Morrison Formation are in the Polacanthidae. These polacanthid ankylosaurs are the rarest dinosaurs in the Morrison fauna.

Ankylosaurs of the Morrison Formation may have had a beak covering the fronts of the upper and lower jaws, a structure used for cropping vegetation during feeding. Like other ornithischians, they may have had cheeks to keep food inside the mouth as it was cropped, chewed briefly, and swallowed. The food may then have undergone microbial fermentation in the gut (Carpenter 1997).

Mymoorapelta maysi

Mymoorapelta was the first ankylosaur from the Morrison Formation and the first Jurassic ankylosaur found in North America (fig. 6.47; Kirkland and Carpenter 1994); it was first identified at the Mygatt-Moore Quarry in western Colorado, and additional ankylosaur specimens (some possibly *Mymoorapelta*) have since shown up at several other sites, including Cactus Park and the Small Quarry in Colorado (Kirkland et al. 1998) and the Hanksville-Burpee and Gnatalie Quarries in Utah. *Mymoorapelta* had short legs and a heavy build and probably weighed about 560 kg (1,232 lb).[9] It walked low to the ground and was herbivorous, so it probably spent much of its time feeding on low-growing plants (fig. 6.48). Studies of stomach contents preserved in an ankylosaur from Australia show that these animals probably were selective browsers, nipping off small pieces of particular parts of soft plants (Molnar and Clifford 2000). Although there is no direct evidence of the diet of *Mymoorapelta* or other Morrison ankylosaurs, we can infer from the Australian specimen a similar feeding mode for these dinosaurs in the Late Jurassic of North America. It wasn't notably fast or agile and really had no defensive weapons, so its main defense mode was the bony armor embedded in the skin and the sharp spines all along its body (fig. 6.47). This defensive armor included

6.48. Restoration of *Mymoorapelta*. Total length ~3.6 m (11. 8 ft). *Color drawing by Thomas Adams.*

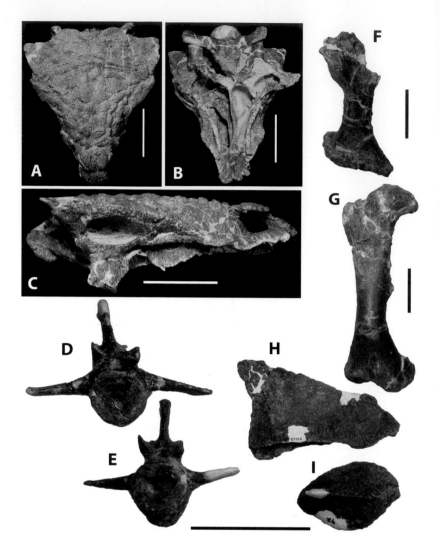

6.49. Elements of the polacanthid ankylosaur *Gargoyleosaurus parkpinorum* (DMNH 27726). (A) Skull in dorsal view. (B) Skull in ventral view. (C) Skull in right lateral view. (D and E) Anterior caudal vertebrae in anterior views. (F) Right humerus. (G) Right femur. (H) Lateral spine. (I) Dorsal dermal plate.

6.50. Skeletal reconstruction of *Gargoyleosaurus*, as mounted at the Denver Museum of Nature and Science.

the shield over the pelvis and two rings of bone around the upper half of the neck just behind the skull.

Gargoyleosaurus parkpinorum

Gargoyleosaurus was named in 1998 on the basis of a skull and partial skeleton from the Bone Cabin Quarry in southeastern Wyoming (fig. 6.49; Carpenter et al. 1998). In general morphology and habits, *Gargoyleosaurus* resembles *Mymoorapelta* (fig. 6.50), but it is larger and differs in some details of the skull and skeleton, including the fact that the neural spines on the caudal vertebrate are not expanded transversely as in *Mymoorapelta*. Measurements of the humerus and femur of *Gargoyleosaurus* suggest that the animal weighed about 754 kg (1,659 lb). *Gargoyleosaurus* also has more elongate vertebral centra than *Mymoorapelta* (Kilbourne and Carpenter 2005). *Gargoyleosaurus* appears to be known now from the Dry Mesa Quarry in Colorado also (Kirkland et al. 2016).

Fleet-Footed Plant Eaters: Heterodontosaurids, Basal Neornithischians, and Ornithopod Dinosaurs

The bipedal ornithischian dinosaurs of the Morrison Formation are an underrated lot. High in diversity and also in numbers during the Late Jurassic in North America but less often preserved, they attract neither the attention nor the praise that are often showered on more flashy herbivores of the time, like *Stegosaurus* or the many huge sauropods. But they were an important group of animals because there were a number of different species (fig. 6.51), because most were likely relatively fast when pursued, and particularly because the small ones were abundant on the Morrison floodplain during the Late Jurassic. They were in fact the most numerous dinosaurs around at the time, if our reconstructions of the ecosystem are anywhere near correct.

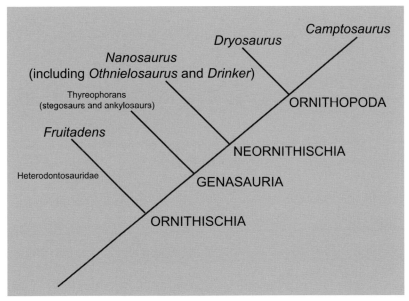

6.51. Relationships of ornithischian dinosaurs of the Morrison Formation. Based on Butler et al. (2008), Boyd (2015), and Galton and Carpenter (2016b).

Ornithopods, heterodontosaurids, and basal neornithischians of the Morrison Formation were bipedal plant eaters of the order **Ornithischia** that had medium- to small-sized heads and short forelimbs with five digits on the hands. Their later relatives among the ornithopods were the more advanced duck-billed dinosaurs of the Cretaceous period, whereas all are to varying degrees related to the **Thyreophora** and Marginocephalia. In several Morrison bipedal plant-eaters, the hind legs were long, and the tibia in most was longer than the femur, suggesting high cursoriality, but recent discoveries of Cretaceous basal neornithischians in burrows suggest that at least some species may have been diggers that lived in dens at times (Varricchio et al. 2007). This behavior is something that Bakker suggested for some *"Drinker"* specimens he found at Como Bluff years ago (Bakker 1998a). The teeth of Morrison ornithopods, heterodontosaurids, and basal neornithischians were small, numerous, and leaf shaped and were set in from the lateral edges of the jaws, indicating the presence of cheeks or skin along the sides of the mouth to keep food inside. All were herbivorous, although it is conceivable that some of the smallest species were omnivorous and also fed on insects.

Heterodontosauridae

FRUITADENS HAGAARORUM

This species from the Fruita Paleo Area in Colorado was identified as a small basal ornithischian similar to *Heterodontosaurus* in having large, canine-like teeth near the front of its jaws and was recognized as such soon after it was discovered in 1976 (Callison 1987). A small lower jaw from the FPA (LACM 128258) contains typical primitive ornithischian teeth similar to those of *Othnielosaurus* but with a single caniniform tooth

6.52. Restoration of the heterodontosaurid dinosaur *Fruitadens*, the smallest adult dinosaur in the Morrison, as exhibited at the Museums of Western Colorado. Note the canine-like tooth of the lower jaw. This small dinosaur was likely omnivorous. *Model by Doyle Trankina (LACM), color pattern by George Callison.*

near the front of the jaw. It was finally officially described as *Fruitadens* after having been variously identified as some type of heterodontosaurid for many years (Butler et al. 2010). *Fruitadens* would have weighed approximately 0.75 kg (1.6 lb) and ate mostly plants and very likely some insects (Campione et al. 2014; Butler et al. 2010, 2012). It was bipedal and the smallest adult dinosaur known from the formation (fig. 6.52; Callison and Quimby 1984). The FPA is the only locality in the Morrison that has so far yielded a heterodont basal ornithischian like *Fruitadens*. This was one of the smallest adult dinosaurs of the Late Jurassic not just in the Morrison but worldwide.

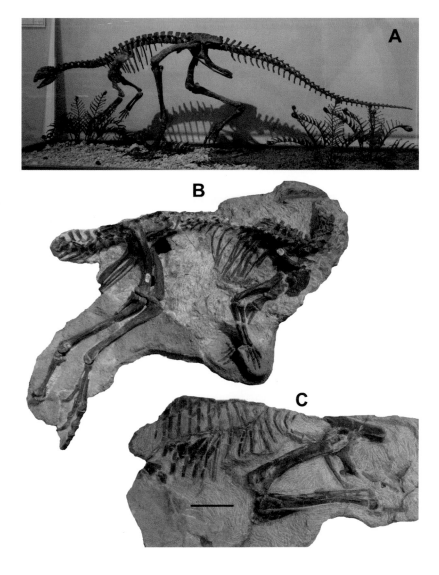

6.53. Small bipedal neornithischians *Nanosaurus* from the Morrison Formation. (A) Cast of a skeleton from northwestern Colorado at the Museums of Western Colorado. (B) Articulated skeleton (BYU 163) from near Willow Springs, Utah. (C) Partial skeleton from the Alcova Reservoir area, Wyoming (UW 24823).

Basal Neornithischians

OTHNIELOSAURUS CONSORS*

Galton renamed this species based on Marsh's original naming of "*Laosaurus consors*," which had been collected from Como Bluff in the 1870s (Galton 2007). The genus name *Othnielosaurus* refers to Marsh's first name and the root for "lizard." The specimens Marsh named "*Laosaurus gracilis*" and "*Nanosaurus (Othnielia) rex*" are referable to *Othnielosaurus consors* (junior synonyms in the language of taxonomy) or are indeterminate (Marsh 1877; Galton 1977, 1983). However, as we will see below, even *Othnielosaurus* itself is likely a junior synonym of *Nanosaurus agilis*. "*Othnielosaurus*" (*Nanosaurus*) was a small animal of about 5 kg (11 lb; fig. 6.53); it was a browser of low-growing vegetation and had small teeth shaped like diamonds with several **denticles** on each of the two upper edges. The skull was small and short with relatively large orbits, and the relatively long legs had a tibia that was notably longer than the femur. When chased, "*Othnielosaurus*" could run away rather quickly. Based on the numbers we have found of this animal in most areas of Morrison Formation exposure, and considering the typically greater abundance of small herbivorous animals in most ecosystems, "*Othnielosaurus*" was likely among the most common of dinosaurs during Morrison times. Long considered an ornithopod and a hypsilophodontid, "*Othnielosaurus*" now appears to be a basal neornithischian outside **Ornithopoda** (Butler et al. 2008; Boyd 2015; Galton and Carpenter 2016b).

A basal neornithischian from the Middle Jurassic of Siberia has been found to have feather-like structures, serving a possible function of insulation, around the head and body (Godefroit et al. 2017), which indicates a possibility that other neornithischians like those from the Morrison Formation had similar integumentary covering.

DRINKER NISTI*

Bakker and others named this small basal neornithischian (Galton and Carpenter 2016b) on the basis of teeth that have cusped ridges along one edge and main tooth denticles composed of three cusplets (Bakker et al. 1990) rather than the simple denticles apparently characteristic of "*Othnielosaurus*." Otherwise, the teeth and skeleton of *Drinker* are similar to those of "*Othnielosaurus*," though generally smaller. Most *Drinker* specimens are from the Breakfast Bench part of Como Bluff, and they may have been burrowing, denning animals (Bakker 1998a). More recent specimens have shown, however, that the tooth types thought to be characteristic of "*Othnielosaurus*" and "*Drinker*" occur in the same, more complete jaws of small bipedal neornithischian specimens from the Morrison Formation.

Nanosaurus was named in 1877 by Marsh based on a specimen from Garden Park, Colorado. Galton and Carpenter (2016b) and Carpenter and Galton (2018) had *Nanosaurus agilis* as a distinct taxon, though others have recently considered it a nomen dubium (Boyd 2015). Carpenter and Galton (2018) reviewed the bipedal ornithischian specimens of the Morrison Formation and concluded that "*Drinker*" and "*Othnielosaurus*" are probably the same animal and that both (and all their specimens) are also likely *Nanosaurus agilis* (fig. 6.53). The original *Nanosaurus* specimen was not spectacular, but enough better specimens have been found (usually referred to "*Othnielosaurus*" or "*Drinker*" in recent years) to allow comparisons that showed that they are all the same small, bipedal plant-eater species. So it appears that the Morrison Formation had only a single species of very common small neornithischian, *Nanosaurus*.

Ornithopods

DRYOSAURUS ALTUS

Dryosaurus ("oak lizard") was a fast, bipedal herbivore (fig. 6.54) and was roughly similar to *Nanosaurus* except that it grew to a significantly larger size and had more advanced teeth with a strong median ridge on the lateral surface. *Dryosaurus* was described by Marsh on the basis of a specimen he originally called *Laosaurus altus*. A second species, *Dryosaurus lettowvorbecki*, from the Late Jurassic Tendaguru deposits of southeastern Tanzania, is closely related but has been returned to its original separate genus, *Dysalotosaurus*. A fully illustrated description of *Dryosaurus altus* also appeared in two publications by Galton (1981b, 1983). My measurements of the femur of YPM 1876, a *Dryosaurus* specimen from Quarry 5 at Como Bluff, suggest that these animals weighed about 114 kg (251 lb) as adults, whereas another estimate, based on a similar method but with a different data set, has suggested a weight of 164 kg (361 lb) for the same individual (Campione et al. 2014). Like *Othnielosaurus* (*Nanosaurus*) and

6.54. Restoration of the ornithopod *Dryosaurus*. Total length ~2.4 m (7.9 ft). *Drawing by Thomas Adams.*

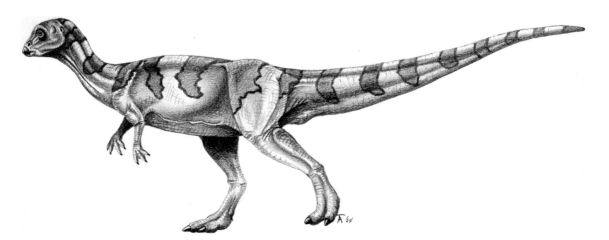

6.55. Skull and lower jaws of the ornithopod dinosaur *Dryosaurus elderae* (CM 3392). Scale bar = 10 cm. *Photo by author. (Courtesy of the Carnegie Institute, Carnegie Museum of Natural History.)*

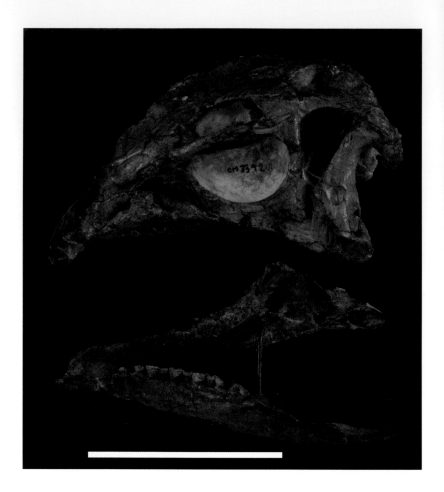

other small ornithopods, *Dryosaurus* fed on the low-growing vegetation that grew on the Morrison floodplain.

Sites near Uravan and Fruita, Colorado, have each yielded hundreds of bones of *Dryosaurus altus* of various ages (Scheetz 1991; Kirkland 1994), and studies of *D. lettowvorbecki* have shown that these ornithopods grew quickly, continuously, and independent of seasonality in their environment (Chinsamy 1995). A baby *Dryosaurus* found at Dinosaur National Monument shows that as these animals grew, the muzzle increased proportionally in length, as did the posterior part of the skull (Carpenter 1994). Like many other vertebrates, the eyes were relatively larger in babies than in adults. Additional partial skeletons of *Dryosaurus* have been found in the Morrison Formation at the Bone Cabin Quarry and the Red Fork of the Powder River in Wyoming and at Lily Park in northwestern Colorado.

DRYOSAURUS ELDERAE

Carpenter and Galton (2018) named the new species *Dryosaurus elderae* from the Carnegie Quarry at Dinosaur National Monument (fig. 6.55). This species is distinguished from *D. altus* by having longer and dorsoventrally shorter cervical vertebrae and a long and low ilium, among other characters.

Marsh (1879b) named *Camptosaurus dispar*, for which Dave Norman (2004) considers there to be numerous synonyms including *C. medius*, *C. nanus*, and *C. browni*. The large type specimen of *C. amplus* from Como Bluff is only a pes (hind foot, and possibly of a theropod), and ?*C. depressus* from the Lower Cretaceous may not be an ornithopod either. The relatively smaller type specimens and referred material of *C. medius* and *C. nanus* contain some unfused vertebral neural arches and probably represent young individuals referable to *C. dispar*. Many individuals of *Camptosaurus* are known from Quarry 13 in the eastern part of Como Bluff.

Camptosaurus ("bent lizard") was a large ornithopod that was the sister taxon to *Iguanodon* and its relatives plus the hadrosaurs (duck-billed dinosaurs); it was therefore more derived than *Nanosaurus* or *Fruitadens* (fig. 6.51). The skull of *Camptosaurus* was longer than those of other Morrison ornithopods and the teeth more robust and tightly arranged along the jaw. The teeth of *Camptosaurus* also had thick median ridges on their lateral sides and denticles along their edges; both were more fully developed than in *Dryosaurus*. The hind foot had three main,

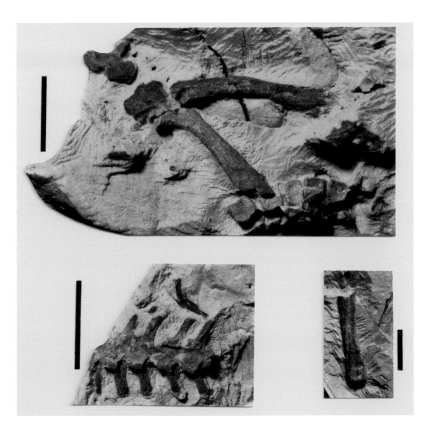

6.56. Partial skeleton of an embryo of the ornithopod dinosaur *Camptosaurus* from Dinosaur National Monument (DINO 15624). *Top*, scapula, coracoid, humerus, and a string of five vertebrae. *Lower left*, vertebral neural arches and transverse processes in ventral view. *Lower right*, tibia and fibula. Scale bars = 1 cm.

6.57. Skeleton of the large ornithopod *Camptosaurus aphanoecetes*, as mounted at the Carnegie Museum. *Photo by author. (Courtesy of the Carnegie Institute, Carnegie Museum of Natural History.)*

weight-bearing digits, but the hand retained five digits. My measurements of Yale Peabody Museum specimens YPM 1877 and YPM 1880 suggest that most adult *Camptosaurus* individuals weighed about 785 to 874 kg (1,731 to 1,927 lb). An average adult was about 6 m (19.7 ft) long, and larger individuals were 7 to 8 m (23 to 26 ft; Erickson 1988). The smallest individual *Camptosaurus* found in the Morrison Formation was a 24-cm-long (9.4 in.) embryo found at Dinosaur National Monument in Utah, though this specimen may be of a separate species of *Camptosaurus* (Chure et al. 1994; see below; fig. 6.56). *Camptosaurus* was an herbivore and fed on low-growing vegetation (fig. 6.57); it may have maintained a quadrupedal stance while feeding. It appears to be more commonly associated with sandstone deposits than are most other dinosaurs, to a degree that suggests it may have been ecologically segregated to some degree from other taxa (Dodson et al. 1980; Foster 2013). Some of its food may have been fairly tough, because the teeth are often well worn. Studies of iguanodontians suggest they may have been able to attain running speeds of close to 25 km (15 mi) per hour.

CAMPTOSAURUS APHANOECETES

Camptosaurus aphanoecetes was named as a new species from Dinosaur National Monument relatively recently (fig. 6.57; Carpenter and Wilson 2008); it was later placed in the new genus *Uteodon* (McDonald 2011) but returned to *Camptosaurus* in part because one of the elements used to establish it as a new genus (a braincase) turned out to belong to *Dryosaurus* instead (Carpenter and Lamanna 2015). *C. aphanoecetes* has shorter cervical vertebrae, longer caudal vertebral neural spines, a more curved than straight scapular blade, and a relatively long, low deltopectoral crest on the humerus, compared with *C. dispar* (Carpenter and Wilson 2008). *C. aphanoecetes* also may have been a frequently quadrupedal animal. Considering the relative age of the Dinosaur National Monument Carnegie Quarry (~150.9 mya), where *C. aphanoecetes* was found, versus even high levels at Como Bluff (Quarry 9 at ~152.5 mya, and most *C. dispar* material is from the lower Quarry 13; Trujillo and Kowallis 2015), it appears likely that *C. aphanoecetes* is at least a couple million years younger than *C. dispar*; the differences between the species may then not be surprising.

The mammals of the Morrison Formation were all mouse- and squirrel-sized creatures weighing about 5 to 150 g (0.2 to 5.3 oz). Most were less than 100 g (3.5 oz). The lower jaws of most species are less than about 25 mm (1 in.) long. These were descendant species of some of the world's first mammals from approximately 75 million years before. Mammals were first found in the Morrison Formation at Como Bluff, Wyoming, in the 1870s and are now known from more than 25 sites. The mammals are as diverse in the Morrison Formation as are the dinosaurs and comprise seven taxonomic groups (fig. 6.58).

Mammals of the Morrison Formation are known mostly from teeth or lower jaws; only a few are known from upper jaw elements, and fewer still have any other skull material known. Fossils of a number of species from the Morrison indicate that the lower jaws of these mammals still retained some of the bones typically occurring in premammalian ancestors. In modern mammals, the lower jaw is composed entirely of the

Jurassic Fur: The Mammals

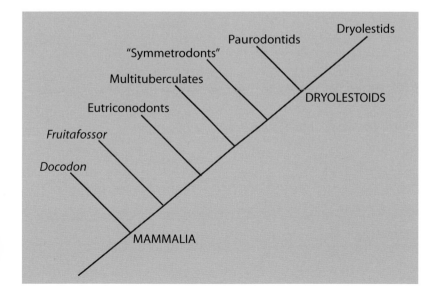

6.58. Relationships of mammalian groups of the Morrison Formation. For reference, modern marsupial and placental mammals are derived from relatives of dryolestoids. Monotremes, as they relate to Morrison taxa on this diagram, may be the sister taxon to dryolestoids and their relatives, or they may be between docodonts and other mammals. Based on Luo et al. (2002), Kielan-Jaworowska et al. (2004), and Luo and Wible (2005).

dentary bone, whereas in reptiles there are several other bones behind the tooth-bearing dentary. In the ancestors of mammals, the bones behind the dentary are reduced in size and through time are slowly lost. Morrison mammals show evidence that splinters of one or two of these bones remained in the lower jaw and thus that these mammals were not quite modern. Some of the bones lost from the lower jaw became part of the middle ear, and those in Morrison mammals still contacting the dentary probably were also involved in the ear to some degree but were somewhat larger and less efficient than those in modern mammals. Like other Mesozoic forms, Morrison mammals were almost certainly furry, as shown by a well-preserved Early Cretaceous mammal skeleton from China (Ji et al. 2002). It is unknown whether mammals of the Late Jurassic had live birth or laid eggs. Either is possible for any or all mammals of the Morrison Formation. Some may have given birth to less-developed offspring, similar to modern marsupial mammals.

Mesozoic mammals in general demonstrated a variety of ecologies from large carnivores (large being medium-dog sized in the case of *Rapenomamus* from the Cretaceous of China) to burrowing insect eaters to gliding mammals to beaver-like species and **omnivores** and insectivores. They also probably were active at a variety of times during the day, though by far most were probably nocturnal (Maor et al. 2017). Mammals even today seem to bear the morphological signals of having spent tens of millions of years during the Mesozoic as mostly nocturnal species (Heesy and Hall 2010; Gerkema et al. 2013), while their synapsid ancestors before the age of dinosaurs seem to show a range of activity preferences among species (Angielczyk and Schmitz 2014) and by the Cenozoic mammals as a group become active during all parts of the day. This suggests that to some degree the time of dinosaurs was one during which mammals remained small and largely nocturnal in order to avoid "antagonistic interactions" with predatory dinosaurs (i.e., getting eaten; Maor et al. 2017). Still, with so many mammal species active at night, there was an opportunity for any that wanted to brave the daylight, and it is likely a few species took advantage of that, perhaps venturing out in the twilight at morning and evening. And certainly the fact that more than a few predatory dinosaurs appear to have been nocturnal too (Schmitz and Motani 2011) suggests that nighttime was not entirely safe and secure for the mammals; so the opportunity may have been worth the somewhat increased risk for some species and possibly even some Morrison mammals were active in the morning and evening or daytime.

Docodonta

Docodonts are a group of Mesozoic mammals with such unusual tooth structure that no one is sure exactly to what other mammalian groups they are related—they left no modern descendants. The lower jaws had seven or eight molars, each rectangular in occlusal view (looking at the chewing surface) and with several bulbous **cusps** of differing heights and

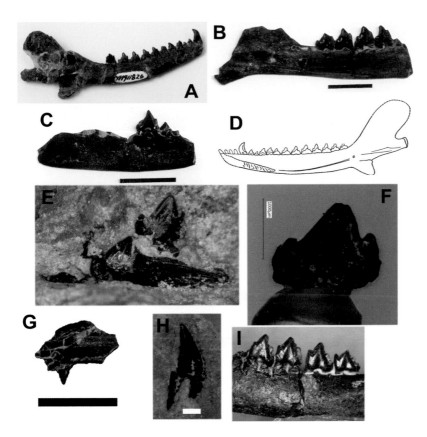

6.59. Elements of the mammal *Docodon* from the Morrison Formation. (A) Complete lower jaw (YPM 11826) from Como Bluff. (B) Partial lower jaw (SDSM 60480) from the Little Houston Quarry, Black Hills. (C) Jaw fragment (SDSM 26911) from Little Houston. (D) Outline of a lower jaw from Simpson (1929). (E) Jaw fragment with a molar and an erupting tooth (SDSM uncataloged), Little Houston. (F) Upper molar from Ninemile Hill, Wyoming (UW specimen). (G) Left maxilla fragment with a canine(?) (MWC specimen), Little Houston. (H) Isolated canine tooth (SDSM uncataloged). (I) Close-up view of four molars of SDSM 60480 in labial view, showing the complex shearing surfaces. Scale bars in B, C, and G = 1 cm. Scale bars in F and H = 1 mm. *Photo in A courtesy of the Peabody Museum of Natural History, Division of Vertebrate Paleontology, Yale University, peabody.yale.edu. Photomicrograph in F by Laura Vietti.*

with distinct shear surfaces between them. The group is represented in the Morrison Formation by the genus *Docodon*.

DOCODON

D. VICTOR

D. APOXYS

The unusual Jurassic mammal *Docodon victor* was named by Marsh in 1880 (fig. 6.59). Other named species include *D. striatus*, *D. affinis*, *D. crassus*, and *D. superus*, this last species consisting of the upper dentitions (Marsh 1880; Simpson 1929). Zofia Kielan-Jaworowska et al. (2004) believed that all of this material was likely referable to *D. victor*, as Farish Jenkins (1969) had previously concluded that *Docodon* may be monospecific, although he did not formally synonymize them. And recently Schultz et al. (2017) formally synonymized Marsh's various species under *D. victor*. *Docodon apoxys* was named recently from several jaws from the Small Quarry at Garden Park, Colorado, that have distal molars reduced in size relative to the more forward molars (Rougier et al. 2015). (Keep in mind that docodonts had up to eight molars in each lower jaw.) Several other species of docodonts have been found in the

Mesozoic outside North America, and the genus probably also occurs in the Early Cretaceous in Great Britain.

Docodon is one of the more common and widespread mammalian genera in the Morrison Formation, occurring at several sites around Como Bluff and at Garden Park, as well as at the Little Houston Quarry in northeastern Wyoming (Foster et al. 2006). Interestingly, it is unknown or very rare at the FPA and the Rainbow Park microvertebrate sites on the Colorado Plateau. Malcolm McKenna and Susan Bell (1997) have listed the **Docodonta** (including *Docodon*) as an outgroup of the **Mammalia**, essentially relegating *Docodon* and its relatives to a status of mammal-like but not true mammals. Some more recent analyses include the docodonts within Mammalia, however.

Docodon also retained, as the lesser part of its lower jaw articulation, a bone called the articular. This bone is the only one involved in the articulation at the back of the lower jaw in reptiles, but in modern mammals, the articulation is formed by the dentary, and what used to be the articular has become part of the middle ear. Like many mammals of its time, *Docodon* had the articular in an "intermediate" position. The fact that docodontids have no modern descendants, however, shows that the trend toward reduction and loss of postdentary bones in the lower jaw and the incorporation of some into the ear was not a stepwise process from one group to the next within mammals as a whole but rather one that occurred in all groups of Mesozoic mammals, even those with no modern representatives. *Docodon* may have had a middle ear that contained only the stapes bone, and a groove in the back part of the lower jaw probably contained, in addition to the articular and angular, the **coronoid** and splenial (Kielan-Jaworowska et al. 2004).

It has been suggested that the Portuguese docodont *Haldanodon* (Lillegraven and Krusat 1991) was semiaquatic or **fossorial** (burrowing), and *Docodon* may have been as well (T. Martin 2005). Their common preservation in freshwater deposits in the Morrison Formation may reflect a semiaquatic mode of life, although it does not confirm it. If they were semiaquatic, certain docodonts may have lived like modern muskrats or desmans, which feed on aquatic plants, vertebrates, and invertebrates (such as insects and freshwater shrimp) but which also forage on land (Macdonald 2001). Middle and Late Jurassic docodonts from China with specializations for arboreal and burrowing modes of life, respectively, indicate the ecological diversity of the group (Luo et al. 2015; Meng et al. 2015). The abundance of burrowing or semiaquatic docodonts of the Morrison Formation, relative to more surface-dwelling forms from other groups, may indicate either that the docodonts were more common or that they were, living close to or in the water, more likely to be preserved in these wet environments that more often bury fossils (Foster et al. 2006; Foster and McMullen 2017). The structure of the skeleton of at least the docodont *Haldanodon* is similar to those of some modern burrowing animals as well; the skull is relatively flat, and the humerus and femur are robust. The truth is, we don't really know yet just how Morrison docodonts lived (Kron 1979; Kielan-Jaworowska et al. 2004); mostly we know that they were unusual animals compared with what we know today.

Docodonts appear to have been omnivores and to have eaten at least some plant material in addition to small invertebrates. The multiple cusps of docodonts have numerous shearing surfaces, which would have been useful in processing small, soft-bodied invertebrates and adult insects, but the teeth overall have relatively shorter, blunter cusps and greater occlusal surface area than those of most other Jurassic mammals, other than multituberculates, and these features probably allowed for greater flexibility of diet and for the inclusion of small seeds and other plant material.

My estimate of the mass of *Docodon* is approximately 142 g (5 oz). This is based on comparisons with the lengths of the lower jaws of modern small mammals and their observed masses (Foster 2009). *Docodon* thus would have been about the same weight as a small squirrel. Its jaw was about 35 mm (1.4 in.) long. As Morrison mammals go, however, *Docodon* was huge. The average adult weight for mammal species in the Morrison Formation was just about 49 g (1.7 oz). *Docodon* was the largest, and the smallest was just 5 g (0.18 oz).

Undetermined Group

FRUITAFOSSOR

This surprising new mammal is based on a nearly complete skeleton found by Wally Windscheffel at the Tom's Place Quarry at the FPA. A partial skeleton has also been found very recently at the Cisco Mammal Quarry in Utah (Davis et al. 2018). The animal is unusual for a mammal in that it has a dentition reduced to simplified peg-like teeth, a robust humerus, and large, spade-shaped claws on the forelimb (fig. 6.60; Luo and Wible 2005). The forelimb appears specifically adapted for digging. The teeth suggest a diet of insects, perhaps termites or ants. As an apparent burrowing insectivore, *Fruitafossor windscheffeli* is the most clearly specialized mammal ever found in the Morrison Formation. Its adaptations

6.60. Skeletal reconstruction of the burrowing, insectivorous mammal *Fruitafossor*. Note the peg-like teeth, spade-like claws on the forefeet, and robust humerus. *Drawing by Mark Klingler; courtesy of the Carnegie Museum.*

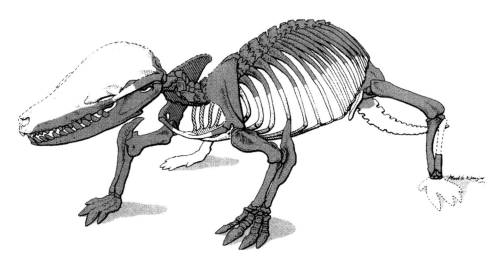

for feeding on termites and ants and for digging are not seen again until the evolution of paleanodonts, armadillos, moles, and aardvarks more than 85 million years later, so this is a prime example of evolutionary **convergence**, the appearance of similar structures in unrelated species as a result of similar functioning and lifestyles. *Fruitafossor* also has the smallest jaw and apparently the lightest mass of any mammal from the Morrison Formation. It probably weighed just 5 g (0.18 oz; Foster 2009). Although it was initially unclear to what group *Fruitafossor* belonged, and although it is still unique in the Morrison Formation, the recent analysis of Huttenlocker et al. (2018) suggested that *Fruitafossor* was near the base of a clade of mammals that eventually led to monotremes (a group today represented only by the duck-billed platypus and echidnas).

Multituberculata

Like the docodonts, multituberculates left no modern descendants. They appeared early in mammalian history and survived the Cretaceous-Tertiary extinction, but they disappeared several tens of millions of years later after the rise of the rodents. The overall construction of their skulls and lower jaws is rodent-like in having deep, robust lower jaws; a reduced number of greatly enlarged and lengthened incisors; a significant gap between the incisors and the next teeth back; and rectangular molars. The multituberculates were unique, however, in possessing one or more premolars in the lower jaw that were enlarged and laterally compressed with diagonal ridges along the sides sweeping back along the tooth. The lower premolars of multituberculates almost look like miniature, half power-saw blades. The molars were reduced to two (on each side in the upper and lower jaws) and were unique in having chewing surfaces consisting of several rows of many small cusps. All cusps were of equal height and were perfectly aligned in anteroposterior rows like ranks of soldiers.[10] Into the grooves between the rows of cusps fit the cusps of the opposing tooth, and the multituberculates had an unusual chewing stroke in which the teeth came together and then the lower jaws slid straight backward.

Multituberculates may have appeared in the Middle Jurassic or possibly the Late Triassic, but the oldest undisputed remains are from the Late Jurassic of Portugal, just before those of the Morrison Formation. These mammals presumably had hair and were thus warm-blooded. Unlike the docodonts and many other Late Jurassic mammalian groups, almost all multituberculates had a lower jaw consisting entirely of the dentary bone, and the other bones seen in docodonts, for example, had moved to the middle ear or been lost. Multituberculates may have had a sprawling limb posture, and various taxa may have burrowed, lived in trees, or progressed by hopping. Unfortunately, not enough of the skeleton of any multituberculate from the Morrison Formation is known to be certain if any had one of these modes of life.

Ctenacodon

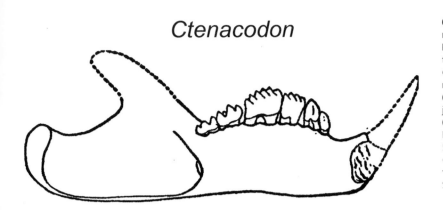

6.61. Multituberculate mammals of the Morrison Formation. *Top,* lower jaw of the multituberculate mammal *Ctenacodon.* Total length ~18 mm (0.7 in.). From Simpson (1929). *Middle,* partial lower jaw of possible *Psalodon* (SDSM 26912) from the Little Houston Quarry. Scale bar = 5 mm. *Bottom,* lower jaw of *Glirodon* (LACM specimen) from the Fruita Paleo Area. Scale bar = 1 cm.

Psalodon

Glirodon

Multituberculates of the Morrison Formation belong to four groups: the Allodontidae, the Zofiabaataridae, a possible Plagiaulacidae, and an indeterminate form (Kielan-Jaworowska and Hurum 2001).

Allodontidae

CTENACODON

Marsh named two species of *Ctenacodon*, *C. serratus* and *C. laticeps*, from Quarry 9 at Como Bluff, Wyoming. Paleontologist G. G. Simpson later named *C. scindens*. *Ctenacodon* is a smaller multituberculate for the Morrison at about 21 g (0.74 oz; Foster 2009) and, like the other multituberculate forms in the formation, has several bladelike premolars increasing in size posteriorly (fig. 6.61).

Ctenacodon and other multituberculates may have lived on the ground surface and in and around bushes. Although some multituberculates from the Morrison may have burrowed, there is no direct evidence of this. The diet of multituberculates has been debated, although it is generally agreed that they must have been herbivorous to some degree. Simpson concluded that "plagiaulacoid" and ptilodontoid multituberculates were largely herbivorous. David Krause's study of *Ptilodus* indicated that this genus and ptilodontoids generally were probably omnivorous. *Ptilodus* (a form from the Cenozoic) is not much larger than the Morrison multituberculate taxa, and although it has a single, larger, bladelike fourth premolar, *Ptilodus* has a fairly similar lower dentition to the Morrison taxa *Ctenacodon* and *Psalodon*, which have four smaller, interlocking, bladelike premolars. Krause also noted that ptilodontoids were probably too small to be entirely folivorous (leaf eating). Mesozoic mammal specialists Bill Clemens and Kielan-Jaworowska (1979) concluded that, with the exception of taeniolabidids, multituberculates were omnivorous and fed on plant material and various invertebrate prey items when available; such omnivory was also the conclusion for a very well-preserved Jurassic multituberculate from China that was of a grade similar to most Morrison Formation multituberculates (Yuan et al. 2013). It is most likely, then, that multituberculates of the Morrison Formation were omnivores and ate a variety of plant material and seeds as well as insects and other small invertebrates (Simpson 1926c; Clemens and Kielan-Jaworowska 1979; Krause 1982; Kielan-Jaworowska et al. 2004).

PSALODON

Psalodon was recognized as a separate form and named by Simpson on the basis of upper jaw material described by Marsh as *Ctenacodon potens* (*P. potens*) and *Allodon fortis* (*P. fortis*). Simpson also named ?*Psalodon marshi* for lower jaw elements probably attributable to this genus (fig. 6.61). *Psalodon* is similar to *Ctenacodon* but is noticeably larger. The type material comes from Quarry 9 at Como Bluff, and a lower jaw probably belonging to *Psalodon* was one of the first mammals found by my crews at the Little Houston Quarry in northeastern Wyoming. *Psalodon* and *Ctenacodon* form a **monophyletic** group within the multituberculates, called the Allodontidae.

Psalodon was the largest of the known multituberculates of the Morrison Formation, weighing about 58 g (2 oz) according to my estimates (Foster 2009). It likely had a similar lifestyle and diet to *Ctenacodon* and other Morrison multituberculates.

Zofiabaataridae

ZOFIABAATAR

This new species of multituberculate (*Z. pulcher*) was named and described on the basis of a complete and relatively robust lower jaw from the Breakfast Bench locality in the eastern part of Como Bluff (Bakker et al. 1990; Carpenter 1998b). This jaw, found in a block of matrix in the lab at the University of Colorado by Carpenter many years ago, is more similar to some Lower Cretaceous forms than it is to other Morrison multituberculates and is from relatively high in the formation. The animal's mass was about 35 g (1.2 oz).

Family Indet.

GLIRODON

This new multituberculate from the FPA and Rainbow Park microvertebrate sites has recently been named and described as *Glirodon grandis* ("grand rodent tooth"; fig. 6.61). *Glirodon* appears to form a monophyletic group with *Bolodon* and two taxa that were previously considered primitive taeniolabidoids, *Eobataar* and *Monobaatar* (Engelmann and Callison 1999). *Glirodon* was the smallest multituberculate in the Morrison Formation at around 17 g (0.6 oz; Foster 2009).

Plagiaulacidae?

MORRISONODON

Morrisonodon brentbaatar was described originally as a species of *Ctenacodon*, with some uncertainty about the genus and family assignment (Bakker 1998a), but it was later recognized to be outside the Allodontidae and possibly a member of the Plagiaulacidae (Kielan-Jaworowska and Hurum 2001). It was given its separate genus name by Hahn and Hahn (2004). *Morrisonodon* is based on a fragment of right maxilla with two upper premolars that appear to be more derived than those of most allodontid multituberculates (Bakker 1998a); it is from the Breakfast Bench locality in the eastern part of Como Bluff. *Morrisonodon*, along with *Zofiabaatar*, may be one of the most derived Morrison multituberculates. It was named after the Morrison Formation and after Wyoming paleontologist and Society of Vertebrate Paleontology auctioneer extraordinaire Brent Breithaupt.

Eutriconodonts get their name from having simple teeth with three cusps in a line. In some forms, the cusps are all the same height, and in others, the center cusp is taller and the two on the sides are shorter and roughly equal in height. Within the Morrison Formation and the Mesozoic in general, they are relatively large mammals, and as seems to be the case with a number of early mammal groups, the eutriconodonts show a mixture of primitive and derived characters in different parts of the skeleton within the same species. Early Cretaceous eutriconodonts from China got just plain gigantic by Mesozoic mammal standards, reaching 13 kg (nearly 30 lb), and were carnivorous, feeding on small vertebrates, including baby dinosaurs. No eutriconodonts from the Morrison Formation were nearly that large or carnivorous, but some may have eaten the occasional small lizard. A well-preserved skeleton of a eutriconodont from China was illustrated by Ji et al. (1999).

The eutriconodonts of the Morrison Formation are of two types: the triconodontids include *Priacodon* and *Trioracodon*, and the "amphilestid" grade group contains *Aploconodon*, *Comodon*, and *Triconolestes*.

PRIACODON

Marsh named *Priacodon robustus* on the basis of a lower jaw, and *P. ferox* from upper and lower dentitions (fig. 6.62). Both species are from the important Quarry 9 at Como Bluff in Wyoming. Simpson named the species

6.62. Jaws of triconodont mammals of the Morrison Formation. *Top, Trioracodon* (YPM 10340) from Como Bluff (courtesy of the Peabody Museum of Natural History, Division of Vertebrate Paleontology, Yale University, peabody.yale.edu). *Bottom, Priacodon.* Total length ~33 mm (1.3 in.). From Simpson (1929).

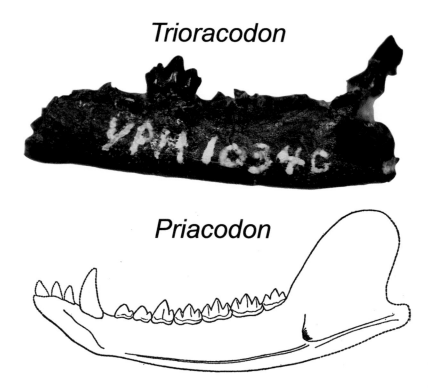

6.63. Restoration of the Morrison mammal *Priacodon*. *Drawing by Thomas Adams.*

P. gradaevus and *P. lulli* from upper dentitions from Quarry 9, and Tom Rasmussen and George Callison described *P. fruitaensis* from a lower jaw (and later referred material) at the FPA in Colorado. Thus, if all described upper dentitions matched one of the lower dentitions, there would be at least three species of *Priacodon* in the Morrison Formation. Studies of the relationships of *Priacodon* consistently place it close to *Trioracodon* and *Triconodon*. *Priacodon* and other Morrison eutriconodonts are relatively large mammals for the formation; my estimate of the mass of *Priacodon* is about 111 g (3.9 oz; Foster 2009).

Eutriconodonts may have been **scansorial** (on the ground but capable of climbing) but were most likely ground dwelling (fig. 6.63). The skeleton was relatively robust and the skull large. Eutriconodonts do not appear to have been particularly fast or agile. The association of eutriconodont fossils with bay and lagoon sediments in some formations, along with tooth comparisons to marine mammals, led some to speculate that eutriconodonts were piscivorous. However, the cusp morphology of these animals indicates that early forms may have fed on invertebrates, including insects, larvae, and worms, and that some of the larger Jurassic eutriconodonts may have become at least partly carnivorous (Jenkins and Crompton 1979; Cifelli and Madsen 1998). The eutriconodonts of the Morrison Formation were ostensibly largely insectivorous but were large enough to have occasionally eaten small vertebrates as well.

TRIORACODON

Marsh described *T. bisulcus* from Quarry 9, although he assigned the species to the previously described genus *Triconodon* (Marsh 1880). Simpson (1928) noticed the difference of the specimen from *Triconodon* and gave it the new generic designation. Simpson also recognized and named an additional species of *Trioracodon* from the Purbeck of England. The teeth of *Trioracodon* are similar to those of *Priacodon* (fig. 6.62), but in the former there are only three molars, whereas *Priacodon* has four. *Triora-codon* and *Priacodon* form a monophyletic group along with *Triconodon*, which is known from England (Rougier et al. 1996). *Trioracodon* weighed approximately 101 g (3.5 oz; Foster 2009) and probably had a diet similar to that of *Priacodon*.

APLOCONODON

Simpson named the small and rare *Aploconodon comoensis* from Quarry 9 on the basis of a jaw fragment with two molars. The molars of *Aploconodon* (along with those of *Comodon* and *Triconolestes*) differ from those of *Priacodon* and *Trioracodon* in that the central cusp is the largest and the two flanking cusps are much smaller (what has been called an "amphilestid" type of eutriconodont tooth structure, after the genus *Amphilestes*). The latter two genera, again, have molars with three cusps of equal height. The molars of *Aploconodon* have particularly small accessory cusps. *Aploconodon* was much smaller than *Priacodon* and *Trioracodon* at approximately 18 g (0.6 oz). This size difference and the disparate molars suggest a different diet for *Aploconodon*, although exactly how it differed is difficult to say. Perhaps individuals of *Aploconodon* fed almost exclusively on insects and other small invertebrates.

COMODON

Comodon started out with the name *Phascolodon*. Simpson (1929) named *Phascolodon gidleyi* from fragments of a jaw, with four molars, from Quarry 9. That name turned out to be preoccupied by that of a single-cell organism named back in 1859, so the new name *Comodon* was proposed. This species has molars of the same style as *Aploconodon*, with accessory cusps that are about half the height of the main cusp. *Comodon* was larger than *Aploconodon* and was average sized for a mammal from the Morrison Formation, weighing about 52 g (1.8 oz).

TRICONOLESTES

This new genus was named by George Engelmann and George Callison (1998); it is a new eutriconodont from Rainbow Park in Dinosaur National Monument, Utah. The molars are of the amphilestid type, but the cusps are recurved, unlike any other Morrison eutriconodont. The mass of *Triconolestes* was probably about 43 g (1.5 oz).

"Symmetrodontans"

"Symmetrodontans" are the rarest and least diverse group of mammals in the Morrison Formation, and they are similarly rare in other formations during the Jurassic. Their tooth structure is distinctive, and the molars each have only three cusps. Unlike in eutriconodonts, however, the three cusps are not in a line parallel to the jaw but rather form a

Amphidon

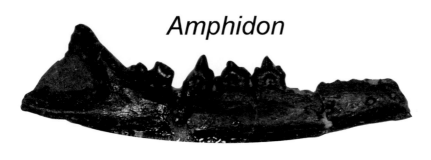

Tinodon

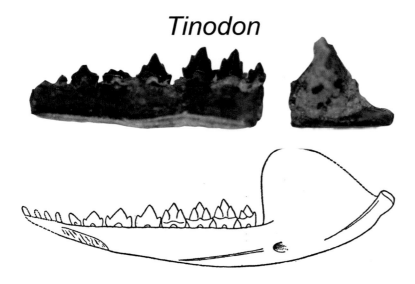

6.64. Jaws of "symmetrodont" mammals of the Morrison Formation. *Top, Amphidon*, Como Bluff. *Middle, Tinodon* jaw, Como Bluff. *Bottom*, Drawing of a complete *Tinodon* jaw. Total length ~28 mm (1.1 in.). *Top* and *middle* courtesy of the Peabody Museum of Natural History, Division of Vertebrate Paleontology, Yale University, peabody.yale.edu. *Bottom* from Simpson (1929).

triangle when viewed from above, with the tallest cusp in the middle and the two smaller cusps offset and flanking it. The relationships of the "symmetrodontans" are unclear, but they appear to be a **paraphyletic** group (some members are more closely related to other groups of mammals than to each other, hence the quotation marks around the name). Some seem to be relatively derived Mesozoic mammals not far from the ancestry of the precursors of modern mammalian groups, and some may be only distantly related and not much closer than docodonts. The two "symmetrodont" taxa from the Morrison retained some of the extra bones of the lower jaws, and one was small.

"Symmetrodontans" may have been scansorial, ground dwelling, or possibly fossorial, although no postcranial elements are known from Morrison forms to help resolve this. A nearly complete symmetrodont from China, however, demonstrates that the forelimbs of these animals were slightly more sprawling than those of most modern mammals (Hu et al. 1997). How this might have affected their life habits is unclear. The teeth of "symmetrodonts" were adapted for slicing and have several shearing surfaces. This was ostensibly for breaking up prey such as soft, small invertebrates and not for chewing plant material. The diet most likely consisted of larvae, worms, and small adult insects (Cassiliano and Clemens 1979; Kielan-Jaworowska et al. 2004).

Amphidon

Simpson named and described *Amphidon superstes* on the basis of a lower jaw from Quarry 9 (fig. 6.64). It is the only known specimen of this species and has obtuse-angled molars that are functionally monocuspid with small, indistinct accessory cusps. The small size of the cusps, however, could be due to wear or poor preservation. *Amphidon* is one of the smallest mammals in the Morrison Formation at just 15 g (0.5 oz). The range of body sizes of the vertebrates inhabiting the Rocky Mountain region during the Late Jurassic is well demonstrated by *Amphidon*, as one of the smallest vertebrates of the time, and the sauropod dinosaur *Brachiosaurus*, one of the largest. *Brachiosaurus* weighed approximately 975,000 times as much as *Amphidon*!

Tinodon

Marsh described *Tinodon bellus* and, later, *T. lepidus*, on the basis of lower jaw material from, once again, that mammal mother lode, Quarry 9 (Marsh 1879c; fig. 6.64). *Tinodon* appears to consist of the lower jaws associated with *Eurylambda*, a genus named from upper jaw fragments. *Tinodon* is now also known from the Early Cretaceous of England and possibly the Callison Quarry at the FPA (Crompton and Jenkins 1967; Ensom and Sigogneau-Russell 2000). My estimate of the mass of *Tinodon* is approximately 72 g (2.5 oz).

Eurylambda*

Simpson (1929) named *Eurylambda aequicrurius* (on the basis of a maxilla with one molar from Quarry 9; he had initially (1925) referred it to *Amphidon*. Simpson also noted that *E. aequicrurius* was a possible upper dentition of *Tinodon*, and Crompton and Jenkins (1967) concluded that this was very likely the case. Don Prothero (1981) seemed to follow this, although he kept the separate names for the sake of clarity in his separate phylogenetic analyses of upper and lower dentitions. Finally, McKenna and Bell (1997) and Rougier et al. (2003) officially listed *Eurylambda* as a junior synonym of *Tinodon*.

Dryolestoidea

Dryolestoids are a group of Mesozoic mammals including the families **Paurodontidae** and **Dryolestidae** (note that not all dryolestoids are dryolestids). They are closely related to *Peramus*, one of the genera that are most closely related to fossil and recent marsupial and placental mammals. The dryolestoids are thus as close to modern mammals as

Paurodon

6.65. Lower jaw of the paurodontid mammal *Paurodon*. Total length ~21 mm (0.8 in.). From Simpson (1929).

we see in the Morrison Formation. The dryolestids and paurodontids have the most developed molars of any Morrison mammals. In the lowers, these consist of three main cusps in the anterior part of the tooth, forming a triangle when viewed from above (a structure known in mammalian tooth terminology as the trigonid). Unlike anything seen in other Morrison mammals, however, both paurodontids and dryolestids have another, much lower cusp coming off the back of the lower molars. In later mammals, including those we have today, this posterior structure is elaborated into a multicusped basin known as the **talonid**. What we see in the dryolestoids of the Morrison Formation, with the well-developed but not elaborate trigonid and what could be considered an incipient talonid, are the beginnings of the modern mammalian tooth structure.[11] Dryolestoids were in a sense the great-aunts and great-uncles of the first marsupial and placental mammals.

The teeth of paurodontids and dryolestids are roughly similar, but the cusps of those of the former group are generally shorter. Clear differences are apparent in the numbers of postcanine teeth in the two groups. Modern marsupials generally have four molars in each side of each jaw, and like us, most placentals have three (two if you've had your wisdom teeth removed). However, whereas most Morrison paurodontids had three to four molars in each lower jaw, dryolestids had seven or eight. This and the fact that the teeth are often tall and sharp cusped really can give dryolestid jaws an almost alien look. Another character that distinguishes dryolestids from paurodontids is that the former have tooth roots that are greatly unequal in size. Whereas paurodontids have oval anterior and posterior tooth sockets of roughly similar size in the jaws, dryolestids have one large, bean-shaped root wrapped around the anterior and lateral sides of a small, circular root.

Paurodontidae

PAURODON

Marsh named *Paurodon valens* on the basis of a jaw from Como's Quarry 9 (Marsh 1887a). This species has a relatively short, robust lower jaw with just six postcanine teeth (fig. 6.65). There is some indication that paurodontids may have been scansorial or arboreal: a complete paurodontid postcranial skeleton from Portugal shows well-developed claws on the digits that may have worked well for climbing (Krebs 1991). On the other hand, the high cusps and many shearing surfaces of the teeth of *Paurodon*, along with the short, robust jaw with few molars, have been compared with the jaws of modern golden moles, possibly indicating a diet of earthworms and insect larvae and suggesting possible fossorial habits (Averianov and Martin 2015). The jaw of *Paurodon* was about 22 mm (0.87 in.) long, and the living animal probably weighed approximately 32 g (1.1 oz; Foster 2009).

A reanalysis of type and referred material of paurodontids has suggested that the lower jaw taxa *Archaeotrigon*, *Araeodon*, and *Foxraptor*, along with the upper jaw genus *Pelicopsis*, are junior subjective synonyms of *Paurodon valens*, in some cases representing ontogenetic (growth) stages of the species (Averianov and Martin 2015).

ARCHAEOTRIGON*

Simpson named *Archaeotrigon brevimaxillus* and A. *distagmus*, both on the basis of lower jaws from Quarry 9. The teeth of *Archaeotrigon* are the same size as those of *Paurodon*, and both animals, if separate, may have been similar in mass. This genus has recently been synonymized with *Paurodon* (see above).

TATHIODON

This slender-jawed mammal was found at Quarry 9 and named *T. agilis* by Simpson (1929). Its molars are roughly similar in size to those of *Archaeotrigon* and *Paurodon*, but the lower jaw is notably more lightly built and is thus likely from an animal that was slightly less heavy (possibly a juvenile). It also appears to match the upper jaw genus *Comotherium* and to possibly be closer to dryolestids than to paurodontids (Averianov and Martin 2015).

ARAEODON*

The specimen on which *Araeodon intermissus* is based was found in 1897 at Quarry 9 by the American Museum of Natural History expeditions led by H. F. Osborn. The specimen was described and named by Simpson 40 years later (Simpson 1937). The jaw, although incomplete, appears to have been shorter in length than that of *Paurodon* and from an animal weighing perhaps 17 g (0.6 oz). Two isolated teeth from the Morrison Formation at Dinosaur National Monument have recently been referred to *Araeodon* (Engelmann and Callison 1998), though the genus has recently been synonymized with *Paurodon* (Averianov and Martin 2015).

FOXRAPTOR

The paurodontid species *Foxraptor atrox* is based on a specimen from the Breakfast Bench locality high in the Morrison at the eastern end of Como Bluff (Bakker et al. 1990). The teeth and probably the jaw were about the same size as in *Paurodon*, so the animals may have been similar in mass, but *Foxraptor* differed from other paurodontids from the Morrison in having five molars.

EUTHLASTUS

Euthlastus cordiformis was named by Simpson (1927) on the basis of an upper jaw fragment with four molars. Like so many other mammals from the Morrison Formation, it was found at Quarry 9. Originally described as a dryolestid, it has since been reidentified as a paurodontid (McKenna and Bell 1997; T. Martin 1999), but whether it represents the upper dentition of a form known from lower jaws or whether it is a distinct species is currently unknown.

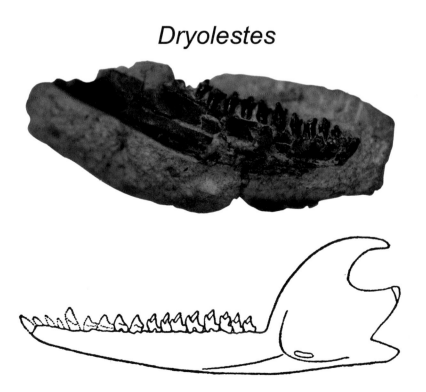

Dryolestes

6.66. Jaws of the dryolestid mammal *Dryolestes*. *Top*, jaw from Como Bluff (YPM specimen; courtesy of the Peabody Museum of Natural History, Division of Vertebrate Paleontology, Yale University, peabody.yale.edu). *Bottom*, drawing of a complete jaw (from Simpson 1929).

From another upper jaw fragment from Quarry 9 (Simpson 1927), this one with three teeth preserved, Simpson named *Pelicopsis dubius*. Simpson described *Pelicopsis* as a probable dryolestid but noted that this mammal probably had only four or five molars and that it may well prove to be a paurodontid instead (Simpson 1929). In recent years, it has in fact been listed as the partial upper dentition of a paurodontid (McKenna and Bell 1997), probably *Paurodon* (Averianov and Martin 2015).

COMOTHERIUM

This genus was found during the AMNH/YPM expedition to Quarry 9 in the late 1960s and named by Prothero on the basis of an upper dentition (Prothero 1981). It was originally described as a dryolestid but has since been identified as a paurodontid (T. Martin 1999) before recently being reidentified as a dryolestid and matched with the lower jaw genus *Tathiodon* (Averianov and Martin 2015).

Dryolestidae

DRYOLESTES

The type specimen of *Dryolestes priscus* (YPM 11820) was the first Jurassic mammal found in North America. Although the genus is known from Quarry 9, the type specimen was collected by Reed in the vicinity of Quarry 5 at Como Bluff in May 1878. Some of the later specimens are in better condition than the type, which contains only one molar. All are lower jaws. The dryolestid genus *"Herpetairus"* was named on the basis of an upper dentition and has since been shown to represent that of *Dryolestes*. *Dryolestes* is larger than most other North American members of the Dryolestidae but is representative of the family in having many (in this case 12) postcanine teeth, including eight molars (fig. 6.66). It also has a relatively slender jaw and molars with tall, pointed cusps. The pair of root openings for each molar is distinctive in *Dryolestes* and other dryolestids in having one large, bean-shaped socket under the anterior end of each molar and a small, circular socket posterior and lingual to that. The lower jaw of *Dryolestes* is about 32 mm (1.3 in.) long, and the whole living creature weighed approximately 104 g (3.7 oz). This is more than five times the weight of the smallest dryolestid from the Morrison Formation. *Dryolestes* is second in abundance in the Morrison Formation only to *Docodon*.

Specimens from the Late Jurassic of Portugal indicate that most dryolestids from this age still possessed coronoid and splenial bones in the lower jaw (Krebs 1971). These are elements that both reptiles and the ancestors of mammals have but that are lost in later mammals. Today we have only one bone, the dentary, as part of our lower jaw. By the Early Cretaceous, dryolestids had lost the coronoid and splenial, but like their

Laolestes

A

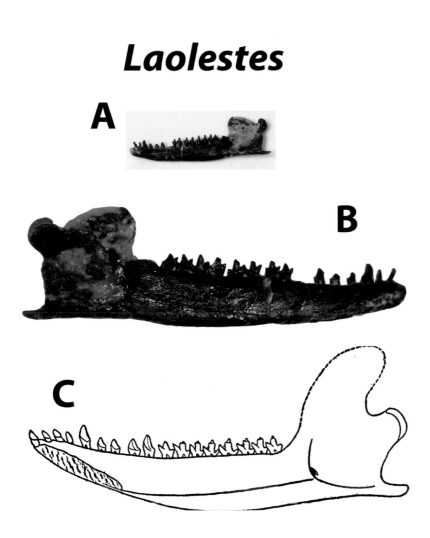

B

C

6.67. Lower jaws of the dryolestid mammal *Laolestes*. (A) Jaw approximately life size. (B) Same jaw in opposite view, enlarged to show detail. (C) Drawing of a jaw. Total length ~27 mm (1.1 in.). A and B courtesy of the Peabody Museum of Natural History, Division of Vertebrate Paleontology, Yale University, peabody.yale.edu. C from Simpson (1929).

Late Jurassic European relatives, *Dryolestes* and other dryolestids of the Morrison Formation probably retained these "extra" lower jawbones of their ancestors. Recall, too, that *Docodon* retained the "reptilian" articular bone in its lower jaw, and *Dryolestes* is much closer to modern mammals than its docodontid Morrison contemporary. If the loss of postdentary bones was a progression that stuck strictly to a phylogenetic sequence, we might expect dryolestids to be considerably more advanced than docodontids and perhaps to have lost all lower jawbones except the dentary. Instead, we see that both groups retained reduced forms of some of these elements; each was probably in the process of a similar but separate trend.

Dryolestids possibly were scansorial or arboreal, but little is known of the postcranial skeleton. They may have been similar in habits to paurodontids, if there are similarities in the skeleton beyond those of the teeth. The noticeably high cusps of the dryolestids indicate a diet of small invertebrates, but the differences in the jaws and numbers of molars between paurodontids and dryolestids suggest some type of dietary separation (Simpson 1933; Kraus 1979).

LAOLESTES

This genus, the third most abundant of the mammals in the formation, is represented by two species (*L. eminens* and *L. grandis*) and was originally identified from Quarry 9 (Simpson 1929). The lower molars of *Laolestes* are unique among dryolestids of the Morrison Formation in that on each, the cusp known as the metaconid is bluntly bifid. The type specimen is a complete lower jaw with most of the teeth, including all the molars (fig. 6.67). This jaw indicates an animal weighing 63 g (2.2 oz). The genus *Laolestes* includes the Morrison taxa "*Melanodon*," which represents the former's upper dentition, and "*Malthacolestes*," which consists of two upper premolars of a young individual that had not yet lost its first set of teeth (T. Martin 1999). Although *Laolestes* is common at Quarry 9 and has also been found at Ninemile Hill, it is unknown from the microvertebrate quarries at Dinosaur National Monument.

AMBLOTHERIUM

This small dryolestid mammal genus was named by Richard Owen in 1871 on the basis of specimens from Dorset, England. In the Morrison Formation, the genus comprised two species named by Marsh and Simpson, but these were later synonymized under Marsh's *Amblotherium gracile*. The dryolestid "*Kepolestes*" was based on a single lower jaw (USNM 2723) from the Marsh-Felch Quarry at Garden Park, Colorado, but it has since proved to be referable to *Amblotherium* (T. Martin 1999). A new, larger species of *Amblotherium* is reported from the Black Hills of Wyoming by Foster et al. (2020). The molar cusps of *Amblotherium* are more pointed and erect than those of *Dryolestes*, and the jaw of *A. gracile* was only about 18 mm (0.71 in.) long, whereas the new species' jaw was closer to 26 mm (1 in.) and had molars nearly 88% larger than those of *A. gracile*. These jaw lengths suggest animals weighing just 19–33 g (0.7–1.2 oz).

MICCYLOTYRANS

Miccylotyrans minimus was found at Quarry 9 and based on an upper jaw with a canine and three molars. Only one specimen of this small dryolestid is known, and little else can be determined about the animal at this time.

Walk and Don't Look Back: The Footprints

As mentioned in another section, footprints are actually common in the Morrison Formation, and there are now more than 70 track sites known in this unit (updated from Foster and Lockley 2006; Lockley et al. 2018).

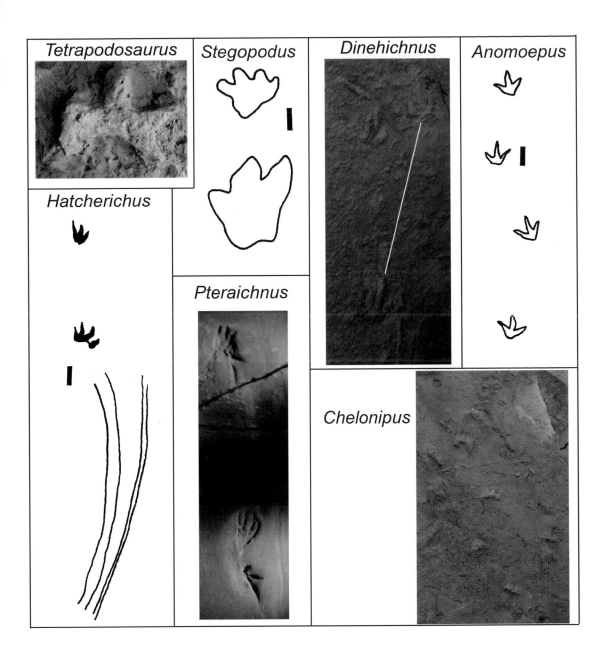

6.68. Ornithischian and nondinosaur track types of the Morrison Formation. *Tetrapodosaurus*, ankylosaur (unnumbered Museums of Western Colorado cast); *Stegopodus*, stegosaur; *Dinehichnus*, small ornithopod (field photo); *Anomoepus*, small bipedal neornithischian; *Hatcherichnus*, neosuchian crocodyliform; *Pteraichnus*, pterosaur (UMNH VP 580, photo by author, courtesy of the Natural History Museum of Utah); *Chelonipus*, turtle (field photo).

Tracks have been found of most of the common types of animals, from a range of dinosaurs and several nondinosaurian groups.

The first documented footprints found in the Morrison Formation were in the southern Black Hills of South Dakota. They were found by George Wieland near Hot Springs and identified by Marsh at Yale in 1899. The original track slabs are in the basement of the Yale Peabody Museum. These were mostly tracks of medium to large theropod dinosaurs, but Marsh identified one as possibly belonging to an ornithopod (likely *Camptosaurus*). New sites showed up every few years until about 1986, when track research took off in the western United States in general.

The types of tracks preserved in the Morrison Formation sample include the following.

Turtles

Turtles are common in the Morrison Formation, so it is in a way surprising that only three sites have been reported that contain footprints of turtles (**ichnogenus** *Chelonipus*). The sites are in Utah north of Lake Powell, northwest of Moab in Utah, and in Colorado National Monument in western Colorado (Foster et al. 1999; Lockley and Foster 2006; Lockley et al. 2018), although a fourth site in the Bighorn Basin of Wyoming contains unidentified tracks that have been interpreted as those of turtles (Jennings et al. 2006). The first site, near Lake Powell, contains more than 20 natural sandstone casts of perhaps two trackways of turtle footprints in the Salt Wash Member. The tracks are about 2 to 3 cm (0.8 to 1.2 in.) across and are closely spaced, indicating that the turtle or turtles progressed at a walking pace across soft mud. The second site, northwest of Moab, contains a few tracks up to 4 cm (1.6 in.) across that are among a small but diverse assemblage of other track types. The third site, in Colorado National Monument, has tracks that appear similar in the size, spacing, and total numbers (fig. 6.68), but the tracks at this site were made in sand.

Lizards

One lizard track has been reported from a sandstone in the Morrison Formation at the FPA in western Colorado (G. Callison, pers. comm., 2001), but I have been unable to relocate this site.

Crocodiles

Crocodile tracks were not reported in the Morrison Formation until quite recently. In 1984, Paul Olsen of Columbia University and Kevin Padian of the University of California at Berkeley proposed that the ichnogenus *Pteraichnus* (which had been described as the trackway of a pterosaur in 1957 by William Lee Stokes) was in fact made by a crocodilian, what were interpreted as impressions of the elongate fourth digit of the hand (the main bones of the wing) of a pterosaur being reinterpreted as the drag marks of the claws of the hand of a crocodilian (Padian and Olsen 1984). This was the first indication of crocodilian tracks in the formation, and for a number of years, most ichnologists agreed with the interpretation. Beginning in the early to mid-1990s, however, new trackways began to emerge that suggested to some that *Pteraichnus* was indeed pterosaurian after all. This split still exists, but in 1994 it became moot in relation to the question of the presence of crocodilian tracks, when uranium miners near Moab, Utah, reported to Martin Lockley (of the University of Colorado at Denver) tracks that were certainly those of a large crocodilian. These

Facing, 6.69. Saurischian dinosaur track types of the Morrison Formation. (A) Track-way of *Parabrontopodus* (sauropods) at the Purgatoire River, Colorado. (B) Trackway of *Brontopodus* (sauropods) at Hidden Canyon Overlook, Utah (Salt Wash Member). (C) Single track of *Parabrontopodus*, Purgatoire. (D) Large theropod dinosaur track of *Hispanosauropus* (allosaurid?) from Copper Ridge, Utah (Salt Wash Member). (E) Large theropod track from Purgatoire. (F) Various tridactyl tracks from North Valley City, Utah (Salt Wash Member). (G) Medium-sized theropod track from Tse Tah Windmill, Arizona (Salt Wash Member). (H) Medium-sized theropod track from Tse Tah Wash, Arizona (Salt Wash Member). (I) Small theropod track from eastern Como Bluff, Wyoming.

were named *Hatcherichnus sanjuanensis* (fig. 6.68), and referred to that name as well was a track originally reported in 1903 from near Cañon City, Colorado (Foster and Lockley 1997). Several years later, I also found *Hatcherichnus*-like tracks in sandstones in the Morrison Formation east of Lander, Wyoming, and Lockley has just recently found more in the Salt Wash Member north of Moab, Utah (Lockley and Foster 2010). The

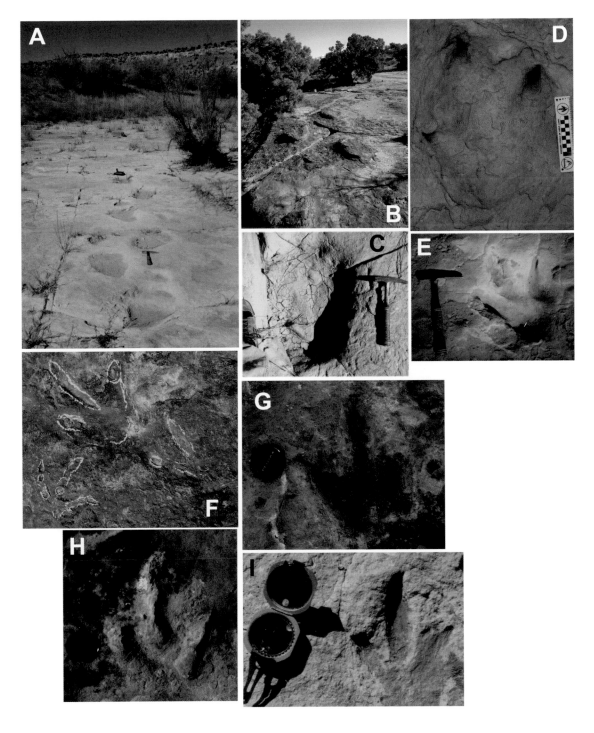

type tracks of *Hatcherichnus* from Utah may be "push-off" marks where only the first three digits of the manus contacted the sediment (digits IV and V are very short in most crocodilians), and where no heel was pressed into the sand in the pes print.

The occurrence of crocodilian tracks in the formation is expected because these animals are nearly as common as turtles at some fossil sites. The only thing that is surprising is that we had not found more of them by now.

Pterosaurs

As mentioned above, the first pterosaur tracks identified from the Morrison Formation were those of a single trackway in the Salt Wash Member in Arizona, found by Stokes in 1957 (fig. 6.68). This specimen was named *Pteraichnus saltwashensis* and is at the Natural History Museum of Utah in Salt Lake City (Stokes 1957; Breithaupt and Matthews 2017, 2018). Numerous pterosaur tracks have been found in the formation since then, mostly in the lower levels and particularly in the area around Alcova, Wyoming (Connely 2006). These tracks show that pterosaurs spent time moving about on sandy substrates and that they progressed often on all four limbs with the wings folded back past their elbows. The identification of *Pteraichnus* as tracks of pterosaurs has not been without its controversies, however; Padian and Olsen (1984) suggest they may be crocodilian, while others believe that Stokes had it right in the first place (Lockley et al. 1995; Mazin et al. 1995; Lockley et al. 2001).

Theropods

There are several general types of theropod tracks that have been found in the Morrison Formation, but few of these have been formally named (fig. 6.69). The large theropod tracks, previously known only as "allosaurid" tracks, are generally 30 to 45 cm (12 to 18 in.) in length and have deep, thick digit impressions; some of these, particularly a trackway at Copper Ridge, north of Moab, Utah, have recently been identified as belonging to the ichnogenus *Hispanosauropus*, an **ichnotaxon** previously known only from the Late Jurassic of Europe (Foster 2015b). They appear consistent with having been made by large theropods such as *Allosaurus* and are found at a number of sites, including the Copper Ridge site in Utah. This intercontinental co-occurrence reiterates the faunal similarities between Europe and North America during the Late Jurassic. Other large theropod tracks from the Purgatoire River site in southeastern Colorado have recently been identified as belonging to the ichnogenus *Megalosauripus* (Schumacher 2018).

Small theropod tracks, also known as "grallatorid" tracks, are 8 to 20 cm (3 to 8 in.) long and are similar in overall shape to the Early Jurassic ichnogenus *Grallator*. The digits are often relatively more slender and pointed than those of the "allosaurid" tracks. They have been found at several sites, but some of the smallest come from sites near Dry Poison Creek and Little Houston Creek in north-central and northeastern Wyoming, respectively (Foster and Lockley 1995, 2006).

"Wide-splay" theropod tracks are known only from the State Bridge site in central Colorado, not far from the Colorado River. The tracks at this site are about 20 cm (8 in.) wide and have a wide divarication between the slender digits. It is unknown what theropod these tracks may belong to.

Sauropods

Sauropod trackways generally consist of pairs of large pes impressions and significantly smaller manus impressions. The manus track is usually oval and often indistinct, but the pes is large, with digit I facing forward and the other digits rotated progressively toward the lateral edge. The claws of the pes were only on digits I–III and were pointed anterolaterally and rotated so that their medial edges pointed somewhat ventrally; there was a large and thick pad behind the metatarsals (Bonnan 2005).

Sauropod tracks in the Morrison Formation are of two types. Those referred to the ichnogenus *Brontopodus* ("thunder foot") are found at the Hidden Canyon Overlook site in Utah and several others. These trackways consist of hind footprints up to 60 cm (24 in.) long, with relatively large front footprints and a wide-gauge track pattern (fig. 6.69). The term "wide gauge" means that the tracks of the right and left sides are relatively widely spaced on either side of the midline, in this case about 10 cm (4 in.) between the midlines of the medial edges of the pes tracks. *Brontopodus* tracks in the Morrison Formation may have been made by camarasaurid or brachiosaurid sauropods (Farlow et al. 1989).

Sauropod tracks of the ichnogenus *Parabrontopodus* are best known from the Purgatoire River site in southeastern Colorado (fig. 6.69), where they exist in multiple parallel trackways. These tracks are about the same size as *Brontopodus*, but the front footprints are relatively smaller. *Parabrontopodus* is a "narrow-gauge" sauropod trackway because the right and left tracks are much closer together (Lockley et al. 1994; Farlow 1992). In fact, the hind footprints of opposing sides often overlap the midline. The tracks belonging to this ichnogenus may have been made by diplodocid sauropods. The gauge of these sauropod trackways indicates something that trackways can tell us about the way the animals moved—something that we hadn't seen from just looking at the skeletons. Look at most of the sauropod skeletons mounted in various museums around the country, and you'll notice that most are standing still in a four-point stance like a forlorn horse braced against a gale. The legs are straight up and down, often with the hind feet having a full meter (3 ft) between their inner toes. Wide-gauge or narrow-gauge sauropod trackways in the Morrison Formation show us that these animals did not walk this way. They may have stood so, but when walking, the legs must have angled slightly inward, and their feet fell close to or just barely on the midline—as opposite as one can get from the old reconstructions of sauropods with sprawled limbs.

Stegosaurs

Stegosaur tracks had not been identified in the Morrison Formation until 1996, when they were reported from near the Cleveland-Lloyd Quarry near Price, Utah (Bakker 1996). Those were hind footprints. The first named stegosaur tracks were a front and back print set from near Arches National Park in Utah, named *Stegopodus* in 1998 (fig. 6.68; Lockley and Hunt 1998). Stegosaurs have five digits on the front feet and three weight-bearing ones on the back. The shape of stegosaur tracks had been predicted (on the basis of the skeletal foot structure) in 1990, so the discovery of tracks that matched several years later was a good example of the interdisciplinary utility of track and bone studies (Thulborn 1990). Since 1998, several more hind footprints of stegosaurs have been identified, and the stegosaur ichnogenus *Deltapodus* (also identified in Europe, Africa, and Asia) has been reported from the Morrison in eastern and southeastern Utah, including a full 11-step trackway (Gierliński and Sabath 2008; Lockley et al. 2017, 2018). At least one more example of *Stegopodus* has been identified near Moab, Utah, as well (Lockley et al. 2017). In general, however, stegosaur tracks are still relatively rare (Lockley et al. 1998b, 2017).

Ankylosaurs

Unknown from the Morrison Formation until just a few years ago, ankylosaur tracks assigned to the ichnogenus *Tetrapodosaurus* have been found recently in Cactus Park (western Colorado) and near a golf course in Grand Junction, Colorado, by Kent Hups and Ken Cart, respectively. As with bone material of ankylosaurs, there was no evidence for decades, and then in the space of the last couple years, there are the tracks, suddenly and in more than one place!

Ornithopods and Bipedal Basal Neornithischians

The bipedal neornithischian tracks known from the Morrison Formation are mostly small, and all are tridactyl imprints with digits generally more widely splayed than in the average theropod track. Tracks referred to the ichnogenus *Anomoepus* (fig. 6.68) are known from western Colorado and eastern Utah (some with scale impressions; Lockley et al. 2018). They lack the hand impressions seen in this ichnogenus in the Early Jurassic of New England. Another ornithopod trackway type was recently named *Dinehichnus* (fig. 6.68) on the basis of several parallel trackways from the Salt Wash Member at Boundary Butte in extreme southern Utah (Lockley et al. 1998a). These tracks have widely splayed digits and show a rotation inward, but they also have distinct separated heel impressions

on each foot. *Dinehichnus* tracks may represent dryosaurid ornithopods, and the parallel nature of their trackways suggests that these dinosaurs were social animals that sometimes moved around in groups.

Although larger tridactyl tracks assigned to possible large ornithopods (*Camptosaurus* would be one possible track maker) have been known from the formation for some time (Marsh 1899), they are rare. These tracks are unnamed, and in addition to the Black Hills they have been found in areas such as the Salt Wash Member south of Green River, Utah.

Mammals(?)

Some apparent burrows in the Morrison Formation in Colorado and Utah have been attributed to small mammals of the time (Koch et al. 2006; Raisanen and Hasiotis 2018). These animals appear to have been digging vertical shafts and more horizontal tunnels, turnarounds, and chambers. If these preserved structures are indeed burrows, it would indicate (as with neornithischian burrows mentioned earlier; Varricchio et al. 2007) that the animals were capable of behavior as complex as much of what we see in digging animals today.

Localities in the Morrison Formation with a significant amount of egg shell preserved are rarer than those with other types of fossil evidence, but some important sites do exist, and collectively, they preserve a range of shell types.

The first site found to contain eggshell in the Morrison Formation was the Young Egg Locality near Delta, Colorado (R. Young 1991). As related in chapter 4, this site was discovered by geologist Bob Young in the mid-1980s and is in the middle Salt Wash Member of the formation. The eggshell from this locality is of the prismatic basic type (named as the **oospecies** *Preprismatoolithus coloradensis*) and may belong to small hypsilophodontid dinosaurs, although this is far from certain. The abundant eggshell at this quarry suggests that it was a nesting site, but the lack

Another Generation: The Eggs

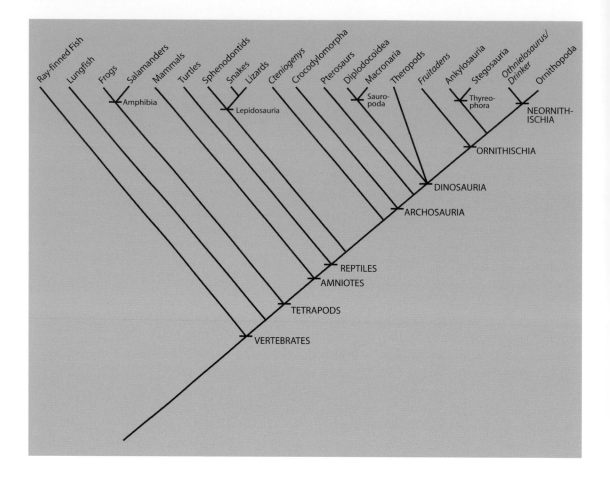

6.71. Relationships of various vertebrate groups known from the Morrison Formation.

of directly associated embryonic material prevents us from being able to be sure of the egg layer's identity. Eggshell belonging to the *Preprismatoolithus* oogenus has also been reported from the Kirkland Locality at the FPA, the Uravan Locality, and the Garden Park site.

There are at least five types of eggshell known from the Morrison Formation as a whole, but none of these types can be assigned with certainty to any one Morrison skeletal taxon (Hirsch 1994; Bray and Hirsch 1998). The quarries such as the Young Locality and Egg Gulch at Garden Park do, however, confirm that Morrison dinosaurs were egg layers and that they used specific nesting sites (fig. 6.70). The eggs from this period appear to have been smaller and thinner than most known from Cretaceous rocks, although the reason for this is not yet clear.

The Morrison Family Tree: Phylogenetic Systematics

How, then, are all these 100 or so types of vertebrates related? We've seen how mammals and several groups of dinosaurs are related within their own clans, but we now should take a look at the overall picture of vertebrates in the formation. Figure 6.71 shows the relationships of the vertebrates of the Morrison Formation in a diagram called a **cladogram**. Unlike traditional trees, a cladogram shows only degrees of relatedness (the order of branching) and implies no direct ancestry. None of the taxa in a cladogram is said to give rise to any others. As you move up the cladogram from the bottom left of the figure, each split (or **node**) represents an evolutionary divergence. These nodes are sometimes labeled with the names of the groups they represent (in fig. 6.71, labeled nodes are indicated with a horizontal line), and every branch, species, or group above that node is a member of that larger group (a clade). Thus, in figure 6.71, the sauropod dinosaurs, for example, are members of the **Saurischia** clade within the Dinosauria, the Dinosauria clade within archosaurs, and the archosaur clade within reptiles.[12] All sauropods are saurischians and dinosaurs and archosaurs, but not all archosaurs are dinosaurs. The labels at the nodes, then, indicate the clade of which every member on every branch above it is a part. Any species or group whose branch is further down the axis to the left is not a part of that clade. For example, the Choristodera is on a branch that intersects the main line above the node labeled "Reptiles" (fig. 6.71) but below that labeled "Archosaurs." This indicates that *Cteniogenys*, the Morrison choristodere, is a reptile but not a true archosaur.[13] It also implies that *Cteniogenys* is more closely related to all archosaurs than it is to other reptiles such as lizards, snakes, and turtles.

Another important thing to remember about cladograms is that the nodes represent divergences from hypothetical common ancestors. As mentioned above, no species are suggested to be directly ancestral to others. That is why nodes are always labeled with clade names and never species names. Even if one species is older than another on a cladogram and the two appear to be closely related, we never go so far as to assume that we have found a direct ancestor in the older species. Given the incomplete nature of the vertebrate fossil record, this caution is well justified.

Each clade is also defined by specific characters of the skeleton that each member shares and that are absent from species outside the clade. For example, all member species (and their ancestors) of the tetrapod clade had four limbs with digits, which distinguishes the group from all other vertebrates (fishes). Similarly, all amniotes (mammals and reptiles) are distinguished from other tetrapods (mainly amphibians) by the presence of an amnion, a membrane surrounding the embryo that allowed the development of shelled eggs and that is absent in amphibians; and all archosaurs are distinguished from other reptiles by several features, including an antorbital fenestra (an opening in the bones of the skull in front of the eyes). Until recently, it appeared clear that the members of two groups of dinosaurs, the Saurischia and the Ornithischia, each had, among several other characters, pelvic structures unique to their clade. This may not be an important character in the relationships of these groups, however, as we will see later. The two pelvic bones that project downward from the fused sacral vertebrae and the ilium (the other pelvic bone) are the pubis and ischium. In sauropods and most theropods (except dromaeosaurs, for example), the pubis projects down and slightly forward and the ischium down and toward the back end.

In ornithischians, however, both pubis and ischium project down and back, with a small, newly developed process projecting anteriorly from the pubis in some cases.[14]

So looking at figure 6.71, we see that lungfish (in the Morrison Formation, several species of *Potamoceratodus* plus *Ceratodus robustus*) are more closely related to tetrapods than they are to other fish. This is also reflected in the fact that the earliest tetrapods appear to have evolved from lobe-finned fish. Lungfish and other lobe-finned fish (plus their tetrapod descendants) form a clade separate from the ray-finned fish (the latter of which, in the Morrison, include *Morrolepis* and *"Hulettia"*). Amphibians, mammals, and reptiles each form separate clades, with mammals + reptiles composing the amniotes. Mammals are distinguished from other Morrison amniotes (reptiles) by the construction of the posterior part of the skull and its single opening behind the orbit on each side. Within the Morrison mammal group, docodontids and multituberculates are unique and probably left no descendants; eutriconodonts, some "symmetrodonts," and dryolestoids (paurodontids + dryolestids) are likely progressively more closely related to modern mammals. No mammals of the Morrison Formation had teeth of what is called the "tribosphenic grade," a form of tooth structure seen in the ancestors of today's marsupial and placental mammal groups. But Morrison dryolestoids are one evolutionary step below the first tribosphenic mammals.

Sphenodontians and lizards form a clade separate from other Morrison reptiles. The archosaurs share a number of features, including the antorbital fenestra, mentioned above, and having teeth set in sockets. The flying pterosaurs are the closest relatives of the dinosaurs but are not actually dinosaurs themselves.

Dinosaurs are defined by several features of the skull and ankle (Sereno 1999) and used to be split into two groups based on hip structure. The relationships of early dinosaurs are no longer so clear; as more taxa belonging to the basal branches of the three main groups have shown up in the fossil record, the relationships among those three groups have become a bit muddied. A reassessment of the basal Theropoda, Sauropodomorpha, and Ornithischia (mostly of the Triassic period) recently suggested that theropods were actually allied with the ornithischians (in the **Ornithoscelida**), rather than with sauropodomorphs (Baron et al. 2017; similar results found by Parry et al. 2017), a major shake-up in the long-standing (and long taken for granted) basic classification of dinosaurs. The differences between apparent evolutionary paths leading to the traditional arrangement (theropods + sauropods) and that leading to the Ornithoscelida (theropods + ornithischians) are relatively slight, with tree length differences of a number of steps but only 1% or so in tree length, so it is difficult to say if this new classification will prove robust. Another team has since reinterpreted the same data set, rescoring some characters based on firsthand knowledge of the specimens, and recovered the traditional theropod-sauropodomorph pairing (Langer et al. 2017), though they admitted that support for it was not particularly strong either. What the Baron et al. (2017) and Langer et al. (2017) papers do show, however, is that we do not really know with much certainty how the theropod, sauropodomorph, and ornithischian lineages are interrelated.

Whether they are closer to sauropods or ornithischians, the theropods of the Morrison Formation (*Allosaurus*, *Ceratosaurus*, and *Coelurus*, for example) are all bipedal carnivorous dinosaurs. The sauropods (e.g., *Camarasaurus*, *Apatosaurus*, and *Diplodocus*), on the other hand, are all large, quadrupedal plant eaters.[15] Within Ornithischia (again, whether they are on their own or allied with the theropods), the armored ankylosaurs such as *Mymoorapelta* and *Gargoyleosaurus* form a clade known as the Thyreophora ("shield-bearers") with the plated and spiked stegosaurs *Stegosaurus* and *Hesperosaurus*. The **Neornithischia** includes the bipedal herbivores *Nanosaurus* and the ornithopods *Dryosaurus* and *Camptosaurus*.

The origins of each group from the Morrison Formation go back well before the Late Jurassic. The first fish appeared during the Cambrian period, although fish of the type seen in the Morrison did not come along until much later in the Paleozoic. Lungfish and amphibians also originated during the Paleozoic, and even the line that eventually led to mammals split from other amniotes in the later Paleozoic, although true mammals did not evolve until the Late Triassic (~225 mya). Still, this means that mammals had been around for some 75 million years by the time those of the Morrison Formation lived—that is more time than has elapsed since the demise of the last dinosaur species. The earliest representatives of the amniote groups (synapsids leading to mammals, early reptiles) split off near the end of the Paleozoic, but most archosaur groups seen in the Morrison came along later—semiaquatic crocodilians at the middle of the Jurassic (the roots of the cursorial crocodilian groups are back in the Triassic), pterosaurs during the Late Triassic, and dinosaurs in the Late Triassic, around the same time as mammals. The oldest choristodere, *Cteniogenys*, is known from Middle Jurassic deposits in Europe, and the group may be no older than Early Jurassic, but other archosauromorphs are known from much older sediments.

Within Dinosauria, primitive theropods are known from the Late Triassic Chinle Group of New Mexico, Arizona, and Utah, and ancestral members of some Morrison theropod groups are known from the Triassic–Jurassic boundary. Prosauropods (a paraphyletic group including probable ancestors of true sauropods) are well known from the Late Triassic and Early Jurassic. True sauropods are known from the Early Jurassic in Africa and several other places, but there have been recent reports of true sauropods in the Late Triassic as well. Primitive ornithischians may have been found in rocks of Late Triassic age in North America, but we haven't been able to confirm this. As we have seen, most immediate groups of Morrison Formation dinosaurs, such as the megalosaurid and ceratosauroid theropods and the ankylosaurs, stegosaurs, and ornithopods, are known from Early to Middle Jurassic rocks in other areas.

The Mess and the Magic: Vertebrate Paleoecology of the Morrison Formation

7

Scientists need reverie. We need long walks and quiet times at the quarry to let the whole pattern of fossil history sink into our consciousness.

Robert Bakker, *The Dinosaur Heresies*, 1986

As we have seen in previous chapters, the Morrison Formation contains an abundance of vertebrate taxa and thousands of fossils, but what do all these individual bones tell us? What do the data actually mean? To determine this, we need to take a detailed look at as many aspects of the Morrison's fossil record as possible. The first step in doing this is to compile lists of as many localities in the formation as possible. From there, one can list all the vertebrate taxa to have been found at each site, and finally, one can list counts of the numbers of bones or skeletons of each species at each site. This requires a lot of digging through collections and catalogs and other papers at a lot of museums and localities around the country and around the world.[1] But the work gives us a total sample from the Morrison Formation, and these census data form the basis of the analysis of the vertebrate fauna of the Morrison Formation that we will look at now. Much of this work builds on that of researchers such as Dale Russell, Peter Dodson, Robert Bakker, Kay Behrensmeyer, Jack McIntosh, and Jim Farlow, all of whom, as we will see, paved the way for much of what is now possible in Morrison Formation paleoecological studies (Dodson et al. 1980; Russell 1989; Farlow 1976, 1987, 1990, 1993; Russell and Farlow in Coe et al. 1987). Supplementing this work with the better understanding of the Morrison climate and physical setting provided by recent multidisciplinary research group studies led by Pete Peterson and Christine Turner, along with other group projects and broad paleoecological studies, only adds to the improved picture of the environment, setting, and paleoecology of the Late Jurassic in North America (Turner and Peterson 2004; Farlow et al. 2010; Noto and Grossman 2010; Whitlock et al. 2018).

One goal of large-scale paleoecological studies is to identify possible temporal (time-related), geographic, or paleoenvironmental variations in the distributions of taxa and guilds. Although guilds were originally defined as groups of related animals that make their living the same way, the term has more recently come to mean simply a group of species (related or not) with a similar ecological function. On the African savanna, for example, lions and cheetahs would be part of the carnivore guild, and zebras and wildebeest would be part of the large-grassland-herbivore guild. Changes in paleoecological structure within a geological formation can be identified by means of the varying numbers of taxa, relative abundances, and biomass of genera or guilds. Guild analysis allows groupings of different, sometimes unrelated taxa united by their similar paleoecological roles and facilitates analysis of the ecological structure of a paleocommunity.

The ecological characterization of each known taxon from the Morrison Formation includes the estimated body mass, the diet or feeding mode, and the locomotion and habitat mode (where an animal lives and how it moves around). For example, the characterization of these three variables for the theropod dinosaur *Allosaurus* would be as follows: (1) a mass of about 1,000 kg (2,205 lb) as an adult; (2) a diet of meat, obviously—a feeding mode of "large carnivore"; and (3) a terrestrial habitat, because it appears best adapted to dry ground and is not, for example, semiaquatic. Two other applicable variables, feeding habitat and shelter habitat, are not included in our look at the Morrison because in almost all cases in which these can be reliably determined, they are the same as the taxon's locomotion and habitat mode. With the characterizations combined with the census data, the changes can then be identified in diversity, relative abundance, and biomass of different guilds through time, across geographic areas, and in different environments.

Where the Vertebrates Are Found

One of the first things it would be interesting to know is the characteristics of the quarries in the Morrison Formation from which vertebrates have been collected. What types of environments do they represent? Which types contain the greatest diversity of animals? To do this, we need to define some categories for environments. Paleoenvironmental categorizations of the Morrison Formation would probably not be defined the same way by any two researchers, but there are some areas of general agreement, and I based my categories on the lithofacies definitions from the work of Dodson and others in 1980. I subdivided some of their lithofacies on the basis of specifics of the taphonomy seen in certain sites, but otherwise the two systems are similar. The categories by which we might classify the paleoenvironments of the quarries in the Morrison Formation are as follows: (1) river channels, (2) sand splays, (3) damp (poorly drained) floodplains, (4) ponds, (5) dry (well-drained) floodplains, and (6) lakes and wetlands (Dodson et al. 1980; Foster 2003a).

River channel deposits are usually composed mostly of sandstone and may represent perennial rivers or ephemeral streams. Representatives of river channel quarries are the Carnegie Quarry at Dinosaur National Monument, the Dry Mesa Quarry, the Hanksville-Burpee Quarry, and the Bone Cabin Quarry. The Gnatalie Quarry near Blanding, Utah, may represent a sandy braided stream. Sand splays are formed during floods as river flow breaks through a levee and spreads a thin fan of sand onto the floodplain. These deposits are different from the river channels themselves in that they are generally thinner, more laterally restricted in extent, and flatter bottomed (channels are often convex down from having scoured into the floodplain). Sand splays contain fewer fossils than channels, at least in the Morrison Formation. Quarries in sand splays include the Cleveland Museum Quarry in Garden Park and possibly the *Dystrophaeus* site in Utah. Damp floodplain deposits consist of green and gray mudstones laid down on the plains, mostly during floods. Some of these muds were probably later covered with low-growing vegetation, and others may have seasonally contained shallow water and thus formed ephemeral ponds. Some damp floodplain deposits include weakly developed paleosols, and others include a high percentage of dark-gray to black, carbonaceous mudstone. Damp floodplain deposits frequently contain vertebrate fossils in the Morrison Formation, and Quarry 10 at Como Bluff (where Bill Reed found Marsh's *Brontosaurus excelsus*) and some of the Breakfast Bench sites are examples. Ponds in our context are close to river channels and may be next to levees, slightly farther out on the floodplain, or may even be abandoned channels themselves. The sediments in pond deposits are mostly silts, fine-grained sands, and claystones. Two types of pond deposits can be distinguished on the basis of the taphonomy of their small vertebrate bones. One kind of pond deposit often contains many disarticulated small bones in thin but dense layers, while the other contains less abundant bones in thicker, less dense accumulations but often with significant articulation of the specimens, including complete skeletons. The microvertebrate sites at Quarry 9 at Como Bluff, the Small Quarry at Garden Park, and the Little Houston Quarry are examples of the first type of pond deposit; the second type is exemplified by the microvertebrate sites at the Fruita Paleontological Area, Rainbow Park, and the Wolf Creek Quarry. Dry floodplain deposits are composed of red to maroon mudstones that represent parts of the Morrison plain that were generally higher above the water table than other muds and were thus better drained of moisture. Ground moisture also evaporated out of the muds. Many dry floodplain deposits are in fact incipient to well-developed ancient soils. Many contain evidence of roots and vegetation in the muds. Fossils are known from a few dry floodplain deposits, including some of Cope's quarries high in the formation at Garden Park and the Twin Juniper Quarry (apatosaur and ceratosaur) near the Mygatt-Moore Quarry. Lake and wetland deposits consist of interbedded gray limestones and gray mudstones. A few vertebrates are known from these types of rocks, including fish, crocodiles, and sauropod dinosaurs at the Purgatoire River site in Colorado, at Garden Park, and at Blacktail Creek in the Black Hills of Wyoming.

A survey of more than 170 quarries in the Morrison Formation indicates that 41% are in damp floodplain deposits and 36% are in river channel deposits. The other quarter of localities are in the other environments (lakes and wetlands, 8%; dry floodplains, 7%; ponds, 6%; sand splays, 2%). These numbers are updated from my earlier work (Foster

2003a) and are similar to results that Sharon McMullen got in a separate study of Morrison fossil occurrences (McMullen 2016). More than three-quarters of quarries in the Morrison Formation are in environments representing river channels and relatively wet parts of the floodplain. To some degree, this distribution depends on how common particular rock types are in the formation, but not entirely. Along the Front Range of Colorado, limestones of the lake and wetland deposits outnumber channel sandstones. But the distribution does not fully reflect the habitats of the animals; rather, burial and preservation potential are higher in rivers and floodplains than in some other areas.

Stratigraphically, nearly two-thirds of quarries in the Morrison Formation occur in the upper half of the formation, at the lower to middle levels of that half. The lower half of the formation contains just 23% of the localities, so we know comparatively little about the first several million years of Morrison time. There are actually a number of bone occurrences low in the formation (in the Salt Wash Member, for example), but few of these have been developed into quarries, in part because of the difficulty of working in a sequence of rock that is largely sandstone benches and cliffs.

Quarries are also relatively sparse near the top of the formation. Most of what we can say with confidence about the vertebrate faunas of the Morrison is really mainly true of the middle to late Kimmeridgian, the early second half of Morrison time. The early years and those near the end of deposition are a little hazy. Indications suggest that there may have been a slightly different fauna during that first half and that some changes were beginning late in Morrison time, but these fossil collections need much improvement before we can be sure of the patterns.

Geographically, quarries range nearly the entire surficial extent of the Morrison Formation, from central New Mexico, Arizona, and Oklahoma in the south to central Montana in the north, and from central Utah east to the Front Range of Colorado and the Black Hills of South Dakota.

Abundances and Diversities

The core of these and other analyses was the database of quarry faunal lists and specimen counts. Compiling the database involved three main, sometimes tedious, steps: (1) listing the known localities, (2) listing the known taxa from each site, and (3) counting specimens of each taxon from each site. The first two could often be done simultaneously. Many papers have been published over the years about specimens from various quarries in the Morrison Formation. Digging through these to compile faunal lists for each site is not as slow as it might seem because most paleontologists share scientific papers freely and hoard their collections like pack rats, and what we don't have, a colleague undoubtedly will. The real work is in the counting of specimens. This can be done only by visiting museums and pulling open drawers. If you're lucky, the collections manager will have a computer printout of their collection, and all you have to do is check the bones against the catalog and figure out what may go with what as one individual. The process of specimen counts takes time. The problem, of course, is what qualifies as one specimen. An isolated bone from one site obviously represents one individual, but what about many disarticulated bones of the same type of animal from one quarry? How many individuals does that actually represent? The goal here is not to just count each species once, no matter how many bones are present, but also not to overestimate a species abundance by counting every bone as a separate individual. We are, after all, trying to get a realistic picture of which species were abundant and which were rare, so being too lax in the counting method doesn't help us.

There are two main methods to count fossils from one site more realistically. One method, known as minimum number of individuals (MNI), adds up how many bones of each skeletal element there are from each species and determines the lowest number of individual animals they represent. For example, in a completely disarticulated sample of 95 caudal vertebrae, 29 ribs, 3 left humeri, and 5 right femurs, all of *Camarasaurus*, we could assume that each bone had come from a different individual, but it is unlikely we really have 132 camarasaurs represented. We don't really know what goes with what, particularly among ribs and caudal vertebrae, but we can be sure that no individual has more than one left humerus or more than one right femur. We therefore assume the MNI represented by the sample, which would be five *Camarasaurus*, on the basis of the femurs, which indicate at least five animals. Obviously, an articulated animal can be counted as a single individual. MNI counts are used when it is likely that at least some of the disarticulated bones in the deposit belong to the same individual and when this likelihood varies among species represented (for example, when sauropod partial skeletons and individual bones are mixed with isolated bones of rare species). The other method is known as number of identified specimens (NIS). This is just a straight count of specimens, and it is used when the characteristics of the quarry suggest that it is unlikely that any two bones in the deposit belong to the same individual and when this is about equally true of each species represented in the deposit (for example, in a deposit consisting entirely of disarticulated and unassociated small vertebrate remains).

The abundances of each species represented at each quarry were determined by whichever was the more appropriate of these techniques for the individual site. Generally, dinosaur sites were counted with the MNI method, and localities with microvertebrates were counted with NIS. Some adjustments had to be made, however. At the Cleveland-Lloyd Quarry in Utah, the counts included standard MNIs of the other animals but only one *Allosaurus* because there are nearly nine times as many allosaurs in the quarry as there are any other type of animal. This unusual deposit might skew the analysis, and so its allosaur sample was reduced to show only the presence of the species. Similarly, at the Howe Quarry, where there are at least a dozen diplodocid sauropods and a number of other, less common animals, only the individuals that have been positively identified so far have been included in the count. Little of the Howe Quarry material has been prepared, and only these specimens were included in the count.

The straight counts are just one step. With raw numbers like these, we can really start to refine the picture. Our first problem with straight counts is that skeletons of small animals get abused by the elements. The skeleton of a small animal that isn't ingested by a carnivore often gets scattered during scavenging, stepped on, rained on, bleached in the sun, cracked, broken down by microorganisms, and otherwise destroyed by different processes in its environment. Large animal skeletons, on the other hand, can sit on a floodplain being hit with the same processes for decades but still be recognizable. Therefore, the fossil record is greatly biased in favor of large animals. The most abundant dinosaur in the Morrison Formation, after all, is *Camarasaurus*, a sauropod that may have weighed about 14,000 kg (30,865 lb). Generally, however, larger living animals are less abundant numerically than small ones in any one ecosystem. We therefore need a way to adjust our counts so that they are a little more realistic. This is tough to do, but paleoecologist Dale Russell calculated just such numbers a few years ago by using an equation based on the work of Kay Behrensmeyer and others in east Africa. Behrensmeyer and coworkers calculated a regression tying in the mass of modern African savanna species with the ratio of the numbers of their skeletons seen on the floodplain to the numbers of individuals of the same species seen living in the same area. The species' weight correlates to its ratio, so that the smaller animals, though numerous in the populations, are less likely to be observed as skeletons and thus less likely to be preserved. Adapting this regression for the Morrison, Russell calculated adjusted estimated population percentages for different dinosaur species on the basis of the raw counts. Larger species came out less numerically abundant compared with smaller ones than they were in the raw count data, but within a similar body size, the raw counts still had an effect. Even though small species came out much more abundant than large species, a rare, small species was still rarer than more abundant small species (Coe et al. 1987; Russell 1989; both based in part on Behrensmeyer et al. 1979).

We can do the same calculations with our counts of Morrison Formation species, building on Russell's pioneering work and using updated numbers and nondinosaurian taxa. These estimated population percentages for species of the Morrison Formation vertebrate fauna will give us a look at what the ecosystem was like and can form the basis for biomass estimates. Total biomass is an important way to look at species abundance and how ecosystems are constructed. A given species may be rare, but if it is also large, it can still have a major influence on the use of resources and transfer of energy within a community. To calculate total biomass percentages of each of the species of the Morrison, we need to multiply each species' adjusted abundance count by its estimated mass

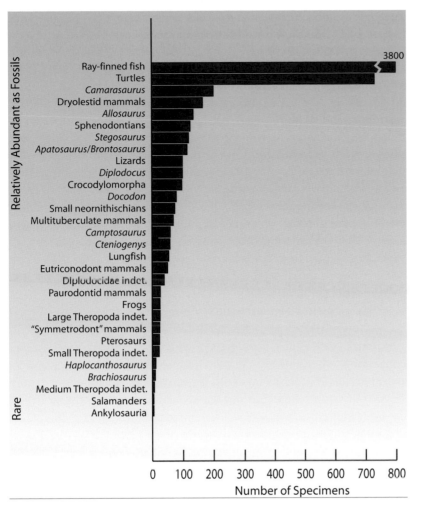

7.1. Abundances of different groups of fossil vertebrates in the Morrison Formation, measured by raw numbers of specimens. Specimens at dinosaur localities with high degrees of articulation or association were counted by MNI; specimens at microvertebrate sites where association was minimal were counted by NIS.

and normalize all the numbers to 100%. This gives the percentage of total biomass represented by each species and can be used to look at the data in a number of ways.

So we've dug through dozens of museum basements and offices, driven out to the middle of nowhere to log data from small, unpublished localities, loaded all the MNI and NIS data into a computer spreadsheet (along with the quarry characteristics), and made adjustments for taphonomic bias and for biomass totals. Now we can start looking at the numbers. We will get to population and biomass percentages later, but for now, one of the first things we can investigate is the straight relative abundances and diversities. If we look at the total sample of vertebrate fossils in the Morrison Formation, we see that by far the most common animals are turtles and fish (fig. 7.1). These two groups outnumber the next most common animal more than two and three to one. The Rodney Dangerfield of sauropod dinosaurs, *Camarasaurus*, is that next most common animal and is both the most abundant dinosaur and the most

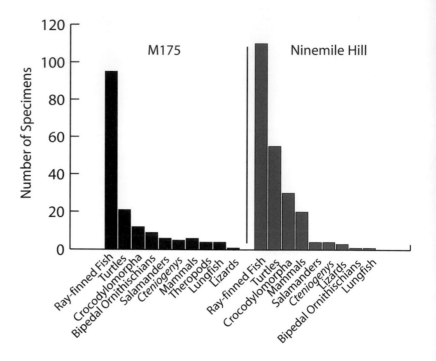

7.2. Abundances of fossil vertebrate groups from two microvertebrate sites in the Morrison Formation of Wyoming, showing great abundance of ray-finned fish and turtles. About half the specimens at each site are fish, and aquatic and semiaquatic taxa combined account for 89% of the samples. Both localities were worked by screen washing soft mudstone matrix at the sites. Counts are NIS. Data updated from Trujillo (1999) and Foster and Heckert (2011).

abundant single genus known from the formation. Although it gets little press in the world of famous dinosaurs, *Camarasaurus* is in fact one of the most important vertebrates of Late Jurassic times.

Interestingly, a group of mammals (the dryolestids) is the fourth most abundant type of vertebrate in the Morrison. This is likely partly the result of the diversity of the group and partly because small animals can be abundant in the quarries at which they are found (no matter how difficult it is to locate them in the first place). The abundance of dryolestid mammals in the sample underscores the critical ecological role of our furry ancestors during the Late Jurassic.

Among the other vertebrate taxa in the Morrison Formation, some of the more common dinosaurs are *Allosaurus*, *Stegosaurus*, *Apatosaurus*, and *Diplodocus*. Other nondinosaurian vertebrates are mixed in among these, and *Camptosaurus* is less common, but all other dinosaurs are rare. Ankylosaurs, not found in the Morrison Formation until 1990, are the rarest dinosaurs.

As we saw in figure 7.1, one of the most abundant groups of vertebrates in the Morrison Formation is the ray-finned fish. What that figure shows is the total, formation-wide count, however. It can also be of value to look at the fish samples from individual sites. Although they are better known for other reasons, usually involving their dinosaurs or other vertebrates, a number of the large quarries in the Morrison Formation contain ray-finned fish fossils. The rare articulated fish found near Cañon City and in Rabbit Valley in Colorado are almost unique occurrences and are found with relatively few other significant fossils in the same layers, but at other sites, huge dinosaurs or terrestrial mammals can be mixed in with fish. For example, Reed's Quarry 9, the Bone Cabin Quarry, the Dry Mesa Quarry, the Small Quarry, the Lindsey Quarry, Stovall's Quarry 8, Rainbow Park, the Little Houston Quarry, Callison's Main Quarry and Tom's Place at the FPA, and the Wolf Creek Quarry all contain fragmentary fish remains among their other vertebrates. Most often preserved are scales, teeth, jaw fragments, and vertebrae. These fish remains are more abundant and more easily found in the Morrison Formation than is often appreciated. In some cases, the best way to find fish, and in relatively abundant numbers, is through screen washing (Foster and Heckert 2011). Collecting soft matrix from sites containing abundant surface fragments of turtles and crocodiles, for example, often leads to the discovery of many more vertebrate types at the site. The soft matrix is put in screen boxes and broken down in water, after which it is dried. The reduced volume of smaller rock pieces is sorted under a microscope, usually revealing numerous small, fragmentary, but identifiable fossils of vertebrates. Mammals, lizards, salamanders, small dinosaurs, and a number of other taxa are commonly identified this way, and at some sites, we find that the most abundant fossils are those of ray-finned fish. At some localities with eight or more species identified, about 50% of the sample are various elements of fish, and the other seven taxa comprise the remaining half of the sample (fig. 7.2). Turtles are also common at these sites, and aquatic and semiaquatic taxa overall consistently account for approximately 89% of the preserved fossils.

At other sites, the presence of fish is indicated indirectly by fossils of certain types of bivalve mollusks, which depend on the gills of fish for the larval stage of their reproductive cycle. These mussels are not uncommon in many channel sandstones of the formation and can be found in quarries with dinosaurs. This is not surprising, given that most vertebrate fossils are preserved in what were freshwater environments 150 million years ago. That turtles and fish are so common in a preserved pond environment into which the terrestrial dinosaurs were washed is not unexpected, but the abundance of aquatic taxa in the Morrison Formation reminds us that perennial water sources were not scarce at the time. And this underscores the importance, to the plants and animals of the Morrison basin, of groundwater and surface runoff from the western mountains. In the subhumid to semiarid climate of the region, rainwater that had not even fallen in this area was still critical to the abundance and diversity of life of the Morrison basin. The results of screen washing have taught us that there is no paucity of sites that contain fish in the Morrison Formation and that within these sites aquatic and semiaquatic taxa can be quite abundant. Clearly, despite apparently high evaporation rates in the basin, the groundwater, western highlands, and seasonal rains kept many (although not necessarily most) ponds, wetlands, and river courses filled with water year-round. And plenty of fish seem to have inhabited these water sources.

Another way to look at abundance is by the number of localities at which a species or group occurs. This helps moderate the effects of

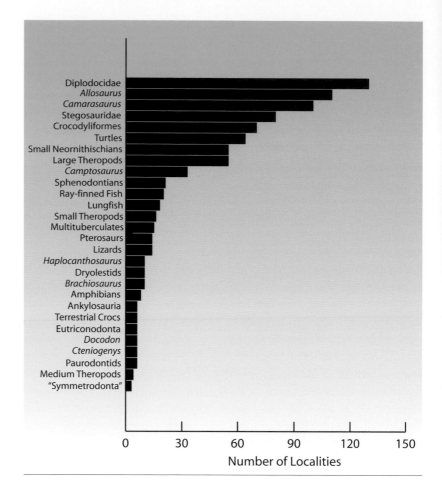

7.3. Abundances of groups of fossil vertebrates in the Morrison Formation, measured by the number of localities at which each occurs.

what happens to be preserved in abundance at particularly rich sites. Viewing the data for vertebrates in the Morrison Formation this way (fig. 7.3), we see again that *Camarasaurus* and *Allosaurus* are among the most abundant taxa. In this case, however, *Allosaurus* is found at more sites, although we saw that *Camarasaurus* is numerically more abundant, indicating that *Allosaurus* is more often found by itself at sites and *Camarasaurus* more with other members of its species. Diplodocids as a group are the most abundant when counting by locality, but in this case, it is because we have combined all species together, including the abundant *Apatosaurus* and *Diplodocus*. *Stegosaurus* demonstrates an abundance similar to that seen in straight specimen counts, and the sauropods *Haplocanthosaurus* and *Brachiosaurus* are relatively rare.

A chart of the diversities of different vertebrate groups in the Morrison Formation (fig. 7.4) shows that the diversity of mammals far exceeds that of all other groups. In fact, there are nearly as many types of mammals in the formation as there are of all four groups of dinosaurs (sauropods, theropods, thyreophorans, and ornithopods) combined. After the mammals, theropod and sauropod dinosaurs are next most diverse, and crocodilians show a surprising array of species for fourth place among

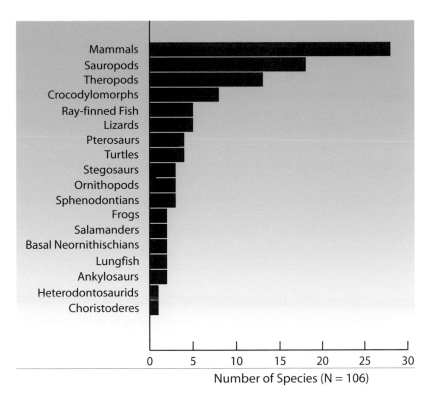

7.4. Diversities of vertebrate groups from Morrison Formation rocks, measured by numbers of species.

the groups. The rich diversity among mammals indicates the degree to which these animals had evolved in the 75 million years the group had existed up to that point. Another 85 million years were yet to go before dinosaurs left the scene and opened the world to serious mammalian diversification, but by the Late Jurassic, mammals had obviously created enough ecological niches for themselves to equal the dinosaurs in species richness, if not in size or flashiness. Their abundance also implies a diversity of prey items and exploited habitats. In order for this many mammals to coexist in the same area, they had to have adopted a variety of lifestyles and had to have had available plenty of the insects, worms, larvae, seeds, and other small vertebrates needed for their respective diets. It is difficult for us to determine the exact lifestyles and diets of individual mammal species of the Morrison Formation, but it is likely that as a group, they partitioned their resources so as to avoid competition with each other as best they could. Some likely fed mostly on insects, others mostly on seeds, and some on small vertebrates at least part of the time; and we seem to have some specialists in eating earthworms (*Paurodon*), in eating termites and ants (*Fruitafossor*), and in living semiaquatically (*Docodon*), so that at least some of this resource partitioning is becoming apparent for Morrison mammals (Averianov and Martin 2015; Luo and Wible 2005; T. Martin 2005). Also, they may have taken shifts, with some being nocturnal or crepuscular (active and feeding at night or in the morning and evening twilight); others may have been diurnal (active and feeding during the day). It was once believed that most Mesozoic mammals were active nocturnally to avoid becoming a dinosaur snack, but there is no reason some mammals would not have been diurnal, because scurrying around at night is no guarantee of safety. After 75 million years of coevolution with mammals, dinosaurs undoubtedly would have figured out that there was good dining at night, and I suspect that some small Late Jurassic theropod dinosaurs had adaptations to hunt in the dark. In particular, *Ornitholestes* had unusually large orbital openings in the skull (see Chure 1998; Carpenter et al. 2005b). This may indicate large eyes, good for seeing mammals and lizards in low light. So day or night may not have made much difference, and mammals may have adapted as much as possible for one or the other. In this way, they would not all be out feeding at once, which would reduce competition. Some mammals of the Morrison Formation may have scampered on the ground surface most of the time, and others may have stayed mostly in trees or bushes, whereas others were burrowers (*Fruitafossor*). *Docodon* may have burrowed or perhaps been semiaquatic. All these were different ways the mammals got by and allowed for a greater diversity than would have been possible if, for example, they all had been strict tree-dwelling, nocturnal insectivores.

The species richness of the various small-vertebrate groups may tell us something as well. The difference in diversity between mammals and other small vertebrates such as frogs and salamanders is interesting, for example. Frogs, salamanders, and lizards all occur together with mammals in the large sample from Quarry 9 at Como Bluff, but there are still only 2 kinds of each of these animals known from the deposit, whereas there are 22 genera of mammals. Amphibians are generally less diverse and abundant than mammals in a lot of modern communities, so perhaps this isn't surprising, but the low abundance (as seen previously) and diversity of lizards in the Morrison Formation are less expected, especially in a semiarid environment. Some of the lizard diversity and abundance would be made up for by the fact that, during Morrison times and unlike in modern communities (outside New Zealand, anyway), sphenodontians probably filled a lizard-like niche. So there may have been eight lizard-like small reptile species (including the five true lizards) and, with about as many species of sphenodontians known as lizards from the fossil sample, twice the abundance. Still, compared with some modern and ancient faunas, this is not particularly rich. One wonders whether the diversity and abundance of small reptiles in the Morrison Formation truly was lower than might be expected.

It has been suggested that modern reptilian biomass and density in Africa are less than in the neotropics because of large herbivorous mammals. Habitat modification by the herbivores is one suggested possible cause of the low reptilian biomass, and the other is the increased diversity of carnivorous animals that mainly feed on the mammalian herbivores but also occasionally feed on reptiles (Owen-Smith 1988). Undoubtedly, in the Morrison Formation, sauropods had a significant impact on the habitat and may have had some effect on small reptilian density and perhaps diversity, if indeed that is one of the causes of the apparent lower modern reptilian densities and if a similar mechanism would have worked in the Jurassic. The modification caused by the large sauropods probably would have been even greater than that caused today by elephants in Africa in which, in severe cases, large areas of woodland can be stripped nearly clean of vegetation.

Another possible cause for the Morrison Formation paleocommunity distribution would parallel the increased-carnivore hypothesis. The relatively high diversity of sauropods (and the associated increased range of herbivore and carnivore body sizes) may have helped allow an increase in theropod diversity. This may have resulted in greater predation on, and thus lower densities and perhaps diversities of, smaller reptiles. Several of the Morrison Formation theropods were fairly small and probably fed on smaller nondinosaurian vertebrates as adults, and juveniles of larger theropod genera likely competed with the smaller species for these smaller items. A young *Allosaurus*, for example, was about the size of adults of theropods such as *Ornitholestes* and *Coelurus* and probably fed on some of the same animals (assuming that *Allosaurus* was not a social animal and that the swift, maneuverable juveniles mainly hunted on their own and did not often feed on large vertebrate carcasses brought down by adults, as young lions do; however, if Bakker is correct that allosaur adults fed the young, then competition with small theropods would be less). So the great abundance and diversity of sauropods would have indirectly had an effect on small reptilian densities by allowing greater diversities and densities of theropod predators and thus greater predation intensities on small reptiles by small theropods (Janzen 1976).

The Upper Cretaceous Djadokhta Formation of China and Mongolia represents a different paleoenvironment from that of the Morrison Formation, but it contains a paleofauna with an ankylosaur, a sauropod, a hadrosaur, and three ceratopsians including *Protoceratops*, as well as the theropods *Tarbosaurus*, *Saurornithoides*, *Oviraptor*, and *Velociraptor*. Preserved with these dinosaurian taxa are 15 genera of lizards. The fact that, in contrast, the 8 lizard-like species in the Morrison Formation are preserved with 10 types of sauropods, 2 ankylosaurs, 2 stegosaurs, and 5 ornithopods may indicate that a less diverse and abundant large herbivore assemblage (in the Djadokhta) indeed allows for a more diverse and abundant community of small reptiles. This assumes, of course, that the differences between the Morrison and Djadokhta are not simply results of abiotic environmental or preservational conditions. The reasonably high total known diversity of vertebrates in the Morrison Formation (90+ species) may also indicate a high degree of habitat heterogeneity, in part because of what we know about plant productivity in modern ecosystems (Rosenzweig 1995). For example, an increase in total plant productivity in a desert environment, changing it to a semiarid grassland, could decrease the original small vertebrate diversity by homogenizing the habitat distribution. Could the vertebrate diversity of the Morrison have been different had the climate been wetter or drier and the level of plant productivity higher or lower? Perhaps. The true vertebrate paleofaunas of the sand dune and interdune paleoenvironments of the Navajo Sandstone of the Early Jurassic in western North America, for example, may have been a little less diverse than that in the Morrison Formation because plant productivity was lower and environments were perhaps more homogeneous in the Navajo. Greater productivity and habitat heterogeneity in the Morrison may have allowed the greater diversity of vertebrates.

Distributions of Guilds

We talked briefly about analyzing the data from the census in terms of guilds in addition to categorizing by taxon. Again, we need to define the guilds and decide to which one any given species belongs. First of all, several important characteristics of a species can be broken down that have an influence on its individual ecology. Three of these are the average adult body mass of the species, its locomotion and habitat mode, and its feeding mode. These categories are some of the most important and definable for most kinds of animals, but they certainly are not all that are possible to assess (Behrensmeyer et al. 1992). (Some characteristics known to be important in living ecosystems may be unmeasurable in paleoecosystems. Much time in paleontology is consumed trying to estimate factors that are obvious and easily measured in living systems.) The average adult body mass influences metabolism, reproduction rate, and a range of other factors in a species' physiology and is important to the ecosystem in determining energy flow within the food web. The trick with fossil species is determining a way to estimate the mass of the animals. Each vertebrate species from the Morrison Formation was assigned to an appropriate weight category on the basis of its mass estimate. These categories were <1 kg, 1–10 kg, 10–100 kg, 100–1,000 kg, 1,000–10,000 kg, and >10,000 kg (1 kg = 2.2 lb). Knowing the accuracy of the mass estimates is difficult, but in some ways, making the estimate methods consistent across the range of species in a group is more important because that will affect the analysis directly, and we really can never know the actual accuracy of our estimates. Consistency in the estimates is the most important because, in this case, it is the mass estimates of the taxa relative to each other that will most affect the analysis.

For this study, we can base our small-vertebrate mass estimates on comparisons with modern, related forms when such are available (fish, frogs, salamanders, sphenodontians, and crocodilians). Most of the mass estimates for dinosaurs that I gave in chapter 6 were calculations based on measurements of well-preserved specimens in museum collections. On the basis of circumferences of the humerus and femur, one can estimate the mass of a quadrupedal dinosaur by using a formula published in 1985 (J. Anderson et al. 1985).[2] Estimates for bipedal animals need only the femur circumference and a slightly different formula. The following two equations give approximate mass estimates for the dinosaurs. For quadrupedal animals, the following equation is used:

$$W = 0.078(C_{h+f}^{2.73})$$

For bipedal animals, the following equation is used:

$$W = 0.16(C_f^{2.73})$$

In these equations, W is the mass (in grams), and C_h and C_f are the humerus and femur circumferences (in millimeters), respectively. These mass estimate computations can be done easily on a hand calculator, and one can usually have an estimate for a species of dinosaur shortly after measuring the specimen. However, this is not the only method of estimating dinosaur masses, and it is certainly not without controversy (see box 7.1). It is one of few, however, that allow you to estimate the masses of bipedal or quadrupedal individuals relatively easily and without needing the entire skeleton to be reconstructed.

Box 7.1—Weighing In

Mass estimation in dinosaurs generally falls into one of two categories: body volume estimation (off scale models) followed by calculation of mass based on an assumed density, or measurement of the robustness of skeletal limb elements and comparison with a range of modern animals' same proportions. Different methods have, historically, contributed to a wide range of mass estimates for the same taxa (see table 1 in Sander et al. 2011).

The method used here for quadrupedal dinosaurs was developed by J. Anderson et al. (1985) and was based on measurements of several dozen species of modern mammals, but because many dinosaur species of the Late Jurassic are larger than elephants, many of the species for which we are calculating estimates are being extrapolated from a pattern of smaller animals. These mammals also have a range of ecologies and locomotor modes and even relatively different metabolic levels. And regardless of limb proportions, the dinosaurs had a variety of skeletal morphologies. So the mass estimates for extinct dinosaurs here are very much ballpark figures. In fact, what has been termed the "traditional" method of dinosaur mass estimation, as exemplified by the J. Anderson et al. (1985) study, has been challenged by some, including Packard et al. (2009), who argued (largely on mathematical grounds) that a nonlinear regression was a better model for the original data set. Cawley and Janacek (2010), however, presented data showing that on modern animals with known masses the Packard et al. (2009) updated formula consistently (and sometimes dramatically) overestimated the masses of small- to medium-sized mammals and may well underestimate the masses of dinosaurs. The matter is far from settled (e.g., Packard et al. 2010), but Cawley and Janacek (2010) also presented data showing that the traditional method (e.g., J. Anderson et al. 1985) more accurately predicted known masses of a wide range of modern mammals. Campione and Evans (2012) updated the traditional method and found that it was reasonably accurate, and Campione et al. (2014) also found that the basic J. Anderson et al. (1985) formula was a "robust, consistent method"; thus, it is used here.

Meanwhile, some recent studies use the body volume method instead (Gunga et al. 2008 on *Giraffatitan*; P. Christiansen and Fariña 2004 on theropods) and produce results somewhat heavier than those produced by the Packard et al. (2009) formula (38,000 kg for *Giraffatitan*) and the traditional method. With the limb element measurement method, Cawley and Janacek (2010) noted that while the modal estimates were more accurate, the range of estimates produced by the traditional method was significant. All of this suggests that, as paleobiologist Jim Farlow once put it, all dinosaurian mass estimates, regardless of the method used, should be taken with "an evaporite deposit's worth" of salt.

In order to put all this into context and perspective, let's lay out what different methods and formulas predict for the masses of some well-known dinosaurs of the Morrison Formation:

| Species | Specimen | C_{h+f} | Mass estimate (kg) by method of | | |
			J. Anderson et al.	Campione and Evans*	Packard et al.
Brachiosaurus altithorax	FMNH 25107	1,604	43,896	50,945	23,954
Diplodocus longus	USNM 10865	1,017	12,657	14,559	8,411
Barosaurus lentus	AMNH 6341	996	11,957	13,747	8,017
Apatosaurus louisae	CM 3018	1,461	34,035	39,411	19,330
Brontosaurus excelsus	YPM 1980	1,291	24,247	28,050	14,549
Stegosaurus stenops	DMNH 1483	739	5,284	6,052	4,039
Gargoyleosaurus parkpinorum	DMNH 27726	362	754	850	784

C_{h+f} = circumference of humerus plus femur (in millimeters); author's direct measurements of specimens.

Mass estimates in kilograms; multiply by 2.2 to convert to pounds.

*Campione and Evans (2012) Equation 1.

You'll notice that among midsized dinosaurs (the smaller ones here) the methods can be quite close, and in fact Packard et al. (2009) predicts a slightly heavier mass for *Gargoyleosaurus*, but among the largest animals there are drastic differences between J. Anderson et al. (1985) and Campione and Evans (2012) on one hand, and Packard et al. (2009) on the other. I have to admit that 8.4 metric tons for an 80-foot-long *Diplodocus* sounds a bit on the light side considering that many estimates for 45-foot-long *Tyrannosaurus* specimens are often in the 7–8 ton range. On the other hand, keep in mind that sauropod specialist Matt Wedel (2005) has suggested that sauropods (at least) may have weighed a good 10% less than we are predicting with these methods, simply due to the degree of pneumaticity in the presacral vertebrae.

But there you have it, a summary of the best that we can do at the moment. And don't forget that salt.

A problem is presented by the mammals of the Morrison Formation. There are mass estimate calculations for small mammals, but most are based on the length or area of the first lower molar. Tooth size and skull size generally correspond to mass and physiology in mammals. Nearly all modern mammals have three or four lower molars, and the size of the first molar correlates well with body size. In Cenozoic mammals, which also have three or four lower molars, such correlations still work. The problem in the Morrison is that, as we have seen, a number of Jurassic mammals had seven or eight molars, and the size of the first of these (or any other) does not necessarily correlate to skull or body size in the same way. The mass estimates for mammals mentioned in chapter 5 are based on measurements of the total length of the lower jaw and comparisons with this measurement in modern small (mostly marsupial) mammals (Foster 2009). Skull size also correlates to body mass, so this estimate should work better for Jurassic mammals than if we tried to measure the teeth.

The locomotion and habitat guild category is assigned to each species based on how it moved around and where it did so. Each species' locomotion and habitat categorization was determined based on comparisons with modern, related taxa or on morphological characteristics. The locomotion and habitat categories were as follows: aquatic for fully water-bound forms (e.g., fish); semiaquatic for animals that spend various amounts of time in water and on land or are dependent on water (turtles, crocodilians, and frogs); terrestrial for animals that spend most of their time on the land surface (dinosaurs); specialized terrestrial for mostly small animals that may be burrowers, part-time climbers, ground surface living, or fully tree living but that, because their postcranial elements are rarely preserved, cannot be further categorized (mammals and lizards); and **aerial** for animals capable of flight (pterosaurs).

Feeding mode is simply a definition of the general diet of a species. The categories we will use for the Morrison Formation include **invertivores**; invertivore/carnivores; omnivores; small, medium, and large carnivores; and low, medium, and high browsers. Invertivores were mainly small forms believed to have fed largely on adult insects, grubs, worms, and other small invertebrates. Animals were assigned to the group on the basis of comparison with related modern groups, such as frogs and lizards; relatively high-cusped, small-crown-area, shearing-type molars in mammals; and the small size of the animals (implying a dependence on a relatively high-protein diet).

Invertivore/carnivores probably fed on invertebrates and some small vertebrates. The types of prey animals used probably differed from one form to another. For example, some aquatic forms may have fed on bivalves, gastropods, and small fish, whereas a terrestrial invertivore/carnivore may have fed on insects and small vertebrates.

Omnivores include forms that appear to have fed consistently on plants and small vertebrates or invertebrates. Determination of members of this group was based on comparison with modern taxa (turtles) or on flat, grinding, or relatively blunt-cusped molars in mammals (multituberculates and docodonts).

Carnivorous groups are those that fed mainly on other vertebrates (at least as adults). Most of these taxa have sharp, sometimes serrated teeth, but in fossil forms for which no skull is known, the presence of sharp claws and the animal's relationship to other carnivorous forms was assumed to imply a similar life habit. Carnivores of the Morrison Formation include theropod dinosaurs, crocodilians, and pterosaurs and were subdivided into small, medium, and large categories on the basis of mass estimates. For our purposes, small carnivores were those under 50 kg (110 lb), medium carnivores were those between 50 and 400 kg (110–881 lb), and large carnivores were those above 400 kg (881 lb). In my early work with the theropods, I'd hoped that they could be subdivided into guilds on the basis of morphological characteristics in a manner similar to University of California–Los Angeles paleontologist Blaire Van Valkenburgh's work on fossil mammalian predators, but too many of the Morrison theropods are missing key skeletal parts (skulls and forelimbs) for this to work just yet. The mass subdivision is at this point the best method for ecologically subdividing the theropod dinosaurs because mass greatly affects prey choice.

The herbivorous species were subdivided into low, medium, and high browsers. Low browsers fed mainly at levels within 2 m (6.5 ft) of the ground surface, medium browsers fed mainly at levels below 4 m (13 ft), and high browsers fed mainly above 4 m (13 ft).

Identifying which of these feeding modes the Morrison vertebrates belonged to was based mainly on their tooth type. Teeth are the key to inferring diet. Unfortunately, Mesozoic herbivores have less differentiated tooth shapes than modern animals. This doesn't mean that they lacked dietary variety or specialization; it may just be that processing of plant material occurred, for many herbivorous dinosaur species, more in the gut than in the mouth. Dinosaurs such as sauropods, ankylosaurs, stegosaurs, and ornithopods have spoon-, peg-, leaf-, or chisel-shaped teeth that are generally agreed to indicate a diet of plants, although it is difficult to determine what type of plants dinosaurs may have fed on the most. The sphenodontid *Eilenodon* is the only nondinosaurian herbivore, although this form may have fed on invertebrates as well.

The estimated browsing heights of the dinosaurs were based on normal head heights of the animals as indicated by relatively complete mounted skeletons. Ankylosaurs, stegosaurs, and ornithopods all were considered low browsers because their head heights were generally

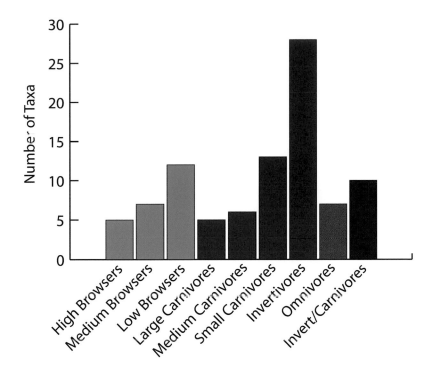

around 1 m (3.3 ft). The ornithopod *Camptosaurus* could have fed a bit higher than 2 m (6.5 ft) but probably stayed closer to the ground. Sauropods were in the medium and high browser categories. Medium-browsing sauropods probably overlapped to a large degree with the low browsers but were able to feed consistently up to 4 m (13 ft) or so. High browsers include brachiosaurids that were probably feeding higher than 4 m (13 ft) some or most of the time. Camarasaurs may have been in this group as well, although recent work suggests their neutral head heights were a bit lower than in brachiosaurids, which were probably at around 4–5 m (13 ft). The very large diplodocids (*Supersaurus, D. hallorum*) were primarily adapted for feeding on low-growing horsetails and small conifers and ginkgoes (with ferns and cycads being rarer menu items) on the open plains, but they could have reached up above 4 m on occasion. The tripodal, high-feeding stance proposed for stegosaurs and diplodocids is a possible but not habitual feeding mode for these animals.

After assigning each Morrison species to its weight, locomotion and habitat, and feeding mode guilds, we can now take a look at the fauna from this perspective, free of taxonomic considerations. If we look at the diversities of the different feeding mode guilds (fig. 7.5), we see that most species by far were invertivores. Most of these were small mammals; few dinosaurs, for example, would have been strictly feeding on insects or other invertebrates, but this again shows the diversity and importance of the small mammals.

Many of the insects eaten by the invertivores would have been her-bivorous, so we can see the beginnings of a food web. Plants as producers begin the chain of energy transfer; herbivorous insects (primary consumers) feed on the plants; invertivores (secondary consumers) eat the insects; and small carnivores, perhaps in the form of small theropods like *Ornitholestes*, feed on the invertivores. Because energy is lost at each link in this chain, there is only a small fraction of the original amount produced available to the upper-level consumers. A high diversity, abundance, and biomass of the lower levels of this network are therefore necessary to sustain the carnivores like the theropods. This is demonstrated by the high number of species and small size of the invertivores and invertivore/carnivores.

Note also that among the herbivores and carnivores, the low browsers and small carnivores are the most diverse. There were probably more plants and more plant biomass close to the ground for herbivores to feed on than there would have been in taller trees, so the diversity of low browsers is expected. Among the carnivores, because there are generally more smaller animals and fewer larger ones in most ecosystems (including both predators and prey), the distribution in the Morrison Formation is typical, except perhaps in the absolute size of the large carnivores and how many of them there were.

Looking at the diversity of the five locomotion and habitat guilds (fig. 7.6), the terrestrial and specialized terrestrial categories are by far the most diverse, mostly because of the species of mammals in the specialized terrestrial category and the large number of dinosaur taxa among terrestrial animals. Although most vertebrate fossils in the Morrison Formation are associated with deposits representing river, pond, and lake environments, most of the diversity is in land animals. Are we simply able to discern species better among these animals? Are the larger terrestrial animals more likely to fossilize even though fewer individuals likely became buried in the appropriate wet environments than did aquatic and semiaquatic animals? Or was diversity truly higher among terrestrial animals than those living in the ponds and rivers? Perhaps there were more ways to make a living in microhabitats on the floodplain than there were for the fish, turtles, and crocodilians in the water, but that seems unlikely. Alternatively, the low aquatic diversity may be a result of the fact that some of the ponds and some streams in the Morrison basin were ephemeral. It is difficult to establish a diverse aquatic community in a pond that exists only part of the year. Most likely the high diversity among terrestrial and specialized terrestrial guilds was due to peculiarities of each: most terrestrial guild members were dinosaurs and thus were large on average and more likely to fossilize, and most specialized terrestrial guild members were mammals, a group that is often species rich and good at partitioning its resources, no matter what time or environment the members happen to live in. Because of the distinctiveness of their teeth, mammals also can be more easily divided by species than can many other groups such as fish. There probably is greater diversity out there among the semiaquatic and aquatic taxa. We just can't recognize it with the material we have thus far.

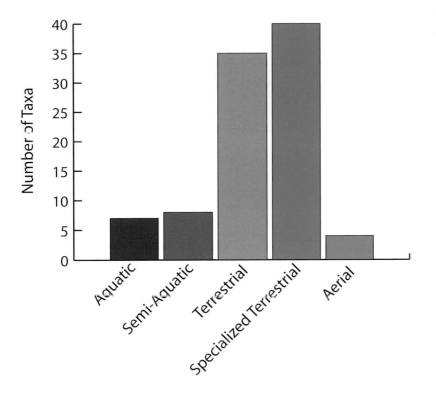

7.6. Diversities of locomotion and habitat guilds in the Morrison Formation.

The Morrison Formation seems to have had an unusually high diversity of large carnivores compared both with today and with other past ecosystems. A number of Mesozoic dinosaur faunas had one large theropod and a significant size gap between it and the second-largest carnivore. The Late Cretaceous–age Hell Creek Formation is a prime (almost exaggerated) example of this pattern. *Tyrannosaurus* probably weighed about 6,000 kg (13,228 lb), but the next largest theropods (a large dromaeosaur, an ornithomimid, and two oviraptorosaurs) were each "only" about 275–400 kg (606–880 lb). All but the dromaeosaur among those theropods lacked teeth, however, and may have been omnivorous, so the next largest pure carnivore after *Tyrannosaurus* at six metric tons and *Dakotaraptor* at a little over half a metric ton was a troodontid that weighed about 35 kg (77 lb). That makes for a gap of more than 5 metric tons (~11,000 lb) between the adult sizes of the two largest theropods of the time! In the Morrison Formation theropod fauna, there was no such gap (Foster et al. 2001). At least three species of Morrison theropods had adult body weights more than any in the Hell Creek except for *Tyranno-saurus*, and seven weighed more than the Hell Creek troodontid. Figure 7.7 compares the weight distributions of the theropod faunas of the two formations on a logarithmic scale. As animals grow in length, their mass does not increase linearly in the same way, and seeing patterns in linearly graphed weight distributions ranging from 6,000 kg to less than 5 kg can be difficult, so the logarithmic scale for this graph will assist us. (Keep

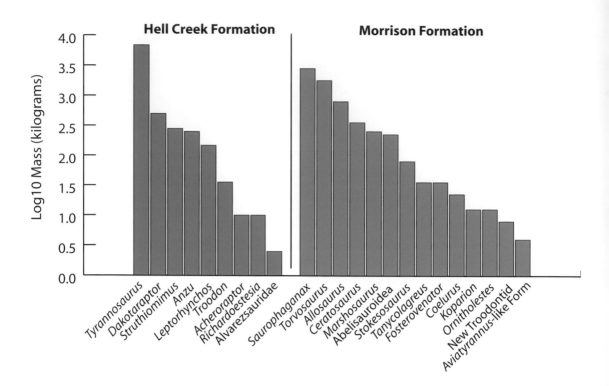

Hell Creek Formation **Morrison Formation**

Log10 Mass (kilograms)

Tyrannosaurus, Dakotaraptor, Struthiomimus, Anzu, Leptorhynchos, Troodon, Acheroraptor, Richardoestesia, Alvarezsauridae, Saurophaganax, Torvosaurus, Allosaurus, Ceratosaurus, Marshosaurus, Abelisauroidea, Stokesosaurus, Tanycolagreus, Fosterovenator, Coelurus, Koparion, Ornitholestes, New Troodontid, Aviatyrannus-like Form

7.7. Masses of theropod dinosaurs from the Hell Creek and Morrison Formations. Notice that the distribution from the Morrison Formation shows greater species diversity, lesser body mass range from smallest to largest, and lesser disparity in mass from one species to the next smaller than is seen in the Hell Creek Formation. Mass on the y-axis is log scale, so that each whole number increase on the axis indicates a 10× increase in weight. Thus, *Tyrannosaurus* weighs more than 10 times as much as next smaller theropod species in the Hell Creek Formation. No such extreme disparity seems to have existed between theropod species in the Morrison Formation.

in mind that on the graph, each increase of a whole number equals an increase of 10× in actual mass in kilograms; also, to convert the y-axis scale to real weights, 1 on the graph is 10 kg, 2 is 100 kg, and 3 is 1,000 kg.)

By just glancing at the graph of the Hell Creek animals on the left and the Morrison on the right, some differences are immediately apparent. First, the overall drop in body size in the Hell Creek theropod fauna is precipitous across the whole range, whereas the size decrease in the Morrison theropod fauna is just as strikingly gradual and steplike. These patterns result from the mass differences between ranked taxa being significantly less in the Morrison than in the Hell Creek, where the least size difference between one species and the next lowest or highest is still a magnitude of almost three times. Second, the Morrison Formation theropod fauna has neither a species as large nor a species as small as in the Hell Creek. Third, the gap between *Tyrannosaurus* and the next largest theropods is more than an order of magnitude (more than 10×), and the gap between *Tyrannosaurus* and the next largest true carnivore is more than two orders of magnitude (more than 100×). In the Morrison distribution, the largest gap between any two taxa is only four times.

Clearly, different ecosystem dynamics were operating in Morrison times compared with Hell Creek times—and indeed most of the Cretaceous. Few studies have addressed these differences from the Jurassic to Cretaceous (Farlow 1993; Foster et al. 2001; Bykowski 2017), and even fewer answers have been identified. If *Tyrannosaurus* dominated its world even more than other large theropods had dominated theirs, what was different during the Late Jurassic that allowed there to be four carnivorous dinosaurs weighing more than 300 kg (661 lb), which is way beyond what is considered to be large among modern carnivores? Perhaps the Hell Creek is an extreme case and the immense gap between *Tyrannosaurus* and troodontids developed because juveniles of the *Tyrannosaurus* population filled the middle size ranges. The theropods of the Morrison Formation, on the other hand, may have been indirectly influenced by the mass distribution of the stegosaurs, sauropods, and ornithopods. Most likely, an unusual abundance and diversity of large herbivorous prey, along with a relatively open habitat, well suited to hunting by large predators, resulted in a wide range of possible specializations in large carnivores and thus a higher diversity of these animals than in other times.

It is also interesting that the largest prey dinosaurs of the Late Cretaceous (hadrosaurs and ceratopsians) were nowhere near the size of the largest dinosaurs of the Late Jurassic, yet *Tyrannosaurus* was much larger than most Late Jurassic carnivores. This appears counterintuitive. Why would gargantuan sauropods not result in behemoth carnivores in the Late Jurassic of North America? (And why would *Tyrannosaurus* be so much larger than it needed to be to either kill hadrosaurs and ceratopsians or chase off other carnivores from kills? But that's a book or two in itself and best left to the Late Cretaceous folks.) Why are Morrison theropods so diverse yet so comparatively small? As we discussed in chapter 6, the theropods may not often have attempted to kill adult sauropods, which may have been simply out of the size range for the carnivores in most cases. But there still were a diversity and abundance of stegosaurs, ornithopods, and young sauropods to sustain a diverse group of moderately large Morrison theropods. It is likely that *Tyrannosaurus* is an unusual case and that this freak of the Late Cretaceous could be misleading when trying to understand most theropod faunas. Many other faunas in the Cretaceous have fewer large herbivorous species than the Morrison and also have a smaller dominant carnivore than does the Hell Creek in *T. rex*. This seems to be the normal situation and *Tyrannosaurus* more of an anomalous Mesozoic carnivore on steroids.[3] Thus, the theropods of the Morrison Formation, it seems, did just fine and were as large as they needed to be. And perhaps there were simply so many large prey dinosaurs around compared with other times in the Mesozoic that there was room for a wide variety of the Late Jurassic carnivores.

If we take a look at the mass distribution of the theropod dinosaurs of the Morrison Formation on a linear scale (fig. 7.8), we see that most were small and some were large. Although the steepening of the curve from smallest to largest appears drastic and indeed represents a wide range of body size, the log-scale distribution in figure 7.7 demonstrated that the size increase amounts between species are quite consistent in magnitude. Although we can barely see it here, the log-scale graph demonstrated that the biggest relative size increase is from *Coelurus* to *Stokesosaurus*, at about three times. We see here (fig. 7.8) and in the log-scale graph that, with a few exceptions, the general pattern is that each larger species is approximately twice as massive as the one below it. The previous

7.8. Mass estimates for theropods of the Morrison Formation in kilograms and on a linear scale. *Allosaurus*, although dominant in terms of abundance, was not the largest carnivorous dinosaur of the time.

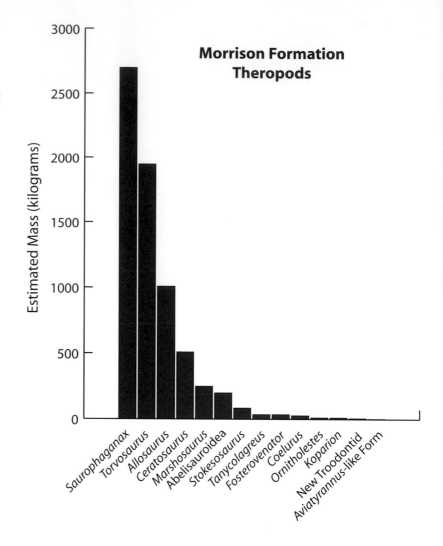

log-scale graph didn't lie—these doublings in size are fairly regular and gradual compared with the tenfold to hundredfold increases between species in the Hell Creek. Most important in figure 7.8 is that unlike the Hell Creek, where the gigantic *Tyrannosaurus* dominated the theropod fauna, in the Morrison Formation, the dominant theropod (in terms of total faunal biomass) is the third largest in adult body size.

Torvosaurus and *Saurophaganax* are, respectively, about two and nearly three times heavier than *Allosaurus* among Morrison Formation theropods, yet *Allosaurus* is the keystone carnivore species of the time. How does a relatively small ("only" 1,000 kg) carnivore dominate the ecosystem? Sheer numbers. If we look at a sample of 194 theropod dinosaur specimens from the Morrison Formation (fig. 7.9), we see that nearly 75% are *Allosaurus*. The next most abundant are *Torvosaurus*, *Coelurus*, and *Ceratosaurus*, and the remaining six species each account for just a few percentage points (or less) of the sample.

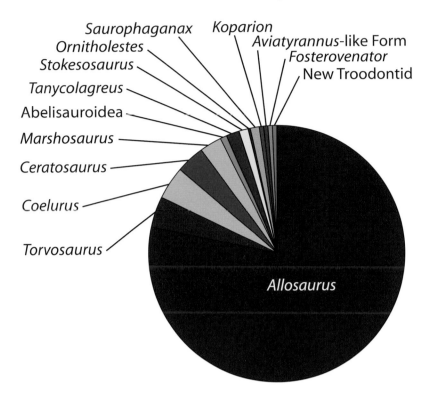

Morrison Formation Theropods (N=194)

Saurophaganax
Ornitholestes
Stokesosaurus
Tanycolagreus
Abelisauroidea
Marshosaurus
Ceratosaurus
Coelurus
Torvosaurus

Koparion
Aviatyrannus-like Form
Fosterovenator
New Troodontid

Allosaurus

7.9. Relative abundances of theropod dinosaur species in the Morrison Formation, showing the numerical dominance of *Allosaurus*. Counts are raw MNI. If these numbers are converted to biomass percentage estimates, *Allosaurus* still is the most abundant theropod.

Certainly, the small theropods were competing to some degree for the small vertebrate prey, and the numerical abundances of these small carnivores are relatively equal. But among the large theropods, why did *Allosaurus* dominate? What were *Ceratosaurus*, *Torvosaurus*, and *Saurophaganax* doing differently (or where were they living) that they are so much less abundant in the rock record? *Saurophaganax* is rare and is known only from the top of the Morrison Formation in Oklahoma and possibly New Mexico; it is also morphologically similar to *Allosaurus*, so it is difficult to say how or if it was ecologically different from its smaller and more abundant cousin. *Ceratosaurus* and *Torvosaurus*, however, are different from *Allosaurus* (particularly in their teeth and forelimbs), and they must have been doing something differently. The three are found together in some of the same deposits and in some that show little sign of transport of skeletons, so we know that they overlapped in time and that they lived in the same areas. But might the rarer *Ceratosaurus* and *Torvosaurus* have preferred habitats where preservation was less likely than those that were frequented by *Allosaurus*? Perhaps. Those two genera are rarely found by themselves, however, and are often found in the same quarries with *Allosaurus*. Where *Ceratosaurus* or *Torvosaurus* occur without *Allosaurus*, they are almost always single specimens, and when the two occur with *Allosaurus*, *Allosaurus* is always more numerous.

There are no multiple-specimen *Ceratosaurus* quarries, for example, with a paucity of allosaurs. Shed teeth at some sites suggest that *Allosaurus* and *Ceratosaurus* fed on some of the same prey at the same sites, but again, *Ceratosaurus* is rare. This suggests to me that the sample found in the Morrison indicates a true rareness of *Ceratosaurus* and *Torvosaurus* as a result of their specialization rather than geographic or habitat-preference separation of the species from *Allosaurus* in the Morrison basin. It is possible that the two rare theropods lived mostly outside the Morrison depositional area and only rarely ventured into the region.

Torvosaurus was quite a bit larger than *Allosaurus*, so even if it preferred habitats that were more destructive to vertebrate skeletons, it still should have had a respectable preservation potential. Considering this, it is not surprising that *Torvosaurus* is one of the more abundant theropods in the Morrison Formation after *Allosaurus*.[4]

The morphological differences among the three large theropods, therefore, must hint at some ecological separation of feeding mode. Among the three, *Allosaurus* is the most evolutionarily derived theropod but appears to be the most conservative in construction. It is moderate in body size, tooth size, and forelimb build. *Ceratosaurus* is slightly smaller on average than *Allosaurus*, yet the skull of *Ceratosaurus* contains teeth that are absolutely larger and more compressed than in *Allosaurus*. *Ceratosaurus* also has a four-digit hand and longer, more gracile forelimbs than *Allosaurus*. *Torvosaurus* is simply by far larger and more robust than *Allosaurus*. It has a giant premaxilla (and skull), a more robust humerus, and a shorter, much more robust ulna. Somehow these animals must have been focusing on slightly different prey than *Allosaurus*, although the three certainly overlapped on occasion. Were they specialized for diets of specific types of dinosaurs—perhaps the large and robust *Torvosaurus* for larger dinosaurs and the smaller, more gracile, but large-toothed *Ceratosaurus* for midsized dinosaurs? These theropods seem to be specialized for something. Or maybe group-hunting allosaurs were going after the largest dinosaurs? I believe not. The midrange size and moderate robustness of *Allosaurus* strike me as adaptations for flexibility of diet. Unfortunately, it is difficult to be sure of much with the Morrison theropods because too many specimens are incomplete and too few species are well known. The numbers of specimens preserved may give us some hint of how these animals fed, however.

In simplified models of ecosystems with a fixed number of herbivorous vertebrate species and a fixed amount of biomass transfer from prey to predator species, several end-member patterns are possible. The main variables for the models are predator (theropod) diversity (number of species), abundance (total biomass of the species population), and degree of feeding specialization (many or few targeted prey species). We set rules for those fixed amounts of biomass transfer (e.g., only two units per herbivore, but these can go to carnivores in any combination; all herbivore energy units must be consumed). If we model the different possible ways to flow the energy (herbivore units) to the predators, several combinations

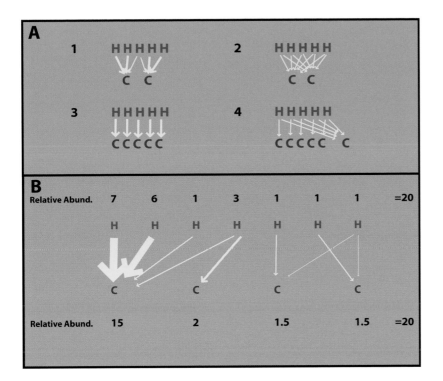

7.10. Simplified models of herbivore-carnivore energy flow. (A) In each of the four models here, each of five herbivorous species (H) has a biomass equating to 2 units of energy to transfer to carnivorous species (C). Herbivores have to give exactly 2, but carnivores can take in 10 total units in any varying amount among themselves, and the number of carnivorous species varies. Thicknesses of the lines are proportional to the number of units represented by each, and energy units are also assumed to reflect postproduction carnivore abundance. Models 1 to 4 are not the only possibilities but represent end members resulting from maximizing or minimizing certain parameters. Models 1 and 2, for example, minimize carnivore diversity, whereas model 1 maximizes carnivore specialization in prey targeting and model 2 minimizes specialization. Models 3 and 4 reflect high carnivore diversity, with model 3 maximizing specialization. When diversities and relative abundances (as reflected by energy units incoming to carnivores) are compared, the model that most closely resembles the Morrison Formation dinosaur fauna is model 4. In this model and in the Morrison, relatively high carnivore diversity is paired with uneven abundance numbers (*Allosaurus* is 75% of sample in the Morrison; the rightmost carnivore species in model 4 takes in 5 units to the others' 1 unit each). Notice that in model 4 there is also a disparity in dietary prey specialization: the lower five carnivores are all specialists, whereas the abundant species on right is a generalist. (B) If we modify model 4 to more accurately reflect the Morrison, we get a similar pattern. In this model, the 7:4 ratio of herbivores to carnivores approximately reflects the relative diversity seen in the dinosaur fauna of the Morrison Formation. Similarly, the abundance/energy unit numbers above herbivores and below carnivores approximate the relative abundances of species in the herbivorous and carnivorous guilds. (The 20:20 ratio of abundance between herbivores and carnivores, of course, is not the case in the formation; this is just necessary in the model because every energy unit needs to be consumed.) Abundance numbers indicate how much energy each herbivore species must give and how much each carnivore species must take in. Again, the lines are proportional to the numbers of units they represent, and we will try to minimize fractioning of whole energy units. Even if we maximize the specialization of the most abundant carnivore, that species still has to feed on more prey species than other carnivores (four prey species, versus one, two, and two for others). These models suggest that, energetically, the abundant Morrison theropod *Allosaurus* may have had to have been a generalist predator that fed on a wide variety of dinosaurian prey.

in the diversity, abundance, and specialization variables result: (1) low predator diversity with equal abundance and high specialization, (2) low diversity with equal abundance and low specialization, (3) high diversity with equal abundance and high specialization, and (4) high diversity with unequal abundance and mixed specialization. If we try a more complex model, incorporating approximate Morrison Formation predator-prey diversity ratios and within-group abundance proportions for both predators and prey, we find that the last of the above-listed simplified models (model 4) most closely matches the situation in the Late Jurassic of North America (fig. 7.10). Reflecting the high carnivore diversity of the complex model and simplified model 4, the record from the Morrison Formation includes at least 11 theropod dinosaur genera compared with 13 to 15 herbivorous dinosaur genera, so theropods are relatively diverse; and the true abundances of the theropod genera are definitely uneven, as we have seen, with *Allosaurus* accounting for nearly 75% of the specimens. In both the simplified and more complex models (fig. 7.10), the most common theropod (*Allosaurus*) must feed on more types of prey species than other theropod taxa in order for the modeled abundances to match those observed in the Morrison Formation. These models suggest that *Allosaurus* was a generalist predator and that other, rare theropod taxa were more specialized. *Torvosaurus* and *Ceratosaurus* probably overlapped to some degree with *Allosaurus* in prey species targeted, but small theropods overlapped only with young *Allosaurus*. It is possible that *Allosaurus* was in fact highly specialized and that it was common simply because its specific prey was. Were *Torvosaurus* and *Ceratosaurus* instead the generalists? Perhaps, but I very much doubt it. What we know of the respective morphologies and relative abundances of these theropods suggests that *Allosaurus* was less selective and *Torvosaurus* and *Ceratosaurus* more specific in what they ate.

A while back, I mentioned the possibility that juvenile tyrannosaurs of the Late Cretaceous filled the role of middle-weight carnivores of the time and that this may be the reason for the big Hell Creek gap in theropod weight distribution. Could a similar process have been occurring among the herbivorous species of the Morrison Formation? Let's graph the diversities of different Morrison Formation species (not just dinosaurs) by their estimated adult mass category and split predators from prey (fig. 7.11). For this graph, only terrestrial taxa were considered, including theropods and terrestrial crocodilians (e.g., *Fruitachampsa*) among predators, and sauropods, ornithopods, ankylosaurs, stegosaurs, lizards, sphenodontians, and mammals among prey. Taxa were counted at the generic level except in the case of the mammals, which were counted by group (e.g., triconodonts) only so that, given the tremendous mammalian diversity, the graph would be more readable.

In figure 7.11, we see that compared with the prey species distribution, the diversities of predator mass categories are relatively even (one to four species across a body size range of 1–10,000 kg [2.2–22,000 lb]). The

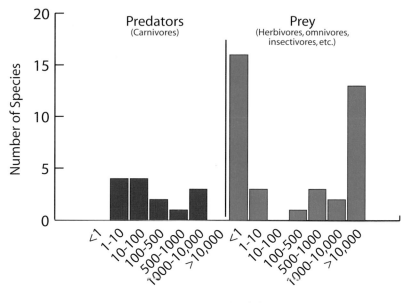

7.11. Diversities of Morrison Formation vertebrate biomass guilds. Predator and prey species are graphed separately. Note the great diversity at very large and very small body sizes in prey species and the gap in diversity of animals with adult body masses in 10 kg to 500 kg categories. This gap may have been caused by a variety of factors, including high predator diversity in similar body sizes driving prey species to larger or smaller sizes, and juveniles of larger species occupying the 10 kg to 500 kg range.

prey distribution is highly uneven, ranging from 1 to 14 species across the full range of body mass categories. Most striking about the prey species graph, however, is the diversity of animals at the very small and very large body sizes and the paucity of species with moderate body weights (10–500 kg [22–1,102 lb]). Sauropods and stegosaurs, which form the bulk of the diversity on the right side of the prey graph, would have had large juvenile populations in just this range much of the year, and this may have prevented the evolution of many dinosaur species with adult body masses of corresponding amounts. Juvenile sauropods, stegosaurs, and ankylosaurs probably would have initially occupied that 10–500 kg range and would have provided food for large theropods, as well as heavy competition for any herbivorous taxa with 10–500 kg adult sizes. Prey taxa less than 10 kg may have been less affected by competition from juveniles of the larger dinosaur taxa.

Also note in figure 7.11 that peak body mass categories for predators are often close to or larger than some of the prey peaks (except for the >10,000 kg prey peak). The predators in each of the size classes probably fed mainly on the prey groups just smaller than them. Thus, the small theropods in the 10–100 kg category (*Coelurus*, *Ornitholestes*, etc.) probably fed mainly on the prey taxa forming the <1 kg and 1–10 kg categories (lizards, sphenodontians, mammals, and small ornithopods). Also, the large theropods in the 1,000–10,000 kg category probably fed mostly on the prey taxa in the 100–500 kg, 500–1,000 kg, and 1,000–10,000 kg categories (*Camptosaurus*, *Dryosaurus*, stegosaurs, etc.). As discussed previously, it is likely that the adult sauropods in the >10,000 kg size class were, in most cases, safe from predation.

Paleoecologist Richard Stucky has suggested that one possible reason for a similar (although less extreme) bimodality of prey size distribution seen in Late Cretaceous vertebrate communities was that habitat modification by large herbivores increased the visual hunting capabilities of predators by opening up the vegetation understory. Thus, there would have been selective pressure limiting the size of the animals living on the forest floor or in the open. This process may have been operating during the Late Jurassic as well, particularly given the possible habitat modification of larger herbivores and generally more open habitat of the Morrison Formation overall. Browsing by the large herbivores of the time could have limited the understory and put pressure on prey taxa to develop either small adult sizes or, in the case of sauropods, extremely large adult sizes to avoid becoming meals. If the 10–500 kg size range was a dangerous one in the Morrison Formation environment, then the prey animals may have stayed mostly below it or occupied it only temporarily as juveniles, growing out of it to larger size as quickly as possible. The predator diversity peak in the 10–100 kg category indicates that certainly the <10 kg prey taxa were not entirely safe, but species in the 10–500 kg range would have been potential prey for nearly all the predators of the time, from the 10–100 kg small theropods and medium-sized theropods like *Marshosaurus* up to the large and common 1,000 kg *Allosaurus*. With predators ranging over all sizes and vegetative cover comparatively light in many areas as a result of herbivore browsing, prey species may have adopted the strategy of either staying small or becoming very large in order to limit their number of potential predators. Ultimately, predator pressure and the previously discussed competition from juveniles of very large species both likely played a role in greatly limiting the number of taxa with midrange adult body sizes. And such body size bimodality, though perhaps most extreme in the Morrison Formation, appears to be a trend among many vertebrate clades and within vertebrate ecosystems since the Triassic–Jurassic boundary (Benson et al. 2018) and not just during the Mesozoic (Rodríguez 1999).

If we compare overall theropod dinosaur diversity throughout the time of deposition of the Morrison Formation, an interesting pattern emerges. Plotting the diversity of theropods as a percentage of total dinosaur diversity for each of six stratigraphic zones in the formation (fig. 7.12), there is a gradual increase in the relative diversity of theropods throughout the lower three-quarters of the unit and then a distinct drop in relative diversity in the top zone. This does not appear to be an artifact of varying sample size, because the upper two levels (5 and 6) are among the largest samples. The drop in theropod diversity from 40% of total dinosaur taxa late in Morrison times to 20% at the end thus appears to reflect a real phenomenon. We do not, however, know what might have caused this drop. In the upper levels, was it just an increase in nontheropod dinosaur diversity such that theropods are simply relatively less diverse? Or was it the actual loss of some theropod species? Neither of these is obvious, but

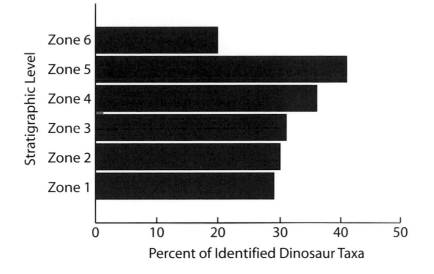

7.12. Relative diversity of theropod dinosaurs from the Morrison Formation by stratigraphic level. Diversity is shown as the percentage of total dinosaur diversity from that level. The apparent drop in diversity in the upper levels remains unexplained.

most of the difference may be in the restriction of a number of small, rare theropod species to single quarries in the middle upper half of the formation. What causes this pattern will be clearer someday, when we have a better sampling of small theropod species from various levels in the formation. The problem, of course, is that these animals are hard to find.

Another way to look at distributions of masses in taxa of the Morrison Formation is through **cenograms.** In this case we will be looking only at herbivorous and omnivorous species, and we will compare the rank abundance distribution of these taxa to some modern and Neogene-age counterpart faunas. Cenograms as a paleobiological tool were pioneered by Legendre (1986) as a means of comparing Cenozoic fossil mammal assemblages to modern environments in order to see if such distributions could tell us something about the habitats of the ancient environments. Cenograms of modern herbivorous mammal faunas differ between humid and dry habitats and between forested and open vegetation. Cenograms historically have not been applied to Mesozoic faunas (but see Farlow et al. 2010), likely in part because it is easier to compare fossil mammal faunas with modern mammal faunas—reptile-mammal comparisons inevitably involve more unknowns. Still, considering the likely intermediate metabolisms of many dinosaurs and that many dinosaurs were filling what are now niches of mammalian species, there may be comparability between some Mesozoic faunas and those of modern settings. If so, what happens if we look at such a distribution for the Morrison Formation? How is it similar to or different from cenograms of mammalian communities?

Again, we are using only herbivorous or omnivorous taxa, and we are excluding aquatic and semiaquatic taxa. We also will not include the flying pterosaurs. We will be ranking the taxa along the x-axis from largest to smallest based on the estimated mass (natural log of the mass in grams),

7.13. Cenogram of herbivorous and omnivorous taxa from Zones 4 and 5 of the Morrison Formation (green); separate rank order plot for strictly carnivorous taxa (red; theropods and terrestrial crocodylomorphs). Taxa are ranked by the natural log of mass in grams (left y-axis), with the approximate mass in kilograms of natural log units (right y-axis). Note the gap from natural log ~10 to ~6.5. Zone assignments and mass estimates are taken from Appendix B.

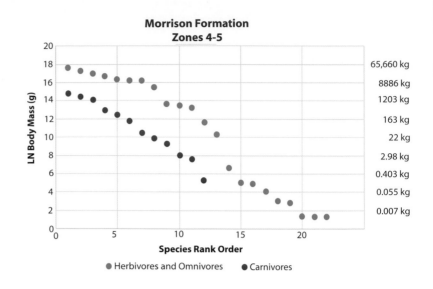

which will be indicated along the y-axis for each species. We also want to make sure that we include only co-occurring taxa, so we will sample taxa from biostratigraphic Zones 4 and 5 within the Morrison.

There are 22 vertebrate species from the Morrison Formation that fit these criteria, and when we graph them in a cenogram (fig. 7.13) we see they range from a top-end natural log ~17.6 down to ~1.4 (strict carnivores are plotted here as a separate line). This is equivalent to a range from about 44,000 kg down to 4 g (~96,800 lb down to ~0.2 oz). Obviously, this is a greater range than any modern or Neogene ecosystem, simply because of the Morrison's almost unparalleled menagerie of truly giant terrestrial herbivores (sauropods). Among the 22 species are herbivorous dinosaurs, an omnivorous heterodontosaurid (*Fruitadens*), docodont and multituberculate mammals (omnivorous), a largely herbivorous sphenodontid (*Eilenodon*), and blunt-toothed and likely omnivorous lizards.

The Morrison Formation cenogram in figure 7.13 consists of a number of large animals forming a line with a relatively shallow slope (sauropods and *Stegosaurus*), a gap between natural log ~16 and ~14, a handful of taxa in a curved line with a steeper slope from ~14 down to ~10 (ankylosaurs, *Camptosaurus*, *Dryosaurus*, and *Nanosaurus*), another gap from natural log ~10 down to ~7, and then a number of taxa from about natural log 6.6 down to 1.4 with an intermediate slope (*Fruitadens*, *Eilenodon*, docodont and multituberculate mammals, and omnivorous lizards).

Major differences of this Morrison cenogram from most modern and Neogene cenograms include a significantly greater range of body sizes, a shallower slope among larger animals than among the smaller group (the opposite is usually the case), and a second gap in larger body sizes around natural log 15 (~3,200 kg, 7,040 lb) — between *Stegosaurus* at about 5,200 kg (11,440 lb) and ankylosaurs at around 800 kg (1,760 lb). On the other hand, the Morrison cenogram is similar to some modern and fossil ones in that it has a gap between smaller and larger body sizes from about

500 g (18 oz) mass up to 8 kg (17.6 lb); this is a feature observed in many cenograms from all ages in open (lightly wooded) environments (Legendre 1986; Travouillon and Legendre 2009). The Morrison cenogram, despite differences in slope between larger and smaller animals, shows an overall pattern similar to tropical savanna and tropical wooded savanna in modern environments (Legendre 1986; based on African settings), which seems to fit predictions of Morrison habitat based on geologic and paleobotanical studies. The pattern among the cenogram data (and even histogram plots of body masses) also most closely resembles the open woodland settings in Australia (Travouillon and Legendre 2009). The Morrison cenogram pattern differs significantly from the patterns demonstrated by modern deserts, on one hand, and closed forests and jungles on the other.

Various factors may influence the differences between Morrison and other cenograms, including differences in vegetation (there were no flowering plants during the Jurassic), metabolic differences between included reptilian taxa and the mammals used in more recent cenograms, and differences between the predatory species of the Morrison and those of other times. As we have seen, there was both a diversity of carnivorous taxa during Morrison times and a close-packed distribution of them in terms of masses. This may have helped reduce the number of medium-sized herbivorous species in the Morrison Formation. Carnivorous taxa that fit the same criteria used for the cenogram (other than being herbivorous or omnivorous, of course) all fit within a natural log range of 15 to 5 (fig. 7.13), which corresponds to the low-diversity medium size ranges for herbivorous and omnivorous species in which there are only six taxa; in contrast, there are eight prey species larger than this and eight smaller than it. The carnivorous taxa occupy a range of sizes that most herbivorous and omnivorous taxa appear to avoid ecologically (Rodríguez 1999). This is a pattern suggested for other, more recent faunas as well (Croft 2001), and it appears to be backed up by patterns on islands, where predators are often very rare and on which larger and smaller animals evolutionarily appear to converge on medium body sizes (J. H. Brown 1995). The greater gap in medium body sizes (larger or more pronounced) seen in cenograms from more open habitats relative to more forested ones may also relate to a possibly greater effectiveness of predators in these open environments (Croft 2001); in light of this, the strong gap in the Morrison cenogram is not surprising. It has been suggested, in fact, that the patterns in cenograms correlate better with carnivore size distribution than with vegetation structure, temperature, or humidity (Rodríguez 1999).

Now let's take a look at some abundance data adjusted by Russell's techniques to estimate population percentages and total biomass percentages. One interesting feature of the Morrison Formation vertebrate paleocommunity is that there are major differences in the abundances of different guilds and ecological categories depending on whether the abundances are measured by the corrected percentage of biomass or by the corrected percentage of individuals. Among the locomotion and

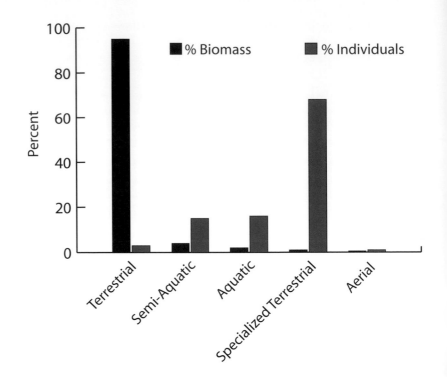

7.14. Relative abundances of Morrison vertebrate locomotion and habitat guilds, comparing the adjusted percentage of individuals to the adjusted total biomass percentage. The original Morrison Formation vertebrate population appears to have been dominated in biomass by terrestrial animals (mostly large, herbivorous dinosaurs), although the numerically most abundant groups of animals were water dwelling (fish, turtles, and crocodiles) and small, terrestrial climbers and burrowers (mostly mammals, lizards, and sphenodontians).

habitat categories, for example (fig. 7.14), terrestrial taxa are the second least abundant animals in terms of numbers, but they account for more than 95% of the biomass. Specialized terrestrial, semiaquatic, and aquatic taxa are most abundant numerically but are almost insignificant in terms of biomass.

In figure 7.15, we can see that most of the biomass of Morrison Formation vertebrate populations was accounted for by the herbivores, as in most ecosystems. Medium and high browsers (mostly sauropods) comprised nearly 75% of the biomass and low browsers (stegosaurs, camptosaurs, etc.) just over 10%. Carnivores, as we will see in more detail soon, were just under 10% of the total biomass, and all others (omnivorous turtles and insectivorous mammals, lizards, and amphibians) comprise the remaining 5%. Remember that these are the adjusted numbers, not the straight, raw data counts. It goes to show how much being a large species can mean, even if there were very few of you, as was the case with the sauropods. Notice, for example, that on the same graph, we have plotted the adjusted percentage of individuals and that approximately 95% of the individuals in the original populations were invertivores, invertivore/carnivores, and omnivores. These species were mostly mammals, lizards, sphenodontians, amphibians, fish, and turtles, and being small, it is not surprising that they were also numerous. But what is surprising is that the herbivorous species, which so dominate the biomass, barely even register in the plot of percentage of individuals. The large body size of herbivorous dinosaurs meant that even though they were not as numerically abundant as other animals, they and their uncounted young were

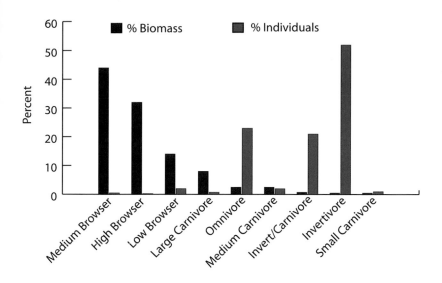

7.15. Relative abundances of Morrison vertebrate feeding mode guilds, comparing the adjusted percentage of individuals to the adjusted total biomass percentage. Browsing dinosaurs dominate the biomass, but small invertivore-carnivores and omnivores were by far more abundant in terms of numbers.

central in the transfer of energy in the ecosystem, particularly as primary consumers. Similarly, in modern megaherbivore communities, elephants and other large herbivorous mammals are outnumbered numerically but dominate in terms of biomass. Populations of white rhinoceros, black rhinoceros, and giraffe in the Umfolozi Game Reserve in southern Africa account for only 7.5% of the herbivorous animals in the region but comprise 50.3% of the biomass density (Owen-Smith 1988). The effects of sauropods and other large herbivorous dinosaurs on the habitat and other species of the Late Jurassic probably were more pronounced but may not have been all that different from what is seen today in megaherbivore communities. Elephants, for example, have a large impact on vegetation in their environment, but the effects in many cases are positive for some smaller herbivorous species. The replacement of woodlands by rapidly growing shrubs after large herbivores have passed through an area can be advantageous to smaller, browsing herbivorous animals (Owen-Smith 1988). Perhaps small plant eaters of the Morrison such as *Nanosaurus* and *Dryosaurus* fed on vegetation that was regenerating after sauropods had fed on woodlands or open fern meadows.

As we have already seen, the sauropods of the Morrison Formation were key elements of the Late Jurassic fauna. One of the most important measures of their effect on the ecosystem is that of their total relative biomass. Even after adjusting their numerical abundance on the basis of taphonomic bias in favor of large specimens, the sauropods alone accounted for approximately 73% of the total vertebrate population biomass of the time and also for nearly 78% of the total dinosaur biomass. Interestingly, the straight minimum numbers of individual counts of specimens in the formation approximate the relative abundances of sauropod species by biomass (Foster 2001b). Measuring by straight counts or by adjusted biomass, the three most abundant sauropods in the Morrison were, in order, *Camarasaurus*, *Apatosaurus*, and *Diplodocus*. Among herbivorous dinosaurs, nearly 40% of the biomass in the population at any one time comprised low-browsing diplodocid sauropods, about 25% of the biomass would have been *Camarasaurus*, and only 6% was high browsers such as *Brachiosaurus*. These numbers stress the importance of sauropods in the overall ecosystem and among the vertebrate fauna and also suggest that herbivores feeding high in the trees were rare indeed. Certainly *Brachiosaurus* was specifically adapted for feeding on high vegetation, but it was likely utilizing a rare resource, and there was room for neither many individuals nor many species to be feeding high in the treetops. More plant biomass and thus more resources were available closer to ground level, and this is reflected in the diversity of medium and low browsers among the sauropods and other dinosaurs.

Nondinosaur Biostratigraphy

Christine Turner and Fred Peterson of the U. S. Geological Survey have attempted to determine the relative positions of many of the quarries in the Morrison Formation (Turner and Peterson 1999), and on the basis of this, they have plotted the occurrence ranges of the dinosaur taxa in the formation. The zonation of taxa from that study was used to specify their Dinosaur Zones 1–4, and early versions of these zones were adopted as stratigraphic subdivisions (Zones 1–6) for my dissertation study (Foster 2003a).[5] As a reference, quarries associated with specific zones include the following: Zone 1, Meilyn Quarry and Cleveland Museum Quarry; Zone 2, Reed's Quarry 13 and Bone Cabin Quarry; Zone 3, Marsh-Felch Quarry; Zone 4, Callison's Main Quarry at Fruita and Dry Mesa Quarry; Zone 5, Reed's Quarry 9 and Carnegie Quarry at Dinosaur National Monument; and Zone 6, Rainbow Park, Breakfast Bench, and Ninemile Hill. On the basis of Turner and Peterson's quarry positions, I plotted those microvertebrate sites and dinosaur sites with nondinosaurian taxa for which good information was available. Not all sites containing nondinosaurian taxa were plotted because many sites contain turtle and crocodilian specimens. Only those that extended the ranges of a turtle or crocodilian taxon were plotted. All known microvertebrate sites were plotted.

A total of 16 higher-level taxonomic groups and 53 taxa were plotted. One feature is that many of the taxa occur at only one site and thus within only one zone. These single-site occurrences include 56.6% of the nondinosaurian taxa, and 37.7% of the nondinosaurian taxa are known only from Como's Quarry 9. Zone 1 is characterized by the appearance of turtles (probably *Glyptops*), the crocodilian *Eutretauranosuchus*, a terrestrial crocodilian, the stegosaur *Hesperosaurus*, the theropod *Allosaurus jimmadseni*, the sauropods *Haplocanthosaurus* and *Dystrophaeus*, and possibly the crocodilian *Amphicotylus*. Particularly in the presence of the dinosaurs *Hesperosaurus*, *A. jimmadseni*, *Dystrophaeus*, and *Haplocanthosaurus*, this lowest level of the Morrison Formation contains a distinctive vertebrate fauna, comprising unique taxa, that is different from that of any other level, more so than the other levels appear to be from each other.

Zone 2 features the first appearances of the actinopterygian fish, the lungfish *Potamoceratodus*, frogs and salamanders, the turtle *Dinochelys*, the sphenodontian *Opisthias*, the lizard *Dorsetisaurus*, the choristoderan *Cteniogenys*, the crocodilian *Amphicotylus* (if not in Zone 1), pterosaurs, the multituberculate *Psalodon*, triconodonts, *Docodon*, and the dryolestids *Amblotherium* and *Dryolestes*.

Zone 3 contains no first or last appearances. Zone 4 is characterized by several first appearances, no last appearances, and a number of single-site taxa. First appearances include the scincomorph *Paramacellodus*, the multituberculate *Glirodon*, the crocodilian *Macelognathus*, and the triconodont *Priacodon*. Single-site taxa include the terrestrial crocodilian *Fruitachampsa* (actually several sites in the FPA), the lizard *Saurillodon* and snake *Diablophis* (also FPA), the triconodont *Triconolestes*, and the pterosaurs *Kepodactylus* and *Mesadactylus*. The sphenodontian *Eilenodon* occurs at two sites in Zone 4.

Zone 5 is characterized by the first appearances of the sphenodontian *Theretairus* and the paurodontid *Euthlastus*; by the last appearances (in the formation) of salamanders, the lizards *Paramacellodus* and *Dorsetisaurus*, the crocodilian *Macelognathus*, and the multituberculate *Psalodon*; and by the only appearances of the crocodilian *Hoplosuchus*, the pterosaurs *Dermodactylus* and *Comodactylus*, the multituberculate *Ctenacodon*, the triconodonts *Aploconodon*, *Comodon*, and *Trioracodon*, the symmetrodonts *Amphidon* and *Tinodon*, the paurodontids, and the dryolestids *Laolestes*, *Melanodon*, and *Miccylotyrans*.

Zone 6 is characterized by a few single-site appearances: the turtles *Dorsetochelys* and *Uluops*, the terrestrial crocodilian *Hallopus*, the multituberculates *Morrisonodon* and *Zofiabaatar*, and the paurodontid *Foxraptor*. Last appearances include actinopterygian fish, lungfish, frogs, the turtle *Dinochelys*, the sphenodontians *Opisthias* and *Theretairus*, lizards, *Cteniogenys*, the crocodilians *Eutretauranosuchus* and *Amphicotylus*, pterosaurs, the multituberculate *Glirodon*, the triconodont *Priacodon*, the docodont *Docodon*, the paurodontid *Euthlastus*, and the dryolestid *Dryolestes*.

Probably none of the higher-level taxonomic groups in this analysis is making a first appearance for real. Most of this is likely a result of taphonomic first appearance. But the evolutionary first appearance of some genera is a distinct possibility, particularly for the more advanced forms high in the formation. The turtle *Uluops* and the multituberculates *Morrisonodon* and *Zofiabaatar* are similar to some Early Cretaceous forms and may in fact be the first of their groups.

The overall pattern of the distribution is similar to that found with the dinosaurs by Turner and Peterson (1999) in that many taxa appear in Zone 2 and diversity remains highest through Zone 5. Unlike the dinosaur ranges, however, 73.9% of the nondinosaur taxa that occur in more than

7.16. Comparison of diversities and sample sizes of nondinosaurian vertebrates for six stratigraphic levels in the Morrison Formation. Not surprisingly, the diversity in each stratigraphic level is to some degree correlated with the number of specimens known from that level. This graph also illustrates the need for more data (i.e., fossils) from the lower half of the Morrison Formation.

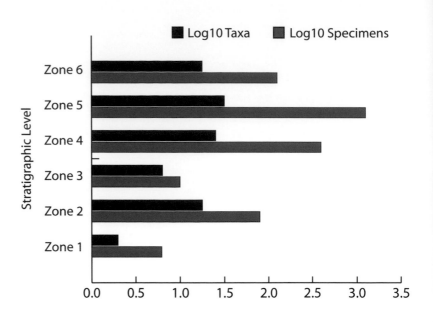

one stratigraphic level are still around in Zone 6. By contrast, only 44.4% of longer-ranging dinosaur taxa make it into Zone 6. This may suggest a greater diversity of nondinosaurian taxa than of dinosaurs in the upper part of the formation. The nondinosaurian diversity pattern, however, seems to be largely dependent on sample size. A log-scale plot of the number of nondinosaurian specimens versus the number of nondinosaurian taxa for each stratigraphic level (fig. 7.16) shows the two numbers to be highly correlated. Stratigraphic levels with larger microvertebrate quarries have more specimens and thus more taxa. But the relative numbers of dinosaurian and nondinosaurian taxa that range into Zone 6 are interesting. If the relative numbers are similar with future collecting, they will lend support to the idea that the upper Morrison was deposited during a time when dinosaur diversity was lower than it had been and the diversity of smaller, nondinosaurian taxa remained relatively high. This pattern was first hinted at by Bakker et al. (1990), who noted a general rarity of large dinosaurs at several localities in the upper Morrison (although certainly some of the dinosaurs that do occur are large). A number of promising sites high in the formation, such as Ninemile Hill and Rainbow Park, should continue to provide additional data to test the idea.

The big caveat here is that as we gather more age data some of these quarries may "jump" from one zone to the next, potentially modifying the above appearance patterns, particularly the middle ones (things in Zones 1 and 6 are likely to remain fairly stable). Problems with regional correlations between quarries (Trujillo 2006; Maidment et al. 2017) mean the zones outlined above are only very generally defined. Recent radiometric results for some of these quarries have shown that sites at rather different stratigraphic levels appear to be the same age (Trujillo and Kowallis 2015). For example, two quarries in Zones 4 and 5 here have returned U-Pb ages that are indistinguishable, but also two sites in Zone 5 have ages suggesting they are 1 million years apart. The shortened story, then, is that we may not have as good a handle on the correlation of quarries in the Morrison as we once thought, so that our hypothesized biostratigraphic patterns are once again a bit tenuous. (For a table of stratigraphic occurrences of Morrison vertebrates, see appendix B.)

Another possible stratigraphic subdivision of the Morrison Formation was proposed recently, one based on sequence stratigraphic and paleomagnetic correlations rather than lithostratigraphic correlation or biostratigraphy (Maidment and Muxworthy 2019). This study found that the formation appears to be younger in the north, something that many researchers have suspected was possible. This of course would mean than localities and taxa found low in the formation in northern Wyoming, for example, would be equivalent to sites more in the middle of the formation (rather than the base) on the Colorado Plateau. These long-distance correlations by stratigraphic sequence (Maidment and Muxworthy 2019), along with refinements based on future work and additional radiometric dates, may allow us to better correlate quarries and thus faunas and begin to decipher changes in the Morrison's paleoecosystem through time.

The specimen count data can also give us a picture of the overall diversity of the formation, regardless of the diversity of different taxonomic groups. One problem of diversity comparisons within a formation is the different number of specimens in each sample. One sample may contain more species than another but may also contain more specimens. Is the higher diversity a result simply of larger sample size? What would be the number of species in this larger sample if it instead had fewer specimens and was the same size as the other, smaller sample? If one could estimate this, different-sized samples could be compared directly to determine whether the samples likely came from similarly diverse original communities. We can do this using a specifically designed computer software algorithm. For this experiment with the Morrison Formation, we will test the samples from five stratigraphic zones in the unit. These samples, from Zones 2 to 6, all have different numbers of specimens and species known. We will be testing to see whether the overall diversity of vertebrates in the Morrison Formation changed through time or whether it was stable. These calculations produce curves of expected diversity versus sample size. If the curves are all similar in shape and within overlapping 95% confidence intervals of the expected diversities at given sample sizes, then we assume that the samples came from original faunas of comparable diversity.

The curves for different, "rarefied" sample sizes for each Morrison stratigraphic level above Zone 1 (fig. 7.17) suggest that the diversity of the vertebrate fauna in the formation was consistent throughout the time of deposition. These data are presented as a general indication of the comparative diversity of the different levels in the Morrison Formation.

General Diversity Trends

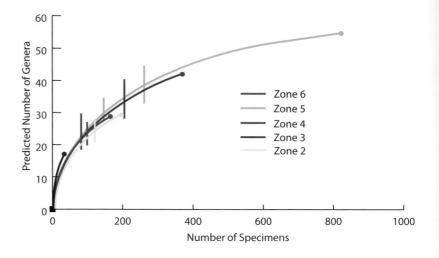

7.17. Rarefaction curves for vertebrate samples from Zones 2 to 6 in the Morrison Formation. The symbol on the right end of each curve represents the known sample size (x-axis) and fossil vertebrate species diversity (y-axis) for that level. The curve from that point down and left approximately represents the estimated diversities expected at smaller sample sizes. Vertical bars represent 95% confidence intervals for each curve; the overlap of most of these bars with adjacent curves suggests no significant difference in diversity across the five stratigraphic levels. Curves based on equations of Tipper (1979), Hurlbert (1971), and Heck et al. (1975). The diversity of Morrison Formation vertebrate species remained essentially constant for several million years.

Zones 4 and 5 seem to be slightly more diverse at smaller sample sizes than other levels, but probably not significantly so. These two zones each contain an unusually diverse site or two (FPA and Quarry 9), and these alone probably account for the slightly higher diversities.

Comparing the diversities of eight of the most productive quarries in the formation also yields similar results. The quarries have produced between 93 and ~4,000 specimens (either MNI or NIS, depending on the characteristics of the site) and have raw species diversities of between 15 and 44 vertebrate taxa. Comparing these by a diversity index that accounts for different sample sizes between sites (the Shannon index and related effective richness; Olszewski 2010), the eight sites have a remarkably similar level of diversity (Foster, unpublished data). And this diversity similarity seems to cut across preservational environments.

Another method of comparing diversity from different sample sizes was developed by E. H. Simpson, who proposed an index on the basis of the number of specimens of each species (n) and the total sample size (N). The Simpson index (SI) ranges between 1 and 0, and the value grows as diversity declines; thus, an SI value of 1 is the upper limit, at which all individuals in the sample belong to one species. Very low numbers result

Table 7.1. Calculated Simpson index (SI)* for the Morrison Formation

Site	SI
Morrison Formation overall	0.060
Morrison Formation stratigraphic level	
Zone 6	0.072
Zone 5	0.081
Zone 4	0.090
Zone 3	0.065
Zone 2	0.070
Zone 1	—

* SI = $\Sigma\,[(n^2 - n) / (N^2 - N)]$.

from large samples in which most species are represented by just one or a few specimens.

The SI value for the Morrison Formation as a whole is 0.060. The SI values for stratigraphic levels in the Morrison range from 0.065 to 0.090 (table 7.1) and suggest a relatively consistent and diverse community from each level. Although all the levels have close SI values and the differences may be insignificant, it is interesting that Zone 4, which contains the FPA and its microvertebrate assemblage, apparently has the lowest diversity. On the other hand, it is also curious that Zone 5, with Quarry 9, is not the most diverse. The Morrison Formation had a high level of vertebrate diversity throughout its time of deposition, and despite large, diverse single quarries in only some levels, the SI value appears to show this.

One point that needs to be stressed yet again about the Late Jurassic world of the giants is that this was a time when land animals were just about as large as they've ever been in Earth's history, but only that upper limit was raised—the size of all species was not shifted up. There were still, as today, many small animals. In fact, there are more small species known from the Morrison Formation than large ones, and as we have already seen, there were about as many types of mammals as there were dinosaurs of all sizes, so the diversity curve was rich in small vertebrates. How small were these small vertebrates? Many were less than 1 kg (2.2 lb), some much less. Some small lizards and frogs probably weighed only a few grams (less than 1 oz). The smallest mammal weighed less than 10 g (<1 oz), and the mammals ranged up to nearly 150 g. The average weight among all Morrison mammal species was about 49 g, and most were between 10 g and 100 g (Foster 2009). This is a size range that many modern mammals still attain as their adult body mass, even though some terrestrial mammal species today are much larger. The diversity at small body sizes has not changed much in 150 million years.

Predator-Prey Ratios

Predator-prey ratios are counts of the relative biomass of carnivorous species from a fauna (in this case represented in the fossil record) compared with that of the fauna as a whole. Carnivorous taxa are themselves counted in the prey category because, obviously, there was little to stop a large carnivorous dinosaur from eating a smaller one when the opportunity arose. If we correct the abundance counts we started with in this study and convert these to biomass, we can easily compare the predators to prey. What does this tell us? It can suggest the relative metabolism of the predator species. Most modern ecosystems contain a higher abundance, density, and percentage of carnivorous reptiles than endothermic carnivorous mammals (Bakker 1972, 1980, 1986; but see also Farlow 1980). This is largely because reptiles require less food per animal per day than does a mammalian (endothermic) carnivore of the same size, and thus a given amount of prey biomass can support a certain number of mammals but a greater number of same-sized reptiles.

The predator–prey ratio technique for paleoecosystems assumes that the populations were steady and that the number of skeletons produced and preserved was roughly equivalent to the secondary productivity of the taxa. The Russell correction method used here (Coe et al. 1987; Russell 1989) should account for the preservational bias against the original numbers of skeletons available. Several years ago, Farlow noted some of the other assumptions involved in the use of predator-prey ratios (Farlow 1993). Most important among these is that one assumes that all productivity of the prey species is consumed by the carnivores, but in that study, Farlow was mainly concerned with large, tyrannosaur-sized theropods and had to discount smaller theropods. In this case, because we are including all theropods as well as nondinosaurian predators, the assumption is less problematic, although it is probably impossible for all prey biomass to be consumed by predators. Other assumptions that Farlow (1980, 1993) noted were relatively even distributions of predator and prey species and that prey species distribution densities were fairly high. Another factor here is the assumption that a given carnivorous taxon consumed relatively even and high amounts of the productivity of all of its utilized prey taxa. Farlow noted that a more even consumption of prey species would result in lower predator-prey ratios than if any predator concentrated more on one of the prey species. Thus, highly diverse, potentially specialized communities of predators could appear to have lower predator-prey biomass ratios than they had in reality.

So predator-prey ratios have been used to test whether past vertebrate ecosystems were more like those containing mammalian predators (low ratios of predators to prey) or more like those containing reptilian predators (with higher ratios but still well below 1—prey almost always outnumber predators; Bakker 1980, 1986). What is the case in the Morrison Formation? It appears that the ratios are well below those demonstrated by typical reptilian-predator ecosystems, but they are still higher than those of mammalian systems. The carnivores, it seems, at least from this evidence, were like neither modern reptiles nor mammals in terms of their metabolism.

The predator-prey biomass ratio for the Morrison Formation as a whole is 8.6%. The values range from 5.1% to 11.9% at various stratigraphic levels in the formation (table 7.2). These percentages were

Table 7.2. Calculated predator-prey ratios for the Morrison Formation

Site	Ratio (%)
Overall biomass ratio	8.6
By stratigraphic level	
Zone 6	6.0
Zone 5	5.6
Zone 4	11.9
Zone 3	8.0
Zone 2	5.1
Zone 1	—

calculated including all vertebrate taxa in the formation; comparisons of predator-prey biomass that use only dinosaur species demonstrate a range from 4.4% to 11.0%, with an average of 6.44%. Interestingly, a predator-prey ratio based on the Morrison Formation's footprint data is quite similar at 8.2%, although considering the very different types of data, the significance of this is difficult to determine, and the similarity of the percentages may be coincidental.

Bakker's predator-prey ratios for general dinosaur communities range from less than 1% to just over 5%, similar to those of modern and fossil mammal communities (Bakker 1972, 1980, 1986). The overall predator-prey ratio in the Morrison (8.6%) is above any of Bakker's ratios but is still at the top of the range demonstrated by later fossil mammal communities. The range of variation within the Morrison Formation by stratigraphic level is from the top end of Bakker's dinosaur range up to the lower end of the vertebrate ectotherm range. If these ratios mean anything about the physiology of the dinosaurs, it is important to remember that the predator-prey biomass for the Morrison includes taxa that are generally agreed to have been ectothermic (crocodilians). This fact would increase the apparent ratio of dinosaurian predators. However, the crocodilians and medium and small theropods combined account for only 1.3% of the Morrison biomass, so the adjusted formation-wide ratio of large theropods would be about 7.3%. The Morrison ratio and range may support the "intermediate" dinosaurian metabolism proposed by some (Reid 1997, 2012), but it is interesting that it is more within the upper range of mammalian ratios than in the lower range of that of ectotherms. It is important to emphasize, however, the previously mentioned assumptions that are involved and the fact that the possible metabolic implications only directly relate to the predators and would not necessarily suggest that all prey species had similar physiologies.

The Morrison Formation predator-prey ratio is not particularly high or low, despite the fact that some of the prey species had adult weights of 20 to 40 times that of the largest common predator and that most of the predator species were in fact relatively small, regardless of popular belief about the great size of all dinosaurs. Of the at least 11 carnivorous dinosaur genera known from the Morrison Formation, just 4 may have weighed more than 1,000 kg (2,204 lb), and 3 of those are comparatively rare; another 3 or 4 theropods were less than 100 kg (220 lb). On the basis of a calculation developed by Farlow (1993), the predicted population densities of the Morrison theropods would range from 5.11 animals per square kilometer for the small *Ornitholestes* to 0.022 for the massive *Saurophaganax*. The most common larger theropod, *Allosaurus*, has a predicted density of 0.061 animals per square kilometer; this translates to a total population of *Allosaurus* in the Morrison basin of about 61,000 individuals at any one time. Adjusting for the different relative abundances of the theropod taxa (remember that *Allosaurus* alone accounts for ~74% of the total number of theropod specimens known from the formation), the greatest densities of animals occurred among *Coelurus*, *Ornitholestes*, and *Allosaurus*. By measure of biomass, *Allosaurus* still dominated the predator population. The 8.6% ratio of predator-to-prey biomass was thus a result of moderate numbers but a large biomass of *Allosaurus* and large numbers but a low amount of biomass contributed by small theropods like *Ornitholestes* and *Coelurus*. This may suggest that *Allosaurus* was the main predator on a variety of medium-sized dinosaurian prey, and considering its relative abundance, *Allosaurus* was, as discussed previously, probably less selective and less specialized in its feeding than other large theropods like *Torvosaurus* and *Ceratosaurus*.

Small theropods, meanwhile, probably concentrated mainly on smaller, nondinosaurian prey, and the total biomasses of both predatory small theropods and small nondinosaurian prey had little effect on these predator-prey ratios. The small theropods could all have fed on a variety of small vertebrates and been less specialized in feeding because there was a great diversity and number of smaller animals, particularly mammals.

Similar estimations of population density for large herbivorous dinosaurs suggest that, depending on the metabolic level that these animals had, common sauropods such as *Camarasaurus* and *Diplodocus* would have had about one adult animal for every square kilometer of floodplain, whereas rarer sauropods of the same size like *Barosaurus* would have had densities of about 0.23 animals per square kilometer (Farlow et al. 2010).[6] This suggests populations of sauropods in the Morrison basin of about 230,000 to 1 million individuals at a time, depending on the genus.

Geographic Distributions

The dinosaurs and other vertebrates of the Morrison Formation are widely distributed and have been found in all eight states in which there are surface exposures of the unit. Fossil vertebrates are found across an expanse of western North America that covers about 1,250 km (775 mi) north–south and 800 km (~500 mi) east–west (Dodson et al. 1980; Foster 2003a). It would seem that across such great distances there should be some regional zonation of the vertebrates. Some should be more common in or even restricted to certain areas of the Morrison basin, but this is not immediately obvious when one first looks over the data. In their study of Morrison Formation paleoecology, Dodson, Behrensmeyer, Bakker, and McIntosh (1980) noted that sauropod dinosaurs occur in nearly all Morrison paleoenvironments in all parts of the rock unit's exposure and that the most common dinosaurs (*Camarasaurus* and *Allosaurus* among them) are found nearly everywhere. Because the north–south extent of the Morrison Formation outcrop is 50% more than that of its east–west width, and because temperature and environmental gradients are more likely in the north–south extent, it might be revealing to check relative abundance data for dinosaurs first across the north–south expanse of the formation. (Environmental gradients probably also occurred east to west, but these were likely less pronounced and caused more by localized groundwater conditions in the basin.) If we group nearly 90 quarries and their respective vertebrate samples into North, Middle, and South zones

Table 7.3. Comparison of abundances of five groups of Morrison Formation dinosaurs across the north–south extent of the formation's outcrop area*

Taxon	South Zone	Middle Zone	North Zone	Total
Allosauridae	22 (15.0)	34 (41.4)	13 (12.6)	69
Camarasaurus	26 (23.1)	57 (63.6)	23 (19.3)	106
Diplodocidae	26 (30.5)	84 (84.0)	30 (25.5)	140
Stegosaurus	14 (14.8)	48 (40.8)	6 (12.4)	68
Bipedal Neornithischia	11 (15.7)	50 (43.2)	11 (13.1)	72
Totals	99	273	83	455*

* Observed values are listed; expected values based on totals in each category are listed in parentheses. Despite a few anomalies, most observed numbers are close to their expected values. $\chi^2 = 15.26$; critical value = 15.51 (at $P = 0.05$); degrees of freedom = 8.

across the extent of the Morrison Formation and statistically compare relative abundances of allosaur, camarasaur, diplodocid, stegosaur, and ornithopod dinosaurs, we can test the likelihood that the distributions of dinosaurs are not homogeneous. Accounting for the total samples of each dinosaur group and from each north–south geographic zone, we can calculate the expected number of specimens of each group in each zone, assuming a homogeneous relative abundance distribution. The total samples in each north–south zone may be different, and within each zone some dinosaur groups may be more common than others, but our expected values are those at which these relative abundances are statistically the same for each zone. By calculating the total deviation of our actual samples from the expected values, we can determine the probability that each zone does not have a similar relative abundance. Table 7.3 shows the results for our Morrison Formation north–south compilation of data. As shown by the χ^2 value for this contingency table being less than the critical value, there is less than a 5% chance that the relative abundances of the dinosaurs in the three zones are not essentially equivalent (Foster 2000). Thus, the dinosaurs of the Morrison Formation were more or less homogeneously distributed, and there were no great differences in the abundances of common taxa across the north–south geographic regions in the basin. There is, of course, that slim chance that the anomalies we see in the data are in fact significant and not in range of the variation we would expect. Notice, for example, that in the North Zone, there are less than half as many *Stegosaurus* specimens as should be expected. Is this by chance, or were these dinosaurs really less common in this part of the Morrison basin? It is still too early to say, but recent finds in Montana (too recent to be incorporated here) suggest that there may be plenty of stegosaurs in the northern reaches of the Morrison after all.

The first hints of possible **provinciality** of dinosaur faunas may be emerging from Montana, where *Suuwassea* was found recently; northern Wyoming and Montana also seem to be yielding an unexpected diversity of the stegosaur *Hesperosaurus*. If this relatively small dicraeosaurid sauropod and the stegosaur were restricted to the slightly cooler and damper areas to the north, it may be the first of a possibly more diverse endemic fauna of dinosaurs in that region (Maidment et al. 2018). However restricted the local animals may have been in Montana, we still find remains of the typical southern Morrison fauna up there as well. There thus does not seem to have been anything about the environment that deterred "regular" Morrison dinosaurs from living in or migrating through these northern reaches of the floodplain.

It would not be surprising if most dinosaur species in the Morrison Formation were widely and similarly distributed across the extent of the formation (Foster 2000; Whitlock et al., 2018). Compared with most terrestrial animals today, a significant percentage of dinosaur species of the time were large and would have had large home ranges, and many may have migrated seasonally. Oxygen isotope ratios of rocks and sauropod and theropod teeth indicate that these dinosaurs (at least from some regions) made trips up to hundreds of kilometers, probably seasonally (Fricke et al. 2011; Bronzo et al. 2017). There may not have been changes in paleoenvironments across the floodplain significant enough to restrict dinosaurs from moving all around the region either. For smaller animals, the environments may have been restricting to some degree, but for large animals capable of significant migration, this would have posed little problem.

Among the small vertebrates, a qualitative look at the data suggests at least two possible patterns. There may still be too few sites with microvertebrates to make reliable comparisons of relative abundances, and too many taxa are still known from only one or two sites, but looking at the geographic distributions of microvertebrates reveals hints that there may be some provinciality among these species. For example, the semiaquatic or burrowing Morrison mammal *Docodon* is the most common single species of mammal in the formation, and it is often the most abundant mammal at the quarries from which it is known, but it is almost unknown from the western part of the Morrison basin. Although it is known from dozens of jaws and occurs even at some sites with only a few mammals in the eastern outcrop area, *Docodon* is represented by just a couple of possible fragments even from the vast microvertebrate collections in the west, particularly at significant sites such as Rainbow Park and the FPA. This distribution appears to represent a genuine paleobiogeographic pattern of greater abundance or even restriction of *Docodon* to eastern parts of the Morrison basin, even across a wide north–south range (Foster et al. 2006).

Why is this? Perhaps *Docodon* preferred the distal parts of the Morrison floodplain further from the source areas of the rivers or within the wetlands of the eastern and northern parts of the basin. One geologic aspect of the areas of the Morrison Formation where *Docodon* is abundant is that these appear to have been areas that during the Late Jurassic had a high water table and multiple ephemeral and perennial lakes and wetlands, many more so than regions to the west (Dunagan and Turner 2004). Maybe this regional environment was more to the liking of *Docodon*, and perhaps this supports the idea of this Morrison mammal being semiaquatic like the modern desman and muskrat (T.

Martin 2005; Foster et al. 2006). There is nothing to rule out its being a burrowing animal, but if the environment was dotted with ponds and wetlands and the water table often high, a semiaquatic and omnivorous existence may have been a widely available approach for many animals, including a mammal.

Another species that seems to show an interesting paleobiogeographic pattern is the small semiaquatic choristodere *Cteniogenys*. Of the nine localities at which this reptile has been found, seven are in the northern regions of the formation. Specifically, six are in Wyoming, one in South Dakota, and one each further south in northeastern Utah and western Oklahoma. Of the nearly 60 specimens of *Cteniogenys* found so far, all but three are in Wyoming, so although it did range down to the southern extent of the formation, *Cteniogenys* was apparently far more abundant in the northern half of the formation (Chure and Evans 1998; Foster and Trujillo 2000). It is difficult to say why this may be, except that perhaps the wetter and slightly cooler (or rather less hot) conditions up north were more what *Cteniogenys* preferred environmentally. A larger sample of microvertebrates from the Morrison Formation will eventually clarify these patterns and will almost certainly refine our interpretations of why they occur.

Finally, although they range throughout the formation and occur in nearly all areas, turtles and semiaquatic goniopholidid crocodilians appear to be much more abundant, or at least more often preserved, in the wet environmental settings of the eastern and northern parts of the Morrison basin (Foster and McMullen 2017).

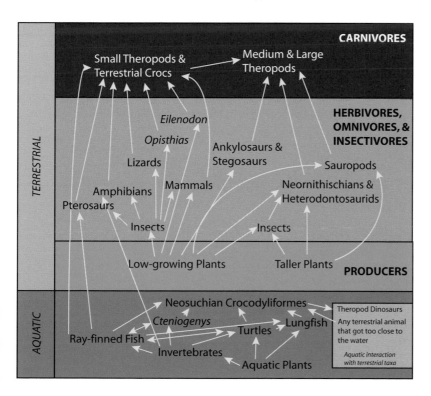

7.18. Simplified food web for the Morrison Formation. Arrows indicate hypothesized feeding and direction of energy flow. Some groups probably also fed on each other, although this is not shown. In the aquatic environment, crocodilians were the top predators and were, along with aquatic invertebrates and fish, the link with the terrestrial system.

We can determine what types of food different vertebrate species were likely eating during Morrison times, but specifically what and who might they have been feeding on? We can guess on the basis of the structure, size, and teeth of the animals and diagram this information. Figure 7.18 presents a simplified hypothetical food web for the Morrison Formation. The first, and in some ways most important, group in this diagram is the plants that convert sunlight into energy that in turn becomes available not only to the vertebrates but to all animals of the time. Feeding on the plants were herbivorous dinosaurs and insects, and then feeding on these were various levels of carnivores.

Presumably herbivorous terrestrial invertebrates (mostly arthropods) served as the main food for most mammals, amphibians, and lizards of the Morrison Formation. Feeding on these vertebrates were predators such as the small theropods (*Ornitholestes*, *Coelurus*, *Koparion*, *Tanycolagreus*) and terrestrial crocodilians. Larger herbivores included the stegosaurs, ankylosaurs, ornithopods, diplodocids, *Haplocanthosaurus*, *Brachiosaurus*, and *Camarasaurus*. These were prey for the next level of larger theropods such as *Allosaurus*, *Ceratosaurus*, *Torvosaurus*, *Marshosaurus*, and *Stokesosaurus*. Aquatic invertebrates (arthropods and mollusks) and plants served as the main food of turtles, actinopterygian fish, *Cteniogenys*, and lungfish; these were fed on by crocodilians and possibly carnivorous dinosaurs. There was probably a fair amount of overlap in diets as well. *Cteniogenys* probably fed on insects and small fish, pterosaurs on fish and small vertebrates when possible, crocodilians on small dinosaurs when the dinosaurs were near the ponds and rivers, and turtles on amphibians and small vertebrates. Docodonts and multituberculates probably fed on seeds and other plant material as well as insects, triconodonts on very small vertebrates as well as insects, and larger theropods on smaller theropods. Both aquatic and terrestrial environments have four trophic levels: producers (plants), primary consumers (herbivores), and two levels of secondary consumers (carnivores). This is typical of most ecosystems. And as in most ecosystems, there surely were plenty of decomposers in the Morrison basin (such as bacteria, and mushrooms and other fungus) to degrade the organic matter not ingested by the herbivores or carnivores.

Although the web contains a minimum of four trophic levels, the number of links can be as high as seven. For example, large theropods such as *Allosaurus* can feed at level 3 when feeding on herbivorous dinosaurs, but seven levels would exist if *Allosaurus* fed on a medium-sized theropod such as *Marshosaurus*, which would result in the following sequence: plants, herbivorous invertebrates, small insectivorous frogs, triconodont mammals, small theropods (perhaps *Ornitholestes*), *Marshosaurus*, *Allosaurus*. Various studies of food webs suggest that communities with shorter, more connected webs are more stable than those with many trophic levels and fewer prey species per predator, although stability studies of complex systems suggest that above a certain number of species, increasing connectedness actually reduces stability

(Rosenzweig 1995; Gotelli and Graves 1996). In this case, stability means independence of species populations from variations in other species. For example, highly connected food webs allow predators to concentrate on other prey when one prey population drops. Thus, fluctuations of one species do not necessarily affect all others. The Morrison Formation web seems to be fairly highly connected and to have an average number of trophic levels. Predatory species have a variety of possible prey species available. The main difference separating this web from those of modern ecosystems is that there is no way to directly observe diet for an ancient species, and thus most features of the web are conjectural.

Evidence from trackways of vertebrates adds another line of evidence to the paleoecological picture in the Morrison Formation. Unlike skeletal material, which is often preserved in paleoenvironments in which the animals did not actually live, tracks provide direct evidence of the areas where dinosaurs and other taxa naturally walked. Two ways to measure the abundance of animals in any one area or level are by the number of trackways and the number of localities at which different types of tracks occur. Probably the more accurate of these is the count of the number of sites, because the trackway counts can be dominated by the census of a single large locality. Figure 7.19 compares the numbers of trackways of several dinosaurian groups and other vertebrates in different paleoenvironments. Floodplain (mudstone) environments preserve sauropod and theropod dinosaur tracks but little else, and the numbers of each are

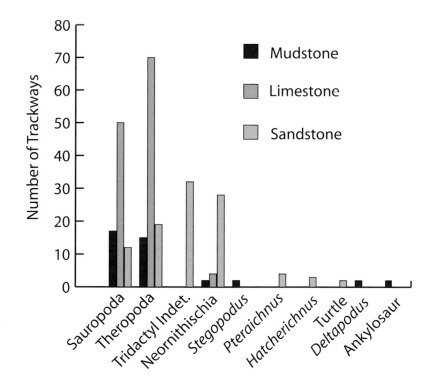

7.19. Abundance of different vertebrate track types by substrate lithology, measured in number of trackways.

not great, probably because of the preservational mode of these types of sites. Most floodplain localities are areas of relatively small sandstone overhangs with several isolated natural sandstone casts of tracks that were originally made into mudstone. Without large areas of exposed bedding planes, these sites rarely contain more than a few tracks. The fact that most lacustrine shoreline (limestone, lake) localities are larger, exposed bedding planes with natural impressions of multiple trackways is seen in the numbers of sauropod and theropod tracks in this category. These numbers are dominated by the large numbers of sauropod and theropod trackways at the Purgatoire River (Lockley et al. 1986) and Como Bluff track sites. The sandstone splays-and-bars paleoenvironment seems to preserve the most diverse assemblage of vertebrate tracks, and in this category, the sauropods and theropods are more evenly abundant than other dinosaurs. Probably the numbers of trackways of sauropods and theropods in lacustrine shoreline settings are unnaturally high compared with those of other taxa because of the two sites mentioned above. The lower diversity of other dinosaurs in the floodplain settings may just be a result of their smaller tracks being harder to recognize as casts.

Figure 7.20 shows a presumably more realistic abundance of the vertebrates in plotting the numbers of sites at which different vertebrate taxa occur in the three paleoenvironmental settings. Sauropods are the most common group in floodplain settings, although theropods are not uncommon there, and sauropods and theropods occur at nearly equal numbers of lacustrine shoreline localities. There are far more localities with sandstone splay-and-bar paleoenvironments, however, and these preserve a greater diversity of track types. Interestingly, the abundance relationships seen in the previous graph are reversed, and theropods occur at twice as many sites as sauropods. Sauropods are thus about as common as ornithopods and indeterminate tridactyl tracks in splay-and-bar paleoenvironments.

The relative abundances of different dinosaur track types in the Morrison Formation (Foster and Lockley 2006; Lockley and Foster 2017) suggest that sauropods may have preferred floodplains and areas near lakes and ponds. The fact that their tracks are far less common in sandstone beds may indicate that they traveled along and near rivers less often, or alternatively, it may simply reflect the fact that many sandstone localities consist of relatively small surface areas of exposed bedding planes, and large sauropods are less likely to have left footprints in these areas than smaller animals. The floodplains on which sauropods walked may have been proximal relative to river channels and may have consisted of soft, wet or hard, dry substrates. The fact that several floodplain sites in the Morrison Formation preserve sauropod tracks with casts of infilled mud cracks both in the sandstone bed overlying the mudstone into which the tracks were made and in the track cast itself indicates that the floodplains on which the sauropods were walking were wet at the time and dried some time later. The tracks the sauropods made in the floodplain mud were often 30–40 cm (12–16 in.) deep, and the muddy floodplain surface

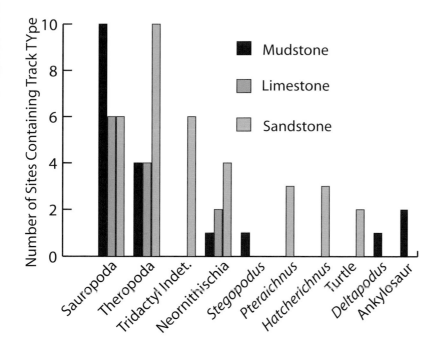

7.20. Abundance of different vertebrate track types by substrate lithology, measured in number of localities.

and the tracks themselves later dried and cracked before being covered with the sand deposit, probably during a flood. Areas where sauropods may have traveled on hard, dry substrates obviously would not preserve trackways, and thus it can only be assumed that the sauropods lived in these areas as well. Sauropods may have in fact lived in all environments, and their common occurrence in lacustrine limestones as well as floodplain muds, and the fact that at least one splay-and-bar site (Hidden Canyon Overlook) contains three parallel sauropod trackways, suggest that these animals ranged across most paleoenvironmental settings of the Morrison Formation.

Still, the distinctly low percentage of track sites in sandstones that contain sauropod tracks is interesting. Considering the preservational similarities between sites in limestone beds and those in sandstone beds (i.e., tracks preserved as impressions; bedding plane exposures), it would appear that both the rarity of sauropod tracks and the diversity of other track types in sandstone deposits are real. What, then, might have kept sauropods away from sandy substrates? If the sandstone deposits were related to river channels, and if plant species were on average larger or plant biomass more densely concentrated near river channels and other water sources, as suggested by paleobotanical studies, then perhaps this density prevented sauropods from penetrating the vegetative cover and traveling along river courses (and in near-channel areas where splays may have been deposited) as often as they did out on the floodplain, where vegetation may have been less dense. With the possibly greater abundance of vegetation near the channel, and with their greater ability to maneuver in denser vegetation, smaller herbivorous species (and the carnivores that fed on them) would then have been relatively more common in near-channel environments. The common occurrence of sauropod tracks in lacustrine shoreline settings may suggest less dense vegetation on the perimeter of the lake. If these lakes were on the flood-plain separated from river channels, they may have fluctuated in water level but would not have flooded the surrounding area like the rivers would have. Therefore, the nearby plant growth may not have been as thick as near the flooding and avulsing channels, although high water tables near the lakes would have allowed significant plant growth. The areas around lakes and ponds, then, may have been more open and more frequently traveled by sauropods. Still, sauropods obviously did spend time near river channels, as evidenced by their tracks, which are found at five sites that indicate splay-and-bar deposits.

Equally interesting is the distribution of theropod tracks. Theropods occur nearly as commonly as sauropods in floodplain and lacustrine shoreline paleoenvironments but are more commonly found in splay-and-bar deposits. Theropods thus seem to have ranged across all environments. If the more common occurrence of theropods in near-channel environments is real (as indicated by tracks), then the theropods may have simply been staying near a greater concentration of prey species.

In the cases of pterosaurs, crocodilians, and turtles, their occurrence in the more sandy environments and not in others may be a result of their smaller size or the greater number of track sites in sandstones, but, at least in the case of pterosaurs, it may reflect a habitat preference. If pterosaurs in fact fed on fish, one would expect to find their trackways in the lime-stones of lacustrine deposits and in sandstones, indicating proximity to water. Crocodilian and turtle tracks should occur in limestones as well. Perhaps, though, there are still too few limestone track localities known. The number of sites with turtle, crocodilian, and pterosaur tracks is still smaller than the total number of track sites in sandstone beds.

A major note of caution here is that the above discussion is based on the following assumptions: (1) the trends seen in distributions of vertebrate tracks are real and not variations that result from the different preservational characteristics of the tracks and the kinds and areas of bed-ding in which they are found; (2) vegetational cover was differentiated among areas of the floodplain basin, with the variations at least generally understood; and (3) a large enough sample of vertebrate tracks is available that real trends can be indicated.

The distribution of track types (as measured by number of localities) by stratigraphic level (fig. 7.21) indicates a decreasing abundance and diversity upward through the formation that is a result of progressively decreasing numbers of localities at higher stratigraphic levels. The strati-graphic zonation here is divided into quarters, estimated to be rough equivalents of the upper and lower Salt Wash and Brushy Basin Mem-bers. The decrease in numbers of sites upward through the formation is probably a result of the manners in which tracks are preserved in the Mor-rison. Most track sites occur either as relatively large, exposed bedding

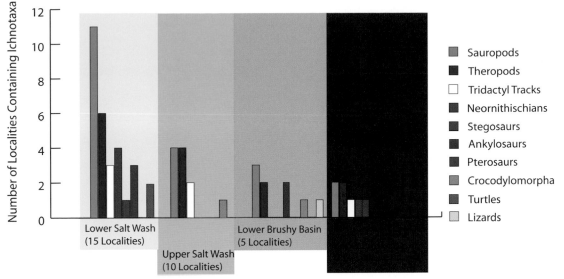

planes of sandstones or limestones or as collections of natural casts under sandstone bench overhangs above mudstone units. The abundance of limestone beds low in the formation in some areas and the abundance of sandstones in the Salt Wash Member mean that a large number of sites occur low in the Morrison. Higher in the formation, in the Brushy Basin Member, thick units of mudstone are common, and sandstone and limestone beds are more rare; thus, fewer track sites occur at this level.

The relative abundances of sauropods and theropods seem to be consistent in most levels, and probably there was little change temporally. The slight variations in abundance and diversity higher in the formation are likely the result of smaller sample sizes. The relative abundances of sauropods, theropods, and ornithopods in the lower Salt Wash equivalent levels are interesting because they are probably the best record we have of abundances of vertebrates from what would probably be Zone 1 of the body fossil record, the zone for which the least skeletal information is available. In some ways, the track record from low in the Morrison fills in for what has so far been missing in the body fossil record. The lower Salt Wash equivalent track record indicates a lower presence of ornithopods than is recorded from body fossils. Interestingly, though, the relative abundances of sauropods, theropods, and ornithopods are closer to those of the straight counts of body fossils than to the corrected count abundances. Larger sauropod tracks are more likely to be found and preserved than smaller tracks, so there may be some preservational bias, but if we compare the trackway counts from the splay-and-bar paleoenvironment, ornithopods are more common than theropods and sauropods, a pattern that more closely resembles a realistic population composition.

384

Many Rivers to Cross: A Late Jurassic Journey across the Morrison Floodplain

8

This is all a dream we dreamed one afternoon long ago

Robert Hunter, "Box of Rain," 1970

Like other people, paleontologists spend a lot of time thinking about various aspects of their work, but in our case, it is partly because we deal with worlds that no longer exist. A significant amount of energy can be spent in the physical labor and logistics of excavation, preparation, and interpretation of fossils, and all of this work ultimately leads toward the goal of the reconstruction of past environments and ecosystems. So by the time we get to the point of being able to stop and ponder the world whose evidence we see when we find and excavate fossils, a lot of time and sweat have already been invested in the project. But envisioning what we are seeing so many years after its living presence is, I suspect, irresistible to all of us. It is sometimes difficult to remember, much less to fathom, that the bizarre animals that we dig up were once integral parts of a very different, almost alien world—but one that existed on the same planet and in areas that are now as unexotic as freeways and backyards, prairies and deserts.

Imagining the world of the Late Jurassic, or any other time, can help bring together many diverse elements and scattered data into a more informative, whole picture of a former ecosystem. As an exercise, it helps improve our overall understanding of what we are seeing in the fossils we have found already. We can imagine taking a trip across the floodplain whose sediments became the Morrison Formation, and we might find ourselves transported back 150 million years to a starting point in what is now Montana. For the trip, we will be moving on foot, with backpacks to carry some gear, and will journey from the mouths of some Morrison rivers to their headwaters. This means we will be traveling from the shore of the ancient seaway of northern Montana and Canada to the south and west, tracing stream and river systems from where they empty into the sea all the way to their origins in the mountains west of the Morrison basin.

We will be traveling about 40 km (25 mi) per day on average, and the trip will take us a little over three months. Making 40 km per day may be a little optimistic because of the terrain. Most of the trip will be over relatively flat land, but forging through vegetation and mud and fording streams could all slow things down considerably.

8.1. Modern relatives of plants seen in the Morrison Formation. (A) Horsetails *Equisetum* growing as ground cover in a wet meadow in Sequoia National Park in the Sierra Nevada Mountains, California. (B) Modern fern with a leaf shape very generally similar to the Morrison's *Cladophlebis*, from Camarillo, California. (C) Modern fern with a leaf shape very generally similar to the Morrison's *Coniopteris*, from Glacier National Park, Montana. (D) Modern cycadophyte *Zamia*, from Palm Desert, California. Some Morrison cycadophytes may have had this general appearance. (E) Podocarpacean conifer *Dacrydium*, New Caledonia. (F) Araucarian *Araucaria bidwilli* (43 m tall; 141 ft) from Camarillo, California. (G) Foliage of the cupressacean conifer *Thuja plicata* (western red cedar), Glacier National Park, Montana, with scaled leaves similar to some Morrison fossils. (H) Leaves of the modern *Ginkgo*, indoors at the Royal Tyrrell Museum of Paleontology, Alberta. Morrison fossil *Ginkgo* leaves often had multiple lobes (fig. 1.29G), but individual trees may have had some leaves similar to those of modern trees. (I) Close-up of foliage from the *Araucaria bidwilli* shown in F, similar to some Morrison fossils. Image in E courtesy of Alan Titus.

From the Beach

If we start off on the shore of the seaway, we find ourselves double-checking our packs and supplies on a flat, wide stretch of sand with a calm sea to the north. The beach is not steep, so small waves roll in from hundreds of meters (yards) offshore. Wading out into the warm water to catch some fresh fish for the early stage of the trip, we see horseshoe crabs moving along the sand in the clear, knee-deep water, and we know that farther out in the deeper parts of the sea live the dolphin-like marine reptiles, the ichthyosaurs, and small squid-like animals. Coming back in with a few fish, we might see some oyster shells in the sand on the beach. As we get ready to move off the beach to start the first leg of the journey, we observe the vegetation south of the shoreline. It is mostly conifers near the sand, but it becomes denser and more diverse to the south. Several hundred meters to the east is the edge of a delta where a stream system empties into the sea. We will begin by following this river to the south.

Our bodies will, we hope, adjust to the atmosphere of 150 million years ago. Although the percentage of oxygen is about the same as today (at about 21%) or slightly higher, the concentration of carbon dioxide is close to three or four times higher. The greater amount of CO_2 not only keeps the planet's atmosphere warmer in general but also might make breathing a little labored for us, at least at first. The higher concentration of CO_2 in the air we breathe in might make it harder for our bodies to get rid of the CO_2 in our blood and perhaps leave us gasping for more oxygen to counteract this. The animals of the time evolved in such conditions and are thus adjusted to it, but for us, plunked down in the middle of a foreign environment, it might take a period of adjustment. Or perhaps we wouldn't adjust, and we'd be working hard for our entire time in the Late Jurassic.

Among our packing duties, we make sure we have all our food packs as a base diet and that each of us has a water purifier. Dive masks are in our packs for exploring water environments. We also take a few minutes to make sure that we have a sufficient and clean supply of predator deterrents: for each of us, a shotgun as an emergency measure for accidental close encounters; a team rifle of standard size for food hunting; and a team semiautomatic rifle for encouraging large theropods to head the other direction, from long distances, when necessary. We consider a large-game gun but ultimately decide that we are trying to run the animals off rather than actually drop them (as much as we'd like to dissect one to learn about its internal anatomy—we'll wait until we come across a carcass). Finally, we lament the lack of any methods by which to fend off the semiaquatic crocodilians; we'll just need to be cautious.

The sky is clear and the temperature in the high 80s ($\sim$30°C) on a midsummer day as we begin walking along the beach to the delta and then turn south so we can walk parallel to the bank of the river. For the next three days, we hike through the riparian woodlands along the banks of the river and occasionally cross wide meadows carpeted with ferns and short, herbaceous cycadophytes. The ground is damp, and while crossing the systems of anastomosing rivers, we notice turtles along the shores, in the water, and on logs.[1] We also need to keep an eye out for the crocodilians that inhabit the rivers. By now, we've noticed that we have to keep food well sealed: small animals have been getting into our packs overnight. On days 4 and 5 of the trip, the main river begins wide meanders east and west, so we keep moving south across the open fern meadows and through scattered woodlands, crossing small streams as we go and now and then fording the main river as it crosses our southerly route. The woodlands are composed mostly of araucarian conifers, sequoias, ginkgoes, and tree ferns. The ground sometimes becomes wet in the open woodlands near the streams, and we can see dark-brown and black soil between the ground-cover ferns, preserving semi-intact fragments of ferns and ginkgo leaves. As we cross another of the large fern meadows, we see several of the long, slender sauropods feeding on the low-growing ground cover. They are probably *Diplodocus*. On the edge of one of the woodlands along a major river, where the trees are more openly scattered, there are a pair of 15 m (50 ft) *Camarasaurus* feeding on branches, leaves, and fronds of the conifers, ginkgoes, and tree ferns.

A week into our trek, we begin to notice a decrease in density and frequency of the woodlands through which we have been passing. The open fern-cycadophyte meadows are becoming larger and the woodlands more scattered. Woodlands seem to be distributed along stream courses, where they are densest, and in lower areas of the floodplain, sometimes around ponds. Some open stands of moderate-sized trees do occur out away from water. In these areas, there is an understory of ferns and bush-like cycadophytes and seed ferns. We've already seen plenty of horsetails, a diversity of ferns, cycads, various conifers, and ginkgoes (fig. 8.1)

On day 8, 280 km (174 mi) into the trip, as we slowly work our way into the interior of the Morrison basin, we get some scattered clouds and a short rain shower in the afternoon. This is short-lived, however, and the temperature is soon back up to what it was. It is still close to 70°F (21°C) at night and quite comfortable camping. Late on the eighth day, we notice several small, robust diplodocids feeding in a fern meadow that we are crossing. They are several hundred meters east of us and appear to be about 13 m (43 ft) long. It is difficult to tell from a distance and without a skeleton visible, but on the basis of their overall shape and size, they may be young *Apatosaurus* or perhaps the dicraeosaurid diplodocoid *Suuwassea*.

On day 9, we turn east-southeast and begin heading through what one day will be southern Montana and northeastern Wyoming.

The day and night temperatures have increased slightly, and we find ourselves drinking more from the rivers than we have in past days. One advantage of the relatively dry atmosphere is that our sweat evaporates quickly and keeps us cool, but we risk dehydration. For food, we would live more on cooked and dried and salted meats and maybe some seeds. We would find Jurassic plants mostly bitter or otherwise unappetizing. The fruits and most of the vegetables we eat today are parts of angiosperms (flowering plants), which didn't exist during Morrison times. The

8.2. An *Apatosaurus* feeding near the edge of an open plain covered by ferns and a scattering of cycadophytes. In Morrison times, such "fern meadows" may have covered much of the landscape between the many woodland areas. Trees of the woodlands were mostly araucarian conifers, *Podozamites*, and ginkgoes. *Drawing by Laura Cunningham © 2005.*

modern relatives of the plants that were around during the Late Jurassic are not very tasty, so their ancient forms probably weren't much better. Few dishes today are made from parts of cycads, ferns, or conifers, so we'd likely not eat much of these as we hiked the Morrison. More likely we'd salt some seeds (although even these would probably be hit and miss) and meat of small vertebrates we'd caught and dried. But perhaps we would have found some way to stomach boiled ginkgo leaves or some of the softer herbaceous cycadophytes from the meadows.

As we move across the plains in this area, we encounter the same mixture of rivers, scattered forested areas, and open meadows as we've seen for days. One thing we have noticed is that the rivers so far are all of only moderate size and their channels laterally restricted. As we cross the sandbars, we notice that the sediment carried in these streams is mostly medium- to fine-grained sand, and it is composed mostly of quartz. There are few other minerals represented in the rounded sand grains. From outcrops of the Morrison Formation, we know that this will change as we get farther south and west—the sediment becomes coarser and the mineralogy of the grains and pebbles more diverse.

For the next three days, we cross woodlands, rivers, and meadows and see several herds of *Camarasaurus* feeding in the open woodlands and small groups or individuals of *Apatosaurus* mostly in the meadows (fig. 8.2). Many of the *Camarasaurus* are 14-ton, 15 m (50 ft) adults, but several young juveniles and subadults are in the groups as well. The juveniles are only a couple of years old, and the subadults are probably less than 10 or 12 years old, although they are close to 11 m (36 ft) long. Several of the adults are older than the others, although they are not much larger. They move a little more slowly and carefully, as some excess arthritic growth has begun in their vertebrae. The youngest camarasaurs, probably less than a year old, are about the size of a cow. Some of the *Apatosaurus* are 24-ton, 21 m (70 ft) monsters, but several young juveniles and subadults are seen as well. We seem to see *Apatosaurus* less often in large groups; sometimes they are on their own (fig. 8.2).

About two weeks into the trip, things continue as they have been: temperatures around 90°F (32°C) during the day, little rain, mostly clear skies, and days spent wading streams, crossing meadows, weaving through woodlands, and crossing the occasional dry wash. We still see sauropods feeding, small groups of ornithopods, and fish, turtles, and crocodiles in the rivers and ponds. We have seen only a few carnivores, however. Those that we have seen have mostly been lounging at the edge of the woodlands, keeping an eye on small neornithischians and ornithopods or small sauropods but seemingly uninterested in pursuing them. On day 14, though, we stumble onto a scene far different from what we've witnessed before. Coming out of a woodland, we emerge on the edge of a meadow, and looking out, we see two adult *Allosaurus* pulling apart the fresh carcass of a juvenile *Diplodocus*. In the distance, the rest of the *Diplodocus* herd is slowly moving away. The allosaurs snap at each other with their jaws, each trying to keep the other away from its side of the carcass. The scene attracts the attention of a circling pterosaur, but because the two carnivores and their meal are right in the way of where we were hiking, we duck back into the woodland, circle around the scene, and continue on.

Ponds and River Crossings

At midafternoon on day 15 of our expedition, we take off our packs and sit in the shade of several tree ferns to rest. As we cool off, we drink some of the water we've been filtering from the streams—we don't want to drink straight from them because the unfamiliar, ancient microorganisms in the water would likely make us sick. We watch several small bipedal herbivores—probably *Nanosaurus*—feeding on the vegetation, and we get up and follow them as they slowly amble deeper into the woodland, eating as they go. We soon come to a pond. It is about 50 m (162 ft) long and 10 m (33 ft) wide and is slightly curved in shape. Its banks are lined with horsetails, ferns, and small cycads, and the woodland surrounding it is made up mostly of tree ferns and small conifers. The muddy banks of the pond are covered with trampled stems of horsetails along with fragments of cones from the trees. The water is quiet and slightly turbid with mud. Although we see little as we first step to the shore, after sitting down and staying quiet for several minutes, we notice a number of turtle shells sticking up above the surface along the opposite shore, the snouts of the animals' heads just barely breaking the water. They are sitting on submerged logs. One turtle moves through the water along the surface close to us but soon ducks under and disappears. There seem to be at least two types of turtles here, as indicated by the shapes of their heads and the slight differences in color patterns.

As we lean out over the water and look down into the shallows, we can just see a few small ray-finned fish darting around the submerged bases of the horsetails. Sitting motionless in the shallows in another part of the pond is a large, chunky fish about the size of a loaf of bread. This one has a boxy head and fins shaped more like paddles than fans—we've spotted our first lungfish of the trip. As we sit by the pond for a while longer, we notice that there is little sound except the buzz of some insects, the occasional chirp or birdlike call of one of the nearby feeding *Nanosaurus*, and, less frequently, the small splash of a fish jumping as it goes after a surface insect in the pond.

The pond is long and narrow. At one end, there are no trees, just bushy cycadophytes, some fresh ferns, and lots of horsetails. As we check this end more closely, we see that the muddy ground near the water's edge is covered with small bones mixed with the plant fragments. As we move away from the pond, we see that many of the horsetails are leaning over near horizontal, pointed toward the pond. Around some of the bushy plant bases are wrapped fern fronds and horsetail stems, and on the edge of the clearing near the woodland, we find a partial skeleton of a sauropod half-buried in mud. After picking up several of the white, slightly weathered bones, we determine that it is an adult *Camarasaurus* skeleton, probably washed in from elsewhere. We conclude that it is probably a section of the floodplain that flows with water only during floods—hence the washed-in sauropod carcass, the carpet of small bones, the knocked-down horsetails, and the bushes wrapped with other plants. Between floods, it is the quiet pond we see this afternoon, and it will remain so until the next event reactivates this part of the floodplain and washes in silt, sand, and bone debris in a relatively short period of time. Back at the pond, we notice the eyes of a small, motionless atoposaurid crocodilian breaking the surface near the far shore; its snout is unusually short compared to those goniopholidids we've been seeing in most rivers and ponds. Near the dry bank of the pond, a very young, downy-feathered *Allosaurus* rests on a log occupied by a sphenodontid (fig. 8.3). Far off to the north in a distant fern meadow, some sauropods are feeding. We haven't yet come to the active river channel we suspect is nearby, but we will soon enough. As we pick up our packs and prepare to move on, we notice something small splash into the water not far away. A small, furry animal begins crossing the pond, its nose held just clear of the surface and its waving tail propelling it smoothly through the brownish-green water. It reaches the other shore, dives briefly, and disappears. After a minute or so, it pops out of the water along the bank and vanishes into a hollow log. It is a mammal, perhaps *Docodon*.

We've got a few more hours to go before we camp, so we start off to the south, passing several more herds of sauropods and small ornithopods as we go. It is still warm at night, and we have no trouble sleeping in the open air by the fire. The next morning, it is pushing 80°F (27°C) not long after sunrise, and as we work our way south, the horizon ahead is a flat plain many miles distant. The landscape is almost entirely flat in every direction, much like being in the middle of California's Central Valley today. In the afternoon, as we cross a large meadow of ferns and cycadophytes, we get another rare shower that lasts nearly half an hour and just begins to soak the ground. We stop to check out a few of the cycadophytes called *Zamites*, which grow in small thickets and occasionally as isolated,

bushy plants. The horsetails seem to be fed on by herds of diplodocid sauropods in the meadows and the conifers and ginkgoes by the *Camarasaurus* along the edges of the woodlands. As we continue across this meadow—one of the largest we've crossed—we pass a lone 24-ton *Apatosaurus* and a herd of about a dozen *Diplodocus*, all feeding on the short vegetation growing close to the ground. We stop to eat a quick snack and drink some of the filtered water we have been carrying since crossing the last stream. We sit and watch the diplodocids. Their long necks, 6 m (20 ft) in length, are slung out almost horizontally from their torsos, sloping just slightly downward. Their heads hover just a meter or so above the ground, and they grab mouthfuls of horsetail or fern material with the teeth at the front of their jaws and pull back, stripping the vegetation; only one or two more quick chews and they swallow, grab another bunch of fern fronds, and repeat the process. As they do, their necks sweep slowly from side to side, covering on a full left-to-right arc about the equivalent of a modern, roomy two-lane roadway. As the sauropods work their way across the plain, they occasionally take a step forward, inching into fresh vegetation every few minutes. Now and then, one of the sauropods will lift its head up several meters to quickly scan the floodplain. We also notice

8.3. A very young *Allosaurus* with proportionately long hind legs and a covering of fine feathers encounters the sphenodont *Opisthias* when both find themselves occupying different parts of the same hollow log. Based on finds at the Little Houston Quarry in northeastern Wyoming. *Painting by Mark Witton © 2015.*

that their skulls rest naturally at nearly a 90-degree angle to their necks most of the time, so that as they feed with their necks out horizontally, their skulls are oriented almost straight down.

For the next four days, our routine continues. We get up and on the move early enough to be moving in the cooler part of the day, stop in midafternoon to rest during the hottest part of the day, and continue the hike for a few more hours as it cools down around sunset. Day in and day out, we cross fern-cycad meadows, scattered woodlands, and a few streams. Sometimes we pass small ponds and mudflats. There are some tree ferns scattered in parts of the meadows, and between the meadows and the dense woodlands close to the rivers, there are often open woodlands with a mixture of shorter trees and bushes and a ground cover of ferns. In a number of areas in the meadows of the floodplain, the ferns are beginning to brown slightly, but we encounter little bare ground and no long, open stretches of dirt. On one of the mornings, we come across an open woodland not far from a river. We cannot actually see the river through the dense woodland of tall conifer trees and ginkgoes, but we know it is there, beyond the levee. To the east as we approach the open woodland is a large fern meadow and a much more distant woodland. To the west are the dense forest and the river. As we come to the open woodland area where the trees are shorter and more widely spaced than along the river, with fewer tall araucarians and sequoias, we notice that many of the tree ferns, ginkgoes, and smaller conifer trees have been nearly stripped of all leafy branches. The ground cover is largely trampled. A number of the trees have been knocked over, and many of the bushy cycadophytes are stripped and damaged as well. But access to the bushy thickets and the ferns and other plants below them is a little better now.

As we move deeper into this impacted area, we see in the distance a herd of *Camarasaurus* slowly making its way southward away from us. With a distant, muffled crack, a small tree fern goes down in the middle of the herd, and several individuals begin feeding on the fronds. We are following in the feeding path of the camarasaurs, and what we are seeing is the result of their work. The trampled and stripped open woodland in which we are standing is no quiet place, however. As we watch the camarasaurs moving off into the distance, we notice a lot of activity around us. Patient observation of the area during a break from our hike reveals to us that the droppings of the passing sauropod herd have attracted numerous flies and dung beetles, and a few of the small basal neornithischians (probably *Nanosaurus*) are moving around the area in twos and threes, snapping up the insects as they also feed on the trampled ferns and partially stripped cycadophyte bushes. We are lucky enough to also catch brief glimpses of tiny mammals and lizards ambushing and eating the beetles and trying unsuccessfully to catch the flies. Meanwhile, as we begin to move out again, we also happen to scare up and see a small multituberculate (perhaps *Ctenacodon*) running back to its cover with a seed it retrieved from one of the damaged cycad plants. As we move south along the sauropod-damaged swath, we also observe several 5 m

(17 ft) long *Camptosaurus* ornithopod dinosaurs feeding on what is left of the fronds and leaves of the knocked-down trees. They also feed on the trampled ground cover ferns and the more bush-like plants.

The real shock comes minutes later, when we hear nearby cracking vegetation and a cowlike honk. We turn to see behind us a 6 m (20 ft) *Allosaurus* crouched over a 4 m (13 ft) *Camptosaurus* that it has caught. From the direction in which the other camptosaurs are warily moving away, it appears that the allosaur lunged out of the denser forest nearer the river and made a short dash to catch the camptosaur before the herbivore could escape. The *Camptosaurus* struggles, but it cannot move much because the 800 kg (1,700 lb) allosaur is on top of it, sharp-clawed hands gripped into the ribcage, one hind foot and much of its weight pinning the hips, and teeth and jaws clamped tightly around the neck. The *Allosaurus* releases the grip with its jaws just for a fraction of a second to bite the camptosaur's neck again, then repeats this process a couple of seconds later. After several more destructive bites, the camptosaur stops moving, and the allosaur begins to feed, ripping off chunks of flesh from the large, muscular thighs, hips, and basal tail. The allosaur pulls the chunks most of the way off, but many pieces are still partly attached, and the allosaur has to back up, snap its neck back, and put its own body weight into the action to yank the pieces loose. As it does, the whole camptosaur carcass also is pulled in short jerks. It is getting dragged slowly across the ground several meters from where it fell. This event will no doubt attract the attention of other terrestrial predators, so, not wanting to risk becoming meals ourselves, we continue south.

Three weeks into our hike, on day 22, we come to a large river that we must cross. We are now in an area that will in our time be southern Wyoming. The river is one of the largest we have encountered, so we stand on the sandy banks and assess the situation before we wade into it. It is approximately 200 m (650 ft) across and is running muddy and slow, perhaps less than 1 m (3.3 ft) per second. We see turtles moving around in the calm, small embayments along the banks; see small ornithopods in the vegetation along the banks; and know of a *Camarasaurus* feeding in the trees behind us who is ignoring our presence, but otherwise the immediate area shows little animal activity. The problem is, we know that there are crocodiles in the rivers and ponds (fig. 8.4), and although we have encountered and have been in plenty of rivers before, this is the largest stream we've had to deal with. The water is an opaque greenish color with cloudy brown mud, so we cannot see the bottom or any crocodiles that may be around. On the basis of the width of the river, the speed of water flow, and the lack of ripples or significant boils or eddies on the surface, we guess that the bottom is probably just sand and gravel—but more importantly, it is also relatively deep and cannot be waded across like so many of the other, smaller rivers we've forded. We will have to swim. And there is no way we can avoid our vulnerability to the crocodiles as we do. We don't have time to build a full-size raft for all of us and our gear, and we can't really help each other much from the

8.4. Life near a tributary stream in southeastern Wyoming during Morrison times. During an afternoon rain shower, a feathered and nearly camouflaged *Ornitholestes* has scavenged an amioid fish from the river and retreated up a log to feed. It looks out across a river populated with *Amphicotylus* crocodilians and *Glyptops* turtles. On the opposite shore, a juvenile *Diplodocus* keeps a wary eye on several predatory *Allosaurus*. The stream leads to a major river channel, and bones of animals such as these eventually end up buried in the sands there. Based on finds at the Bone Cabin Quarry near Medicine Bow, Wyoming. *Painting by Todd Marshall © 2004.*

shore if we swim singly, so our only defense is to swim as a group and cross our fingers. After making small rafts of vegetation to float our packs across and lashing these together, we wade in and are quickly up to our waists. Only a few more steps and we are in water over our heads. We begin swimming, kicking with our feet and side-paddling with one hand, the other hand grabbing the float for our packs. After a number of long minutes, we get to the opposite shore, drag our gear onto the sand, and begin unlashing our packs.

As we clean vegetation from our packs and secure them on our backs, we catch a two-second glimpse of a small, bipedal, brightly feathered animal dashing through the vegetation between us on the sandbar and the dense forest to our west. It appears to have been chasing something small, but we can't see what. Moments later, we glimpse it again as it pauses behind a scattering of horsetails, trying to relocate its prey. It is only about 1 m (3.3 ft) high and has long, slender back legs, a long tail, a small torso, and a short, boxy head that is probably less than 15 cm (6 in.) long. Its arms are relatively short but have long hands with sharp claws. Although we can't see its teeth from this distance, from the body form and the sharp claws, we can infer that this is a small carnivorous theropod. As the animal stands still, it turns its head quickly left, then right, then down in short, almost jerking motions quite unlike the slow smoothness of the sauropod dinosaurs we see out on the floodplains. Suddenly the small theropod jams its snout down into the vegetation like a modern heron snatching a fish, and when it comes up, we see the flailing tail of a squirming mammal sticking out of the end of the theropod's mouth. A few quick bites (small toss-up-and-catch motions, really), and the theropod has worked the mammal to the back of its mouth and swallowed it. The predator opportunistically grabbed an unfortunate mammal that got too close, and we'll never know what animal it was chasing when we first spotted it. As the small theropod pauses for a few seconds after its meal, we notice it is a relatively brightly colored animal but with a subtle pattern to it. It looks almost furry. We notice, however, that there are what appear to be short feathers along the edge of the forearm and the top of the neck and head. We can't see all of the tail, but the feathers we do see suggest that the end of the tail may have such structures as well. We also can infer from the arm and head feathers that the rest of the body is covered not in fur but in short, downy feathers. There are a number of small theropods that this beast could be, we realize as it dashes off again into the forest. Perhaps it is an *Ornitholestes*.

As we prepare to leave the sandbar to head through the forest and out onto the floodplain, we hear a splash and a small chorus of birdlike screeches from the opposite shore. We turn to see four small *Nanosaurus* trotting across a sandbar and disappearing into the vegetation on the other side of the river as a 2.5 m (8 ft) crocodile (probably an *Amphicotylus*) slowly slides empty mouthed from the sandbar into the river. The small neornithischians will move somewhere else to get a drink. We duck into the vegetation and leave the river, remembering that the sandbar onto which the crocodile just lunged is the same one we swam from minutes before.

The next day, we are several hours into our daily hike and the temperature is climbing toward 89°F (32°C) on a sunny day with blue sky and patchy clouds when we see, off to the west, several *Stegosaurus*. These are easy to identify, of course, because of all their unique ornamentation—the large plates and long tail spikes. We see them moving about in the middle of a huge fern meadow far from the large river we crossed yesterday. We've not come across any other surface water, so the ground here is fairly dry, and the plants are noticeably shorter. The main thing we note about the stegosaurs, however, is that the skin enclosing the large plates all along the back is brightly colored in many individuals. The stegosaurs also move at a slow walk as they feed, cropping ferns and other small plants with the beaks of their long heads. Their necks are shorter than those of the diplodocid sauropods, so they can't move their heads very far laterally, but they do "mow" their way through the meadows.

On day 23, we cross a small, knee-deep stream in the morning, and then, just before lunch, we arrive at an elongate pond tucked into a lightly wooded area paralleling a bend in a nearby large river. The pond is surrounded by low-growing vegetation and scattered trees, and because it looks like a nice place to rest and eat lunch, we take the rest of the day off and break here, picking out a large tree fern under which to drop our packs. In the heat of midday, not much is moving around the pond.

After eating, we wade in to cool off and notice a few fish jumping and breaking the surface, but not much else is going on. We slog through the mud along the shore of the pond, working our way through thick stands of horsetails growing in ankle-deep water, but we don't see much, so we start heading back to our tree. Then we spot a small, lizard-like animal, about 30 cm (1 ft) long, as it slides off the shore into the water. With its tail snaking through the water to propel it, the animal disappears into the dense stand of horsetails in slightly deeper water. It is small and went by quickly, so it is hard to say what it was, but from its noticeably long, narrow skull and general size, it may have been the choristodere *Cteniogenys*. We come out of the shallow, heavily vegetated water along the shore and relax in the shade of the tree for the rest of the afternoon. Around dusk, we spot another small, lightly feathered theropod dinosaur patrolling the land around the pond, but this one has a different color pattern and shorter feathers. It also is slightly larger and has relatively much longer legs than the previous one from near the large river two days before. Perhaps it is *Coelurus*. Why it is stalking this area becomes clear as darkness falls and the chirping of crickets and the croaking of frogs start up all around the pond. It is no longer quiet, but in a way the familiar sounds are soothing in a foreign world. We knew there were at least relatives of crickets in the Morrison Basin, but as far as we know, we haven't found any trace of crickets themselves. We know they live in what will become Europe, however, so as their chorus starts tonight, we are not surprised by it. The frogs are animals that we knew were there even though we hadn't seen them; their fossils are known in the future. But we now know that they not only look much like modern frogs but also make noise like them. Although we haven't seen them, several species of mammals are probably out hunting insects also, and these mammals may be a favorite part of the diet of the small theropod we saw a while back. As we work around camp in the evening, we spot one small salamander squirming through the damp, fallen vegetation back toward the pond.

The next morning, the clouds are still pink with dawn when we spot a small, furry mammal with a triangular skull and an elongate body crossing a rotten log near camp (fig. 8.5). This mammal has a long, slightly bushy tail like a chipmunk, not like the thin tail of a rat or mouse. The mammal is out looking for insects, and its small nose and whiskers twitch as it smells the air. It will head for cover as the morning warms and the sun gets high because this mammal's species, perhaps a dryolestid, is active mainly in the evenings and mornings. A short time later, as we prepare to move out for the day, we see another chunky *Docodon* mammal cruise out of the water among the horsetails and waddle ashore, heading into the vegetation. We see it hop after a small lizard that it sees at the last second, but the lizard scampers away, and the *Docodon* continues into the ferns, probably searching for a morning meal of insects. The *Docodon* moves less gracefully on land than does the possible dryolestid, and it has a larger body and more ratlike tail. After looking at the small

footprints the *Docodon* left in the mud along the shore, we head off for another day of hiking.

We are now 573 miles (928 km) from our starting point and are nearing what will one day be the Colorado-Wyoming border. As we work our way farther south, both day and night temperatures are getting hotter. It is now sometimes more than 95°F (35°C) during the day and around 75°F (24°C) at night. There is very little rain and few clouds, and as we head south-southeast, we cross more of the fern-cycad meadows and woodlands we have gotten so used to. The meadows are interesting in that there is an incredible diversity of fern species and a number of types of herbaceous cycadophytes. We have only hints of many of these in the future fossil record. In some of the meadows, we count a couple dozen species of ferns and several low-growing cycadophytes, as well as one or two types of the bushy cycadophytes. Most of the "typical" cycads (those with palmlike fronds and thick trunks like the modern *Cycas*) are restricted to the open woodlands and the thicker forests near the rivers and ponds. The ferns and cycadophytes in the meadows seem to be thick and green after the rainy season and hang on through the dry season, although by its end, they are quite brown in many areas. Plenty of small animals live in burrows in the meadows, feeding on the insects and seeds. And of course, the meadows are prime feeding grounds for the herbivores of the Morrison, particularly the sauropods.

8.5. Dawn in southern Wyoming, 152 million years ago. Two Morrison mammals, *Docodon* and *Dryolestes*, meet on a log while out foraging one morning during the rainy season. Also living near the horsetail-lined pond are lizards and frogs. A *Camarasaurus* feeds in the trees and other vegetation around the pond, while on the open plain in the distance a *Diplodocus* is feeding on ferns. Based on finds at Quarry 9 at Como Bluff, Wyoming. *Painting by Donna Braginetz © 2005.*

By day 25, the heat is getting to us. There is little water and no shade as we move across the large, open meadows, so we stick closer to the woodlands now, even if it means taking roundabout routes around the meadows. And we still avoid hiking in the middle of the afternoon. We continue to filter our water from the perennial streams. Finding them has not been a problem, but we are beginning to notice that some of the smaller streams are now just sandy, dry washes. Crossing them makes us feel the heat that much more. Our progress slows a bit in this period now, too. Our rest breaks for water and shade are a little longer, and winding our way through the woodlands is more time consuming.

As we move through the woodlands, we make some more observations—things we hadn't noticed when we were spending more time in the meadows. There are small, bushy-tailed mammals similar to the one we saw several days ago near the pond, but these live mostly in the trees of the woodlands, feeding on insects. When we get the opportunity, we dissect one of these mammals after lunch. From the tooth and jaw characteristics, we decide that the arboreal mammals are some type of dryolestid. We also dissect a small, burrowing mammal that one of the team members has trapped; it has teeth like the dryolestid but only four molars rather than the seven in the dryolestid, and its stomach contains remains of recently consumed earthworms. We decide it is one of the paurodontids, probably *Paurodon* itself. And there are plenty of insects for the dryolestids to feed on. We find this out the hard way. The air is still hot in the woodlands, but in the shade of many of the trees, swarming clouds of gnat-like insects and many small, biting flies drive us crazy as we hike. Again, we didn't know these were here from any fossil evidence, but their presence is hardly surprising. Insects have been around in large numbers seemingly forever. We also encounter, at various points during our trip, dragonflies, mayflies, wasps, cockroaches, water skaters, locusts, cicadas, giant water bugs, and beetles.

On day 29, we continue to see the usual cast of dinosaurs in the woodlands and beyond, out in the fern meadows. But we are seeing more small animals now in this environment, with its many small habitats and hiding places. As we cross a particularly densely vegetated and damp part of the thick forest fairly far from any river, we notice a small salamander weaving its way around, under, and over the wet ground litter. By its color pattern and body form, we see it is a different species than the one we saw several days ago in camp, headed to the pond. This one is less elongate and seems to be less aquatic than the previous one. It may even live in burrows in the forest floor.

As we emerge from one of the woodlands, we spot another rodent-like multituberculate. This one is larger and may be a species that we have not seen before, perhaps *Psalodon*. Although we suspect it also eats insects, it is unlikely that we would spot it actually catching one. Instead, we see it scampering around the edge of the open woodland, searching for seeds among the cycadophytes. We also spend some time in this woodland taking a closer look at a type of tree we've seen a number of times but never stopped to truly observe. This is a *Podozamites*, a type of conifer quite different from the sequoias and the other araucarians we also see on the trip. *Podozamites* has long, flat, symmetrical leaves different from the needles and scales of most conifers; it is similar to types of modern trees, including two from New Zealand and Japan known as the kauri (*Agathis*) and *Nageia*, respectively. The other araucarian conifers we see remind us of the species we remember from modern times. As we step out from the edge of another woodland one morning a month into the trek, we see a small lizard sunning itself on a log. The lizard's perch is the last sign of trees, dead or alive, for quite a way. We are approaching a large fern meadow.

As we get beyond the lizard's log, we immediately come across a sauropod skeleton bleaching in the sun. The bones are dull white, and their surfaces are cracking slightly; most bones are half buried in the dirt, and some have ferns and cycadophytes growing around them. Some bones have shallow, circular pits in their surfaces, as if someone with a power drill started boring into the bones in dozens of places but never went in more than a few millimeters. The bones are scattered over an area about the size of a house. We see vertebrae, ribs, a pelvic element, and a few limb bones. On the basis of the vertebrae, it appears to be a *Camarasaurus*. It has likely been here for several years, perhaps a couple of decades. Sauropod bones do not decay quickly. The meat is long gone. It was probably quickly consumed by carnivorous dinosaurs, other scavengers, microorganisms, and the dermestid beetles that made the shallow, circular pits we see in some of the bones.

We leave the *Camarasaurus* skeleton and wade into the meadow. It is large and wide and there is no avoiding it, so we will try to cross it before the hottest part of the day. We can see a distant woodland on the horizon, however. And scattered across the meadow are several small ponds—something we've not seen much of recently. The landscape is beginning to subtly change yet again.

Beyond the large meadow, we pass more small ponds, and we continue through more woodlands. We see a few mudholes also; some near the ponds are damp, but others are dry and dusty. All have been trampled by dinosaurs. A storm system is moving through now, and the skies are partly cloudy. The temperature has dropped to about 85°F (29°C) during the day as we head straight south. We step out of a dense growth of ground cover, bushes, and a few tree ferns and come to the largest lake we've yet seen. It is several kilometers across and appears to be relatively shallow, and several slender diplodocid sauropods are walking in the dark-gray mud along the shore. They are several meters from the water on one side and the thick plant growth that lines the beach on the other. As we watch the sauropods and their lengthening trackways, we note that the animals

Wetlands and Lakes

walk with their left and right legs surprisingly close together. The hind legs are angled just slightly inward down from the hips so that the left and right footprints overlap along their medial edges.

We work our way down the beach toward the water, crossing the trackways of the sauropods. Soon after, we sink into ankle-deep mud as we reach the water. Small waves lap the shore of this lake, and looking closely, we see that in the shallowest water along the beach, the mud forms pinhead-sized balls that roll back and forth in the gentle wave action. The rest of this day and next are spent in much the same manner. We cross meadows and woodlands, and we also occasionally pass small ponds. We stop at one pond that is about 1,000 m (0.6 mi) across. The floodplain immediately around the pond is covered in ferns, tree ferns, and cycads that are noticeably greener and thicker than at this time of the season farther out on the plains. The margin of the pond is thick with horsetails that also grow in the water, similar to their modern counterparts and to modern cattails. We wade out into the pond, and our feet quickly sink into deep mud on the bottom. We have to push to get through the horsetails to the more open water, and when we get there, we find that we can still stand up. The water out here is still not deep. We put on dive masks and duck under the surface for a short swim. The water is relatively clear away from our footfalls in the muddy bottom; we can see more than a few meters ahead of us. Much of the bottom out in the open part of the pond is covered with what appears to be seagrass. After pulling up a bunch of the strands, we surface and inspect it closely. We find a few pinhead-sized reproductive structures, which look like tiny cartoon beehives. We realize that the "grass" is actually charophyte algae, and its reproductive structures we have found are preserved in the future rocks of the Morrison Formation. The subsurface charophyte meadow covers much of the bottom of the pond, and we find them in many of the other ponds as well.

We dive down again. We explore further and notice abundant freshwater snails among the charophyte algae. Attached to a few small rocks on the bottom of the pond is a small freshwater sponge, a primitive animal that filters food out of the water. This is the first time we've seen a whole Morrison sponge. We usually find only their spicules preserved in limestone beds. We also see a few tiny animals in the water that look like brine shrimp with small, clear clamshells draped over their backs. Small fish dart in and out of the charophyte algae and horsetails near the bottom as we head back to shore. As we slog through the bottom mud while working our way out of the water, we notice several turtles swimming along the surface near the shore, and on the far side of the pond, we now see some small crocodiles returning to the water from a short exploration of the land around the pond. We dry off and prepare to resume our hike.

Around camp that night, not far from a river, we notice ant-like insects, which we've had around fairly often during the trip. In several places, we've also found what appear to be termites. This time at our camp, however, for the first time, we also locate plants with several damaged leaves that show evidence of having been chewed by herbivorous insects.

On day 33, we awake to find overcast skies. It begins to rain in the morning as we start out, and we end up hiking most of the day in a light drizzle. We are now in an area not far from what will eventually be Colorado's Front Range and the cities of Denver and Colorado Springs. The terrain is still flat and consists of the same fern meadows, woodlands, small rivers, and scattered ponds that we've encountered for days now. We notice that some of the wetlands that we pass, although they contain standing water, are nearly totally choked with horsetails and aquatic plants, so they look more or less like flooded meadows. Some of the wetlands on higher parts of the floodplain have no standing water and are actually dry, but the vegetation is still greener and more thickly grown than in the areas around them. These dry wetlands look like shallow, rounded meadows.

We travel another day through similar terrain and come to a large river about 100 m (328 ft) across. We have emerged from the vegetation lining this river at a large, horseshoe-shaped bend, and across from us, on the inside of the curve, is a wide, open sandbar. The cutbank drops away from our feet at the edge of the river straight down into the water. Here, the vegetation has been undercut, and some of the roots hang in the air about 2 m (6.6 ft) above the river. We lower ourselves down this cutbank, drop into the river, and begin the swim across, again keeping an eye out for crocodiles. On the other shore, we take a few minutes to explore the sandbar. We find a few smaller dinosaur bones embedded in the bar, and some plant litter has been washed into some areas of it too. Some ripple marks line the sand parallel to the water, and there are tracks of tiny vertebrates in the fine sand in this area. These may be mammal tracks, but all we can be sure of is that they are of small animals with four or five toes.

Low Hills

We take a break, preparing to set off across an open woodland on this far side of the river. The woodland has a lot of tree ferns, small ginkgoes, and young *Brachyphyllum* trees. As we rest, we briefly see a large theropod with a horned snout charge a 2.5 m (8 ft) long, beaked ornithopod, which ducks into dense brush and avoids the predator. The rare horned theropod is almost certainly *Ceratosaurus*, and the ornithopod seems to be an adult *Dryosaurus*. As we begin the hike from the river, we are now heading west. After several hours, we stop to camp, and as the sun sets, it appears that the terrain ahead is mostly similar to what we've been hiking through for several days. The next day, we find that there are meadows, wetlands, and scattered woodlands all crossed by a few streams, and we travel through these all day. By late afternoon, through the clear, dry air, we can just make out some low, dark hills on the horizon. Although we get a little closer by evening, we still cannot make out any details, but we should get to them tomorrow. As we sit in camp at dusk, yet another small group of *Nanosaurus* steps slowly through some nearby cycadophyte bushes, feeding on ferns and other plants as they go. We frequently see these little neornithischians as we travel, and they seem to be relatively undisturbed by our presence. Every so often, one of them will pause to look up and keep an eye on us for a number of seconds, but then they go back to feeding. As long as we are just sitting with our packs, the *Nanosaurus* are not too concerned. The evening insects have been irritating recently, particularly early on in this stretch, when we made the mistake of camping too close to a wetland.

The next morning, as we are preparing to move out, a large pterodactyloid pterosaur glides silently overhead and disappears beyond some conifers to the southwest. Several hours later, as we continue west after crossing a small stream, we find ourselves following several small, forking, dry creeks. We pick one of these and begin tracing it upstream. It is mostly dry, but we can tell it was damp after a rain several days ago, and it is mostly sandy until we work our way several miles farther up. As we follow the dry wash closer to one of the low, dark hills that we first saw yesterday, we notice that the sand in the bed of the wash is turning to pea-sized gravel and is now largely pink in color. Finally, on a gentle slope at the base of one of the hills, the gravel is about olive sized and is all pink, like a salmon steak. We begin hiking into one of the low hills we first spotted yesterday. Their dark appearance from a distance came from the fact that the hills are covered in short conifer trees and hardy cycadophyte brush, the only large plants that will grow in the thin, tough soil. Shortly after reaching the high point of this hill, we see that the conifer-topped hills are small in area and only a couple tens of meters in elevation above the surrounding floodplains. The conifer-covered hills are low and rolling, and the pink bedrock is exposed in numerous spots on their slopes. The pink color comes from potassium feldspar, a mineral common in granites—in this case a 1,000-million-year-old granite that will one day erode down to form Colorado's Pikes Peak and much of the surrounding rock. At this time, however, only a few low hills of it are exposed above the Morrison plain, and its pink gravel is washing down into some of the surrounding streams, the lithified remains of which will one day be studied by geologists and will be found to contain the telltale pink potassium-feldspar-rich rock.

As we stand atop the only real hill we've encountered in 36 days of travel, we can't help but take a long break to look out over the terrain we have ahead of us to the west. The plain is crossed by a number of perennial and dry streams, none particularly large, and is still dotted with wetlands and a scattering of woodlands. As the summer has worn on, the meadows have begun to take on a much more brownish tinge. And in the next several miles, there are about a half dozen more of the dark, hilly "islands" in the floodplain like that on which we now stand. As we start down the hill through the conifers and cycadophytes, we scare a small mammal living in the brush, and it scampers away. We turn west and continue hiking for the next two days. We pass a *Stegosaurus* skeleton in a dry wash at one point and a small group of *Camptosaurus* feeding on some low tree ferns at another.

We are now 1,531 km (945 mi) into our journey. It is late summer, the daytime temperature consistently hovers around 97°F (36°C), and there is rarely any cloud cover and no precipitation. As we continue west, we now see fewer and fewer wetlands, and the woodlands are becoming more widely spaced. The vegetated meadows between are drier and browner and less densely grown over than they had been earlier in the trip and farther north. Some patches of the meadows are now bare ground. The soil is dry and beginning to crack in some areas as well; no longer do we occasionally get bogged down in mud. We try to move as much as possible along the permanent river courses so that we can stay in the shade of the riparian forests, but where these rivers don't travel west or northwest, we have to cross dry washes and open meadows and hop from one perennial river to another. Because it is getting so hot and dry and because there is no shade there, crossing the meadows is tough, and we often rest in the shade of the trees both before and after traversing the meadows. And we are filtering a lot of drinking water out of the rivers.

One afternoon, while some of us are taking a break in the shade of the riparian woodlands, we are suddenly startled by four loud gunshots in quick succession. Sitting up, we know some of our party, on a trek down to the nearest river to cool off and refill on water, have gotten in trouble. But we do not know exactly where they are. We hear distant shouting, another blast from one of our weapons, and a lot of cracking branches and crunching forest litter. Our colleagues appear, winded and thirsty, one bleeding badly from a mangled right arm, and as a group, we retreat to the densest part of the forest, all shotguns now drawn.

As we rehydrate in our protective thicket and wrap up the limb of our injured expedition member, the water-retrieval party tells us what happened. As they came upon the river and the sandy clearing along its banks, they stepped a little too abruptly out of the vegetation and startled a pair of allosaurs that were feeding on a *Dryosaurus* carcass. Angry and protective of their kill, the allosaurs charged the group, and two members fired their shotguns at the carnivores as all expedition members scrambled back into the vegetation and began a full-tilt run through the forest, snapping branches as they went. The shotgun blasts stalled the larger theropod, but the unhit juvenile kept coming into the forest a short way and snapped onto the arm of one of the group after she stumbled and thrust her shotgun at the theropod to fend it off. Another blast from the shotgun of a different member of the group drove off the juvenile *Allosaurus* and allowed all in our group to retreat safely. As we finish up our work in the retreat area and prepare to move out (very cautiously now), we know we will need to keep an eye on our colleague's injury.

The Dry Stretch

As we cross the meadows of this region, we still see groups of 10 to 12 *Diplodocus* and smaller groups or individuals of *Apatosaurus* feeding as best they can on the horsetails, slowly drying ferns, and herbaceous cycadophytes out on the plains, but we no longer stop for long to observe them—not because they are any less majestic but because we don't want to linger long in the sun. We sometimes see the sauropods cooling off each day in the rivers or in small water holes dug in the dry streambeds as well as in the shade of the open woodlands, but otherwise, when feeding out in the meadows, they seem to be, for hours at a time, essentially impervious to the heat, probably because of their massive bulk, which insulates them from both heat and cold. These diplodocids do seem to be sticking closer to the rivers now, however. Perhaps this is in part to be close to the water sources and in part because they are beginning to feed more on riparian vegetation as the meadows dry late in the season.

We are now crossing dry washes more frequently than having to wade or swim flowing rivers. The flowing streams are in fact becoming somewhat rare, and we find ourselves rationing our own bottles of filtered river water, hoping we can stay in the shade and hydrated as much as possible until we find the next perennial stream. At the dry washes, we often see water holes dug into the streambed by dinosaurs—holes that do not have to be all that deep to reach the water table. Several types of larger dinosaurs are often gathered at and near these holes, including sauropods (who may have dug the holes with their clawed feet), stegosaurs, and ornithopods. No doubt theropods are lurking nearby. As we travel along the dry washes, we sometimes see dinosaur skeletons partly buried in the sand, and on close inspection of the dried, white bones, we see that many have been marked by theropod dinosaur teeth and by larvae of dermestid beetles that had been feeding on the flesh and bored just into the surface of the bone. We also spot two large theropods in one day: first a *Ceratosaurus* feeding on the carcass of a sauropod, and then a desperate *Allosaurus* try to solo attack a large adult *Camarasaurus*. It gets whacked by the tail and knocked down.

Late in the afternoon of day 42, we sit down to take a break along the bank of a curving perennial stream. The meadows are dry now, and ferns are much less common than they were at any previous point of the trip. We decide to camp here and rest for the afternoon. As we sit quietly and watch the surroundings, we notice that we are on a wooded floodplain and levee next to a bend in the river. Across the bank, beyond the opposite levee, is a muddy area that appears to be the remains of a pond from earlier in the season. We move little and don't talk, and as it turns to evening, various animals, which probably hid when we first dropped our packs, begin to reappear. In the next few hours, we spot a number of animals: a small lizard, a lizard-like animal that is probably a sphenodontian, several scampering mammals that move so fast that we don't get much chance to identify them, a distant *Stegosaurus*, and a small, long-legged terrestrial crocodilian moving along the bank. We see it only for a second and are not even sure what we saw at first. It looks vaguely like a scaly cat, and from its size and proportions, we guess that it was one of the Morrison's several cursorial crocodilians. We also see a pair of small *Nanosaurus*-like animals feeding on insects and plants downstream a ways. We'd guess they were young *Nanosaurus* except that they have a different color pattern from either the young or adult *Nanosaurus* that we have seen. When we view them with binoculars, we see that in fact they are not *Nanosaurus* at all but rather heterodontosaurids, as indicated by the enlarged, canine-like tooth near the front of each jaw. These must be the tiny plant- and insect-eating dinosaur *Fruitadens*, whose jaws and other elements have been found at the Fruita Paleontological Area in western Colorado.

The injury to our friend from the allosaur is becoming problematic. Although it took a while to stop the bleeding and deal with all the tissue damage, those problems are now less of an issue than the infection she has developed. Some of the bacteria in the mouth of the allosaur were transferred into the punctures and slashes, caused some serious inflammation, and impeded the healing process. Our expedition doctor provides heavy-duty antibiotics to our injured colleague. We have no way of knowing whether these will work on whatever forms these Late Jurassic bacteria are, but we have no choice, and we are optimistic that because no microorganisms from this age have encountered designed antibiotics, the drugs should be effective.

The next morning, we see to our east a large sauropod feeding in the conifer trees. By the shape of the skull, long, angled neck, deep chest, long forelimbs, and short tail, we recognize it as the first *Brachiosaurus* we have seen on the journey. We watch it for several hours before resuming our hike. It feeds mostly on conifers like sequoias and *Pagiophyllum* and large tree ferns, and less on shorter plants. It moves slowly among the trees along the edge of the riparian forest, moving its neck up and down, left and right, selectively browsing the trees. It is an adult and very large. Unlike the diplodocids and camarasaurs, the brachiosaur has such long limbs that a person could walk underneath its belly practically without hitting her head.

As we move out after an enjoyable time watching the *Brachiosaurus*, we are beyond the river on the drier part of the floodplain when we pass a tiny, conical hill covered in small pebbles. The ant-like insects living in it, which we see on the ground, are going about their business with determination. As we continue by, however, we catch a glimpse of three or four small loads of dirt being flung out of a small burrow near the base of the insects' hill. We haven't seen what animal it was that dug the burrow, but in the next few minutes, several more loads of dirt are pushed out of the opening. Soon the hill comes alive as a good portion of the population begins to appear, scrambling out of various exits. One small insect even comes back out of the larger animal's burrow—apparently the unseen burrower has accessed the insects' subterranean gallery and is feasting while as many insects as can flee the carnage. We wait a while to see the burrower, but apparently it is still feeding, so we resume our

8.6. Life in a small Morrison Formation lake. The fish (*left to right*) cf. *Leptolepis, Morrolepis schaefferi,* and *"Hulettia" hawesi* swim in the shallow waters of the lake while on the ripple-marked bottom a crayfish approaches a sunken branch from a *Podozamites* tree. Based on finds in Rabbit Valley, western Colorado. *Painting by Donna Braginetz © 2005.*

hike. As we walk, we decide the burrower was likely a mammal, most probably *Fruitafossor*.

We haven't gone more than an hour when we come to a small wetland. Here, washed up on the shore, we see several small silver-and-black-striped fish, which are, as best as we can determine, *"Hulettia" hawesi*. In the shallows of the small pond, we see a crayfish moving along the bottom (fig. 8.6). Several hours later, we come upon an interesting scene. Among the conifers of an open woodland out on the floodplain is an area probably little more than 200 m (660 ft) across that is surrounded by a thick growth of horsetails, ferns, and herbaceous cycadophytes. The plants are greener than in other areas, and in the center of the ring of plants and trees is a muddy area where several *Allosaurus* seem to be pulling apart a decaying sauropod carcass (fig. 8.7). The theropods rip flesh off several pelvic and limb elements and occasionally play tug-of-war with each other over a piece of the dinosaur. We can't tell exactly what sauropod it is, but from some of the bones and the coloring, it appears to be an *Apatosaurus*. Looking closer—but keeping our distance from the predators—we see that several sauropod bones are already trampled in the mud. Some of these appear to be from other, older carcasses. Around and in the muddy area are abundant standing and trampled horsetail plants, and fragments of other plants and wood seem to be mixed into the mud as well. Although this is the first mud we've seen in a long time, it has

not rained recently, and other than a single, small puddle near the center of the mud pit, there is no standing water. The greenery of the area and the dampness of the ground are likely the result of a high water table in this area, and the promise of water seems to keep a number of wary sauropods in the area in case the theropods do eventually leave. We keep an eye on the *Allosaurus* ourselves as we go around the muddy area and its surrounding growth. About an hour later, we spot our first ankylosaurian dinosaur of the trip. The low-slung, spiked, and armored animal moves among the riparian woodlands near a dry riverbed with several water holes and patiently picks off leaves from young conifers, sapling ginkgoes, and ferns and cycadophytes.

It is getting late. We camp on a gravelly sandbar along one of the few flowing rivers. It is warm overnight, almost hard to sleep, but the river keeps things cooler than they could have been. The next morning, we start off again, this time to the north. There is still no rain and no clouds, and the daytime temperatures are often close to 102°F (39°C). It is the hottest part of the late summer. One day we read the temperature in the shade at 109°F (43°C), but thankfully, that is only for a brief time in the late afternoon, and only for one day. We're moving slower now, mainly because we're taking frequent shade breaks, and we make our best progress in the early morning and evening, resting in the shade of the trees for a long time at lunch. The dinosaurs of the plains are really staying close to the few low but flowing rivers, what drying ponds are left, and the dry washes where sauropods have dug down to the water table, creating holes filled with turbid, brown water. In the shrinking ponds and in the

8.7. At a damp and muddy but drying ephemeral waterhole, an *Allosaurus* feeds on the carcass of an *Apatosaurus* that it has just found, concentrating on the large muscles around the pelvis and base of the tail, and in the process gouging deep tooth marks in the ischium bone of the sauropod. Bones of other apatosaurs litter the mudflat, and two other *Allosaurus* are arriving on the scene. While feeding on sauropod carcasses over many years at the site, generations of allosaurs will shed teeth that will drop into the mud and be preserved along with the sauropod bones, eventually to be found by paleontologists. Based on finds at the Mygatt-Moore Quarry in western Colorado. *Painting by Todd Marshall © 2006.*

water holes, turtles and crocs have retreated as well. In one case, we see several small herds of dinosaurs standing around a severely reduced pond flat, trying to work their way in toward the water, while the remaining shallow pond is packed with what appear to be logs but are in fact the densely compacted numbers of several ponds' crocodile populations, all now in one place. We come to a low perennial river and refill our water bottles. This area is a refuge as well, and it is active with dinosaurs along the banks. Although the crocodiles here have it better than in the pond we just crossed and are not compacted into one place, we know that they are still in the river, and so we cautiously wade across it.

As we cross one of the nearly brown meadows between woodlands, we notice that the areas around the rivers and the few areas of damp ground are about the only places still with significant greenery. As we stand in the meadow and stop to take a drink from our bottles, we see a herd of *Diplodocus* moving east. The relatively large group is kicking up a low-hanging cloud of tan dust as it passes us. This is not the first group we've seen on the move, and we will see several more, but for the first time, we realize that all the groups were moving east. Are they perhaps traveling east to get to the wetland-rich areas we passed two weeks ago?

Shadows and Tall Trees

8.8. A large diplodocine sauropod walks along a river sandbar one morning during Morrison times, leaving deep tracks in the ripple-marked sand surface. Based on finds at the Copper Ridge Dinosaur Tracksite near Arches National Park, Utah. *Artwork by Brian Engh © 2016.*

As we camp that night not far from one of the shrinking ponds, we again hear a chorus of frogs after sunset, and the next morning, we catch a glimpse of two *Allosaurus* teaming up to stalk a nearly adult *Apatosaurus* in one of the open woodlands. In the late morning of day 46, we come to a large perennial river about 200 m (660 ft) across. It appears to be down from 300 m (990 ft) earlier in the summer. Like the other areas with rare water that we've seen, there are a lot of dinosaurs here. Crocodilians and turtles can be seen occasionally on the shore and in the water. We don't try to cross this one. We watch through the binoculars as there is a brief three-way fight among an *Allosaurus*, a *Ceratosaurus*, and a third species of slightly smaller theropod (perhaps *Marshosaurus*) at the site of a sauropod carcass on a sandbar across the river and downstream several hundred

meters. The small groups of sauropods move up and down the expanded sandbars of the shrunken river (fig. 8.8), switching between feeding along the levees and getting water from the river or water holes. In the shallows of the river along one sandbar, we notice freshwater bivalves.

We turn southwest from the large river and then south, moving back onto the floodplain. In the late afternoon, we see a rare set of thunderstorm clouds moving in from the west, but they seem to have little power. Indeed, they drop nothing more than wisps of **virga** over the flats, shade us for a few minutes, and are almost gone by the time the sun is low on the horizon. After breakfast the next morning, we are moving south and are making good progress across a large, open expanse in the cooler part of the day when we spot three large diplodocids in the distance feeding on the low-growing greenery along a riparian woodland. They look much like *Diplodocus* except that their color pattern (basic as all sauropods are) is different and they have a noticeably longer neck. In fact, the neck alone is more than 6 m (20 ft) long. These appear to be the rare diplodocid sauropod *Barosaurus*. For the next three days, we cross terrain similar to what we've been passing through for the last week. Out on the plains, we see occasional individuals and small groups of *Stegosaurus* and *Camptosaurus*; along the sides of the woodlands (fig. 8.9), we spot a few *Dryosaurus* feeding on the low-growing tree ferns, seed ferns, trees, bushes, and cycadophytes. We see a few unidentified small theropods in the open woodlands, and we continue to see groups of our friends, the *Nanosaurus*, almost every day. We also see an amazing sight when a

8.9. During a relatively dry season, many species of Morrison dinosaurs are attracted to a shrinking but still-flowing braided river. Apatosaurs, stegosaurs, camptosaurs, diplodocines, camarasaurs, allosaurs, and brachiosaurs all feed and access water. *Artwork by Brian Engh © 2016.*

very large, very bulky theropod we've not seen before charges out of the riparian woodland one day and catches a 7.5 m (25 ft) long subadult *Camarasaurus*. The large theropod, probably *Torvosaurus*, wrestles with the sauropod for some time but eventually subdues it. We had no idea such a large predator could be so close to us without our knowledge. Stunned, we move on through the woods.

On day 51, we find ourselves moving across similar terrain to what we have been in for a few days, but now things are even drier. There are still dry washes with water holes, and there are wetlands, but most of them have dried up and are mud cracked. There are a few remnants of ponds, shrunk to almost nothing, but fewer animals live in them; some of the ponds appear to have just inches of saline water in them, so the vegetative cover of the area is much less thick as well. The water table in this area seems to be high, but the water is quickly evaporating, and few areas have standing water. What water is there is often alkaline, and in evaporated ponds, a saline crust is left behind. Around midday, we start seeing a column of what appears to be dark ash rising tens of thousands of feet into the atmosphere, probably several hundred miles to the west. The dark cloud from the volcano doesn't seem to move from our distance, but over the minutes and hours, we watch it billow up, spreading like an anvil and starting to blow east.

As we move through a lightly wooded area, walking single file in twos and threes but each group scattered somewhat laterally, one of our colleagues jumps and reports that, among the ferns and horsetails he was moving through, he caught a brief glimpse of a snake slithering away under the vegetation. He may have seen a *Diablophis*. But after our questioning, he can't confirm whether or not it had vestigial limbs. We keep moving, albeit a bit more cautiously.

Soon the woods become a bit denser and the conifers soar ever taller. Within a few minutes we are in a forest of araucarians more than 38 m (125 ft) tall, with an abundance of ferns, tree ferns, and cycads in the shade around us. As we walk and crack branches under our boots now and then, we notice that the forest floor here, even in this dry period, is mostly soft, damp, and covered with rotting organic material; the forest is quiet except for the sounds of our steps and breathing as we climb over logs (some more than a meter in diameter), and the occasional buzz of an insect or chirp of an unseen small dinosaur. Eventually we emerge into another more open area with shorter ginkgo trees scattered among a fern meadow, and soon we come upon a small lake. The ferns and ginkgos along the shore are dense, and in the water we notice a relatively large aquatic bug swimming just over the decomposing vegetation on the bottom, perhaps hunting for small snails.

In camp that night we are happy to hear a positive report from our doctor on the condition of our injured expedition member. The antibiotics are working. Her injured arm is looking much better under the bandages, with much less inflammation and no more red streaks under her skin. The arm should heal well, although she will have some permanent damage to the muscles and a number of scars.

We wake up the next morning to fluttering ash coming down. Our camp, packs and all, and the surrounding landscape are coated in a dusting of ash. The sky is gray, and visibility is low. The sounds around camp as we prepare to move out are muffled, like those on a foggy day or after a heavy snow. We finally get going and head southwest. As we move across the gray landscape, we see no evidence of animals except for the trackway in ash of a lone small tridactyl dinosaur. Later in the day, visibility improves, and the sky slowly begins to clear. The fine ash has gotten into everything in our packs, which makes dinner in camp that night less than ideal. By the next morning, the sky has cleared, and some of the ash has begun to blow away from some areas of the landscape and from some vegetation. We move throughout the morning, and things continue to clear. In the afternoon, we see a distant heavy thunderstorm up ahead. An hour or so later, we pause at a dry wash before crossing. We notice a trickle of water coming down the sand, and we watch as, for the next 10 minutes, the amount of water gradually increases. Within another 5 minutes, the wash is full with a muddy torrent of rushing water. We decide to wait it out. The flash-flood water is also choked with ash, as well as small branches and other plant debris. We follow the channel downstream as the water gradually recedes. The floodwaters have spilled into a wide basin and now fill several shallow ponds with brown water, but they have broken through a levee in one area, and a mud-ash slurry has spilled into a formerly mud-cracked pond bed. When we check out this mud deposit, we find that it is mixed with abundant pebbles and contains several fragments of dinosaur bones and one complete limb bone, and many of these are beetle bored. We also notice, half buried in the ash, the fresh carcass of a 60 cm (2 ft) long *Eilenodon*, a large, robust sphenodontian herbivore. The mud also contains wood fragments, sticks and twigs, and some leaves. The storm passes to the north of us as we get moving again toward the southwest. After several more hours of hiking, we find a place to camp and get set up. The sunset tonight is spectacularly red as a result of the residual ash in the atmosphere.

The Drought Breaks

The next day, as we continue southwest, it gets up to 97°F (36°C) during the day. It is mostly clear, but variable cloudiness will begin soon enough and continue for the next few days. We are in the same general terrain as previous days, and the vegetation is still largely brown in the meadows, but the extremely dry, mud-cracked barren flats are behind us. We cross a nearly dry braided river with only two of its channels flowing at a trickle; the rest of the river is now just sandbars for the hundred meters or so across its width (fig. 8.10). The sand, however, is still damp from the day of the thunderstorm, which undoubtedly had this braided river flowing briefly as well. On one of the sandbars in the braided river is a stretch

8.10. An injured *Allosaurus* limps across a wet river sand bar as a turtle works its way across the same surface. Based on finds at the Copper Ridge Dinosaur Tracksite near Arches National Park, Utah. *Artwork by Brian Engh © 2016.*

Facing top, 8.11. A scene in eastern Utah during Morrison times. Two *Brontosaurus* stroll along a shallow stream during the early rainy season. Two pterosaurs skim the surface of the stream, grabbing fish. A *Stegosaurus* moves along the opposite shore. Lining the mudflat along the stream are horsetails, ferns, cycads, and ginkgoes. Several weeks earlier, rains had soaked the muds in the foreground and right distance, and sauropods and small theropods left footprints in the mud in the left and center foreground. Sunny, hot weather has since formed desiccation cracks in the dried mud and footprints, and soon another rainy cycle will flood the stream and wash a sheet of sand over much of the area, filling in the mud cracks and footprints and preserving them for the next 150 million years. Based on the Dalton Wells East track site near Arches National Park, Utah. *Painting by Todd Marshall © 2004.*

of three parallel sauropod trackways, which disappear into the vegetated banks to the west of us. The hind-foot tracks of each trackway are about 60 cm (2 ft) long and about 10 cm (4 in.) deep. The front foot tracks are shaped like lima beans and are a bit smaller. The trackways are separated laterally by about 1.5 m (5 ft) and indicate animals moving together in the same direction.

We cross the rest of the braided river, wading through water only about boot deep in the ripple-marked flowing sections, and continue southwest. Later in the day, we come across a partial *Apatosaurus* skeleton half buried in soil and with some dry, brown vegetation growing around much of it. It has probably been here for many years, and the bones' surfaces are dried and flaking just slightly. The next day, as we push on in the same direction, we notice some significant cloud cover to the west. At one point, we stop to inspect several crisscrossing trackways of theropods and sauropods in hard, brown mudstone in an open flat not far from a river. The tracks are much deeper than those on the braided river sandbar were, and the entire surface of the mudflat is baked hard and cracked, even the sides and bottoms of the tracks themselves (fig. 8.11). The flat is only about 30 m (99 ft) across and is surrounded by a scattering of seed ferns, tree ferns, ginkgoes, araucarians such as *Brachyphyllum*, and bushy cycadophytes. In the distance are a pair of what we surmise to be the apatosaurine sauropods *Brontosaurus*; they are slightly less robust than the *Apatosaurus* we have been seeing (though every bit as large), and their color patterns are distinguishable from those of their cousins.

Several hours later, as we cross a dry fern meadow, it begins to rain gently. It rains off and on all night and is still going the next morning as we move out. By lunchtime, the meadows we cross are beginning to get a bit muddy, so we move into the woodlands to try to get some drier ground. In the evening, before we stop to camp, the mud is beginning to slow our progress, and it rains that night as well. We get a little bit of sun the next day, but once again, nighttime brings steady rain. At one point early in

8.12. Battle in the marsh. During the Morrison wet season, a subadult *Apatosaurus* falls victim to an opportunistic and adventurous group of young *Allosaurus*. Based on finds at the Mygatt-Moore Quarry, western Colorado. *Artwork by Brian Engh © 2016.*

8.13. At a full-flowing river during the rainy season, several *Camarasaurus* swim across the channel while a few goniopholidid *Amphicotylus* lurk below. Bones from skeletons that ended up in the channel during previous dry seasons lie buried or partially buried in the sand on the bottom of the channel. On shore, a theropod lurches its head back to swallow a chunk of meat from a sauropod carcass. Based on finds at the Mill Canyon Dinosaur Bone Trail north of Moab, Utah. *Artwork by Brian Engh © 2016.*

the predawn darkness, it pours for 20 minutes. We can no longer stay dry in camp, and for the first time in almost two months, we begin to feel chilled at night. The next morning, we slog on through a light drizzle, keeping our distance from what appears to be an apatosaur in distress at the claws of several *Allosaurus* (fig. 8.12).

The other herbivorous dinosaurs we see along the way go about their business of feeding and seem oblivious to the rain, although we suspect that they must be much less stressed now that there are many puddles and some of the streams are beginning to flow again. And the rain has brought out frogs and salamanders by the hundreds from small ponds that hardly existed just days ago. The rise in the rivers has interfered with us a little, as we tried to get out of the mud by moving along the sandy stream courses, but this was difficult because previously dry streams are now flowing and the perennial rivers are rising.

On day 58, we find that crossing rivers is getting to be a problem because there are simply more of them now and they are becoming swollen with the runoff (fig. 8.13). We move upstream parallel to those flowing from the southwest. We also notice that we have almost imperceptibly begun to travel up a gentle slope to the southwest, and the streams and rivers are just a bit more gravelly than they had been. It is still raining off and on, sometimes heavily, but there are occasional breaks in the clouds when we have some sun. And once or twice we have been able to see the horizon to the west, and we've noticed that there are low hills and distant mountains ahead of us—the first real topography we've seen on the trip and the only hills we've seen in weeks. By the next day, the continuing rain is making the mud even worse. We slog slowly through it, and the rivers are getting swollen. In the afternoon, we are moving parallel to a river on the edge of a woodland and not far from a fern meadow when we suddenly notice water now flowing around our mud-encrusted boots. In a few moments, we realize that the levee has been breached by the river not far away, and water is spilling out onto the floodplain. The water continues past us and begins flooding the meadow, while around us the water rises. The entire woodland around us is now under water almost up to our knees, then close to our waists. We quickly climb nearby medium-sized trees. We sit in the branches and watch as brown, swirling water surrounds us on all sides. After several hours, it becomes clear that we will not be able to come down before tomorrow and will thus have to spend the night in the trees. We hardly sleep.

By the next morning, we are exhausted and sore, and the water is barely flowing. The entire plain around us is flooded, and although the current does not seem to be dangerous, the water is still too high for us to move, so we wait it out again. Plant debris floats by frequently, and at one point, we watch as a slightly bloated *Camptosaurus* carcass drifts slowly through the flooded woodland. The large sauropods are big enough to wade through the water up to their chests. Later in the day, the sky clears to patchy clouds, and by dawn the next morning, the water has gone down enough for us to come down and begin moving again to the southwest. Although wading through the water is strenuous, it beats the cramping and soreness of sitting in a tree for hours, so we are happy to be on the move again—and to see some sun. Eventually we get out of standing water and

8.14. A *Ceratosaurus* and a *Stegosaurus* square off while in the foreground shartego-suchid crocodylomorphs (*Fruitachampsa*) raid a nest of *Dryosaurus* eggs and a snake (*Diablophis*) attempts to drop in on the burrowing mammal *Fruitafossor*. Based on finds at the Fruita Paleontological Area, western Colorado. *Sketch by Brian Engh © 2017.*

Illustration by Brian Engh : dontmesswithdinosaurs.com

onto muddy ground; we never would have thought we'd be happy to see mud, but we are. Crossing the streams is a bit dangerous now—many are still high. And at the numerous flood ponds on the landscape, the carcasses of drowned dinosaurs are attracting those carnivorous dinosaurs that survived, so we make a point of staying clear of those areas. We come to a fresh *Camarasaurus* carcass half buried in a crevasse-splay sand, and several small, feathered theropods and pterosaurs are beginning to pick at it. We don't linger long; we get on our way before larger carnivores show up. Now that the water level is down, the exposed mud is the perfect medium in which to preserve footprints, and today, we see large numbers of small animal tracks. Some are lizard tracks, others mammals. At one of these track sites, we see a decent-sized triconodont mammal scamper by with the carcass of a small sphenodontid in its mouth. We come to a fairly large river that we'd like to cross, but we can't because it is still far too high and dangerous. As we prepare to go the long way around, we notice the body of an *Allosaurus* floating by in the muddy current.

Over the next few days, things dry out a bit, and the plants and animals are recovering. While resting at midday, our expedition artist makes a quick sketch from his lunch spot a hundred or so meters to our south (fig. 8.14). The scene causes him to retreat back to where the rest of the group is.

Reaching the Hills

We are now 1,428 mi (2,313 km) into our journey. It is day 61. The terrain is changing significantly. After following the rivers upstream the last few days, we notice that some are beginning to branch from the single larger rivers we've been following into more, smaller, and steeper creeks as we begin to enter the foothills. We are working our way up a gravelly stream surrounded by forest. The stream is wide and shallow and is part of an alluvial fan coming out of the foothills. The streams contain only rounded pebbles and almost no sand. The next day, as we follow the gravelly stream, we pass through a notch in the foothills and enter the mountains. The terrain is significantly steeper here, and the creek rushes down at us with much greater speed. The rocks in the creek have increased in size from cobbles to small boulders, and the creek's water roars as it splashes down its rocky course. The mountains are covered in araucarian and sequoia conifers, plus ginkgoes and tree ferns, and the forest here is denser than anything we've experienced before (fig. 8.15). Logs and rocks in the forest are covered with moss in many places.

On day 64, we pass a rock outcropping, something we haven't seen since the pink granite around day 38. This outcrop consists of a wide, 10 m (33 ft) high prominence of bedded gray limestone. On closer inspection, we see that it contains fossils of horn corals, which indicate that it is a Paleozoic limestone, already during the Late Jurassic more than 100 million years old. It has been uplifted in these Late Jurassic mountains and now is being slowly eroded and washed down to the Morrison plain. The next day, we enter an area of the mountains that was affected by the

volcanic eruption a couple weeks ago; it received a far thicker coating of ash than we've seen before, and although rains have washed the ash off most of the ground and vegetation, gray ash still chokes some parts. As we get farther into the mountains, the temperatures have cooled to about 82°F (28°C) during the day. Even the highest parts of the mountains are only moderate in elevation. There is no snow in these mountains, and as we keep winding upstream, the creek gets smaller and steeper, the forest even more dense, and the rain showers more frequent. After the rains, the normally clear water of the creek flows light brown. Around midmorning the next day, the clouds clear a little, and off to the northwest, we see the peak of a volcano several miles off. The mountain looms ahead of us as we move for the rest of the day. It is a relatively low-lying volcano, and it is quite broad in its basal diameter.

Day 67 dawns clear—the first totally clear day in some time—and we start off on the last leg of our trip. For the next several hours, we hike up toward the base of the volcano, passing through the dense green forest and along the cascading creeks until we finally reach our destination. We drop our packs and sit down to rest, having finally achieved our goal. From the ocean shore a 1,594-mile walk away, we've traveled more than two months, and here in the mountains, looking out over the Morrison Basin, we are rewarded with a view of the plain far to the east. The slopes of the volcano north of us are bare, a result of the heavy load of ash and the destructive power of the eruption. The trees and forest around us are still intact because the eruption blew mainly north and east, knocking down trees for several miles in those directions. As we sit and rest, taking

8.15. In an opening in the dense forest of the foothills, a *Ceratosaurus* looks out on the heavily vegetated and misty scene near a clear and relatively fast-flowing river. *Painting by Karen Foster-Wells © 2006.*

in the view of the plain and the volcano's slopes, we see an unusual small dinosaur in the forest nearby. It appears to be a bipedal ornithischian, but it has a short tail, grasping hands and feet, long arms, and a beaked *Dryosaurus*-like head, and it is feathered. Watching it feed, we determine that it is partly a seed-eating animal, and we are slightly surprised when we see it quickly scramble up a tree. Is it arboreal? Perhaps. This is an animal of which we have absolutely no fossil record in the Morrison Formation, of course, but apparently it lived in the western mountains. It illustrates the unique potential of the mountains west of the basin to have held a fauna different from that of the floodplain. We simply have no idea what kinds of animals lived in the mountains along the Morrison Formation's southern and western margins. It is entirely possible that there were dinosaurs and other creatures in the mountains that we cannot now imagine simply because they were so different from their cousins preserved in the Morrison Formation sediments.

As we sit in the mountains and reflect on our trip, we are struck by a number of things. For one, it was even warmer and flatter than we'd expected—or perhaps it just seemed that way after more than 1,500 miles. We also remark on the large number of small neornithischians among the dinosaurs; the familiar insects such as wasps, crickets, termites, beetles, and annoying flies and gnats; the diversity of small animals such as fish, amphibians, lizards, and mammals; the relative abundance of plants on the floodplains and near the rivers, considering the infrequent rainstorms we encountered; and the abundance of rain when it did come. We also feel lucky to have had only one truly unpleasant encounter with carnivores over a period of more than two months and to have had our colleague survive the bite of a juvenile allosaur. The color patterns on some of the dinosaurs are also hard to forget: the feathered small theropods that looked almost like large birds, the bright plates of the *Stegosaurus*, and the dusky but distinct colors of different species of sauropods. We also remember the large, lumpy, polygonal scale patterns and low midline spines of so many of the large dinosaurs, giving them a much more ornamented appearance than we had expected.

When we return to our own time, we'll no doubt be exhausted from the journey, but we will also know that we have experienced one of the most spectacular times in the history of life on Earth. And it will be difficult not to want to go back and see more.

9.1. The sauropod dinosaur *Dystrophaeus viaemalae* strolls up a small sandy braided stream at sunrise one morning 157 million years ago, at the very beginning of Morrison times. Based on data collected at J. S. Newberry's quarry in southeastern Utah. Recent finds have not yet indicated exactly to what sauropod group *Dystrophaeus* belongs, but they have shown that it is likely not a diplodocoid. *Painting by Brian Engh © 2018.*

Epilogue: The Morrison Fauna in World Context

What can we learn from the Morrison Formation and its vertebrate fauna? What is this temporally distant world of the Late Jurassic (fig. 9.1) saying to those that will listen? Perhaps the greatest value in the formation's vertebrate fossil record is in illustrating the great variety of morphologies, ecologies, physiologies, and body sizes that are possible in vertebrate animals. Certainly, these four aspects are all directly related to each other, but there are, in the fossil vertebrate species of the Morrison Formation, unique examples in each of the above categories that demonstrate adaptations beyond what we see today. Among the unique morphologies seen in the Morrison are the incredibly long necks of the sauropods; the large skulls and sharply clawed hands of the bipedal theropods; the long-legged, cursorial, terrestrial crocodylomorphs; the spikes and plates of stegosaurs; the retained postdentary bones in the lower jaws of some mammals; and the pleurocoels ("side hollows") of the presacral vertebrae of the sauropods and theropods, suggesting a nearly avian pneumatic air-sac system in large dinosaurs. These are all structures either unseen or not developed nearly to the same degree in modern vertebrates. Ecologically, today we do not see nearly the numbers of large herbivore or carnivore species in any terrestrial community. The concept of what constitutes a large carnivore in our world is quite different from the Late Jurassic, and it makes *Torvosaurus* or *Saurophaganax* seem almost ludicrously gigantic. Our largest modern terrestrial herbivores are barely larger than some of the Morrison carnivores. Although no well-preserved Morrison sauropods appear to have been heavier than our modern blue whale (although some were approximately as long), they were much larger—both longer and heavier—than any terrestrial herbivores before or since the Late Jurassic.[1] The "land-whale" sauropods are arguably the most majestic example of how different a terrestrial herbivore can be—nothing in the modern world parallels a 47-ton, terrestrial, plant-eating reptile with a physiology that probably was not entirely "reptilian."

Dinosaur studies suggest, as we have seen earlier, that whereas most animals today are either ectothermic or endothermic, dinosaurs at least were probably intermediate in their metabolic rate and general physiology. Again, we see in the vertebrates of the Morrison Formation another way of doing things compared with modern animals. And among Morrison dinosaurs, there likely were a variety of intermediate metabolic rates. Although what we see today are "reptilian" or "mammalian" metabolisms, and although we initially assumed that the options were similar for dinosaurs, it appears instead that most Morrison dinosaurs were reptiles that were not exactly like either mammals or modern reptiles in their life-time growth rates or their apparent resting metabolic rates. The body-size range of Morrison Formation vertebrates, from the 5-g (0.18 oz) *Fruitafossor windscheffeli* up to the 47 metric ton (103,617 lb) *Camarasaurus supremus*, covers a full 6.9 orders of magnitude—the largest Morrison dinosaur is 9.4 million times heavier than the smallest mammal! Body-size ranges in most modern ecosystems aren't quite that dramatic, although in Africa one could see a difference of 6 orders of magnitude and about 1.4 million times between a 7-metric-ton elephant and a 5-g insectivore.

Of course, there are also similarities between the vertebrates of the Late Jurassic and those of modern times, and these can be equally striking. From the Morrison vertebrates, we see that modern turtles, some crocodilians, salamanders, frogs, and lizards are little changed ecologically from what they were 150 million years ago. Like today, large vertebrates then were mostly herbivorous browsers, although the teeth and the cropping and chewing styles of the dinosaurs were significantly different during the Late Jurassic. We also see convergence in Morrison vertebrates—that is, unrelated animals adapting to the same conditions or habits with similar but not necessarily homologous structures. The peg-like teeth and robust humerus of *Fruitafossor* are much like the same elements in armadillos and moles, respectively, and likely evolved for similar reasons, in this case feeding on insects and digging. But it is really the armadillos, aardvarks, and moles of today that are imitating *Fruitafossor*, not the other way around. The Morrison mammal developed a lifestyle of digging after subterranean insects 150 million years earlier. We also may be seeing a certain amount of convergence in the robust seed- and insect-eating lower jaws of Morrison multituberculates and modern rodents; in the small, lightweight, and long-limbed carnivorous terrestrial crocodylomorphs of the Morrison and small, modern feline carnivores; and in the long necks and long forelimbs of both *Brachiosaurus* and the modern giraffe. The Morrison Formation vertebrate fauna thus gives us an appreciation both for the aspects of modern species that are either unchanged from the Late Jurassic or are, in a sense, being revisited and for the forms, habits, and physiologies that vertebrate species can adopt that we might not have imagined possible based on our modern world.

Unfortunately, we probably cannot learn much about extinction from the Morrison Formation because if there is an extinction in the terrestrial faunas of North America near the Jurassic–Cretaceous boundary, it is not recorded, at least in the Rocky Mountain region. If the story of the terrestrial Jurassic–Cretaceous transition in North America were a play, the third act would get cut short midway through. We have no idea how it ends. That is because, unlike at the end of the Cretaceous, where we have a preserved and continuous record of an extinction, there is a large gap of time between the top of the Morrison Formation and the base of the overlying Lower Cretaceous rocks, and this gap occurs everywhere between the two. It really does not matter much if the top of the Morrison Formation ranges just slightly into the early part of the Cretaceous period (and most data suggest that it does not) because there is nothing in the fossil record of the formation to suggest major faunal turnover near the top. There are hints that some Cretaceous-grade forms were appearing high in the Morrison within the established groups (multituberculates, for example), but there are no major disappearances, and high in the formation we still see *Camarasaurus*, *Apatosaurus*, and other forms typical of the earlier millennia of Morrison time. So there doesn't seem to have been any significant faunal or floral event at the end of the Morrison era, regardless of whether it coincided with the Jurassic–Cretaceous boundary. That boundary definition is based on changes in the marine invertebrate record in Europe, so even if the time interval were preserved in the sediments of the Morrison Formation, it may not have left a significant mark in the fossil record of the land vertebrates.

On the other hand, as alluded to above, most recent radiometric dates from the Morrison Formation of the Colorado Plateau suggest that the upper layers of the unit are still several million years older than the end of the Jurassic, so the Jurassic–Cretaceous boundary may be represented in the time gap between the Morrison and the Lower Cretaceous rocks and thus not even be preserved in the rock record. What we do know is that the vertebrate fauna of the Morrison Formation continued without significant change up to near the top of the formation, and then there is a 10- to 20-million-year gap represented by a single surface in the outcrops around the western United States. Above this, immediately on top of the Morrison Formation, is Lower Cretaceous rock approximately 125 million years old that preserves a much different group of dinosaurs. A lot can happen in 20 million years. Even though the Gaston and Dalton Wells Quarries, which occur in the Early Cretaceous Cedar Mountain Formation near Moab, Utah, both are relatively close above the contact with the top of the Morrison Formation, their dinosaur faunas are different from the Late Jurassic unit. Ankylosaurs, rare in the Morrison, are relatively common in the Cedar Mountain (e.g., *Gastonia*); sauropods still occur in the Cedar Mountain, but the fauna is less diverse and is dominated by basal macronarians (e.g., *Moabosaurus*) and lacks diplodocoids entirely; iguanodontids have appeared; and gone are most of the large theropods, replaced by a few coelurosaurian theropods and the moderately large dromaeosaurid *Utahraptor*. Although ankylosaurs are abundant, stegosaurs are gone, as is the high diversity of small basal neornithischians.[2]

Something clearly changed over those 10–20 million years, but whether it was a gradual or a dramatic event, we may never know because the right-age rocks are simply missing from between the Morrison and overlying formations. So there is no "what happened to the dinosaurs of the Morrison Formation?" story that we can realistically investigate. Instead, we must understand the 7 million years or so for which we do have a record of the dinosaurs and other vertebrates of the Late Jurassic. This forms the basis of possible interesting comparisons with other times during the Mesozoic. The Morrison Formation fauna does demonstrate something about diversity. There are more than 100 fossil vertebrate species in the formation, and those are just the ones we have found so far. Very likely there are many more that we have not yet found, particularly among the small vertebrates, or that lived in the area but were rare and unlikely to fossilize. The known vertebrate record probably just hints at the diversity that likely was there. And the recent finds or naming of *Diablophis*, *Fruitafossor*, *Kaatedocus*, *Galeamopus*, and *Hesperosaurus* show that even after more than 140 years of exploration in the Morrison, it is still possible to find entirely new types of mammals that we never suspected were there and to identify new types of dinosaurs and other reptiles in the formation. Could the true diversity of vertebrates that lived on the Morrison floodplain have been twice the number we have found so far? There is no way to know, but it is possible.

The Morrison also helps show, through its paleontological and geological records, the differences in the world and its environments through time. Differences in the Rocky Mountain region between the Late Jurassic and today include Morrison times having a regional area that was essentially flat, a higher sea level, a bit more precipitation, more vegetation fed by a high water table (but certainly no jungle), warmer temperatures year round and at higher latitudes (with probably no ice caps and no snow in the western mountains), and similar atmospheric oxygen but about three times the atmospheric carbon dioxide concentration. All these individual elements have varied through geologic time, and although they can be interconnected and influence each other, they do not necessarily change in lockstep. Oxygen and CO_2 levels often affect each other but can vary independently, as can temperature. So there is really nothing that we can call a "normal" climatic condition in Earth history. There have been times in the past both a bit warmer and much colder than Morrison times, so the Morrison, warm as it was compared with today, was not near the atmospheric extremes reached, for example, during the Cretaceous or the Eocene, when the earth was probably the hottest it has ever been (at least since very early on), or recently (the past 2 million years or so), when it has been unusually cold compared with most of Earth history. And Morrison times were much more like today in their atmospheric oxygen level, especially compared with most of the Paleozoic era and near the end of the Permian into the Triassic period, when the amount of atmospheric oxygen dropped to well below what it is now.

Ice ages are rare in Earth history, and because we are currently in one, our time is a bit unusual. There were periods of significant glaciation late in Precambrian time and during the Permian; the latest started just under 2 million years ago. But other than that, most of geologic history has been comparatively warm. There were no major periods of glaciation during the entire Mesozoic, and probably for much of that time, there were no polar ice caps. Most of the Cenozoic was warm (sometimes very warm), and even the cooldown that began millions of years ago did not lead to a cycle of ice ages until relatively recently. It is also important to remember that although we are currently in an interglacial period during which it has been comparatively warm, over the last 1.8 million years of the current ice ages, we have gone through many cycles of cold-wet, then warm-dry. The last cold cycle ended 10,000 years ago or so, and previous cold glacial cycles in our current ice age series may have been more severe than the most recent one. However, several of the previous interglacial warm periods in the last 1.8 million years were a bit warmer than it has been during the last 10,000 years. Although it may have been alternatively much colder and much warmer than today at different times just in the last 2 million years, all of that time has been unusual in Earth history in being cooler than average. In that sense, although there has been no "typical" Earth climate, what we see in the world of the Late Jurassic may give us a better idea of the less extreme times in the past.

The Morrison Formation fauna also could teach us something about evolutionary change in ancient vertebrate lineages and ecological change in communities through 7 million years of the Mesozoic, but we still need more data from the full stratigraphic range of the formation. Much of what we do know comes from the middle upper half of the unit, and too little comes from the lower half. The Morrison's vertebrate record also illustrates what we don't know. It comprises probably almost exclusively taxa from the basin itself. We have no idea what species lived in the mountains west and south of the Morrison Formation or how they differed. And we have no indication of the types of vertebrates that lived in eastern North America at that time. The earliest North American history of the vertebrate groups known from the Morrison Formation is also unclear because the Early and Middle Jurassic records from the continent consist mostly of footprints. The only good window to a North American Early Jurassic community is the body fossil record of the Kayenta Formation in Arizona, but it is too brief, too geographically restricted, and still too small to really fill in the history of the groups we know from the Morrison Formation.

From these limitations, we can already see some directions that Morrison Formation paleontological research may head in the future. The lower half of the formation is in need of serious exploration, and there is no reason more sites shouldn't be there; it has simply been less frequently worked in the past, probably in part because fossil material can be harder to find in it and because, given the number of thick, cliff-forming sandstones in the Salt Wash Member, for example, the fossil layers that are found can be difficult to work. But more data from the lower Morrison would be helpful. With those data, we may be able to answer questions such as the following: (1) Are *Allosaurus fragilis* and *A. jimmadseni* truly separated chronostratigraphically (as they appear to be lithostratigraphically), and do they perhaps represent ancestor-descendant? (2) How did the *Haplocanthosaurus* sauropod fauna of the lower Morrison differ ecologically from that of the upper half? (3) What can *Dystrophaeus* (fig. 9.1) tell us about the origin of eusauropods in North America? (4) What more can we learn about *Hesperosaurus mjosi*, and did it overlap in time with *Stegosaurus stenops*? (5) Did the small-vertebrate faunas of the upper and lower halves of the formation differ in composition or ecology? (6) Did the apparently closely related sauropods *Diplodocus* and *Barosaurus* coexist throughout Morrison time (and thus originate before the Morrison), or is their most recent common ancestor perhaps a third species preserved near the base of the formation? (7) Are there paleobiogeographic differences among other small vertebrates in the formation, and what can these reveal about the Late Jurassic ecosystems? Or are there patterns among the dinosaurs that we have so far missed? These are just a few questions. There are many events that may have occurred during the time of deposition of the Morrison. Their patterns may be preserved in the formation, but in order to see these better—or at all—we need to further explore the lower levels of the Morrison Formation, and we need to develop some stratigraphic control leading to correlation of the ages of major quarries. This latter goal is what may be in sight with the blending of quarry positions relative to local stratigraphic sections (Turner and Peterson 1999) with long-distance correlations based on stratigraphic sequences, radiometric dates, and paleomagnetic correlations (Trujillo and Kowallis 2015; Maidment and Muxworthy 2019).

Similarly, the uppermost levels of the Morrison Formation could use more data. Are the large animals like *Saurophaganax*, *Maraapunisaurus fragillimus*, and *Camarasaurus supremus* at the top of the formation in Oklahoma and Colorado just anomalies, or is there a true late-Morrison increase in body size among the dinosaurs? And what of Bob Bakker's suggestion of an increase in abundance of small vertebrate species near the top of the formation? Could an apparent late-Morrison increase in humidity have allowed an increase in the total vertebrate biomass that could be supported in the basin, and did this result in larger dinosaurs and more small animals? Was the increased mass added at the ends of the body-size spectrum? There also are some Cretaceous-grade vertebrate groups appearing near the end of Morrison time, but these are known mainly from the Como Bluff and Dinosaur National Monument regions, and filling in the record in other areas will help clarify what was going on before the curtain came down on the Late Jurassic in North America. The latest Morrison is an interesting interval and should be looked at more intensively.

Other areas of research in the Morrison Formation that will improve our understanding of the formation include studying microvertebrates, finding more plant localities, and certainly finding more insect body fossils. Microvertebrates are important because they have the potential to tell us more per pound about environments, ecosystems, distributions, and abundances than dinosaurs do, and they have been underutilized in the Morrison so far. We really have good plant macrofossil records only from one site in Montana, one in Utah, and two in Colorado. We need to better understand what was going on at the time among the plants of the basin. Among invertebrates, although we have good records of freshwater clams and snails, have found trace fossils of insects, and even have crayfish fossils and caddis fly cases preserved in the Morrison Formation, it would be good to have more body fossils of insects from the unit, beyond the tantalizing wing from Temple Canyon, Colorado, and partial insect from southeastern Utah.

The Morrison Formation's vertebrate fauna gives us one of the best windows we have into the world of one Mesozoic ecosystem. As much as we know, and as much time and hard work by many researchers as it has taken to get to this point, by summarizing it, we now only see more clearly how much there may be yet to learn from the formation's fossil record. In our process of trying to understand the ancient world, we necessarily categorize the animals we find by species, a concept taken from modern biological studies and based on simple and definable criteria. Although the fossil species of past times and the zoological species of modern environments are often defined on slightly different criteria, each is meant to delineate the same idea: populations unique in their reproductive isolation from others. And although species defined that way are certainly real and properly categorized as such, thinking of fossil species purely as defined units separate from others obscures one of the most inspirational elements of the history of life—its continuity over hundreds of millions of years. The fragmentary nature of the vertebrate fossil record and the division of geologic time into distinct periods and epochs add to this false sense of fossil species as isolated biological "islands" in a four-dimensional sea of space and time. By far the majority of vertebrate species are known only from one or a few specimens from one time and from one or a few localities somewhere on the globe. Certainly, some species were endemic to small areas, were short lived (even if successful), or were rare in their communities, but others were likely more common, wide ranging, and long lived than is apparent from the fossils we have. And all of these species were part of a larger, more long-lived history of life.

Parallel to our use of the species concept, and despite the limitations of the fossil record, we need to remember that the history of our phylum, Chordata, has been a continuum of hundreds of millions of generations—an unbroken series of parents and offspring from the first free-swimming relatives of *Pikaia* and *Metaspriggina* to every vertebrate alive today. In that sense, the successive species along the way and the geologic time periods used to order them, although real in that they are based on concrete characteristics of fossils and rocks, are imposed concepts that obscure the obvious continuity of life. Although we may see in the fossil record only one or a few individuals of a few species, we know that countless other individuals and many other species occurred along the lines from the first chordates to the modern array of vertebrates. Every individual vertebrate today is connected in a real way through millions of individuals and hundreds of successive species back through every one of the last 500 million years to the first chordates. The history of vertebrates is our history, even as individuals. All of it in some way shaped us, from the influence of the first tetrapods on our basic limb structure to the influence of the first placental mammals on how many molars we have. We think of ancient species with significant gaps between them because that is the reality of the record we have, but in fact, as humans, even we have ancestors, not just species but individuals, in every one of the time periods throughout history.

Our structure goes back to the primates that lived alongside the pantodonts and early horses of the Eocene epoch and to the first primates that appeared near the time the dinosaurs disappeared. Our direct ancestors, and those of many modern mammals, were among the first placental mammals that appeared early in the Cretaceous. Our genes and those of all today's mammals—zebras, the platypus, weasels—come down from mammals that lived somewhere in the world during the Late Jurassic. Back in the Permian, our ancient relatives were nonmammalian synapsids, reptilian in general appearance. And tracing back generation to generation far enough, we would eventually end up among the early chordates of the Cambrian period, sharing the world with the trilobites. Every vertebrate alive today could trace its history similarly, eventually converging with all of the rest of us back in the Cambrian.[3] Reptiles would trace back a different path of three-hundred-millionth great-grandparents, perhaps, but would merge with our mammalian line in the Paleozoic, and from then back, we would find we had many of the same relatives. And every one of these modern vertebrate species would trace a series of generations right through the Late Jurassic. Their Late Jurassic relatives might have looked quite similar to them, or they might have been quite different, but the relatives would be there. Somewhere in the Late Jurassic world, perhaps in some cases in the Morrison basin, lived the direct ancestors of the modern lizard living in the woodpile, ancestors of the frog living in the meadow, ancestors of the crocodiles in the Nile, ancestors of the birds perched in the tree, ancestors of the koalas in the eucalyptus, ancestors of the salamanders in the forest, and ancestors of the turtles in the pond. The Morrison Formation, as a window to the Late Jurassic world, is therefore a look back to one time in the history of today's fauna. It is a snapshot of where we came from—one of the few that are available out of many, many times and places that have changed through vertebrate history. It is a part of our story in a direct sense, and because of that, we can feel a certain connection to the Morrison fauna. We're a part of it too.

Appendix A: Vertebrate Fossil Localities

This section details the main data used in this study from Morrison Formation fossil vertebrate localities, updated from Foster (2003a). Abbreviations are as follows:

Main collection method: QU = Quarrying operations
SU = Surface collection
SW = Screen washing

Data collection: PO = Personal observation of museum and field specimen collection records and data
RE = References such as published papers and unpublished reports
VS = Field visitations of quarries and field areas
WK = Working the site personally for a significant period
PC = Personal communication from fellow researchers involved

Paleoenvironment: CH = River channel
OS = Crevasse splays
FPP1 = Wet floodplains
FPP2 = Wet floodplains, carbonaceous mudstone
FPW = Dry floodplains
PD1 = Overbank pond, dense microvertebrates
PD2 = Overbank pond, rare microvertebrates
LK = Lake deposits

Counting method: MNI = Minimum number of individuals
NIS = Number of identified specimens

Estimated stratigraphic level: Zones 1–6, as in chapter 7.

Geographic tiers are in North, Middle, and South thirds across the latitudinal extent of the formation and West, Central, and East thirds across the longitudinal extent.

Notes: A = Adult; J = Juvenile; H = Hatchling (very young juvenile)
Only vertebrate taxa are listed; for invertebrates, plants, and trace fossil taxa, see references in the text.

Arizona

Navajo County

PIÑON LOCALITY

Main collection method: SU
Data collection: PC, RE, PO
Stratigraphic level: —
Paleoenvironment: —
Geographic tiers: South, West

Taxa	MNI
Apatosaurus	1

Colorado

Delta County

YOUNG EGG SITE

Main collection method: SU, QU
Data collection: RE, WK, PO
Stratigraphic level: Zone 1
Paleoenvironment: FPW
Geographic tiers: Middle, West

Taxa	MNI
Shartegosuchid new species	1
Chelonia	1
Theropoda	1

Fremont County

CLEVELAND QUARRY

Main collection method: QU
Data collection: PO, VS, RE
Stratigraphic level: Zone 1
Paleoenvironment: OS
Geographic tiers: South, Central

Taxa	MNI
Haplocanthosaurus	1
Eutretauranosuchus	1
Glyptops	1
Theropoda	1

CLEVELAND II

Main collection method: QU
Data collection: PO
Stratigraphic level: —
Paleoenvironment: —
Geographic tiers: South, Central

Taxa	MNI
Haplocanthosaurus	1

COPE'S NIPPLE

Main collection method: QU
Data collection: PO, VS, RE
Stratigraphic level: Zone 6
Paleoenvironment: FPW
Geographic tiers: South, Central

Taxa	MNI
Camarasaurus supremus	6
Allosaurus	2
Apatosaurus	1
Maraapunisaurus	1
Stegosaurus	1
Amphicotylus	1
Glyptops	1

COPE QUARRY 8 (THE FORT)

Main collection method: QU
Data collection: RE
Stratigraphic level: Zone 6
Paleoenvironment: —
Geographic tiers: South, Central

Taxa	MNI
Camarasaurus supremus	1

COPE QUARRY 12

Main collection method: QU
Data collection: PO, RE
Stratigraphic level: Zone 6
Paleoenvironment: —
Geographic tiers: South, Central

Taxa	MNI
Diplodocus (*Amphicoelias*)	1

COPE'S MYSTERY LOCALITIES

Main collection method: QU
Data collection: RE
Stratigraphic level: —
Paleoenvironment: —
Geographic tiers: South, Central

Taxa	MNI
Glyptops	2
Camarasaurus	2
Apatosaurus	1
Camptosaurus	1
Stegosaurus	1

DEWEESE QUARRY

Main collection method: QU
Data collection: PO, VS
Stratigraphic level: Zone 4
Paleoenvironment: FPP1
Geographic tiers: South, Central

Taxa	MNI
Diplodocus	1
Glyptops	1
Theropoda	1
Crocodylia	1

EGG GULCH

Main collection method: SU, QU
Data collection: PO, VS
Stratigraphic level: Zone 3
Paleoenvironment: FPP1, LK
Geographic tiers: South, Central

Taxa	MNI	
Dryosaurus	1	FPP1
Crocodylia	1	LK

ERIC'S TOOTH

Main collection method: SU
Data collection: PO, WK
Stratigraphic level: —
Paleoenvironment: —
Geographic tiers: South, Central

Taxa	MNI
Allosaurus	1
Glyptops	1
Crocodylia	1

FELCH QUARRY 2

Main collection method: QU
Data collection: RE, VS
Stratigraphic level: Zone 3
Paleoenvironment: CH
Geographic tiers: South, Central

Taxa	MNI
Allosaurus	1
Camarasaurus	1
Diplodocus	1

GARDEN PARK GENERAL

Main collection method: —
Data collection: PO
Stratigraphic level: —
Paleoenvironment: —, CH
Geographic tiers: South, Central

Taxa	MNI	Notes
Allosaurus	1	
Glyptops	2	
Stegosaurus	1	J
Nanosaurus	1	
Camarasaurus	1	CH

GREEN ACRES GENERAL

Main collection method: SU
Data collection: PO, VS
Stratigraphic level: —
Paleoenvironment: —
Geographic tiers: South, Central

Taxa	NIS
Glyptops	11
Ornithopoda	2
Theropoda	1
Crocodylia	1
Eilenodon	1

GREG'S BONE

Main collection method:	—
Data collection:	RE
Stratigraphic level:	—
Paleoenvironment:	—
Geographic tiers:	South, Central

Taxa	MNI
Stegosaurus	1

JENNINGS AND JOHNSON

Main collection method:	SU
Data collection:	PO, VS, RE
Stratigraphic level:	Zone 6
Paleoenvironment:	OS
Geographic tiers:	South, Central

Taxa	MNI
Hallopus	1

KENNY'S *STEGOSAURUS*

Main collection method:	—
Data collection:	RE
Stratigraphic level:	Zone 3
Paleoenvironment:	—
Geographic tiers:	South, Central

Taxa	MNI
Stegosaurus	1

KESSLER QUARRY

Main collection method:	QU
Data collection:	PO
Stratigraphic level:	Zone 2
Paleoenvironment:	FPP1
Geographic tiers:	South, Central

Taxa	MNI
Stegosaurus	1
Allosaurus	1
Glyptops	1
Amphicotylus	1
Ornithopoda	1

LINDSEY QUARRY

Main collection method:	QU
Data collection:	PO, RE
Stratigraphic level:	Zone 4
Paleoenvironment:	PD2
Geographic tiers:	South, Central

Taxa	NIS	Notes
Actinopterygii	2	
Anura	1	
Glyptops	11	
Opisthias	1	
Crocodylia	1	
Allosaurus	3	2A, 1J
Diplodocus	1	
Camarasaurus	2	1A, 1J

LUCAS'S SITE

Main collection method:	QU
Data collection:	RE, PO
Stratigraphic level:	Zone 6
Paleoenvironment:	CH
Geographic tiers:	South, Central

Taxa	MNI
Amphicotylus lucasii	2

MARSH-FELCH QUARRY

Main collection method:	QU
Data collection:	RE, PO, VS
Stratigraphic level:	Zone 3
Paleoenvironment:	CH
Geographic tiers:	South, Central

Taxa	MNI	Notes
Potamoceratodus	2	
Glyptops	1	
Amphicotylus	1	
Eutretauranosuchus	1	
Allosaurus	3	
Ceratosaurus	2	
Coelurus	1	
"*Elaphrosaurus*"	1	
Camarasaurus	4	2A, 2J
Apatosaurus	3	2A, 1J

Diplodocus	3	2A, 1J
Brachiosaurus	1	
Haplocanthosaurus	2	
Stegosaurus	3	
Nanosaurus	2	
Dryosaurus	1	
Camptosaurus	1	
Docodon	1	
Amblotherium (*Kepolestes*)	1	

MEYER SITE 1

Main collection method:	—
Data collection:	PO, RE
Stratigraphic level:	Zone 1
Paleoenvironment:	—
Geographic tiers:	South, Central

Taxa	MNI
Amphicotylus	1
Allosaurus	1

MEYER SITE 2

Main collection method:	—
Data collection:	PO, RE
Stratigraphic level:	Zone 1
Paleoenvironment:	—
Geographic tiers:	South, Central

Taxa	MNI
Torvosaurus	1

MEYER SITE 3

Main collection method:	—
Data collection:	PO, RE
Stratigraphic level:	Zone 3
Paleoenvironment:	—
Geographic tiers:	South, Central

Taxa	MNI
Diplodocus	1

MEYER'S PUMP HOUSE LOCALITY

Main collection method:	—
Data collection:	PO
Stratigraphic level:	—

Paleoenvironment:	—
Geographic tiers:	South, Central

Taxa	MNI
Allosaurus	1

SAUROPOD QUARRY

Main collection method:	QU
Data collection:	PO, RE
Stratigraphic level:	Zone 1
Paleoenvironment:	FPP1
Geographic tiers:	South, Central

Taxa	MNI
Diplodocidae indet.	1
Theropoda	1
Chelonia	1

SMALL QUARRY

Main collection method:	QU
Data collection:	PO, VS, RE
Stratigraphic level:	Zone 4
Paleoenvironment:	PD1
Geographic tiers:	South, Central

Taxa	MNI	NIS
Actinopterygii		30
Potamoceratodus		7
Caudata		1
Glyptops		72
Dinochelys		5
Chelonia indet.		21
Opisthias		3
Lacertilia		2
Amphicotylus		14
Kepodactylus	1	
Abelisauroidea indet.	1	
Theropoda indet.	2	
Camarasaurus	1	
Apatosaurus	1	
Stegosaurus	1	
Ankylosauria	1	
Dryosaurus	1	
Docodon		4
Dryolestidae		1
Mammalia indet.		2

TEMPLE CANYON
(MULTIPLE EXCAVATIONS)

Main collection method: QU
Data collection: PO, VS, RE
Stratigraphic level: Zone 1
Paleoenvironment: LK
Geographic tiers: South, Central

Taxa	MNI
Actinopterygii indet.	3
Amiidae	5
Pholidophoriformes	5
Coccolepidae	1
Semionotidae?	3
Halecomorpha	1
Potamoceratodus guentheri	1
Anura indet.	1
Glyptops	1

TIM'S EGG SITE

Main collection method: —
Data collection: PO
Stratigraphic level: —
Paleoenvironment: —
Geographic tiers: South, Central

Taxa	MNI
Glyptops	1

VALLEY OF DEATH

Main collection method: QU, SU
Data collection: PO, VS, RE
Stratigraphic level: Zone 3
Paleoenvironment: —
Geographic tiers: South, Central

Taxa	MNI
Nanosaurus	1
Glyptops	1
Crocodylia	1
Diplodocidae	1

WEBSTER PARK

Main collection method: —
Data collection: PO, RE

Stratigraphic level: —
Paleoenvironment: —
Geographic tiers: Middle, Central

Taxa	MNI
Camarasaurus	1

WILSON CREEK

Main collection method: —
Data collection: RE
Stratigraphic level: —
Paleoenvironment: —
Geographic tiers: South, Central

Taxa	MNI
Allosaurus	1

Grand County
MIDDLE PARK

Main collection method: SU
Data collection: PO, RE
Stratigraphic level: —
Paleoenvironment: —
Geographic tiers: Middle, Central

Taxa	MNI
Allosaurus	1

RADIUM LOCALITY

Main collection method: QU
Data collection: PO
Stratigraphic level: —
Paleoenvironment: FPP1
Geographic tiers: Middle, Central

Taxa	MNI
Allosaurus	1

Gunnison County
CABIN CREEK QUARRY

Main collection method: QU
Data collection: RE, PC
Stratigraphic level: Zone 2
Paleoenvironment: FPP1
Geographic tiers: South, Central

Taxa	MNI
Apatosaurus	1

BLUE MESA QUARRY

Main collection method: QU
Data collection: VS, RE, PO
Stratigraphic level: —
Paleoenvironment: PD2
Geographic tiers: South, Central

Taxa	MNI
Apatosaurus	1
Allosaurus	1
Crocodylia	1

NORTH BEACH CURECANTI

Main collection method: SU
Data collection: WK, RE, PO
Stratigraphic level: Zone 4?
Paleoenvironment: CH
Geographic tiers: South, Central

Taxa	MNI
Ceratosaurus	1
Allosaurus	1
Diplodocidae indet.	1
Diplodocinae indet.	1
Camarasaurus	1

SOUTH BEACH CURECANTI

Main collection method: SU
Data collection: WK, RE, PO
Stratigraphic level: Zone 4?
Paleoenvironment: CH
Geographic tiers: South, Central

Taxa	MNI
Goniopholididae indet.	1
Sauropoda indet.	1
Ornithischia?	1

Jefferson County

INTERSTATE 70 ROADCUT

Main collection method: —
Data collection: PO, VS

Stratigraphic level: —
Paleoenvironment: —
Geographic tiers: Middle, Central

Taxa	MNI	Notes
Allosaurus	2	1A, 1J

LAKES QUARRY 1 AT MORRISON

Main collection method: QU
Data collection: RE, VS
Stratigraphic level: Zone 4
Paleoenvironment: CH
Geographic tiers: Middle, Central

Taxa	MNI
"Goniopholis" felix	1
Allosaurus	1
Apatosaurus	1

LAKES QUARRY 5 AT MORRISON

Main collection method: QU
Data collection: PO, RE, VS
Stratigraphic level: Zone 4
Paleoenvironment: CH
Geographic tiers: Middle, Central

Taxa	MNI
Diplodocus	1
Stegosaurus	1

LAKES QUARRY 8 AT MORRISON

Main collection method: QU
Data collection: RE, VS
Stratigraphic level: Zone 4
Paleoenvironment: CH
Geographic tiers: Middle, Central

Taxa	MNI
Apatosaurus	1
Diplodocus	1

LAKES QUARRY 10 AT MORRISON

Main collection method: QU
Data collection: RE, PO, VS
Stratigraphic level: Zone 5

Paleoenvironment: CH, FPP2
Geographic tiers: Middle, Central

Taxa	MNI	Notes
Apatosaurus	1	CH
Apatosaurus	2	FPP2
Allosaurus	1	FPP2

Larimer County
HORSETOOTH QUARRY

Main collection method: QU
Data collection: RE
Stratigraphic level: —
Paleoenvironment: —
Geographic tiers: Middle, Central

Taxa	MNI
Camarasaurus supremus	1
Allosaurus (*Epanterias*)	1

MASONVILLE LOCALITY

Main collection method: —
Data collection: RE
Stratigraphic level: —
Paleoenvironment: —
Geographic tiers: Middle, Central

Taxa	MNI
Allosaurus	1

Las Animas County
LAST CHANCE QUARRY

Main collection method: QU
Data collection: RE, PC, PO
Stratigraphic level: —
Paleoenvironment: FPP1
Geographic tiers: South, East

Taxa	MNI	Notes
Theropoda indet.	1	
Apatosaurus	1	
Diplodocus	1	J
Camarasaurus	1	

PURGATOIRE RIVER

Main collection method: QU
Data collection: PO, VS
Stratigraphic level: —
Paleoenvironment: LK
Geographic tiers: South, East

Taxa	MNI
Camarasaurus	1

VILLEGREEN

Main collection method: —
Data collection: PO
Stratigraphic level: —
Paleoenvironment: —
Geographic tiers: South, East

Taxa	MNI
Camarasaurus	1
Amphicotylus	1

Mesa County
BOLLAN QUARRY

Main collection method: QU
Data collection: PO, VS
Stratigraphic level: Zone 3
Paleoenvironment: CH
Geographic tiers: Middle, West

Taxa	MNI
Stegosaurus	1
Allosaurus	1

BOLLAN SNAKE SKIN QUARRY

Main collection method: QU
Data collection: PO, VS
Stratigraphic level: Zone 3
Paleoenvironment: CH
Geographic tiers: Middle, West

Taxa	MNI
Diplodocinae indet.	1
Camarasaurus?	1

CACTUS PARK BYU QUARRY

Main collection method: QU
Data collection: PO, RE
Stratigraphic level: Zone 4
Paleoenvironment: —
Geographic tiers: Middle, West

Taxa	MNI	Notes
Apatosaurus	4	1A, 3J
Camarasaurus	1	

HUPS ANKYLOSAUR QUARRY

Main collection method: QU
Data collection: PO, RE, VS
Stratigraphic level: Zone 5
Paleoenvironment: CH
Geographic tiers: Middle, West

Taxa	MNI
Mymoorapelta	1
Amphicotylus	1
Allosaurus	1
Sauropoda indet.	1

HUPS QUARRY 3A

Main collection method: QU
Data collection: PO, VS
Stratigraphic level: Zone 3
Paleoenvironment: FPP1
Geographic tiers: Middle, West

Taxa	MNI
Apatosaurus	1
Camarasaurus	1
Allosaurus	1

DOMINGUEZ-JONES QUARRY

Main collection method: QU
Data collection: RE, PO
Stratigraphic level: Zone 4
Paleoenvironment: FPP1
Geographic tiers: Middle, West

Taxa	MNI
Camarasaurus	1

DRY MESA QUARRY

Main collection method: QU
Data collection: PO, VS, RE
Stratigraphic level: Zone 4
Paleoenvironment: CH
Geographic tiers: South, West

Taxa	MNI	NIS	Notes
Actinopterygii	1	2	
Potamoceratodus	1		
Amphibia	1	1	
Glyptops	1		
Dinochelys	1		
Lacertilia	1	1	
Eilenodon	1		
Eutretauranosuchus	2		
Mesadactylus	3	34	
Ceratosaurus	1		
Torvosaurus	3		
Allosaurus	4		
Marshosaurus	1		
?*Coelurus*	1		
?*Ornitholestes*	1		
Diplodocus	5		3A, 2J
Apatosaurus	2		
Supersaurus	1		
Camarasaurus	5		2A, 2J, 1H
Brachiosaurus	2		
Stegosaurus	1		
Gargoyleosaurus	1		
Dryosaurus	2		
Camptosaurus	2		

EAST TRAIL THROUGH TIME

Main collection method: QU
Data collection: PO, RE, VS
Stratigraphic level: Zone 4
Paleoenvironment: CH
Geographic tiers: Middle, West

Taxa	MNI
Allosaurus	1
Diplodocus	1
Camptosaurus	1
Goniopholididae indet.	1
Diplodocidae indet.	1

Eriksen Ceratosaur

Main collection method: QU
Data collection: PO, VS
Stratigraphic level: Zone 4
Paleoenvironment: CH
Geographic tiers: Middle, West

Taxa	MNI
Ceratosaurus	1

FPA General

Main collection method: QU, SU
Data collection: PO, VS, RE
Stratigraphic level: Zone 4
Paleoenvironment: —
Geographic tiers: Middle, West

Taxa	NIS	Notes
Opisthias	5	
Parviraptor	1	
Lacertilia	2	
Serpentes	1	
Amphicotylus	2	1A, 1J
Fruitachampsa	6	
Pterosaur	1	
Theropoda	4	
Coelurosaur	3	
Camarasaurus	3	2A, 1J
Diplodocus	1	
Apatosaurus	1	
?Brachiosaurus	1	
Stegosaurus	1	
Priacodon	1	
Triconodonta	1	
Symmetrodonta	1	
Dryolestidae	1	
Paurodontidae	1	
Mammalia	4	

Callison Quarry (FPA Main Quarry 4)

Main collection method: QU
Data collection: PO, WK, RE
Stratigraphic level: Zone 4
Paleoenvironment: PD2
Geographic tiers: Middle, West

Taxa	NIS
Chelonia	1
Opisthias	3
Eilenodon	1
Sphenodontia	4
Diablophis	3
Dorsetisaurus	1
Anguimorpha	4
Lacertilia	19
Fruitachampsa	13
Macelognathus	1
Amphicotylus	3
Coelurosaur	3
Theropoda	2
Camarasaurus	1
Diplodocidae	1
Neornithischia indet.	2
Glirodon	1
Tathiodon?	1
Symmetrodonta	1
Dryolestidae	4
Mammalia	5

FPA Tom's Place

Main collection method: QU
Data collection: PO, WK, RE
Stratigraphic level: Zone 4
Paleoenvironment: PD2
Geographic tiers: Middle, West

Taxa	NIS
Actinopterygii	31
Opisthias	11
Sphenodontia	3
Paramacellodus	1
Saurillodon	2
Parviraptor	2
Anguimorpha	3
Lacertilia	3
Amphicotylus	1
Ceratosaurus	1
Theropoda	4
Diplodocidae	1
Fruitafossor	1
Triconodonta	1
Mammalia indet.	1

FPA Fruitadens Pit

Main collection method:	QU, SU
Data collection:	PC, VS
Stratigraphic level:	Zone 4
Paleoenvironment:	FPP1
Geographic tiers:	Middle, West

Taxa	NIS
Fruitadens	1

FPA Quarry 2

Main collection method:	—
Data collection:	PO
Stratigraphic level:	Zone 4
Paleoenvironment:	—
Geographic tiers:	Middle, West

Taxa	MNI
Priacodon	1
Paurodontidae	1

FPA Quarry 6

Main collection method:	—
Data collection:	PO
Stratigraphic level:	Zone 4
Paleoenvironment:	—
Geographic tiers:	Middle, West

Taxa	MNI
Theropoda	1

FPA CrocTurtleClam

Main collection method:	—
Data collection:	PO
Stratigraphic level:	Zone 4
Paleoenvironment:	—
Geographic tiers:	Middle, West

Taxa	MNI
Glyptops	1

FPA Little Blue

Main collection method:	—
Data collection:	PO
Stratigraphic level:	Zone 4

Paleoenvironment:	—
Geographic tiers:	Middle, West

Taxa	NIS
Opisthias	2
Fruitachampsa	2

FPA Mammal Site

Main collection method:	—
Data collection:	PO
Stratigraphic level:	Zone 4
Paleoenvironment:	—
Geographic tiers:	Middle, West

Taxa	MNI
Eilenodon	1

FPA Stego Knoll

Main collection method:	QU
Data collection:	PO
Stratigraphic level:	Zone 4
Paleoenvironment:	—
Geographic tiers:	Middle, West

Taxa	MNI
Stegosaurus	1

HOLT QUARRY

Main collection method:	QU
Data collection:	RE, VS
Stratigraphic level:	Zone 4
Paleoenvironment:	CH
Geographic tiers:	Middle, West

Taxa	MNI
Allosaurus	1
Stegosaurus	1

JONES HOLE

Main collection method:	QU
Data collection:	RE, VS
Stratigraphic level:	Zone 4
Paleoenvironment:	CH
Geographic tiers:	Middle, West

Taxa	MNI
Allosaurus	1
Camptosaurus	1

KINGS VIEW QUARRY

Main collection method: QU
Data collection: WK, PO
Stratigraphic level: Zone 4
Paleoenvironment: CH
Geographic tiers: Middle, West

Taxa	MNI	Notes
Glyptops	1	
Opisthias	1	
Goniopholididae indet.	1	
Mesadactylus	1	
Theropoda indet.	1	
Camarasaurus	1	J
Stegosaurus	1	J

KOIZUMI ALLOSAUR

Main collection method: QU
Data collection: PO, VS
Stratigraphic level: Zone 4?
Paleoenvironment: CH
Geographic tiers: Middle, West

Taxa	MNI
Allosaurus	1

LIZ'S *CAMARASAURUS*

Main collection method: QU
Data collection: PO, RE, VS
Stratigraphic level: Zone 5
Paleoenvironment: CH
Geographic tiers: Middle, West

Taxa	MNI
Camarasaurus	1

MYGATT-MOORE QUARRY

Main collection method: QU
Data collection: PO, WK, RE
Stratigraphic level: Zone 4

Paleoenvironment: PD2
Geographic tiers: Middle, West

Taxa	MNI	Notes
Morrolepis	1	LK deposit
Hulettia	1	LK deposit
cf. *Leptolepis*	1	LK deposit
Caudata indet.	1	
Opisthias	1	
Goniopholididae indet.	1	
Allosaurus	6	
Ceratosaurus	1	
Apatosaurus	5	4A, 1J
Camarasaurus	3	1A, 2J
Diplodocus	2	1A, 1J
Mymoorapelta	2	
Nanosaurus	1	

RABBIT VALLEY AMNH

Main collection method: —
Data collection: PO
Stratigraphic level: —
Paleoenvironment: CH
Geographic tiers: Middle, West

Taxa	MNI
Glyptops	1

RABBIT VALLEY NESTING SITE

Main collection method: —
Data collection: RE
Stratigraphic level: —
Paleoenvironment: —
Geographic tiers: Middle, West

Taxa	MNI
Dryosaurus	1

RIGGS QUARRY 12

Main collection method: QU
Data collection: RE
Stratigraphic level: —
Paleoenvironment: —
Geographic tiers: Middle, West

Taxa	MNI
Camarasaurus	1

RIGGS QUARRY 13

Main collection method:	QU
Data collection:	PO, VS, RE
Stratigraphic level:	Zone 4
Paleoenvironment:	CH
Geographic tiers:	Middle, West

Taxa	MNI
Brachiosaurus	1

RIGGS QUARRY 14

Main collection method:	QU
Data collection:	RE
Stratigraphic level:	—
Paleoenvironment:	—
Geographic tiers:	Middle, West

Taxa	MNI
Camarasaurus	1

RIGGS QUARRY 15 (DINOSAUR HILL)

Main collection method:	QU
Data collection:	PO, VS
Stratigraphic level:	Zone 4
Paleoenvironment:	FPW
Geographic tiers:	Middle, West

Taxa	MNI
Apatosaurus	1

SINBAD SITE

Main collection method:	QU
Data collection:	PO
Stratigraphic level:	—
Paleoenvironment:	—
Geographic tiers:	Middle, West

Taxa	MNI	Notes
Diplodocus	1	J

SPLIT ROCK CHANNEL

Main collection method:	QU
Data collection:	PO, WK, VS, RE
Stratigraphic level:	Zone 5
Paleoenvironment:	CH
Geographic tiers:	Middle, West

Taxa	MNI
Allosaurus	1
Nanosaurus	1
Apatosaurus	1
Camarasaurus	1

TWIN JUNIPER QUARRY

Main collection method:	QU
Data collection:	PO, WK, VS
Stratigraphic level:	Zone 4(?)
Paleoenvironment:	FPW
Geographic tiers:	Middle, West

Taxa	MNI
Apatosaurus	1
Ceratosaurus	1

WOLNY SITE

Main collection method:	QU
Data collection:	RE, PO
Stratigraphic level:	—
Paleoenvironment:	CH
Geographic tiers:	Middle, West

Taxa	MNI
Allosaurus	1
Stegosaurus	1

Moffat County

CALICO GULCH

Main collection method:	QU
Data collection:	PO, PC
Stratigraphic level:	—
Paleoenvironment:	—
Geographic tiers:	Middle, West

Taxa	MNI
Camarasaurus	1
Diplodocus	1

HEADQUARTERS AREA — DINOSAUR NATIONAL MONUMENT

Main collection method: —
Data collection: PO
Stratigraphic level: —
Paleoenvironment: —
Geographic tiers: Middle, West

Taxa	MNI
Potamoceratodus	1

HOMESTEAD QUARRY

Main collection method: QU
Data collection: RE, PO
Stratigraphic level: Zone 2
Paleoenvironment: OS
Geographic tiers: Middle, West

Taxa	MNI
Piatnitzkysaurid, new	1
Dryosaurus	1
Amphicotylus	1

LILY PARK

Main collection method: QU
Data collection: RE
Stratigraphic level: —
Paleoenvironment: —
Geographic tiers: Middle, West

Taxa	MNI
Torvosaurus	1
Dryosaurus	1

MF AMPHITHEATER

Main collection method: SU
Data collection: WK, PO
Stratigraphic level: Zone 4
Paleoenvironment: CH
Geographic tiers: Middle, West

Taxa	MNI
Camarasaurus	1
Allosaurus	1
Stegosaurus	1
Camptosaurus	1

WITHERELL QUARRY

Main collection method: QU
Data collection: PO
Stratigraphic level: —
Paleoenvironment: FPP1
Geographic tiers: Middle, West

Taxa	MNI
Ceratodus robustus	2
Camarasaurus	1
Nanosaurus	1
Theropoda	1
Crocodylia	1
Diplodocidae	1

WOLF CREEK QUARRY

Main collection method: QU
Data collection: PO, RE
Stratigraphic level: Zone 2
Paleoenvironment: PD2
Geographic tiers: Middle, West

Taxa	MNI	NIS
Actinopterygii		1
Anura		2
Caudata		1
Potamoceratodus		1
Glyptops		5
Dinochelys		1
Chelonia		4
Opisthias		4
Dorsetisaurus		1
Lacertilia		3
Amphicotylus		1
Crocodylia		3
Allosaurus	1	
Camarasaurus	1	
Diplodocidae	1	
Stegosaurus	1	
Mammalia		1

Montezuma County
MCELMO CANYON

Main collection method: —
Data collection: PO, RE
Stratigraphic level: —
Paleoenvironment: CH
Geographic tiers: South, West

Taxa	MNI
Allosaurus	1

YELLOWJACKET CANYON

Main collection method: —
Data collection: PO
Stratigraphic level: —
Paleoenvironment: —
Geographic tiers: South, West

Taxa	MNI
Apatosaurus	1

Montrose County
POTTER CREEK QUARRY

Main collection method: QU
Data collection: RE
Stratigraphic level: Zone 4
Paleoenvironment: FPP1
Geographic tiers: South, West

Taxa	MNI
Brachiosaurus	1
Theropoda	1

SCHEETZ LOCALITY

Main collection method: SU, SW
Data collection: RE, PO
Stratigraphic level: Zone 4
Paleoenvironment: FPP1
Geographic tiers: South, West

Taxa	MNI	Notes
Actinopterygii	1	
Dryosaurus	8	A, J, H
Multituberculata	1	

Chelonia	1
Opisthias	16
Eilenodon	2
Squamata	3
Cteniogenys	1
Pterosauria	1
Theropoda	1
Sauropoda	1
Crocodylia	1
Eilenodon	1

Pitkin County
GORDON-BRAMSON-BROTHERS QUARRY

Main collection method: QU
Data collection: WK, PO, RE
Stratigraphic level: Zone 1
Paleoenvironment: FPP1
Geographic tiers: Middle, Central

Taxa	MNI
Haplocanthosaurus	1

Saguache County
LOS OCHOS MINE

Main collection method: —
Data collection: PC
Stratigraphic level: —
Paleoenvironment: CH
Geographic tiers: South, Central

Taxa	MNI
Allosaurus	1

Summit County
HEENEY SITE

Main collection method: —
Data collection: PO
Stratigraphic level: —
Paleoenvironment: —
Geographic tiers: Middle, Central

Taxa	MNI
Allosaurus	1

Montana

Big Horn County

HORSE COULEE / BEAUVAIS CREEK

Main collection method:	QU
Data collection:	RE, PO
Stratigraphic level:	—
Paleoenvironment:	—
Geographic tiers:	North, West

Taxa	MNI
Diplodocus	1

Carbon County

MOTHERS DAY QUARRY

Main collection method:	QU
Data collection:	RE, PC, PO
Stratigraphic level:	—
Paleoenvironment:	CH
Geographic tiers:	North, West

Taxa	MNI	Notes
Allosaurus	1	
Diplodocus	15	J
Stegosaurus	1	
Theropoda indet.	1	

RATTLESNAKE RIDGE QUARRY

Main collection method:	QU
Data collection:	RE, PC
Stratigraphic level:	—
Paleoenvironment:	FPP2
Geographic tiers:	North, West

Taxa	MNI	Notes
Suuwassea	1	
Allosaurus	1	J

Cascade County

YUREK QUARRY

Main collection method:	QU
Data collection:	PC, PO
Stratigraphic level:	—
Paleoenvironment:	—
Geographic tiers:	North, West

Taxa	MNI
Stegosauridae indet.	1

Fergus County

LITTLE SNOWY MOUNTAINS

Main collection method:	QU
Data collection:	PC, PO
Stratigraphic level:	—
Paleoenvironment:	FPP1
Geographic tiers:	North, West

Taxa	MNI
Camarasaurus	1
Theropoda indet.	1
Stegosauridae indet.	1

Gallatin County

T & J QUARRY

Main collection method:	QU
Data collection:	PC, RE, PO
Stratigraphic level:	—
Paleoenvironment:	LK
Geographic tiers:	North, West

Taxa	MNI	Notes
Apatosaurus	2	2J
Camarasaurus	1	
Allosaurus	1	

Park County

O'HAIR QUARRY (STRICKLAND CREEK)

Main collection method:	QU
Data collection:	PC, PO, RE
Stratigraphic level:	—
Paleoenvironment:	FPP1
Geographic tiers:	North, West

Taxa	MNI
Diplodocus	3
Apatosaurus	2
Allosaurus	1
Hesperosaurus	1
Sphenodontid indet.	1

Wheatland County

WITTECOMBE'S RANCH

Main collection method: —
Data collection: PO, RE
Stratigraphic level: —
Paleoenvironment: —
Geographic tiers: North, West

Taxa	MNI
Potamoceratodus	1
Chelonia	1
Amphicotylus	1
Allosaurus	1
Camarasaurus	1
Diplodocus	1
Stegosaurus	1

New Mexico

Cibola County

ACOMA SITE

Main collection method: —
Data collection: RE
Stratigraphic level: Zone 2
Paleoenvironment: FPW
Geographic tiers: South, West

Taxa	MNI
Allosaurus	1

PETERSON QUARRY

Main collection method: QU
Data collection: RE, PO
Stratigraphic level: Zone 5
Paleoenvironment: CH
Geographic tiers: South, West

Taxa	MNI
Glyptops	1
Camarasaurus	1
Diplodocidae	1
Allosauridae	1

BONEY CANYON NORTHEAST

Main collection method: —
Data collection: RE
Stratigraphic level: Zone 5
Paleoenvironment: —
Geographic tiers: South, West

Taxa	MNI
Camarasaurus	1

CONCHO SPRINGS

Main collection method: —
Data collection: RE
Stratigraphic level: Zone 4
Paleoenvironment: FPW
Geographic tiers: South, West

Taxa	MNI
Stegosaurus	1

SUWANEE PEAK

Main collection method: —
Data collection: RE
Stratigraphic level: —
Paleoenvironment: —
Geographic tiers: South, West

Taxa	MNI
Allosaurus	1

Guadalupe County

BULL CANYON

Main collection method: —
Data collection: RE
Stratigraphic level: Zone 1
Paleoenvironment: —
Geographic tiers: South, Central

Taxa	MNI
Stegosaurus	1

McKinley County

BLUE PEAK

Main collection method:	—
Data collection:	RE
Stratigraphic level:	—
Paleoenvironment:	—
Geographic tiers:	South, West

Taxa	MNI
Apatosaurus	1

Quay County

QUAY COUNTY LOCALITIES

Main collection method:	—
Data collection:	RE
Stratigraphic level:	—
Paleoenvironment:	—
Geographic tiers:	South, East

Taxa	MNI
Allosaurus	1
Camptosaurus	1

Sandoval County

HAGAN BASIN

Main collection method:	—
Data collection:	RE
Stratigraphic level:	—
Paleoenvironment:	—
Geographic tiers:	South, Central

Taxa	MNI
Camarasaurus	3

SAN YSIDRO 1

Main collection method:	QU
Data collection:	RE, PO
Stratigraphic level:	Zone 5
Paleoenvironment:	FPW
Geographic tiers:	South, Central

Taxa	MNI
Camarasaurus	2(?)
Allosaurus	1

SAN YSIDRO 2

Main collection method:	QU
Data collection:	RE
Stratigraphic level:	—
Paleoenvironment:	—
Geographic tiers:	South, Central

Taxa	MNI
Diplodocus	1

SEISMOSAURUS QUARRY

Main collection method:	QU
Data collection:	RE, VS, PO
Stratigraphic level:	Zone 4
Paleoenvironment:	CH
Geographic tiers:	South, Central

Taxa	MNI
Diplodocus (*Seismosaurus*)	1
Theropoda	1

Union County

EXTER

Main collection method:	—
Data collection:	RE
Stratigraphic level:	—
Paleoenvironment:	—
Geographic tiers:	South, East

Taxa	MNI
Allosaurus	1
Sauropoda	1
Ornithopoda	1

Oklahoma

Cimarron County

STOVALL QUARRY 1

Main collection method:	QU
Data collection:	RE, VS
Stratigraphic level:	Zone 5
Paleoenvironment:	FPP1
Geographic tiers:	South, East

Taxa	MNI	Notes
Saurophaganax	2	

Apatosaurus	4	1A, 3H	
Camarasaurus	2	1A, 1H	
Brachiosaurus	1		
Stegosaurus	1		
Camptosaurus	1		

STOVALL QUARRY 3A

Main collection method: QU
Data collection: RE
Stratigraphic level: Zone 4
Paleoenvironment: FPP1
Geographic tiers: South, East

Taxa	MNI
Stegosaurus	1

STOVALL QUARRY 5

Main collection method: QU
Data collection: RE, VS
Stratigraphic level: Zone 4
Paleoenvironment: FPP1
Geographic tiers: South, East

Taxa	MNI
Diplodocus	1

STOVALL QUARRY 6

Main collection method: QU
Data collection: RE
Stratigraphic level: Zone 4
Paleoenvironment: FPP1
Geographic tiers: South, East

Taxa	MNI
Apatosaurus	1
Barosaurus	1

STOVALL QUARRY 8

Main collection method: QU
Data collection: RE, PO
Stratigraphic level: Zone 4
Paleoenvironment: LK
Geographic tiers: South, East

Taxa	MNI	Notes
Actinopterygii	1	
Potamoceratodus	2	
Glyptops	1	
Cteniogenys(?)	1	
Amphicotylus	2	1 *"G." stovalli*, 1 sp.
Theropoda	1	
Ornithopoda	2	

STOVALL QUARRY 9

Main collection method: QU
Data collection: RE
Stratigraphic level: Zone 4
Paleoenvironment: FPW
Geographic tiers: South, East

Taxa	MNI
Apatosaurus	1

South Dakota
Fall River County
PARKER'S MYSTERY LOCALITY

Main collection method: —
Data collection: PO
Stratigraphic level: —
Paleoenvironment: CH
Geographic tiers: North, East

Taxa	MNI
Camarasaurus	1

Lawrence County
FULLER'S 351

Main collection method: QU
Data collection: PO, VS, RE
Stratigraphic level: Zone 6 (?)
Paleoenvironment: FPP1
Geographic tiers: North, East

Taxa	MNI	Notes
Allosaurus	1	
Apatosaurus	2	1A, 1J
Camarasaurus	1	

Meade County

AMNH QUARRY 1

Main collection method:	QU
Data collection:	PO, PC
Stratigraphic level:	—
Paleoenvironment:	—
Geographic tiers:	North, East

<u>Taxa</u>	<u>MNI</u>
Camarasaurus	1

AMNH QUARRY 2

Main collection method:	QU
Data collection:	PO, PC
Stratigraphic level:	—
Paleoenvironment:	—
Geographic tiers:	North, East

<u>Taxa</u>	<u>MNI</u>
Diplodocus/Barosaurus	1

BEAR BUTTE

Main collection method:	—
Data collection:	PO, VS
Stratigraphic level:	Zone 6 (?)
Paleoenvironment:	FPW
Geographic tiers:	North, East

<u>Taxa</u>	<u>MNI</u>
Apatosaurus	1

NORTH OF PIEDMONT

Main collection method:	—
Data collection:	RE, PO
Stratigraphic level:	—
Paleoenvironment:	—
Geographic tiers:	North, East

<u>Taxa</u>	<u>MNI</u>
Potamoceratodus	1

PIEDMONT BUTTE

| Main collection method: | QU |
| Data collection: | PO, RE, VS |

Stratigraphic level:	Zone 5 (?)
Paleoenvironment:	FPP1
Geographic tiers:	North, East

<u>Taxa</u>	<u>MNI</u>
Barosaurus	1
Sauropoda	1
Theropoda	1

USNM QUARRY

Main collection method:	QU
Data collection:	PO, PC
Stratigraphic level:	—
Paleoenvironment:	—
Geographic tiers:	North, East

<u>Taxa</u>	<u>MNI</u>
Camarasaurus	1

WONDERLAND NORTH

Main collection method:	—
Data collection:	PO, VS
Stratigraphic level:	Zone 4 (?)
Paleoenvironment:	FPW
Geographic tiers:	North, East

<u>Taxa</u>	<u>MNI</u>
Diplodocus	1

WONDERLAND QUARRY

Main collection method:	QU, SU
Data collection:	PO, VS
Stratigraphic level:	Zone 4 (?)
Paleoenvironment:	FPP1
Geographic tiers:	North, East

<u>Taxa</u>	<u>MNI</u>	<u>NIS</u>
Allosaurus	1	
Aviatyrannus-like form	1	
Barosaurus	1	
Camarasaurus	1	
Glyptops		20
Cteniogenys		1
Goniopholididae		7

Utah

Emery County

AARON SCOTT QUARRY

Main collection method: QU
Data collection: RE
Stratigraphic level: —
Paleoenvironment: LK
Geographic tiers: Middle, West

Taxa	MNI
Chelonia	1
Opisthias	1
Goniopholididae indet.	1
Allosaurus	1
Diplodocinae indet.	1
Camarasaurus	1

AFTON NELSON

Main collection method: QU
Data collection: PO, PC
Stratigraphic level: —
Paleoenvironment: —
Geographic tiers: Middle, West

Taxa	MNI
Stegosaurus	1
Theropoda	1
Dryosaurus/Camptosaurus	1

AGATE BASIN

Main collection method: QU
Data collection: PO, PC
Stratigraphic level: —
Paleoenvironment: —
Geographic tiers: Middle, West

Taxa	MNI
Ceratosaurus	1

CLEVELAND-LLOYD QUARRY

Main collection method: QU
Data collection: RE, VS, PO
Stratigraphic level: Zone 5
Paleoenvironment: FPP1
Geographic tiers: Middle, West

Taxa	MNI
Glyptops	2
Amphicotylus	1
Ceratosaurus	1
Allosaurus	46
Torvosaurus	1
Stokesosaurus	2
Marshosaurus	3
Camarasaurus	3
Diplodocus	1
Barosaurus(?)	1
Stegosaurus	4
Camptosaurus	5

CLEVELAND-LLOYD AREA

Quarry 2

Main collection method: QU
Data collection: PO
Stratigraphic level: —
Paleoenvironment: CH
Geographic tiers: Middle, West

Taxa	MNI
Allosaurus	1

Quarry 5

Main collection method: —
Data collection: PO
Stratigraphic level: —
Paleoenvironment: FPP1
Geographic tiers: Middle, West

Taxa	MNI
Glyptops	1
Crocodylia	1
Allosaurus	1

Quarry 15

Main collection method: —
Data collection: PO
Stratigraphic level: —
Paleoenvironment: FPP1
Geographic tiers: Middle, West

Taxa	MNI
Chelonia	1
Amphicotylus	1
Allosaurus	1

Quarry 23

Main collection method: QU
Data collection: PO, VS
Stratigraphic level: Zone 4 (?)
Paleoenvironment: FPP1
Geographic tiers: Middle, West

Taxa	MNI
Glyptops	1
Crocodylia	1
Allosaurus	1
Apatosaurus	1
Stegosaurus	1

Quarry 67

Main collection method: —
Data collection: PO
Stratigraphic level: —
Paleoenvironment: —
Geographic tiers: Middle, West

Taxa	MNI
Stegosaurus	1

DUKE SITE

Main collection method: QU
Data collection: PO
Stratigraphic level: —
Paleoenvironment: CH
Geographic tiers: Middle, West

Taxa	MNI
Stegosaurus	1

FERRON QUARRY

Main collection method: QU
Data collection: VS, PO, PC
Stratigraphic level: Zone 2 (?)
Paleoenvironment: CH
Geographic tiers: Middle, West

Taxa	MNI
Allosaurus	1

GREEN RIVER QUARRY

Main collection method: QU
Data collection: RE
Stratigraphic level: Zone 6
Paleoenvironment: FPP1
Geographic tiers: Middle, West

Taxa	MNI
Allosaurus	1
Camarasaurus	1
Stegosaurus	1

HATT RANCH

Main collection method: —
Data collection: PO
Stratigraphic level: Zone 1 (?)
Paleoenvironment: CH
Geographic tiers: Middle, West

Taxa	MNI
Theropoda	1

MUSSENTUCHIT QUARRY

Main collection method: QU
Data collection: PO
Stratigraphic level: —
Paleoenvironment: FPP1
Geographic tiers: Middle, West

Taxa	MNI
Chelonia	1
Crocodylia	1
Allosaurus	1
Diplodocus	1
Camarasaurus	1

SAN RAFAEL

Main collection method: —
Data collection: PO
Stratigraphic level: —
Paleoenvironment: FPP1
Geographic tiers: Middle, West

Taxa	MNI
Camarasaurus	1

SAND BENCH

Main collection method:	QU
Data collection:	PO, RE
Stratigraphic level:	Zone 3
Paleoenvironment:	—
Geographic tiers:	Middle, West

Taxa	MNI
Ceratosaurus	1

66 QUARRY

Main collection method:	QU
Data collection:	PO
Stratigraphic level:	—
Paleoenvironment:	FPP1
Geographic tiers:	Middle, West

Taxa	MNI
Allosaurus	1
Apatosaurus	1

WILLOW SPRINGS QUARRY

Main collection method:	QU
Data collection:	PO, RE
Stratigraphic level:	—
Paleoenvironment:	CH
Geographic tiers:	Middle, West

Taxa	MNI	Notes
Allosaurus	1	J
Nanosaurus	1	

PETERSON SITE

Main collection method:	QU
Data collection:	PO
Stratigraphic level:	—
Paleoenvironment:	CH
Geographic tiers:	Middle, West

Taxa	MNI
Torvosaurus	1

Grand County

ARCHES SAUROPOD QUARRY

Main collection method:	QU, SU
Data collection:	WK, PO
Stratigraphic level:	Zone 6
Paleoenvironment:	FPP1
Geographic tiers:	Middle, West

Taxa	MNI
Apatosaurus	1

CISCO MAMMAL QUARRY

Main collection method:	QU
Data collection:	RE, VS, PC
Stratigraphic level:	Zone 4?
Paleoenvironment:	PD2
Geographic tiers:	Middle, West

Taxa	MNI
Actinopterygii	1
Parviraptoridae indet.	1
Fruitadens?	1
Morganucodonta indet.	1
Fruitafossor	1
Glirodon	1
Eutriconodonta indet.	1
Dryolestes	1
Paurodontidae indet.	1

FLOY JUNCTION

Main collection method:	QU
Data collection:	PO
Stratigraphic level:	—
Paleoenvironment:	—
Geographic tiers:	Middle, West

Taxa	MNI
Camarasaurus	1

TOM'S SAUROPOD

Main collection method:	QU
Data collection:	PO, WK
Stratigraphic level:	Zone 4?
Paleoenvironment:	FPW
Geographic tiers:	Middle, West

Taxa	MNI
Apatosaurus(?)	1

MILL CANYON BONE TRAIL

Main collection method:	QU
Data collection:	RE, VS
Stratigraphic level:	—
Paleoenvironment:	—
Geographic tiers:	Middle, West

Taxa	MNI
Allosaurus	1
Camarasaurus	1

MILL CANYON QUARRY

Main collection method:	QU
Data collection:	PO, PC
Stratigraphic level:	Zone 4
Paleoenvironment:	FPW
Geographic tiers:	Middle, West

Taxa	MNI
Brontosaurus	1

WESTWATER

Main collection method:	SU
Data collection:	PO
Stratigraphic level:	—
Paleoenvironment:	—
Geographic tiers:	Middle, West

Taxa	MNI
Eilenodon	1

San Juan County
BROWN'S HOLE QUARRY

Main collection method:	QU
Data collection:	PO, PC
Stratigraphic level:	—
Paleoenvironment:	—
Geographic tiers:	South, West

Taxa	MNI
Camarasaurus	1

GNATALIE QUARRY

Main collection method:	QU
Data collection:	PC, VS, WK
Stratigraphic level:	Zone 4
Paleoenvironment:	CH
Geographic tiers:	South, West

Taxa	MNI
Theropoda	1
Diplodocinae	1
Camarasaurus	1
Mymoorapelta	1

NEWBERRY-BARNES QUARRY

Main collection method:	QU
Data collection:	WK, PO, RE, PC
Stratigraphic level:	Zone 1
Paleoenvironment:	CH/OS
Geographic tiers:	South, West

Taxa	MNI
Dystrophaeus	1

Uintah County
CHEVRON'S MORRISON NORTH

Main collection method:	—
Data collection:	PO
Stratigraphic level:	—
Paleoenvironment:	CH
Geographic tiers:	Middle, West

Taxa	MNI
Camarasaurus	1

CHEVRON'S MORRISON SOUTH

Main collection method:	—
Data collection:	PO
Stratigraphic level:	—
Paleoenvironment:	FPW
Geographic tiers:	Middle, West

Taxa	MNI
Camarasaurus	1

CHEW RANCH

Main collection method: QU
Data collection: PC
Stratigraphic level: —
Paleoenvironment: CH, FPP1
Geographic tiers: Middle, West

Taxa	MNI	Notes
Apatosaurinae indet.	1	CH
Allosaurus	1	FPP1
Stegosaurus	1	

DINOSAUR NATIONAL MONUMENT QUARRY

Main collection method: QU
Data collection: PO, VS, RE
Stratigraphic level: Zone 5
Paleoenvironment: CH
Geographic tiers: Middle, West

Taxa	MNI	Notes
Opisthias	1	
Glyptops	10	
Dinochelys	1	
Amphicotylus	3	
Ceratosaurus	1	
Torvosaurus	1	
Allosaurus	8	5A, 3J
Diplodocus	29	15A, 14J
Apatosaurus	16	11A, 4J, 1H
Barosaurus	5	
Camarasaurus	22	16A, 6J
Haplocanthosaurus?	2	
Stegosaurus	14	11A, 3J
Camptosaurus	7	4A, 3J
Dryosaurus	4	2A, 1J, 1H

DINOSAUR NATIONAL MONUMENT
Apatosaurus Quarry

Main collection method: QU
Data collection: PO
Stratigraphic level: Zone 5
Paleoenvironment: CH
Geographic tiers: Middle, West

Taxa	MNI
Apatosaurus	1

DNM 15

Main collection method: —
Data collection: PO
Stratigraphic level: Zone 5
Paleoenvironment: CH
Geographic tiers: Middle, West

Taxa	MNI
Apatosaurus	1

DNM 5

Main collection method: —
Data collection: PO
Stratigraphic level: Zone 2
Paleoenvironment: CH
Geographic tiers: Middle, West

Taxa	MNI
Allosaurus	1
Dryosaurus	1

DNM Numbered Localities

Main collection method: —
Data collection: PO, RE
Stratigraphic level: —
Paleoenvironment: —
Geographic tiers: Middle, West

Taxa	MNI
JM 60	
Glyptops	1
Amphicotylus	1
DNM 120	
Goniopholis	1
Diplodocus	1
DNM 141	
Camarasaurus	1
Camptosaurus	1
DNM 131	
Allosaurus	1
DNM 172	
Mammalia	1
DINO 11	

Amioid 1
DNM 1
  Amphicotylus 1
  Allosaurus 1
  Torvosaurus 1
DNM 3
  Allosaurus 1
  Camarasaurus 1
DNM 7
  Amphicotylus 1
  Allosaurus 1
DNM 6
  Glyptops 1
  Amphicotylus 1
  Opisthias 1
  Allosaurus 1
  Camarasaurus 1
  Mammalia 1
DNM 307
  Opisthias 1
  Amphicotylus 1
DNM 360
  Glyptops 1
DNM 375
  Schilleria 1
  Cteniogenys 1
  Caudata 1

Rainbow Park Locality 94

Main collection method: SW, QU
Data collection: PO, VS
Stratigraphic level: Zone 6
Paleoenvironment: FPP1
Geographic tiers: Middle, West

Taxa	NIS
Actinopterygii	4
Glyptops	7
Opisthias	7
Lacertilia	4
Serpentes	1
Amphicotylus	5
Allosaurus	3
Koparion	1
Diplodocus	3
Camarasaurus	3

Stegosaurus	3
Dryosaurus	1
Camptosaurus	2
Docodon	1
Multituberculata	3
Mammalia	9

Rainbow Park Locality 96

Main collection method: SW, SU
Data collection: PO, VS
Stratigraphic level: Zone 6
Paleoenvironment: FPP1
Geographic tiers: Middle, West

Taxa	NIS
Actinopterygii	1
Anura	8
Rhadinosteus	10
Enneabatrachus	2
Discoglossidae?	1
Glyptops	1
Opisthias	24
Theretairus	1
Lacertilia	23
Amphicotylus	1
Eutretauranosuchus	1
Camarasaurus	1
Stegosaurus	1
Multituberculata	3
Priacodon	1
Triconolestes	1
Euthlastus	2
Dryolestes	1
Mammalia	10

Soft Sauropod Quarry

Main collection method: QU
Data collection: PO, VS
Stratigraphic level: Zone 5
Paleoenvironment: —
Geographic tiers: Middle, West

Taxa	MNI
Apatosaurus	1

Douglass Draw Quarry

Main collection method: —
Data collection: PO, RE
Stratigraphic level: Zone 6
Paleoenvironment: FPP1
Geographic tiers: Middle, West

Taxa	MNI	Notes
Camptosaurus	1	H

Nielsen Draw

Main collection method: QU
Data collection: PO, RE
Stratigraphic level: —
Paleoenvironment: CH
Geographic tiers: Middle, West

Taxa	MNI
Hoplosuchus	1

Allosaurus jimmadseni *Site (DNM 116)*

Main collection method: QU
Data collection: PO
Stratigraphic level: Zone 1
Paleoenvironment: CH
Geographic tiers: Middle, West

Taxa	MNI
Allosaurus	1

DNM Marshosaurus

Main collection method: QU
Data collection: PO, RE
Stratigraphic level: —
Paleoenvironment: —
Geographic tiers: Middle, West

Taxa	MNI
Marshosaurus	1

JENSEN-JENSEN QUARRY

Main collection method: QU
Data collection: PO, RE
Stratigraphic level: Zone 2

Paleoenvironment: CH
Geographic tiers: Middle, West

Taxa	MNI
Apatosaurus	1
Camarasaurus	1
Brachiosaurus	1
Stegosaurus	1

UTAH FIELD HOUSE MCSTEGO QUARRY

Main collection method: QU
Data collection: RE, PC, PO
Stratigraphic level: Zone 2
Paleoenvironment: CH
Geographic tiers: Middle, West

Taxa	MNI
Stegosaurus	1

Wayne County

BRACHIOSAUR GULCH QUARRY

Main collection method: QU
Data collection: WK, PO
Stratigraphic level: Zone 1
Paleoenvironment: FPP1
Geographic tiers: South, West

Taxa	MNI
Brachiosaurus	1

HANKSVILLE-BURPEE QUARRY

Main collection method: QU
Data collection: PO, VS, PC
Stratigraphic level: Zone 4
Paleoenvironment: CH
Geographic tiers: South, West

Taxa	MNI
Allosaurus	1
Diplodocus	2
Apatosaurus	1
Camarasaurus	1
Mymoorapelta	1
Dryosaurus	1

HANKSVILLE LIZARD

Main collection method: QU
Data collection: PO, PC
Stratigraphic level: —
Paleoenvironment: —
Geographic tiers: South, West

Taxa	MNI
Squamata indet.	1

Wyoming

Albany County

AARG SITE

Main collection method: —
Data collection: RE, PC
Stratigraphic level: —
Paleoenvironment: CH
Geographic tiers: Middle, Central

Taxa	MNI
Potamoceratodus	1
Chelonia	2

ADB SITE

Main collection method: —
Data collection: RE, PC
Stratigraphic level: —
Paleoenvironment: CH
Geographic tiers: Middle, Central

Taxa	MNI
Stegosaurus	1

AMNH 222 QUARRY

Main collection method: QU
Data collection: PO, RE
Stratigraphic level: —
Paleoenvironment: FPP1
Geographic tiers: Middle, Central

Taxa	MNI
Apatosaurus	1

AMNH 223 QUARRY

Main collection method: QU
Data collection: PO, RE
Stratigraphic level: —
Paleoenvironment: FPP1
Geographic tiers: Middle, Central

Taxa	MNI
Diplodocus	1

AMNH 550 QUARRY

Main collection method: QU
Data collection: PO
Stratigraphic level: —
Paleoenvironment: FPP1
Geographic tiers: Middle, Central

Taxa	MNI
Apatosaurus	1

BDRAW SITE

Main collection method: —
Data collection: RE, PC
Stratigraphic level: —
Paleoenvironment: CH
Geographic tiers: Middle, Central

Taxa	MNI
Goniopholididae indet.	1
Sauropoda indet.	1

BECKY SITE

Main collection method: —
Data collection: RE, PC
Stratigraphic level: —
Paleoenvironment: FPP2
Geographic tiers: Middle, Central

Taxa	MNI
Nanosaurus	13

BERNICE QUARRY

Main collection method: QU
Data collection: RE, PC
Stratigraphic level: Zone 6

Paleoenvironment: CH
Geographic tiers: Middle, Central

Taxa	MNI
Uluops	1
Apatosaurus	1
Priacodon	1
Morrisonodon	1

BERTHA QUARRY

Main collection method: QU
Data collection: RE, PO
Stratigraphic level: Zone 2
Paleoenvironment: FPP1
Geographic tiers: Middle, Central

Taxa	MNI
Goniopholididae indet.	1
Allosaurus	1
Brontosaurus	1
Dryosaurus	1

BLUE NIMBUS SITE

Main collection method: —
Data collection: RE, PC
Stratigraphic level: —
Paleoenvironment: FPP1
Geographic tiers: Middle, Central

Taxa	MNI
Potamoceratodus	3
Goniopholididae indet.	4
Triconodonta indet.	1

BONE CABIN QUARRY

Main collection method: QU
Data collection: PO, RE, VS, PC
Stratigraphic level: Zone 2
Paleoenvironment: CH
Geographic tiers: Middle, Central

Taxa	MNI	NIS	Notes
Actinopterygii	1	1	
Glyptops	4	14	
Dinochelys	1	14	
Chelonia indet.	2		

Amphicotylus	5	
Crocodylia	3	
Pterosaur	1	
Allosaurus	7	6A, 1J
Coelurus	1	
Ornitholestes	1	
Camarasaurus	9	7A, 2J
Barosaurus	1	
Apatosaurus	7	4A, 3J
Diplodocus	5	5A
Galeamopus	1	
Stegosaurus	11	9A, 2H
Gargoyleosaurus	1	
Dryosaurus	3	
Camptosaurus	5	
Mammalia indet.	2	

BORIS QUARRY

Main collection method: QU
Data collection: PO
Stratigraphic level: —
Paleoenvironment: —
Geographic tiers: Middle, Central

Taxa	MNI
Apatosaurus	1
Allosaurus	1
Ankylosauria	1

BREAKFAST BENCH MAMMAL

Main collection method: —
Data collection: RE
Stratigraphic level: Zone 6
Paleoenvironment: FPP2
Geographic tiers: Middle, Central

Taxa	MNI
Foxraptor	1

BRONTE QUARRY

Main collection method: —
Data collection: RE, PC
Stratigraphic level: —
Paleoenvironment: FPP2
Geographic tiers: Middle, Central

Taxa	MNI
Chelonia	2
Goniopholididae indet.	1
Allosaurus	1
Neornithischia indet.	2
Mammalia indet.	1

BROWN'S QUARRY A

Main collection method:	QU
Data collection:	RE
Stratigraphic level:	—
Paleoenvironment:	—
Geographic tiers:	Middle, Central

Taxa	MNI
Stegosaurus	1

BROWN'S QUARRY B

Main collection method:	QU
Data collection:	RE
Stratigraphic level:	—
Paleoenvironment:	—
Geographic tiers:	Middle, Central

Taxa	MNI
Stegosaurus	1

BROWN'S QUARRY C

Main collection method:	QU
Data collection:	RE
Stratigraphic level:	—
Paleoenvironment:	—
Geographic tiers:	Middle, Central

Taxa	MNI
Allosaurus	1

BROWN'S QUARRY D

Main collection method:	QU
Data collection:	RE
Stratigraphic level:	—
Paleoenvironment:	—
Geographic tiers:	Middle, Central

Taxa	MNI
Allosaurus	1

BROWN'S QUARRY G

Main collection method:	QU
Data collection:	RE
Stratigraphic level:	—
Paleoenvironment:	—
Geographic tiers:	Middle, Central

Taxa	MNI
Camarasaurus	1

CAM BENCH QUARRY

Main collection method:	—
Data collection:	RE
Stratigraphic level:	Zone 4
Paleoenvironment:	FPW
Geographic tiers:	Middle, Central

Taxa	MNI
Ceratosaurus	1
Allosaurus	1
Camarasaurus	1
Camptosaurus	1

CAM 1993

Main collection method:	—
Data collection:	RE
Stratigraphic level:	Zone 2
Paleoenvironment:	CH
Geographic tiers:	Middle, Central

Taxa	MNI
Camarasaurus	1
Diplodocus	1

CASSIOPEIA QUARRY

Main collection method:	—
Data collection:	RE
Stratigraphic level:	—
Paleoenvironment:	—
Geographic tiers:	Middle, Central

Taxa	MNI
Allosaurus	1

CLAW QUARRY

Main collection method: —
Data collection: RE, PC
Stratigraphic level: —
Paleoenvironment: FPP1/CH
Geographic tiers: Middle, Central

Taxa	MNI
Amioidea	1
Ceratodus robustus	6
Potamoceratodus	17
Chelonia	2
Opisthias	1
Goniopholididae indet.	20
Ceratosaurus	1
Allosaurus	1
Theropoda indet.	3
Diplodocus	1
Camarasaurus	1
Nanosaurus	2
Othnielosaurus	1
Dryosaurus	1
Zofiabaatar	2

CONVENTION SITE

Main collection method: —
Data collection: RE
Stratigraphic level: —
Paleoenvironment: —
Geographic tiers: Middle, Central

Taxa	MNI
Nanosaurus	2

DEAD RABBIT HILL

Main collection method: SU
Data collection: RE, PC
Stratigraphic level: Zone 2
Paleoenvironment: —
Geographic tiers: Middle, Central

Taxa	MNI
Potamoceratodus	1
Glyptops	1
Docodon	1

DEBRA SITE

Main collection method: —
Data collection: RE, PC
Stratigraphic level: —
Paleoenvironment: FPP1
Geographic tiers: Middle, Central

Taxa	MNI
Chelonia	1

DELTA T

Main collection method: —
Data collection: RE, PC
Stratigraphic level: Zone 6
Paleoenvironment: —
Geographic tiers: Middle, Central

Taxa	MNI
Ceratodus	1
Dinochelys	1
Amphicotylus	1
Multituberculata	1

DOUBLE DIP SITE

Main collection method: —
Data collection: RE, PC
Stratigraphic level: —
Paleoenvironment: FPP1
Geographic tiers: Middle, Central

Taxa	MNI
Diplodocus	1

EOS SITE

Main collection method: —
Data collection: RE, PC
Stratigraphic level: —
Paleoenvironment: FPP2
Geographic tiers: Middle, Central

Taxa	MNI
Megalosauridae	1

E. P. THOMPSON QUARRY

Main collection method: —
Data collection: RE, PC
Stratigraphic level: Zone 4
Paleoenvironment: CH
Geographic tiers: Middle, Central

Taxa	MNI
Diplodocus	1

FREDLIN SITE

Main collection method: —
Data collection: RE, PC
Stratigraphic level: —
Paleoenvironment: FPW
Geographic tiers: Middle, Central

Taxa	MNI
Allosaurus	1
Camarasaurus	1

HAYSTACK QUARRY

Main collection method: —
Data collection: RE
Stratigraphic level: —
Paleoenvironment: —
Geographic tiers: Middle, Central

Taxa	MNI
Squamata	1
Theropoda	1
Apatosaurus	1
Camarasaurus	1

JAPATH SITE

Main collection method: —
Data collection: RE, PC
Stratigraphic level: —
Paleoenvironment: FPP1
Geographic tiers: Middle, Central

Taxa	MNI
Ceratosaurus	1
Sauropoda indet.	1

JEFF P QUARRY

Main collection method: QU
Data collection: RE
Stratigraphic level: Zone 5
Paleoenvironment: CH
Geographic tiers: Middle, Central

Taxa	MNI
Camarasaurus	1

JOSEPH QUARRY

Main collection method: —
Data collection: RE, PC
Stratigraphic level: —
Paleoenvironment: CH
Geographic tiers: Middle, Central

Taxa	MNI	Notes
Goniopholididae indet.	1	
Brontosaurus	1	J

JULIE SITE

Main collection method: —
Data collection: RE, PC
Stratigraphic level: —
Paleoenvironment: FPP2
Geographic tiers: Middle, Central

Taxa	MNI
Nanosaurus	1

KAT SITE

Main collection method: —
Data collection: RE, PC
Stratigraphic level: —
Paleoenvironment: FPP1
Geographic tiers: Middle, Central

Taxa	MNI
Camarasaurus	1

LOUISE QUARRY

Main collection method: QU
Data collection: RE

Stratigraphic level: Zone 4
Paleoenvironment: FPP1
Geographic tiers: Middle, Central

Taxa	MNI	Notes
Allosaurus	2	
Edmarka	1	
Camarasaurus	1	
Apatosaurus	1	H
Stegosaurus	1	J

MATT SITE

Main collection method: —
Data collection: RE, PC
Stratigraphic level: —
Paleoenvironment: FPW
Geographic tiers: Middle, Central

Taxa	MNI
Camarasaurus	1

MEL'S FEMUR

Main collection method: —
Data collection: RE, PC
Stratigraphic level: —
Paleoenvironment: CH
Geographic tiers: Middle, Central

Taxa	MNI
Allosaurus	1
Diplodocus	1

MOJO SITE

Main collection method: —
Data collection: RE, PC
Stratigraphic level: —
Paleoenvironment: FPW
Geographic tiers: Middle, Central

Taxa	MNI
Camarasaurus	1

MORTON SITE

Main collection method: —
Data collection: RE, PC
Stratigraphic level: —

Paleoenvironment: FPW
Geographic tiers: Middle, Central

Taxa	MNI
Camarasaurus	1

MUMMY QUARRY

Main collection method: —
Data collection: RE
Stratigraphic level: —
Paleoenvironment: —
Geographic tiers: Middle, Central

Taxa	MNI
Allosaurus	1
Apatosaurus	1
Ankylosauria	1

NAIL QUARRY

Main collection method: QU
Data collection: PO, RE
Stratigraphic level: Zone 4
Paleoenvironment: FPP1
Geographic tiers: Middle, Central

Taxa	MNI	Notes
Goniopholididae indet.	1	
Allosaurus	2	
Torvosaurus (*Edmarka*)	3	
Brontosaurus	2	
Diplodocus	2	J
Camarasaurus	2	1A, 1J
?*Brachiosaurus*	1	
Stegosaurus	1	

NASA QUARRY

Main collection method: QU
Data collection: PC, PO
Stratigraphic level: —
Paleoenvironment: OS
Geographic tiers: Middle, Central

Taxa	MNI
Chelonia	2
Cteniogenys	1
Goniopholididae indet.	1

Allosaurus	1
Diplodocus	1
Camarasaurus	1
Stegosaurus	1
Camptosaurus	3

Taxa	MNI	Notes
Allosaurus	2	
Camarasaurus	6	4A, 2J
Diplodocus	1	
Mammalia	1	

OKIE QUARRY

Main collection method: —
Data collection: RE
Stratigraphic level: —
Paleoenvironment: —
Geographic tiers: Middle, Central

Taxa	MNI
Diplodocus	1

OPK SITE

Main collection method: —
Data collection: RE, PC
Stratigraphic level: —
Paleoenvironment: CH
Geographic tiers: Middle, Central

Taxa	MNI
Ceratodus robustus	1
Potamoceratodus	4
Chelonia	2
Triconodonta indet.	1

PAT M QUARRY

Main collection method: —
Data collection: RE
Stratigraphic level: Zone 4
Paleoenvironment: FPP1
Geographic tiers: Middle, Central

Taxa	MNI
Camarasaurus	1

REED'S QUARRY 1

Main collection method: QU
Data collection: PO, RE
Stratigraphic level: Zone 5
Paleoenvironment: FPP1
Geographic tiers: Middle, Central

REED'S QUARRY 2

Main collection method: QU
Data collection: RE
Stratigraphic level: Zone 5
Paleoenvironment: FPP1
Geographic tiers: Middle, Central

Taxa	MNI
Apatosaurus	1

REED'S QUARRY 3

Main collection method: QU
Data collection: PO, RE
Stratigraphic level: Zone 5
Paleoenvironment: FPP1
Geographic tiers: Middle, Central

Taxa	MNI	Notes
Allosaurus	2	
Camarasaurus	5	5J
Stegosaurus	1	
Dryosaurus	1	

REED'S QUARRY 4

Main collection method: QU
Data collection: RE
Stratigraphic level: Zone 5
Paleoenvironment: FPP1
Geographic tiers: Middle, Central

Taxa	MNI
Allosaurus	1
Camarasaurus	1
Apatosaurus	1
Barosaurus	1
Stegosaurus	1

REED'S QUARRY 6

Main collection method:	QU
Data collection:	RE
Stratigraphic level:	Zone 5
Paleoenvironment:	—
Geographic tiers:	Middle, Central

Taxa	MNI
Amphicotylus	1

REED'S QUARRY 8

Main collection method:	QU
Data collection:	RE
Stratigraphic level:	Zone 4
Paleoenvironment:	PD1
Geographic tiers:	Middle, Central

Taxa	MNI
Chelonia	1
Amphicotylus	1
Allosaurus	1
Coelurus	1
Camarasaurus	1
Diplodocus	1
Stegosaurus	1

REED'S QUARRY 9

Main collection method:	QU, SW
Data collection:	PO, RE, VS
Stratigraphic level:	Zone 5
Paleoenvironment:	PD1
Geographic tiers:	Middle, Central

Taxa	NIS	Notes
Actinopterygii	254	
Potamoceratodus	24	
Anura	5	Incl. *Enneabat.*
Caudata	3	
Glyptops	79	
Dinochelys	62	
Opisthias	47	
Theretairus	1	
Dorsetisaurus	3	
Paramacellodus	14	
Cteniogenys	30	
Amphicotylus	4	

Taxa	NIS	Notes
Eutretauranosuchus	1	
Macelognathus	1	
Comodactylus	1	
Allosaurus	1	
Coelurus	4	
New theropod?	1	
Camarasaurus	1	
Stegosaurus	1	
Nanosaurus	5	
Dryosaurus	1	
Camptosaurus	1	
Docodon	69	
Ctenacodon	28	
Psalodon	18	
Phascolotheridium	1	
Aploconodon	1	
Trioracodon	13	
Priacodon	22	
Tinodon	15	Incl. *Eurylambda*
Amphidon	2	
Araeodon	1	
Paurodon	2	
Archaeotrigon	9	
Tathiodon	1	
Euthlastus	1	
Pelicopsis	1	
Dryolestes	57	Incl. *Herpetairus*
Amblotherium	35	
Laolestes	39	
Melanodon	13	Incl. *Mathacolest.*
Comotherium	2	
Miccylotyrans	1	
Chelonia indet.	169	
Sphenodontia indet.	6	
Lacertilia indet.	7	
Crocodylia	13	
Pterosaur indet.	6	
Multituberculata indet.	4	
Triconodonta indet.	4	
Symmetrodonta indet.	1	
Paurodontidae indet.	4	
Dryolestidae indet.	13	
Mammalia indet.	78	

REED'S QUARRY 10

Main collection method: QU
Data collection: PO, RE, VS
Stratigraphic level: Zone 5
Paleoenvironment: FPP1
Geographic tiers: Middle, Central

Taxa	MNI
Brontosaurus	1

REED'S QUARRY 11

Main collection method: QU
Data collection: RE
Stratigraphic level: Zone 5
Paleoenvironment: FPP1
Geographic tiers: Middle, Central

Taxa	MNI
Apatosaurus	1
Stegosaurus	1
Mammalia	1

REED'S QUARRY 13

Main collection method: QU
Data collection: PO, RE
Stratigraphic level: Zone 2
Paleoenvironment: CH
Geographic tiers: Middle, Central

Taxa	MNI	Notes
Glyptops	1	
Amphicotylus	1	
Coelurus	3	
Camarasaurus	4	3A, 1J
?Brachiosaurus	1	
Diplodocus	1	
Stegosaurus	14	12 A, 2J
Dryosaurus	1	
Camptosaurus	17	11A, 6J

REED'S QUARRY 14

Main collection method: QU
Data collection: RE, PO
Stratigraphic level: Zone 5

Paleoenvironment: FPP1
Geographic tiers: Middle, Central

Taxa	MNI
Allosaurus	1

REED'S QUARRY R

Main collection method: QU
Data collection: PO
Stratigraphic level: —
Paleoenvironment: —
Geographic tiers: Middle, Central

Taxa	MNI	Notes
Allosaurus	2	
Camarasaurus	4	2A, 2J
Apatosaurus	2	1A, 1J
Diplodocus	1	

REGINA QUARRY

Main collection method: —
Data collection: RE
Stratigraphic level: —
Paleoenvironment: —
Geographic tiers: Middle, Central

Taxa	MNI
Nanosaurus	1

ROCO SITE

Main collection method: —
Data collection: RE, PC
Stratigraphic level: —
Paleoenvironment: CH
Geographic tiers: Middle, Central

Taxa	MNI
Dryosaurus	1

SEAN SITE

Main collection method: —
Data collection: RE, PC
Stratigraphic level: —
Paleoenvironment: FPP1
Geographic tiers: Middle, Central

Taxa	MNI
Amioidea	1
Potamoceratodus	1
Dorsetochelys	1
Chelonia indet.	2
Opisthias	1
Goniopholididae indet.	11
Pterosauria indet.	1
Coelurus(?)	1
Zofiabaatar	1
Triconodonta indet.	1
Mammalia indet.	1

SHEEP CREEK QUARRY D(3)

Main collection method:	QU
Data collection:	RE, VS, PO
Stratigraphic level:	Zone 4
Paleoenvironment:	LK
Geographic tiers:	Middle, Central

Taxa	MNI
Diplodocus	3
Stegosaurus	2

SHEEP CREEK QUARRY C

Main collection method:	QU
Data collection:	RE
Stratigraphic level:	—
Paleoenvironment:	LK
Geographic tiers:	Middle, Central

Taxa	MNI	Notes
Camarasaurus	2	1A, 1H
Apatosaurus	1	

SHEEP CREEK QUARRY D

Main collection method:	QU
Data collection:	RE, PO
Stratigraphic level:	Zone 4
Paleoenvironment:	LK
Geographic tiers:	Middle, Central

Taxa	MNI
Apatosaurus	1
Camarasaurus	1
Stegosaurus	1

SHEEP CREEK QUARRY E

Main collection method:	QU
Data collection:	PO, RE, VS
Stratigraphic level:	—
Paleoenvironment:	LK
Geographic tiers:	Middle, Central

Taxa	MNI	Notes
Brontosaurus	3	2A, 1H
Camarasaurus	1	H
Stegosaurus	1	

SHEEP CREEK QUARRY F

Main collection method:	QU
Data collection:	RE
Stratigraphic level:	
Paleoenvironment:	LK
Geographic tiers:	Middle, Central

Taxa	MNI
Apatosaurus	1

SHEEP CREEK QUARRY G

Main collection method:	QU
Data collection:	RE
Stratigraphic level:	—
Paleoenvironment:	LK
Geographic tiers:	Middle, Central

Taxa	MNI
Stegosaurus	1

SHEEP CREEK QUARRY J

Main collection method:	QU
Data collection:	RE
Stratigraphic level:	—
Paleoenvironment:	LK
Geographic tiers:	Middle, Central

Taxa	MNI
Apatosaurus	1
Stegosaurus	1

SHEEP CREEK QUARRY K

Main collection method: QU
Data collection: RE
Stratigraphic level: —
Paleoenvironment: LK
Geographic tiers: Middle, Central

Taxa	MNI
Camarasaurus	1

SHEEP CREEK QUARRY 4

Main collection method: QU
Data collection: RE
Stratigraphic level: —
Paleoenvironment: LK
Geographic tiers: Middle, Central

Taxa	MNI
Apatosaurus	1

STEGO 99

Main collection method: QU
Data collection: RE, PO
Stratigraphic level: —
Paleoenvironment: CH
Geographic tiers: Middle, Central

Taxa	MNI
Glyptops	1
Gonopholididae indet.	1
Theropoda	1
Stegosaurus	1

TRUCK SITE

Main collection method: —
Data collection: RE, PC
Stratigraphic level: —
Paleoenvironment: FPP1
Geographic tiers: Middle, Central

Taxa	MNI
Potamoceratodus	1
Chelonia	2
Goniopholididae indet.	1
Ceratosaurus	1

Allosaurus	1
Megalosauridae indet.	1
Nanosaurus	1

TWO LEG SITE

Main collection method: QU
Data collection: RE, PC
Stratigraphic level: Zone 6
Paleoenvironment: FPP1
Geographic tiers: Middle, Central

Taxa	MNI
Diplodocus	1

WEEGE BOYS QUARRY

Main collection method: QU
Data collection: RE, VS
Stratigraphic level: Zone 4
Paleoenvironment: —
Geographic tiers: Middle, Central

Taxa	MNI
Stegosaurus	1

WKAT SITE

Main collection method: —
Data collection: RE, PC
Stratigraphic level: —
Paleoenvironment: FPP1
Geographic tiers: Middle, Central

Taxa	MNI
Ceratosaurus	1
Apatosaurus	1

ZANE QUARRY

Main collection method: QU
Data collection: RE, VS
Stratigraphic level: Zone 2
Paleoenvironment: LK
Geographic tiers: Middle, Central

Taxa	MNI
Theropoda	1
Camarasaurus	1

Taxa	MNI
Apatosaurus	1
Diplodocus	1
Stegosaurus	1
Ankylosauria	1

Albany or Carbon County
COMO GENERAL

Main collection method: —
Data collection: PO
Stratigraphic level: —
Paleoenvironment: —
Geographic tiers: Middle, Central

Taxa	MNI
Pterosaur	1
Allosaurus	1
Apatosaurus	1
Stegosaurus	1
Dryosaurus	1
Camptosaurus	1
Ctenacodon	1
Psalodon	1
Laolestes	1

COPE'S COMO QUARRY 1

Main collection method: QU
Data collection: PO
Stratigraphic level: —
Paleoenvironment: —
Geographic tiers: Middle, Central

Taxa	MNI
Glyptops	1
Stegosaurus	1

COPE'S COMO QUARRY 3

Main collection method: QU
Data collection: PO
Stratigraphic level: —
Paleoenvironment: —
Geographic tiers: Middle, Central

Taxa	MNI
Allosaurus	1

COPE'S COMO QUARRY 4

Main collection method: QU
Data collection: PO
Stratigraphic level: —
Paleoenvironment: —
Geographic tiers: Middle, Central

Taxa	MNI
Allosaurus	1
Camarasaurus	1
Ornithopoda	1

COPE'S COMO QUARRY 5

Main collection method: QU
Data collection: PO
Stratigraphic level: —
Paleoenvironment: —
Geographic tiers: Middle, Central

Taxa	MNI
Apatosaurus	1

Big Horn County
BIG AL QUARRY

Main collection method: QU
Data collection: RE, PO, VS
Stratigraphic level: Zone 2
Paleoenvironment: CH
Geographic tiers: North, Central

Taxa	MNI	Notes
Pterosaur	1	
Allosaurus	1	
Apatosaurus	1	
Diplodocus	1	J
Camarasaurus	1	J

BIG BUTTE QUARRY

Main collection method: QU
Data collection: RE
Stratigraphic level: Zone 6
Paleoenvironment: FPP1
Geographic tiers: North, Central

Taxa	MNI
Apatosaurus	1
Stegosaurus	1

HOWE QUARRY (BROWN'S HISTORIC)

Main collection method: QU
Data collection: RE, PO, VS
Stratigraphic level: Zone 2
Paleoenvironment: FPP1
Geographic tiers: North, Central

Taxa	MNI	Notes
Allosaurus	1	
Apatosaurus	1	
Diplodocus	6	1J
Barosaurus	1	
Kaatedocus	1	
Camarasaurus	4	3A, 1J
Camptosaurus	1	

HOWE-STEPHENS QUARRY

Main collection method: QU
Data collection: RE, VS, PC
Stratigraphic level: Zone 2
Paleoenvironment: CH
Geographic tiers: North, Central

Taxa	MNI	Notes
Ceratodus robustus	1	
Allosaurus	1	
Diplodocus	1	
Barosaurus	1	
Camarasaurus	2	
?Brachiosaurus	1	
Stegosaurus	1	
Hesperosaurus	1	
Nanosaurus	2	
Dryosaurus	2	

HOWE-SCOTT QUARRY

Main collection method: QU
Data collection: RE, PC
Stratigraphic level: Zone 2
Paleoenvironment: FPP1
Geographic tiers: North, Central

Taxa	MNI	Notes
Galeamopus	1	
Hesperosaurus	1	

LITTLE BUTTE QUARRY

Main collection method: QU
Data collection: PO, RE
Stratigraphic level: Zone 5
Paleoenvironment: FPP1
Geographic tiers: North, Central

Taxa	MNI
Diplodocus/Barosaurus	1

RED CANYON RANCH QUARRY

Main collection method: QU
Data collection: RE, PC, VS
Stratigraphic level: Zone 2
Paleoenvironment: CH
Geographic tiers: North, Central

Taxa	MNI
Stegosaurus	1
Neornithischia	1

SMITHSONIAN QUARRY 1

Main collection method: QU
Data collection: PO
Stratigraphic level: —
Paleoenvironment: —
Geographic tiers: North, Central

Taxa	MNI
Stegosaurus	1

Carbon County

BEER MUG LOCALITY

Main collection method: SU
Data collection: PO, PC
Stratigraphic level: —
Paleoenvironment: —
Geographic tiers: Middle, Central

Taxa	MNI
Diplodocinae indet.	1

CHUCK'S PROSPECT

Main collection method: —
Data collection: RE, PO, PC
Stratigraphic level: Zone 2
Paleoenvironment: —
Geographic tiers: Middle, Central

Taxa	NIS
Amioid	1
Glyptops	9
Cteniogenys	1
Amphicotylus	2
Multituberculata	1
Dryolestes	1
Amblotherium	1
Mammalia indet.	2

DRYOLESTES QUARRY

Main collection method: —
Data collection: RE, PC
Stratigraphic level: Zone 5
Paleoenvironment: FPP2
Geographic tiers: Middle, Central

Taxa	MNI
Dryolestes	1

DYER RANCH QUARRIES 1 AND 2

Main collection method: —
Data collection: RE
Stratigraphic level: —
Paleoenvironment: —
Geographic tiers: Middle, Central

Taxa	MNI
Apatosaurus	1
Diplodocus	1

FREEZEOUT HILLS GENERAL

Main collection method: —
Data collection: PO, RE
Stratigraphic level: —
Paleoenvironment: —
Geographic tiers: Middle, Central

Taxa	MNI
Amphicotylus	1
Allosaurus	1
Camarasaurus	1
?Brachiosaurus	1
Diplodocus	2
Camptosaurus	1
Mammalia indet.	1

FREEZEOUT HILLS QUARRY 4

Main collection method: QU
Data collection: PO
Stratigraphic level: —
Paleoenvironment: —
Geographic tiers: Middle, Central

Taxa	MNI
Camarasaurus	1

FREEZEOUT HILLS QUARRY 6

Main collection method: QU
Data collection: PO
Stratigraphic level: —
Paleoenvironment: —
Geographic tiers: Middle, Central

Taxa	MNI
Allosaurus	1
Apatosaurus	1

FREEZEOUT HILLS QUARRY L

Main collection method: QU
Data collection: RE
Stratigraphic level: —
Paleoenvironment: —
Geographic tiers: Middle, Central

Taxa	MNI
Camarasaurus	1

FREEZEOUT HILLS QUARRY N

Main collection method: QU
Data collection: RE, PO
Stratigraphic level: —
Paleoenvironment: CH
Geographic tiers: Middle, Central

Taxa	MNI
Glyptops	1
Goniopholididae indet.	1
Allosaurus	1
Camarasaurus	4
Apatosaurus	3
Haplocanthosaurus	1
Stegosaurus	2

FREEZEOUT HILLS QUARRY O

Main collection method:	QU
Data collection:	RE
Stratigraphic level:	—
Paleoenvironment:	CH
Geographic tiers:	Middle, Central

Taxa	MNI
Glyptops	1
Crocodylia	1
Apatosaurus	1
Stegosaurus	1

LAKES'S QUARRY 1A

Main collection method:	QU
Data collection:	PO, RE
Stratigraphic level:	Zone 2
Paleoenvironment:	CH
Geographic tiers:	Middle, Central

Taxa	MNI
Allosaurus	1
Camarasaurus	1
Diplodocus/Barosaurus	1
Nanosaurus	1
Camptosaurus	1

MEILYN QUARRY

Main collection method:	QU
Data collection:	RE, VS
Stratigraphic level:	Zone 1
Paleoenvironment:	CH
Geographic tiers:	Middle, Central

Taxa	MNI
Allosaurus	1
Hesperosaurus	1
Ankylosauria indet.	1

MATT QUARRY

Main collection method:	QU
Data collection:	RE, VS
Stratigraphic level:	Zone 1
Paleoenvironment:	FPP1
Geographic tiers:	Middle, Central

Taxa	MNI
Chelonia	1
Crocodylia	1
Theropoda	1
Dryosaurus	1

NINE MILE CROSSING

Main collection method:	QU
Data collection:	PO, RE
Stratigraphic level:	—
Paleoenvironment:	FPP1
Geographic tiers:	Middle, Central

Taxa	MNI
Apatosaurus	1

NINEMILE HILL

Main collection method:	SW, QU, SU
Data collection:	VS, PO, PC, RE
Stratigraphic level:	Zone 6
Paleoenvironment:	FPP1
Geographic tiers:	Middle, Central

Taxa	NIS
Actinopterygii	100
Potamoceratodus	1
Caudata	5
Glyptops	17
Dinochelys	4
Chelonia indet.	33
Cteniogenys	5
Lacertilia	4
Sphenodontia	1
Crocodylia	30
Allosaurus	1
Camarasaurus	1
Diplodocus	1
Nanosaurus	1
Ctenacodon	1

Psalodon	1
Multituberculata	4
Docodon	5
Dryolestes	1
Laolestes	1
Mammalia Indet.	7

REED'S QUARRY 1 1/2

Main collection method: QU
Data collection: RE, PO
Stratigraphic level: Zone 4
Paleoenvironment: FPP1
Geographic tiers: Middle, Central

Taxa	MNI
Allosaurus	1

REED'S QUARRY 5

Main collection method: QU
Data collection: RE, PO
Stratigraphic level: Zone 5
Paleoenvironment: FPP1
Geographic tiers: Middle, Central

Taxa	MNI
Dermodactylus	1
Theropoda	1
Camarasaurus	1
Diplodocus	1
Dryosaurus	1
Camptosaurus	1

REED'S QUARRY 7

Main collection method: QU
Data collection: PO, RE
Stratigraphic level: Zone 5
Paleoenvironment: FPP1
Geographic tiers: Middle, Central

Taxa	MNI	Notes
Amphicotylus	2	1A, 1J
Dryosaurus	1	

REED'S QUARRY 12

Main collection method: QU
Data collection: PO, RE
Stratigraphic level: Zone 5

Paleoenvironment: FPP1
Geographic tiers: Middle, Central

Taxa	MNI
Actinopterygii	1
Chelonia	1
Opisthias	1
Cteniogenys	1
Amphicotylus	1
Allosaurus	1
Coelurus	1
Fosterovenator	1
Camarasaurus	1
Diplodocus	1
Stegosaurus	1

Converse County
DOUGLAS QUARRY

Main collection method: QU
Data collection: PO, PC, VS
Stratigraphic level: —
Paleoenvironment: CH, FPP1
Geographic tiers: North, Central

Taxa	MNI	Notes
Supersaurus	1	CH
Hesperornithoides	1	FPP1

Crook County
BLACKTAIL CREEK

Main collection method: SU
Data collection: VS, PO
Stratigraphic level: —
Paleoenvironment: LK
Geographic tiers: North, East

Taxa	MNI
?Camarasaurus	1

BOBCAT PIT

Main collection method: QU
Data collection: RE
Stratigraphic level: —
Paleoenvironment: FPP1
Geographic tiers: North, East

Taxa	MNI
Camarasaurus	2
Brachiosaurus	1

DILLON'S CORNER

Main collection method: QU
Data collection: WK, PO
Stratigraphic level: Dinosaur Zone 3B (?)
Paleoenvironment: FPW
Geographic tiers: North, East

Taxa	MNI	Notes
?Barosaurus	1	DZ 3B Upper?
Theropoda	1	DZ 3B Lower?
Glyptops	1	DZ 3B Lower?

HADLEY'S HILL

Main collection method: SU
Data collection: WK, PO
Stratigraphic level: Zone 5 (?)
Paleoenvironment: CH
Geographic tiers: North, East

Taxa	MNI
Glyptops	1
Crocodylia	1
Theropoda	1
Sauropoda	1
Ornithopoda	1

LIGHTNING ROD BUTTE

Main collection method: SU
Data collection: WK, PO
Stratigraphic level: Zone 3 (?)
Paleoenvironment: CH
Geographic tiers: North, East

Taxa	MNI
Glyptops	1
Nanosaurus	1

LITTLE HOUSTON QUARRY

Main collection method: QU
Data collection: WK, PO

Stratigraphic level: Zone 2 (?)
Paleoenvironment: PD1
Geographic tiers: North, East

Taxa	MNI	NIS	Notes
Actinopterygii		32	
Anura		1	
Caudata		1	
Potamoceratodus		5	
Glyptops		7	
Dinochelys		45	
Paramacellodidae		1	
Opisthias?		2	
Cteniogenys		6	
Theriosuchus		1	
Crocodylia indet.		27	
Allosaurus	3		2A, 1H
Camarasaurus	6		5A, 1J
?Apatosaurus	1		
Diplodocinae indet.	1		
Nanosaurus	4		
Dryosaurus	1		
?Camptosaurus	1		
Docodon		6	
Allodontidae indet.		1	
?Triconodonta		1	
Amblotherium		1	
Dryolestidae indet.		1	

MIA LOCALITY

Main collection method: QU
Data collection: PO, VS
Stratigraphic level: Zone 2 (?)
Paleoenvironment: FPP1
Geographic tiers: North, East

Taxa	MNI
Diplodocinae indet.	1
Theropoda	1

MILE 175

Main collection method: SW, SU
Data collection: WK, PO
Stratigraphic level: —
Paleoenvironment: FPP1
Geographic tiers: North, East

Taxa	NIS
Actinopterygii	4,198
Potamoceratodus	9
Caudata	6
Dinochelys	10
Glyptops	11
Squamata	1
Cteniogenys	8
Goniopholididae?	67
Theropoda	8
Nanosaurus	10
Docodon	7
Ctenacodon	1
Multituberculata	6

WAUGH QUARRY

Main collection method: QU
Data collection: PO
Stratigraphic level: —
Paleoenvironment: FPP1
Geographic tiers: North, East

Taxa	NIS
Allosaurus	1
Camarasaurus	2
Diplodocinae	1
Stegosaurus	1

Hot Springs County
WARM SPRINGS RANCH BS QUARRY

Main collection method: QU
Data collection: VS, PO, RE
Stratigraphic level: Zone 6
Paleoenvironment: FPP1
Geographic tiers: North, West

Taxa	MNI
Camarasaurus	3
Allosaurus	1
Diplodocidae	1

WARM SPRINGS RANCH I QUARRY

Main collection method: QU
Data collection: VS, PC
Stratigraphic level: Zone 6

Paleoenvironment: FPP1
Geographic tiers: North, West

Taxa	MNI
Stegosaurus	1
Camptosaurus	1

WARM SPRINGS RANCH RB QUARRY

Main collection method: QU
Data collection: VS, PO
Stratigraphic level: Zone 5
Paleoenvironment: FPP1
Geographic tiers: North, West

Taxa	MNI
Allosaurus	1
Apatosaurus	1
Diplodocus	2
Camarasaurus	1

WARM SPRINGS RANCH BB QUARRY

Main collection method: QU
Data collection: PO, VS
Stratigraphic level: Zone 5
Paleoenvironment: CH
Geographic tiers: North, West

Taxa	MNI	Notes
Diplodocidae	1	J

Johnson County
BUFFALO QUARRY

Main collection method: QU
Data collection: RE
Stratigraphic level: Zone 1
Paleoenvironment: —
Geographic tiers: North, Central

Taxa	MNI
Hesperosaurus	1

ELK MOUNTAIN

Main collection method: QU
Data collection: RE
Stratigraphic level: —

Paleoenvironment: —
Geographic tiers: North, Central

Taxa	MNI
Diplodocus	1
Dryosaurus	1

KAYCEE SOUTHWEST

Main collection method: QU
Data collection: RE
Stratigraphic level: —
Paleoenvironment: —
Geographic tiers: North, Central

Taxa	MNI
Allosaurus	1

POISON CREEK QUARRY

Main collection method: QU
Data collection: PO, VS, RE
Stratigraphic level: Zone 2
Paleoenvironment: FPP1
Geographic tiers: North, Central

Taxa	MNI	Notes
Allosaurus	1	
Theropoda indet.	1	
Camarasaurus	4	3A, 1J
Apatosaurus	5	4A, 1J
Diplodocus	5	2A, 3J
Haplocanthosaurus	2	
Stegosaurus	2	
Camptosaurus	2	

RED FORK POWDER RIVER QUARRY A

Main collection method: QU
Data collection: RE
Stratigraphic level: —
Paleoenvironment: —
Geographic tiers: North, Central

Taxa	MNI
Galeamopus hayi	1

RED FORK POWDER RIVER QUARRY B

Main collection method: QU
Data collection: RE, PO
Stratigraphic level: —
Paleoenvironment: —
Geographic tiers: North, Central

Taxa	MNI	Notes
Allosaurus	1	
?Camarasaurus	4	3A, 1J
?Apatosaurus	1	
?Diplodocus	1	
Haplocanthosaurus	1	

SHERIDAN COLLEGE
ALLOSAURUS QUARRY

Main collection method: QU
Data collection: VS, PO
Stratigraphic level: Zone 2
Paleoenvironment: CH
Geographic tiers: North, Central

Taxa	MNI
Allosaurus	1
Cteniogenys	1
?Nanosaurus	1

SHERIDAN COLLEGE QUARRY 1

Main collection method: QU
Data collection: PO, VS
Stratigraphic level: Zone 2
Paleoenvironment: FPP1
Geographic tiers: North, Central

Taxa	MNI	Notes
Allosaurus	1	
Camarasaurus	2	1A, 1J
Apatosaurus	1	
Diplodocus	1	
Stegosaurus	1	

SHERIDAN COLLEGE QUARRY 2

Main collection method: QU
Data collection: PO, VS
Stratigraphic level: Zone 2

Paleoenvironment: FPP1
Geographic tiers: North, Central

Taxa	MNI
Camarasaurus	1
Theropoda	1

Natrona County
ALLEN'S ALLO

Main collection method: —
Data collection: RE
Stratigraphic level: Zone 2
Paleoenvironment: CH
Geographic tiers: North, Central

Taxa	MNI
Allosaurus	1

COTTONWOOD CREEK QUARRY 1

Main collection method: —
Data collection: RE, VS, PO
Stratigraphic level: Zone 2
Paleoenvironment: CH
Geographic tiers: North, Central

Taxa	MNI
Nanosaurus	1

COTTONWOOD CREEK QUARRY 2

Main collection method: —
Data collection: RE, VS
Stratigraphic level: Zone 3
Paleoenvironment: CH
Geographic tiers: North, Central

Taxa	MNI
Camarasaurus	2

COTTONWOOD CREEK QUARRY 3

Main collection method: —
Data collection: RE, VS
Stratigraphic level: Zone 6
Paleoenvironment: CH
Geographic tiers: North, Central

Taxa	MNI
Diplodocus	1

HENRY QUARRY

Main collection method: QU
Data collection: PC, PO
Stratigraphic level: —
Paleoenvironment: —
Geographic tiers: North, Central

Taxa	MNI
Diplodocinae indet.	1

REED'S ALCOVA QUARRY

Main collection method: QU
Data collection: RE, PO
Stratigraphic level: —
Paleoenvironment: —
Geographic tiers: North, Central

Taxa	MNI
Alcovasaurus longispinus	1

Niobrara County
LANCE CREEK

Main collection method: —
Data collection: RE
Stratigraphic level: —
Paleoenvironment: —
Geographic tiers: North, East

Taxa	MNI
Dyslocosaurus (Diplodocoidea indet.)	1

Park County
PAINT CREEK

Main collection method: SU
Data collection: PO
Stratigraphic level: —
Paleoenvironment: —
Geographic tiers: North, West

Taxa	MNI
Camarasaurus	1

Washakie County
DANA QUARRY

Main collection method:	QU
Data collection:	RE
Stratigraphic level:	—
Paleoenvironment:	—
Geographic tiers:	North, Central

Taxa	MNI
Ceratosaurus	1
Allosaurus	1
Torvosaurus	1
Diplodocinae	1
Camarasaurus	1
?Brachiosaurus	1
Nanosaurus	1
Camptosaurus	1
Hesperosaurus	1

Weston County
KU QUARRY

Main collection method:	QU
Data collection:	PC
Stratigraphic level:	—
Paleoenvironment:	FPP1
Geographic tiers:	North, East

Taxa	MNI
Chelonia	4
Diplodocus	1
Camarasaurus	1

Morrison Formation taxon	Stratigraphic zone range	Mass estimate (kg)
Morrolepis	Zone 5	0.0607
Coccolepid B	Zone 1	~0.0607
Amiidae	Zones 1–4	~0.1576
Pycnodontoidea	Zone 4	~0.0300
Pholidophoriformes	Zone 1	0.0115
Semionotidae indet.	Zone 1	~0.0300
"*Hulettia*" *hawesi*	Zone 5	0.0084
cf. *Leptolepis*	Zone 5	0.0197
Actinopterygii indet.	Zones 1–6	~0.0222
Potamoceratodus	Zones 1–6	~2.9
Ceratodus robustus	Zones 2–6	~6.3
Enneabatrachus	Zone 5	~0.0230
Rhadinosteus	Zone 6	~0.0230
Pelobatidae	Zones 5–6	~0.0230
Anura indet.	Zones 2–4	~0.0230
Iridotriton	Zone 6	~0.0160

Appendix B: Morrison Formation Taxa

The table shows approximate stratigraphic range, mass estimate (sauropod estimates here not reduced 10% due to pneumaticity as elsewhere in text), specimen count (MNI or NIS, with a total sample of N = 7509), and sample ratio.

Colored groupings indicate the following:

dark blue = aquatic taxa (N = 4583)
light blue = semiaquatic taxa (N = 952)
orange = terrestrial nonmammalian micro- and mesovertebrates
        and aerial taxa (N = 320)
green = dinosaurs (N = 1082)
tan = mammals (N = 572).

Sample ratio is each taxon's count divided by the respective colored subgroup sample size.

Taxa are arranged approximately systematically except for Crocodylomorpha, which is split between the light-blue and orange subgroupings.

Mass estimate notes (specimens used and sources of methods)	Specimen count	Sample ratio (count/group total)
MWC 440; based on Allan (1982) regression of *Salvelinus fontinalis* (brook trout)	3	0.00065
From *Morrolepis*	1	0.00022
Scaled up from *Morrolepis*	267	0.05826
Average of other small taxa	1	0.00022
DMNH 63734; based on Allan (1982)	5	0.00109
Average of other small taxa	3	0.00065
MWC 5564; based on Allan (1982)	5	0.00109
MWC 3722; based on Allan (1982)	1	0.00022
DMNH 58009; based on Allan (1982)	4,200	0.91643
DMNH 40179 skull scaled up on *Neoceratodus* for total body length; based on Allan (1982)	87	0.01898
MWC 5162 jaw scaled up on *Neoceratodus* for total body length; based on Allan (1982)	10	0.00218
Estimate based on average *Rana temporaria*	3	0.00315
Estimate based on average *Rana temporaria*	10	0.01050
Estimate based on average *Rana temporaria*	1	0.00105
Estimate based on average *Rana temporaria*	12	0.01261
Scaled down from *Ambystoma tigrinum*	1	0.00105

Morrison Formation taxon	Stratigraphic zone range	Mass estimate (kg)
Caudata indet.	Zones 2–6	~0.1260
Glyptops	Zones 1–6	~6
Dinochelys	Zones 2–6	~6
Uluops	Zone 6	~6
"Dorsetochelys"	Zone 6	~6
Testudinata indet.	Zones 1–6	~6
Theriosuchus	Zone 2?	2.2
Amphicotylus	Zone 2–6	201
Eutretauranosuchus	Zones 1–6	45
"Goniopholis" felix	Zone 4	~45
Goniopholididae indet.	—	~45
Cteniogenys	Zones 2–6	~0.0815
Opisthias	Zones 2–6	0.0098
Theretairus	Zones 5–6	~0.0078
Eilenodon	Zone 4	~0.1636
Rhynchocephalia indet.	Zones 2–6	0.0078
Dorsetisaurus	Zones 2–5	~0.0080
Paramacellodus	Zones 4–5	~0.0040
Saurillodon	Zone 4	~0.0040
Schillerosaurus	Zone 5?	~0.0040
Squamata indet.	Zones 2–6	0.0040
Diablophis	Zone 4	~0.2000
Parviraptoridae indet.	Zone 4?	~0.2000
Hallopus	Zone 6	~2
Macelognathus	Zones 4–5	~2
Hoplosuchus	Zone 5	0.023
Fruitachampsa	Zone 4	~3
New shartegosuchid	Zone 1	~2
Mesadactylus	Zone 4	~1.5
Kepodactylus	Zone 4	~8.6
Dermodactylus	Zone 5	~1.5
Comodactylus	Zone 5	~8.6
Harpactognathus	Zone 2	~6
Pterosauria indet.	Zones 2–6	~5
Ceratosaurus	Zones 2–6	438
Fosterovenator	Zone 5	~20
Abelisauroidea?	Zone 4	~35
Torvosaurus	Zones 3–5	~1,950
Marshosaurus	Zones 3–5	~250
New piatnitzkysaurid	—	~200

Mass estimate notes (specimens used and sources of methods)	Specimen count	Sample ratio (count/group total)
Estimate based on average *Ambystoma tigrinum*	15	0.01576
Estimate based on average *Chelydra serpentina*	279	0.29307
Estimate based on average *Chelydra serpentina*	127	0.13340
Estimate based on average *Chelydra serpentina*	2	0.00210
Estimate based on average *Chelydra serpentina*	1	0.00105
Estimate based on average *Chelydra serpentina*	265	0.27836
MWC 5625 (proxy for skull length); based on equation in Farlow (unpublished)	1	0.00105
AMNH 5782; based on equation in Farlow (unpublished)	68	0.07143
BYU 17628; based on equation in Farlow (unpublished)	6	0.00630
Estimate based on *Eutretauranosuchus*	1	0.00105
Estimate based on *Eutretauranosuchus*	96	0.10084
Estimated snout-vent length of 15cm; based on Meiri (2010)	64	0.06723
Skull DINO 16454 scaled up to snout-vent length based on FPA skeleton and Meiri (2010)	134	0.41875
Estimate based on Rhynchocephalia indet.	2	0.00625
Estimate scaled up from Rhynchocephalia indet.	6	0.01875
Snout-vent length of FPA field specimen; based on Meiri (2010)	24	0.07500
Estimate based on Squamata indet. and larger dentary	5	0.01563
Estimate based on Squamata indet.	15	0.04687
Estimate based on Squamata indet.	2	0.00625
Estimate based on Squamata indet.	1	0.00313
Snout-vent length of UMNH 18329; based on Meiri (2010)	79	0.24688
Estimate based on Cisco Mammal Quarry dentary doubled to get approximate skull length, then skull length to mass taken from Gans (1961)	3	0.00938
Estimate based on Cisco Mammal Quarry dentary doubled to get approximate skull length, then skull length to mass taken from Gans (1961)	1	0.00313
Estimate based on general size comparison	1	0.00313
Estimate based on general size comparison	2	0.00625
Farlow (unpublished) based on CM 11361	2	0.00625
Scaled down from *Felis catus* based on skull length	21	0.06562
Scaled down from *Felis catus* based on skull length	1	0.00313
Estimate from 1 m wing span (Wellnhofer 1991); based on Roderick et al. (2017) bird data	4	0.01250
Estimate from *Diomedea epomophora*, comparable wingspan of ~2.5 m	1	0.00313
Estimate from 1 m wing span; based on Roderick et al. (2017) bird data	1	0.00313
Estimate from *Diomedea epomophora*, comparable wingspan of ~2.5 m	1	0.00313
Estimated from skull fragment size	1	0.00313
Average estimate	13	0.04062
MWC 1; based on J. Anderson et al. (1985)	18	0.01663
Estimate based on general size	1	0.00092
Estimate based on *Tanycolagreus* size	1	0.00092
Estimate from Foster (2003a)	15	0.01386
Estimate from Foster (2003a)	6	0.00554
Estimate based on *Marshosaurus*	1	0.00092

Morrison Formation taxon	Stratigraphic zone range	Mass estimate (kg)
Allosaurus fragilis	Zones 4–6	1,388
Allosaurus jimmadseni	Zones 1–4	760
Saurophaganax	Zone 5	2,720
Ornitholestes	Zone 2 (–5?)	16
Coelurus	Zones 2–5	11
Tanycolagreus	Zone 2 (–5?)	36
Stokesosaurus	Zone 5 (and 2?)	~132
Aviatyrannus-like form	Zone 2?	~3.5
Koparion	Zone 6	~3
Hesperornithoides	—	~3
Theropoda indet.	Zones 1–6	~500
Dystrophaeus	Zone 1	~10,000
Haplocanthosaurus priscus	Zones 1–3	11,592
Haplocanthosaurus delfsi	Zone 1	~17,200
Diplodocus	Zones 2–6	12,657
Galeamopus	Zone 2	13,784
Kaatedocus	Zone 2	~11,000
Barosaurus	Zones 2–5	11,957
Supersaurus	Zone 3	~40,200
Apatosaurus	Zones 2–6	34,035
Brontosaurus	Zones 2–5	24,247
Maraapunisaurus	Zone 6	~28,400
Diplodocidae indet.	Zones 2–6	~11,000
Camarasaurus spp.	Zones 2–5	18,502
Camarasaurus supremus	Zone 6	~47,000
Brachiosaurus	Zones 1–4	43,896
Stegosaurus	Zones 2–6	5,284
Alcovasaurus	—	~5,000
Hesperosaurus	Zones 1–2	~5,000
Mymoorapelta	Zones 4–5	562
Gargoyleosaurus	Zones 2–4	754
Ankylosauria indet.	Zones 1, 4	~650
Fruitadens	Zone 4	0.750
Nanosaurus (incl. *Othnielosaurus* and *Drinker*)	Zones 2–6	28.4
Dryosaurus	Zones 1–6	114
Camptosaurus	Zones 2–6	830
Neornithischia indet.	Zones 1–6	~5
Docodon	Zones 2–6	0.1407
Fruitafossor	Zone 4	0.0062
Ctenacodon	Zone 5 (2?)	0.0205
Psalodon	Zones 2–5	0.0583
Glirodon	Zones 4–6	0.0168
Zofiabaatar	Zone 6	0.0350
Morrisonodon	Zone 6	0.0205

Mass estimate notes (specimens used and sources of methods)	Specimen count	Sample ratio (count/group total)
MWC 2850; based on J. Anderson et al. (1985)	141	0.13031
MOR 693; based on J. Anderson et al. (1985)	13	0.01201
Chure (1995)	2	0.00185
AMNH 619; based on J. Anderson et al. (1985)	2	0.00185
YPM 2010; based on J. Anderson et al. (1985)	12	0.01109
TPII 2000–09-29; based on J. Anderson et al. (1985)	1	0.00092
Scaled from femora of *S. langhami* (Benson 2008), which is linearly 56% larger (447 kg); based on J. Anderson et al. (1985)	3	0.00277
Scaled down from *S. langhami* (Benson 2008), which is 5 times larger linearly (125 times the volume)	1	0.00092
Tooth only; estimate from other small Morrison theropods	1	0.00092
Estimate from other small Morrison theropods	1	0.00092
Includes mostly indeterminate large theropods	40	0.03697
Estimate from scale down of *Diplodocus*	1	0.00092
Based on FHPR 1106; using J. Anderson et al. (1985)	11	0.01017
Estimate scaled up from *H. priscus*	1	0.00092
USNM 10865; based on J. Anderson et al. (1985)	125	0.11553
CM 662; based on J. Anderson et al. (1985)	3	0.00277
Estimate based on *Diplodocus*; no limb material	1	0.00092
AMNH 6341; based on J. Anderson et al. (1985)	14	0.01294
Estimate scaled up from *Barosaurus*	2	0.00185
CM 3018; based on J. Anderson et al. (1985)	119	0.10998
YPM 1980; based on J. Anderson et al. (1985)	10	0.00924
Estimate scaled from *Diplodocus* and *Rebbachisaurus*	1	0.00092
Estimate based on *Diplodocus*	22	0.02033
SDSM 351; based on J. Anderson et al. (1985)	197	0.18207
Estimate scaled up from *Camarasaurus* sp.	8	0.00739
FMNH 25107; based on J. Anderson et al. (1985)	17	0.01571
DMNH 1483; based on J. Anderson et al. (1985); MWC 81 close at 5,212 kg based on same equation	119	0.10998
Estimate based on femur size and lack of humerus	1	0.00092
Estimate, no limb material	5	0.00462
Based on similar-sized juvenile *Gastonia* at Museum of Moab; no prepped femora	5	0.00462
DMNH 27726; based on J. Anderson et al. (1985)	2	0.00185
Estimated average	3	0.00277
From Butler et al. (2010)	2	0.00185
SDSM 30490; based on J. Anderson et al. (1985)	49	0.04529
YPM 1876; based on J. Anderson et al. (1985)	33	0.03050
YPM 1877; based on J. Anderson et al. (1985)	63	0.05822
Estimate	10	0.00924
From Foster (2009)	91	0.15909
From Foster (2009)	2	0.00349
From Foster (2009)	30	0.05245
From Foster (2009)	20	0.03496
From Foster (2009)	3	0.00524
From Foster (2009)	3	0.00524
From Foster (2009)	1	0.00175

Morrison Formation taxon	Stratigraphic zone range	Mass estimate (kg)
Multituberculata indet.	Zones 2–6	~0.0300
Triconolestes	Zone 4	0.0429
Aploconodon	Zone 5	0.0180
Comodon	Zone 5	0.0520
Priacodon	Zones 4–6	0.1105
Trioracodon	Zone 5	0.1009
Triconodonta indet.	—	~0.0646
Amphidon	Zone 5	0.0152
Tinodon	Zone 5	0.0723
Symmetrodonta indet.	—	~0.0435
Araeodon	Zone 5	0.0165
Archaeotrigon	Zone 5	0.0318
Euthlastus	Zones 5–6	~0.0318
Foxraptor	Zone 6	0.0318
Paurodon	Zone 5	0.0318
Comotherium	Zone 5	~0.0318
Pelicopsis	Zone 5	~0.0318
Tathiodon	Zones 4–5	0.0318
Paurodontidae indet.	—	~0.0318
Amblotherium	Zones 2–5	0.0192
Dryolestes	Zones 2–6	0.1037
Kepolestes	Zone 3	0.0192
Laolestes	Zones 5–6	0.0630
Melanodon	Zone 5	~0.0630
Miccylotyrans	Zone 5	~0.0510
Dryolestidae indet.	—	~0.0510
Mammalia indet.	—	~0.0270

Mass estimate notes (specimens used and sources of methods)	Specimen count	Sample ratio (count/group total)
Estimated average of multituberculate species	22	0.03846
From Foster (2009)	1	0.00175
From Foster (2009)	1	0.00175
From Foster (2009)	1	0.00175
From Foster (2009)	25	0.04371
From Foster (2009)	13	0.02273
Estimated average of triconodont species	11	0.01923
From Foster (2009)	2	0.00349
From Foster (2009)	15	0.02622
Estimated average of symmetrodont species	3	0.00524
From Foster (2009)	1	0.00175
From Foster (2009)	9	0.01573
Estimate based on *Paurodon*; no dentary (Foster 2009)	3	0.00524
From Foster (2009)	1	0.00175
From Foster (2009)	2	0.00349
Estimate based on *Paurodon*; no dentary (Foster 2009)	2	0.00349
Estimate based on *Paurodon*; no dentary (Foster 2009)	1	0.00175
From Foster (2009)	2	0.00349
Estimate	6	0.01049
From Foster (2009)	37	0.06468
From Foster (2009)	61	0.10664
From Foster (2009)	1	0.00175
From Foster (2009)	40	0.06993
Based on *Laolestes*	13	0.02273
Estimate average of dryolestids; no dentary (Foster 2009)	1	0.00175
Estimate average of dryolestids	20	0.03496
Average for Morrison mammals in Foster (2009)	128	0.22378

Glossary

accommodation space Vertical space for sediment to be deposited in a terrestrial or marine system, often caused by a tectonic drop in the basement rock creating a relative increase in sea level.

acetabulum The hip socket formed by the ilium, pubis, and ischium bones.

Actinopterygii The group name for the ray-finned, bony fish.

aerial Related to the ability to fly.

Agnatha General group name for the jawless fish.

amnion A fluid-filled inner membrane in which an embryo is enclosed; in some tetrapods, this membrane and its fluid allow shelled eggs that can be laid on dry land.

amniote A tetrapod whose embryos are enclosed in an amnion; collectively, reptiles, birds, and mammals are amniotes.

amphibian Member of the class Amphibia including frogs, salamanders, and caecilians; the name refers generally to nonamniote tetrapods, including the many large fossil amphibians.

anapsid General term for the skull condition in amniotes in which there are no temporal fenestrae; turtles and many early reptiles have the anapsid condition.

anastomosing channel Part of a river in which the interconnecting channels are relatively fixed and do not migrate laterally. The banks of anastomosing rivers are often heavily vegetated and do not erode easily.

angular A bone along the bottom of the lower jaw, behind the tooth-bearing dentary.

ankylosaur Member of the Ankylosauria, group name for the heavily armored dinosaurs with boxy, triangular skulls and thick, bony scutes embedded in the skin; named after the club-tailed *Ankylosaurus*. In the Morrison Formation, *Mymoorapelta* and *Gargoyleosaurus* are ankylosaurians.

antorbital fenestra A large opening in the skull just in front of the eyes; typical of archosaurs, including dinosaurs.

Anura Group name for the frogs.

appendicular skeleton Portion of the skeleton including the bones of the limbs and pectoral and pelvic girdles.

aquatic Living in water full-time, as with plants or animals.

arboreal Living in the trees.

archosaur Member of the Archosauria, group name for certain diapsid reptiles sharing characters including, among others, an antorbital fenestra. Crocodiles, pterosaurs, dinosaurs, and birds are archosaurs; lizards and snakes are not.

articular Bone of the lower jaw in reptiles that forms the jaw joint with the quadrate bone in the skull. The articular has been incorporated into the ear in mammals.

astragalus The primary ankle bone in dinosaurs, usually considerably larger than the calcaneum. The astragalus fits tightly against the tibia in most dinosaurs, and the bones of the foot articulate against it. Other tetrapods have astragali, but they are usually one of several ankle bones and are thus less important than they are in dinosaurs.

autopod The distal major skeletal subdivision of a tetrapod limb (hand or foot); developed from endochondral bone.

avulsion The process by which rivers occasionally change course by breaking through a levee and establishing flow through a new part of the floodplain. This is different from cutting off a single bend and rejoining the river's own channel, thus shortening its course. Avulsion at the upstream end of a delta system often creates a new active delta.

axial skeleton Portion of the skeleton including the vertebrae and associated smaller bones.

bedload Sediment transported along the bottom of a river or stream.

bentonite Volcanic ash that has been altered to clay minerals, including montmorillonite, often mixed with mud in floodplain or marine sediments.

biomass The collective mass of all individuals of a defined population of animals or plants; the group can be defined on the basis of species, guild, or other criteria as necessary. Often used to compare relative percentages of groups.

bradymetabolic Characterized by a relatively slow metabolism.

braided stream (or river) A stream or river with many interconnected channels that are characterized by fluctuating flow levels and a relatively high sediment load. Braided rivers also tend to have a high width-to-depth ratio, with the individual channels ("braids") migrating around sandbars within the overall riverbed. High flow often covers the whole river, whereas low flow exposes bars that are sandy to gravelly; the bars may contain vegetation or may be entirely barren.

browser An herbivorous animal that selectively picks foliage from trees, bushes, and low-growing plants.

calcaneum The smaller ankle bone in most dinosaurs; also present in other tetrapods.

caliche A type of pedocal in which the calcium carbonate is well developed into a very hard layer in the soil. Caliche is characteristic of relatively dry environments and restricted to modern soils.

carnivore An animal that feeds on the meat of vertebrates.

carpals Collective term for the bones of the wrist.

cartilage A relatively soft tissue that serves as connective material in the skeleton or as a precursor to bone in embryos. Sharks have skeletons composed entirely of cartilage.

cartilaginous Consisting of cartilage.

caudal rib A laterally projecting bony process attached to the centrum in caudal vertebrae; fused to the centrum in adults.

caudal vertebrae Vertebrae from the tail.

Caudata Group name for the salamanders.

cenogram Graph of rank order by body mass for herbivorous and omnivorous taxa in an ecosystem, most often used to compare modern and extinct mammalian faunas and thought to relate to vegetation structure, precipitation, and predator distributions.

centrum (pl. centra) The spool-shaped main body of a vertebra.

cervical rib An anteroposteriorly projecting bony process coming off each side of the cervical vertebrae.

cervical vertebrae The vertebrae from the neck.

channel sandstone A lens- to sheet-shaped sandstone bed preserving sediments of an ancient river.

Chelonia Group name for the turtles. See also Testudinata.

chert Microcrystalline quartz sometimes formed during geochemical replacement of limestones.

chevron Any of the fork-shaped bones embedded in the muscle of the tail, hanging below the caudal vertebrae and articulating with them at approximately the position of the intervertebral disc.

Chordata Group name for the chordates.

chordate An animal possessing a postanal tail, notochord, and dorsal hollow nerve cord.

Choristodera Group name for the reptiles called champsosaurs, all of which are extinct; includes *Cteniogenys* from the Morrison Formation.

clade A group of species descended from a single ancestral species.

cladogram A branching diagram showing the inferred phylogenetic relationships of species.

clast An object being transported in a sedimentary system, usually larger than the surrounding matrix. Pebbles, clay balls, bone fragments, and teeth can be clasts in the bedload of rivers, for example; also used for larger elements in sedimentary rocks.

coal A highly (almost purely) carbonaceous sedimentary rock composed mostly of compacted plant material.

convergence An evolutionary process by which distantly related animals develop structures similar in appearance and function to each other; the fins and tapered snouts of dolphins (mammals) and ichthyosaurs (reptiles) are one often-cited example.

convergent boundary Geologic term for a boundary between plates of the earth's crust in which the plates move toward each other; in many cases in which oceanic crust and continental crust are converging, the oceanic plate "dives" under the continental plate. These types of boundaries are behind the earthquakes and volcanoes of the Pacific Rim.

coracoid A small bone positioned at the anterior end of the shoulder blade and forming part of the articulation with the humerus of the upper arm.

coronoid A bone of the lower jaw, posterior to the dentary; may be more than one.

cranial skeleton The bones of the skull.

crepuscular Active during the twilight hours of the morning or evening; crepuscular animals include some cats, rabbits, ferrets, and mice.

crevasse-splay A thin sheet of sediment spread in a fan shape from a break in a levee in a river system.

Crocodylia Group name for modern alligators and crocodiles and descendants of their common ancestor; no crocodylomorphs of the Morrison Formation (small terrestrial forms or goniopholidids) are true Crocodylia.

Crocodylomorpha Name for a larger group including modern crocodiles and alligators plus their more distantly related and more ancient relatives; the forms from the Morrison Formation are crocodylomorphs but are not true Crocodylia (crocodiles + alligators).

crossbedded Having bedding deposited at an angle to horizontal, often indicating current flow. Beds dip in the approximate direction of flow.

cusp A cone-shaped projection on a tooth (usually in mammals); cusps may be tall and pointed or short and blunt.

dentary The tooth-bearing bone of the lower jaw.

denticle A small bump on a tooth, usually along a ridge and more often in reptiles.

diapsid General term for the skull condition in amniotes in which there are two temporal fenestrae; lizards, snakes, sphenodontids, crocodilians, dinosaurs, and birds have the diapsid condition.

Dinosauria Group name for the archosaurs of the orders Saurischia and Ornithischia (or Saurischia and Ornithoscelida); Dinosauria is defined on the basis of a number of characters of the skull, pelvis, and ankle.

diversity Count of the number of species in a defined group. Groups may be defined on the basis of higher-level taxonomic categories, guilds, and so on.

Docodonta Group name for the extinct mammals with uniquely multicusped teeth and related to *Docodon*.

dorsal rib A rib of the dorsal series of vertebrae, between the neck and pelvis.

dorsal vertebrae Vertebrae between the neck and sacrum; equivalent to the thoracic and lumbar vertebrae in mammals.

Dryolestidae Group name for the mammalian family that includes mammals closely related to *Dryolestes*. Characterized by having seven or eight molars with prominent cusps and a single posterior cusp but not a fully developed talonid basin; also characterized by unique molar roots.

Dryolestoidea Group name for the Dryolestidae and Paurodontidae and related mammals. Among mammals from the Morrison Formation, the dryolestoids are the most closely related to the ancestors of modern marsupial and placental mammals this is as close as Morrison mammals got to being "modern."

eolian Relating to windblown deposits, such as sand dunes.

Eutriconodonta Group name for the mammals possessing molars with cusps aligned in a single front-to-back row; either the three main cusps were all of similar height, or the central cusp was tallest and the anterior and posterior cusps were shorter and subequal to each other.

external naris The opening in the skull for the nostrils.

femur The upper bone in the hind leg; positioned between the hip and the knee joint.

fermentative endothermy High body temperatures resulting from the digestion of large amounts of plant material.

fibula The outer and more slender bone of the lower part of the hind leg; positioned between the knee joint and the ankle.

floodplain The flat-lying area in a river system away from the river channel; usually flooded during very high water.

frontal Paired bone along the top front part of the skull; positioned roughly between the nose and the eyes (often just above the front part of the eyes).

formation A mappable unit of (usually) sedimentary rock characterized by easily recognizable and consistent lithologies (rock types).

fossorial Burrowing; living underground.

genus A group of closely related species; the next higher taxonomic level above species.

guild A group of species, related or not, that make their living in a similar manner; often grouped by feeding categories but can be grouped any number of ways.

hemal arch See chevron.

herbivore An animal that feeds on plant material.

homeothermic Characterized by maintaining a constant body temperature.

humerus The upper bone in the front leg; positioned between the shoulder and the elbow joint.

ichnogenus/ichnotaxon A type of trace fossil equivalent to genus rank (or of any rank in the case of ichnotaxa, such as ichnospecies, the lowest taxonomic level for naming trace fossils such as footprints); the taxonomy of footprints and other trace fossils is entirely separate from that of body fossils.

ilium The bladelike bone on either side of the sacrum; the lower portion of the ilium forms the top of the hip socket (acetabulum), and the top part of the ilium is an attachment point for much of the muscle of the pelvis and hind leg.

inertial homeothermy Maintaining a relatively constant body temperature through large size and a relatively low surface-to-volume ratio.

invertivore An animal that feeds on invertebrates such as insects, worms, mollusks, and insect larvae.

ischium A paired bone in the lower and posterior part of the pelvis; articulates with the ilium above and the pubis anteriorly to form the hip socket (or acetabulum).

jugal A bone on the lower external part of the skull behind the maxilla and behind and below the orbit.

junior synonym A name lacking priority because it was assigned after another, for animals that turn out to be the same at whatever taxonomic level is being discussed; thus, *Seismosaurus* (1991) has been suggested as a possible junior synonym of *Diplodocus* (1878).

lacrimal A skull bone that borders the upper anterior border of the orbit. In *Allosaurus* and some other theropod dinosaurs, the lacrimal includes a small hornlike structure; in humans, the tear ducts pass through the lacrimal.

limestone A sedimentary rock composed mostly of calcium carbonate, usually precipitated by microorganisms in shallow marine or freshwater lake environments.

Mammalia Group name for the mammals. The general definition of a mammal among modern vertebrates centers around warm-bloodedness, live birth, hair, and mammary glands, but because direct evidence of most of those traits does not actually preserve in fossils in most cases, fossil mammals must be defined on the basis of the skeleton, and here the lines get a little more blurred; some primitive mammals, by one researcher's definition, may be "Mammaliformes" by another's.

maxilla The main tooth-bearing bone of the upper jaw in the skull. In many fish, amphibians, and reptiles, it is not the only tooth-bearing element of the skull; the vomer, palatine, and pterygoid bones also contain teeth in some cases. The maxilla is the most important in having the main teeth that oppose those of the dentary of the lower jaw.

meandering river A river whose usually single channel migrates laterally across its floodplain through multiple horseshoe-shaped bends, cutting into old floodplain deposits on the outsides of curves and depositing sand on the insides of curves.

member A stratigraphic unit of sedimentary rock within a formation that can be distinguished from other levels of the formation on the basis of dominant lithology.

metacarpals The long bones of the hand, between the wrist and fingers.

metatarsals The long bones of the foot, between the ankle and toes.

monophyletic Condition of a taxonomic group of organisms in which all species are descended from a single common ancestor. Obviously, in all natural groups this is the case. In contrast, taxonomic groups that are paraphyletic include some but not all descendants of the common ancestor. For example, because the ancestor of all birds was a reptile, Class Reptilia is paraphyletic unless all of Class Aves (birds) is included within it; a monophyletic Reptilia must include the birds.

mosaic evolution Change within a species or group in which different parts of the skeleton are modified at different rates.

mudstone Very fine-grained sedimentary rock consisting of silt- and clay-sized particles; usually blocky rather than platy (as in shales).

multituberculate Member of the Multituberculata, group name for the extinct mammals possessing molars with several rows of small cusps (all of the same height), large, rodent-like incisors, and enlarged, bladelike premolars.

nasal A paired bone from the front of the skull just behind the premaxilla and external naris and anterior to the frontal.

Neornithischia A group of ornithischians including such Morrison forms as *Nanosaurus*, "*Drinker*," and the ornithopods *Dryosaurus* and *Camptosaurus*, but not the stegosaurs, ankylosaurs, or *Fruitadens*.

neural arch Bone fused (in adults) onto the top surface of the centrum of a vertebra, enclosing the neural canal.

neural canal Tunnel-shaped passage to the top of the centrum of a vertebra, and through the neural arch, that in life contains the spinal cord.

neural spine A long projection from the top of the neural arch; in many dinosaurs, the anterior and posterior surfaces of the neural spines are highly rugose, and in life, these were attachments for ligaments holding the vertebral column together.

niche partitioning The division of one or more limiting resources among the species that share them.

nocturnal Active at night.

node A branching point on a cladogram.

occipital condyle A hemispherical projection of bone at the back of the skull, on which the skull articulates with the neck; the occipital condyle in most dinosaurs is just below the foramen magnum, where the spinal cord exits the skull.

omnivore An animal that eats both plants and other animals.

oospecies A species of fossil egg; taxonomy in fossil egg studies is, like with footprints, separate from that of bones.

orbit The opening in the skull for the eye.

Ornithischia Group name for one of the two orders of dinosaurs in traditional classification, characterized by having a backward-projecting pubis and by possessing a predentary bone. In the Morrison Formation, dinosaurs such as *Stegosaurus*, *Mymoorapelta*, *Camptosaurus*, and *Dryosaurus* are ornithischians; all are herbivorous.

Ornithopoda Group name for the bipedal, plant-eating neornithischians such as *Dryosaurus*, *Camptosaurus*, *Iguanodon*, and the later (and larger) duck-billed dinosaurs.

Ornithoscelida Group name for the combined Ornithischia and Theropoda in the new classification of Baron et al. (2017).

Osteichthyes Group name for the bony fish, including ray fins and lungfish together; does not include sharks or jawless fish.

osteoderm Any flat to bulbous bone plate embedded in the skin; most commonly found in crocodilians, ankylosaurs, and aetosaurs.

overbank Deposits or the environment away from a river channel, particularly in low-lying areas.

paedomorphosis The presence in a descendant species of characteristics more typical of juveniles of its ancestor species. This type of evolutionary change may be achieved by speeding up the timing of sexual maturity or by slowing down other development of the individual; in either case, the result is sexual maturity being attained while the animal still has "juvenile" characteristics.

palatine A paired bone medial to the maxilla in the skull; often forms most of the roof of the mouth and may contain teeth.

paleofauna A group of usually extinct animal species known from a fossil assemblage within a defined area.

paleoflora A group of usually extinct plant species known from a fossil assemblage within a defined area.

paleosol An ancient soil horizon preserved in the rock record.

paleosol carbonate A type of paleosol in the rock record in which the calcium carbonate is well developed into a very hard layer; characteristic of relatively dry environments.

paraphyletic Condition of a taxonomic group of organisms in which not all species descended from the common ancestor are included; see also monophyletic.

parietal Paired bone from the top and back of the skull.

Paurodontidae Group name for mammals of the Dryolestoidea that are distinguished from the Dryolestidae by having only three or four molars (versus seven or eight in dryolestids); individual molars of the paurodontids and dryolestids may be quite similar though the roots are very different. Paurodont means "few teeth."

pedocal Soil characterized by the accumulation of calcium carbonate; characteristic of relatively dry environments.

phalanx (pl. phalanges) Any of the individual bones of the fingers or toes.

phylogenetic systematics Classification of species based on their evolutionary affinities; this is as opposed to a traditional taxonomy-based classification.

poikilothermic Characterized by having a varying body temperature.

postcranial skeleton Collective name for all the bones posterior to the skull.

postorbital A bone of the skull behind and on the upper border of the orbit.

predentary A bone at the front of the lower jaw in ornithischian dinosaurs, usually forming the lower half of a beak.

prefrontal A paired bone of the top of the skull anterior to (not surprisingly) the frontal bone.

premaxilla A paired, often tooth-bearing bone at the very front of the skull, anterior to the maxilla and nasal; forms the anterior border of the external naris. In mammals, the premaxilla contains the incisor teeth.

procoelous Descriptive term for a vertebra in which the anterior face of the centrum is concave and the posterior is convex. This is the condition of cervical and dorsal vertebrae in many species of lizards and crocodilians; it is the opposite of opisthocoelous, in which the anterior face is convex and the posterior concave (typical of cervical and anterior dorsal vertebrae of many saurischian dinosaurs). The meaning of procoelous is easy to remember because it basically means "front hollow."

provinciality A characteristic of ecosystems in which different faunas or floras occur in different geographic areas, as opposed to homogeneity, in which separate areas are similar.

Pterodactyloidea Subgroup of the pterosaurs characterized by relatively elongate wings and skulls and shortened tails; some forms also have no teeth.

pterosaur General name for any member of the flying reptile group Pterosauria, whose wings are made of an elongate fourth digit (or "ring finger") of the hand; closely related to but outside the Dinosauria.

pterygoid A bone of the base of the skull along the midline; approximately at the back of the roof of the mouth or the top of the throat.

pubis A paired bone in the lower and anterior part of the pelvis; articulates with the ilium above and the ischium posteriorly to form the hip socket (or acetabulum). Also an attachment point for muscles of the lower abdomen, and a bone close to which passes the main nerve of the hind leg.

quadrate A bone of the lateral and back end of the skull; articulates with the lower jaw at a joint with the articular bone. In reptiles, this articulation consists of pulley-shaped processes in the quadrate; in mammals, the quadrate has become part of the middle ear.

quadratojugal A bone between the jugal in front and the quadrate in back; often near the bottom edge of the lower temporal fenestra in diapsid reptiles.

radius One of the bones of the lower part of the limb in the foreleg.

resource partitioning See niche partitioning.

rhamphorynchoids Members of the Rhamphorhynchoidea, a paraphyletic subgroup of the pterosaurs characterized by relatively short wings and long tails.

sacral vertebra One of the vertebrae fused with others in the pelvis; some dinosaurs have five fused sacral vertebrae as adults.

saltation In geology, the process of a sand grain, pebble, etc., bouncing along the bed of a river or along the surface of a sand dune; or a mode of locomotion in which an animal hops.

sandstone A sedimentary rock in which most grains are sand sized; may be very fine-grained up to pebbly and may be cemented with clay minerals, calcite, silica, etc., in pore spaces between the sand grains.

Sarcopterygii Group name for the lungfish and other bony fish with fleshy limb bases (as opposed to full ray fins).

Saurischia Group name for one of the two orders of dinosaurs in traditional classification, characterized by having a forward-projecting pubis (downward in dromaeosaurid theropods); consists of the (mostly) carnivorous theropods and the herbivorous sauropods. In the classification of Baron et al. (2017), Saurischia consists only of sauropodomorphs and herrarasaurids, but not Theropoda.

Sauropoda Subgroup of the Saurischia consisting of large, quadrupedal dinosaurs with long necks and tails; related to the prosauropods and more distantly to the theropods. All sauropods were herbivorous. Sauropods include *Brachiosaurus*, *Apatosaurus*, *Diplodocus*, and many other genera.

scansorial Adapted for climbing but not necessarily fully arboreal in habits; squirrels might be considered scansorial.

scapula The main bone of the shoulder blade; the anterior end of the scapula contains part of the socket for the humerus.

sedimentary rock Rock formed by the solidification grains of minerals or rock fragments of sizes ranging from clay and silt up to boulders. These grains are deposited in layers by natural processes on the earth's surface and later turned to rock by heat and pressure applied as a result of deep burial.

sedimentation The process of depositing the grains of minerals and rock fragments that may become sedimentary rock.

semiaquatic Living both on land and in the water. Some semiaquatic species are otherwise terrestrial animals that spend much of their time in water (e.g., crocodiles and turtles), whereas others are dependent on water for reproduction and skin health and thus stay close to water sources, even though they are also adept at moving on land (e.g., many salamanders and frogs).

shale Very fine-grained sedimentary rock with grains of silt and clay size; distinguished from mudstone by its more platy (and less blocky) character in outcrop.

species A distinct and reproductively isolated population of organisms, whose individuals can produce viable and fertile offspring only with each other.

Sphenodontia Group name for the relatives of the modern *Sphenodon* (the tuatara of New Zealand), closely related to but outside the lizards; characterized by features of the skull, including having teeth attached to the top of the jaw (as opposed being attached to the inside of it, as in lizards). Large subgroup of Rhynchocephalia.

splenial A bone of the lower jaw that lies behind and inside of the tooth-bearing dentary; sometimes projects forward along the inside of the jaw to near the symphysis, with parts of the dentary above and below.

Squamata Group name for the lizards, snakes, and their relatives.

squamosal A bone of the back of the skull; in mammals, the lower jaw articulates with the squamosal.

Stegosauria Group name for the armored dinosaurs related to *Stegosaurus*; usually characterized by plates and spikes. Other stegosaurs include *Kentrosaurus* and *Hesperosaurus*.

subsidence The sinking of a basin or other landscape area; often results in the deposition of sedimentary rock.

surangular A bone of the upper part of the lower jaw between the jaw articulation and the dentary.

suspended load Very fine-grained sediment carried almost continuously within the water column in a river or other current and not traveling along the bed. Suspended loads eventually settle out in still water.

"symmetrodontans" Members of "Symmetrodonta," a probably paraphyletic group of mammals characterized by having teeth with cusps arranged in a triangle (when viewed from above) with a tall central cusp and shorter lateral cusps of roughly equal height.

symphysis The point at which the left and right lower jaws meet.

synapsid Group name for the amniotes that all have a single, lower temporal fenestra. Mammals are synapsids, along with the sail-backed "reptile" *Dimetrodon* and its relatives.

tachymetabolic Characterized by having a fast metabolism.

talonid A short or rectangular basin, often bordered by one or more cusps, on the posterior ends of lower teeth in mammals.

temporal fenestra An opening in the skull of an amniote, lateral to or above the braincase, associated with muscle attachments.

terrestrial Living on land.

Tetrapoda Group term for four-limbed vertebrates that generally live on land; includes the amphibians and amniotes. Some tetrapods are secondarily semiaquatic (crocodiles, turtles, etc.) or aquatic (seals, whales, etc.).

Theropoda Subgroup of the Saurischia (or Ornithoscelida) consisting of small to large, bipedal, mostly carnivorous dinosaurs. All carnivorous dinosaurs were theropods, but therizinosaur theropods were omnivorous or herbivorous. Because the ancestor of birds was a theropod, birds are also members of the Theropoda—living dinosaurs.

Thyreophora Collective group name for the stegosaurs and ankylosaurs, the main armored dinosaurs; name means "shield-bearers."

tibia The main weight-bearing bone of the lower part of the hind leg; equivalent to the shinbone.

tooth plate Name for the large, flat or ridged, and pitted individual teeth of a lungfish; there are only two upper and two lower tooth plates in the lungfish skull.

transform boundary Geologic term for a boundary in the earth's crust in which plates move laterally past each other; the San Andreas Fault in California is a transform boundary between the Pacific Plate and the North American Plate, with the former sliding north past the latter.

transverse process The laterally projecting parts of the neural arch in the vertebrae. The transverse processes generally project from the base of the neural spine.

ulna One of the bones of the lower part of the forelimb between the elbow and wrist; usually more robust than the radius.

unconformity A surface between two layers of rock, representing a significant period of time missing from the rock record.

ungual The most distal phalanx in a finger or toe; often a claw.

vertebrate A chordate possessing a spinal column of bone or (rarely) cartilage.

virga Precipitation that evaporates before reaching the ground.

vomer A very thin, paired bone of the front of the skull between the roof of the mouth and the nose.

zygapophysis (pl. zygapophyses) Paired, usually oval articulation processes on the anterior and posterior parts of the neural arch; these form points where successive vertebrae articulate, in addition to the contacts between faces of the centra.

Notes

1. Rainbow Country

1. The latest part of the Precambrian has recently been named the Ediacaran period.

2. Most of North America's west coast is a convergent boundary. Convergence is occurring along the coast of Mexico and from northern California up along the Northwest and British Columbia to Alaska. The section along most of southern and central California is actually a **transform boundary** formed by the San Andreas Fault, which slides laterally, sending Los Angeles up toward San Francisco.

3. The modern volcanoes of the west coast of North America are much younger but are caused by the same process.

4. These stumps were found by Moab-area naturalist Fran Barnes.

5. 4,300 cfs is average for late summer on the Colorado River, and 300 cfs is as bad as it got during the dust bowl drought of 1934. Higher flow along the Colorado at this point is about 50,000 cfs.

6. This lowered sea level left England connected to Europe at the time, and teeth of mammoths have thus been dredged up from the bottom of the English Channel and from other continental shelves.

2. Setting the Stage

1. *Syntarsus* (*Megapnosaurus*) has recently been synonymized with *Coelophysis*, the African species being *C. rhodesiensis*. The species identity of the Kayenta form would be *C. kayentakayae*.

3. The Start of It All

1. Thunderstorms are probably the most common and dangerous of the on-the-job hazards in paleontology. The other hazards listed here are all field adventures that colleagues or I have "enjoyed" in various parts of North America. Get into the more remote corners of the globe and things get even more interesting.

2. The stories date back at least 300 years but probably began when the first Native Americans arrived from Asia at the end of the Pleistocene. Some legends involve giant lizards or other large creatures and major floods (Adrienne Mayor, pers. comm., 2004; Mayor 2005).

3. For the better part of two decades, the Morrison Formation was known by a variety of names; among the most common was Marsh's "Atlantosaurus Beds." The formation was finally officially named by Cross (1894) after the town near which Lakes found his fossil material and was better defined two years later by Eldridge (1896).

4. In fact, *D. carnegii* has been proposed to be made the genotype for *Diplodocus* instead of *D. longus*.

5. What became known as Bone Cabin Quarry may have been worked briefly by others as early as the 1880s. The first AMNH crew member to notice the site was Walter Granger in August 1897, near the end of the season.

6. It is unclear how the reporter found out about the discovery; but given that Reed, except for contacting Marsh in 1877, had never before or after 1898 sought much publicity for his finds, it is unlikely that Reed was publicity hounding. More likely the reporter got Reed's name from someone back east and contacted him.

7. Letters from Riggs to Director Skiff and other museum personnel from July 26 and September 9, 1900, plus other correspondence, in the Museum of Western Colorado Loyd Files Research Library.

8. The San Rafael Swell is an elongate uplift between Green River and Ferron (east to west) and Price and just north of Hanksville (north to south) in Utah; one crosses the swell driving west from Green River along Interstate 70.

4. Renaissance

1. In dinosaur paleontology, there sometimes can be a war of superlatives, not just with *Supersaurus* and *Seismosaurus* (*Diplodocus*) *hallorum* in the Morrison and (elsewhere) *Argentinosaurus*, *Patagotitan*, and other sauropods but now with theropods as well. I find it hard to get overly excited about what new animal was a little bigger than the previous champion, even when everyone agrees—and mostly because even if *Supersaurus*, for example, has now been dethroned several times over, all these large sauropods were almost absurdly gigantic. When dinosaurs get much over 27 m (90 ft) or over 40 metric tons (88,000 lb), it is so much beyond anything we know for land animals today that the title of longest or heaviest loses much of its meaning—they're all big.

5. Fins, Scales, and Wings

1. This name meaning is not as foreboding as is sounds; it is named after a canyon near the FPA.

2. This character may be present in some specimens of *Eutretauranosuchus*.

6. Gargantuan to Minuscule

1. Small false orcas sometimes nip at sperm whales for meat, and the sperm whales regularly survive. See Paul (1998) and Holtz et al. (2004).

2. Paleontologists have a habit of using similar names that mean specific yet different things; it is a result of the rules of taxonomy and group names being based on species names in many cases, often inherited from previous decades. In this case, the endings of each word are the key. It isn't as confusing as it sounds, but it takes a little getting used to. This is a totally separate issue from names that are simply similar yet refer to totally different groups. Anapsids (reptiles) versus anaspids (early fish) is one of my favorite combinations that confused me early on as a student.

3. Osborn and Mook (1921) also allow that *Amphicoelias* and *Diplodocus* may be the same.

4. I once visited Bill Wahl and his crew at the site during excavation of the specimen, and we got run off by some rather proximate lightning. That was one record-setting hill descent.

5. Emanuel Tschopp and I have discussed this and understand each other's taxonomic boundary placements so that in a sense we can converse in either "language" when it comes to apatosaurine relationships.

6. *Ultrasaurus* was a double name (it had already been used for what was thought to be a related form from Korea), so Jensen's would have had to have been changed anyway; it sometimes is, in the form of *Ultrasauros*, as proposed by George Olshevsky. But as explained, the type material likely belongs to *Supersaurus* anyway.

7. *Hallucigenia* is a tiny Cambrian onychophoran invertebrate fossil from the Burgess Shale of Canada that had pairs of tall spikes all along its body.

8. The "thagomizer" cartoon appeared on May 27, 1982; the term was made official by its inclusion in the glossary of the first edition of the edited paleontology book *The Complete Dinosaur*.

9. My estimate is based on measurements of a similar-sized Early Cretaceous, subadult *Gastonia* because the type material of *Mymoorapelta* has a humerus but no femur and the Cactus Park specimen of *Mymoorapelta* has an unprepped femur but no humerus.

10. The structure of the molars gave the multituberculates their name, which means essentially "many tubercles."

11. Yes, even we humans have talonids; they've simply been brought forward in each tooth and fused into the rest of our rather nondescript and squarish molars.

12. Reptiles as defined here include birds. Birds are also archosaurs; otherwise, Reptilia would be paraphyletic.

13. *Cteniogenys* is actually classified as an "archosauromorph."

14. Because there are always exceptions, the dromaeosaurid theropods (small, carnivorous dinosaurs such as *Velociraptor*) actually developed a backward-projecting pubis similar to but less extreme than that of ornithischians, but dromaeosaurids are still clearly saurischians.

15. All sauropods were herbivorous; most theropods (except the therizinosaurs, oviraptorosaurs, and ornithomimids of the Cretaceous) were fully carnivorous.

7. The Mess and the Magic

1. Some original specimens from the Morrison Formation are on display in Switzerland, England, and Japan, for example.

2. Also note that for sauropods these mass estimates were then reduced by 10% to account for extensive skeletal pneumaticity (Wedel 2005). J. Anderson et al. (1985) is not the only mass estimate method available, of course, but it and related techniques are the ones that are most easily performed on a variety of species. Computer simulations of volume and mass are useful, but few taxa are known well enough to use that method for a whole fauna, and circumferences of the femur and humerus are difficult to obtain for every species as it is.

3. *Giganotosaurus* and *Spinosaurus* are as big as or larger than *Tyrannosaurus rex* and are associated, respectively, with the sauropods *Argentinosaurus* and *Paralititan*.

4. Keep in mind that I'm including *Edmarka* in *Torvosaurus* here. Certainly if *Edmarka* is in fact a separate species, the abundance of *Torvosaurus* goes down slightly, but I just don't see as strong differences between the two theropods as Bakker does. Good, old-fashioned difference of opinion. Still, if we think of *Torvosaurus* in this case as a general category for "large megalosaurids," they are second only to *Allosaurus*.

5. As a general reference, Zone 1 extends from the base of the Morrison Formation up to about the middle of the Salt Wash Member and its equivalents, Zone 2 extends from there up to the base of the Brushy Basin Member and its equivalents (just slightly into it), Zones 3 and 4 account for approximately the lower third of the Brushy Basin Member and equivalents, and Zones 5 and 6 account for the upper two-thirds of the Brushy Basin and equivalents.

6. These numbers assume somewhat higher metabolic rates for the animals; densities would be considerably higher assuming reptilian metabolic rates.

8. Many Rivers to Cross

1. Remember from chapter 1 that anastomosing rivers have stable, unmigrating, interconnected channels with often vegetated banks.

Epilogue

1. I'm not counting *Maraapunisaurus fragillimus* here because it was a fragmentary neural arch and spine only and is now lost, but it may have been nearly as heavy as a blue whale.

2. Ankylosaurs were also to become diverse in the Early Cretaceous; stegosaurs were gone from North America in the Early Cretaceous but not from some other areas. Small ornithopods were still around through the end of the Cretaceous, but the ornithopod shift was toward iguanodontids and hadrosaurs.

3. No sooner had I finished the first draft of this chapter for the first edition several years ago than I came across Richard Dawkins's *The Ancestor's Tale* (2004). In it, he describes such a backward journey through human biological history to our earliest vertebrate (and, ultimately, single-celled) ancestors. And each relative has a unique story to tell.

References

Additional Reading Related to the Morrison Formation and Late Jurassic

Bakker, R. T. 1986. *The Dinosaur Heresies*. William Morrow, New York, 481 pp.

Barthel, K. W., Swinburne, N. H. M., and Conway Morris, S. 1990. *Solnhofen: A Study in Mesozoic Palaeontology*. Cambridge University Press, Cambridge, 233 pp.

Brett-Surman, M. K., Holtz, T. R., and Farlow, J. O., eds. 2012. *The Complete Dinosaur*. 2nd ed. Indiana University Press, Bloomington, 1,112 pp.

Brinkman, P. D. 2010. *The Second Jurassic Dinosaur Rush: Museums and Paleontology in America at the Turn of the Twentieth Century*. University of Chicago Press, Chicago, 345 pp.

Colbert, E. H. 1984. *The Great Dinosaur Hunters and Their Discoveries*. Dover, New York, 283 pp.

Hallett, M., and Wedel, M. J. 2016. *The Sauropod Dinosaurs: Life in the Age of Giants*. Johns Hopkins University Press, Baltimore, 320 pp.

Kohl, M. F., and McIntosh, J. S. 1997. *Discovering Dinosaurs in the Old West: The Field Journals of Arthur Lakes*. Smithsonian Institution Press, Washington, 198 pp.

Maier, G. 2003. *African Dinosaurs Unearthed: The Tendaguru Expeditions*. Indiana University Press, Bloomington, 380 pp.

Martin, T., and Krebs, B. 2000. *Guimarota: A Jurassic Ecosystem*. Dr. Friedrich Pfeil, Munich, 155 pp.

Paul, G. S. 2016. *The Princeton Field Guide to Dinosaurs*. 2nd ed. Princeton University Press, Princeton, NJ, 360 pp.

Psihoyos, L. 1994. *Hunting Dinosaurs*. Random House, New York, 267 pp.

Russell, D. A. 1989. *The Dinosaurs of North America: An Odyssey in Time*. NorthWord, Minocqua, WI, 239 pp.

Sampson, S. D. 2009. *Dinosaur Odyssey: Fossil Threads in the Web of Life*. University of California Press, Berkeley, 332 pp.

Wallace, D. R. 1999. *The Bonehunters' Revenge: Dinosaurs, Greed, and the Greatest Scientific Feud of the Gilded Age*. Houghton Mifflin, New York, 366 pp.

Scientific Literature Cited

Ague, J. J., Carpenter, K., and Ostrom, J. H. 1995. Solution to the *Hallopus* enigma? *American Journal of Science* 295:1–17.

Alderton, D. 1988. *Turtles and Tortoises of the World*. Facts on File, New York, 191 pp.

Allan, J. D. 1982. The effects of reduction in trout density on the invertebrate community of a mountain stream. *Ecology* 63:1444–1455.

Allen, A. 1996. Morrison Formation stratigraphy between the classic Como Bluff and Thermopolis areas, Wyoming. *In* Paleoenvironments of the Jurassic, *Tate Museum Guidebook* no. 1, Casper, WY, 19–26.

Allen, E. R. 2010. Phylogenetic assessment of goniopholidid crocodyliforms of the Morrison Formation. *Journal of Vertebrate Paleontology* (suppl.) 30:52A.

———. 2012. Analysis of North American goniopholidid crocodyliforms in a phylogenetic context. Unpublished master's thesis, University of Iowa, Iowa City.

Amaral, D. B., and Schneider, I. 2017. Fins into limbs: Recent insights from sarcopterygian fish. *Genesis* DOI:10.1002/dvg.23052.

Anderson, J. F., Hall-Martin, A., and Russell, D. A. 1985. Long-bone circumference and weight in mammals, birds and dinosaurs. *Journal of Zoology, London (A)* 207:53–61.

Anderson, O. J., and Lucas, S. G. 1995. Base of the Morrison Formation, northern New Mexico. *New Mexico Geology* 17:44–53.

———. 1996. The base of the Morrison Formation (Upper Jurassic) of northwestern New Mexico and adjacent areas. *In* Morales, M., ed., *The Continental Jurassic, Museum of Northern Arizona Bulletin* 60:443–456.

———. 1998. Redefinition of Morrison Formation (Upper Jurassic) and related San Rafael Group strata, southwestern U. S. *Modern Geology* 22:39–69.

Andres, B., Clark, J., and Xu, X. 2014. The earliest pterodactyloid and the origin of the group. *Current Biology* 24:1011–1016.

Angielczyk, K. D., and Schmitz, L. 2014. Nocturnality in synapsids predates the origin of mammals by over 100 million years. *Proceedings of the Royal Society B* 281:20141642.

Arcucci, A., Ortega, F., Pol, D., and Chiappe, L. 2013. A new crocodylomorph from the Morrison Formation (Late Jurassic: Kimmeridgian–Early Tithonian?) from the Fruita Paleontological Area, western Colorado, USA. *Journal of Vertebrate Paleontology*, Program and Abstracts, 2013, 79–80.

Armstrong, H. J., and Perry, M. L. 1985. A century of dinosaurs from the Grand Valley. *Museum Journal*, Museum of Western Colorado 2:4–19.

Ash, S. R., and Tidwell, W. D. 1998. Plant megafossils from the Brushy Basin Member of the Morrison Formation near Montezuma Creek Trading Post, southeastern Utah. *Modern Geology* 22:321–339.

Averianov, A. O., and Martin, T. 2015. Ontogeny and taxonomy of *Paurodon valens* (Mammalia, Cladotheria) from the Upper Jurassic Morrison Formation of USA. *Proceedings of the Zoological Institute of the Russian Academy of Sciences* 319:326–340.

Baars, D. L. 2000. *The Colorado Plateau: A Geologic History*. University of New Mexico Press, Albuquerque, 254 pp.

Bader, K. S., Hasiotis, S. T., and Martin, L. D. 2009. Application of forensic science techniques to trace fossils on dinosaur bones from a quarry in the Upper Jurassic Morrison Formation, northeastern Wyoming. *Palaios* 24:140–158.

Bakker, R. T. 1971. Ecology of the brontosaurs. *Nature* 229:172–174.

———. 1972. Anatomical and ecological evidence of endothermy in dinosaurs. *Nature* 238:81–85.

———. 1978. Dinosaur feeding behaviour and the origin of flowering plants. *Nature* 274:661–663.

———. 1980. Dinosaur heresy–dinosaur renaissance: Why we need endothermic archosaurs for a comprehensive theory of bioenergetic evolution. *In* Thomas, R. D. K., and Olson, E. C., eds., *A Cold Look at the Warm-Blooded Dinosaurs*, American Association for the Advancement of Science, Selected Symposium 28:351–462.

———. 1986. *The Dinosaur Heresies*. William Morrow, New York, 481 pp.

———. 1996. The real Jurassic Park: Dinosaurs and habitats at Como Bluff, Wyoming. *In* Morales, M., ed., *The Continental Jurassic, Museum of Northern Arizona Bulletin* 60:35–49.

———. 1998a. Dinosaur mid-life crisis: The Jurassic–Cretaceous transition in Wyoming and Colorado. *New Mexico Museum of Natural History and Science Bulletin* 14:67–77.

———. 1998b. Brontosaur killers: Late Jurassic allosaurids as sabre-tooth cat analogues. *Gaia* 15:145–158.

———. 2017. Stegosaur martial arts: A Morrison Formation allosaur stabbed and killed by *Stegosaurus*. *Journey to the Jurassic: Exploring the Morrison Formation*, 10th Founders Symposium Program and Abstracts, Western Interior Paleontological Society, 26–29.

Bakker, R. T., and Bir, G. 2004. Dinosaur crime scene investigations: Theropod behavior at Como Bluff, Wyoming, and the evolution of birdness. *In* Currie, P. J., Koppelhus, E. B., Shugar, M. A., and Wright, J. L., eds., *Feathered Dragons*, Indiana University Press, Bloomington, 301–342.

Bakker, R. T., and Mossbrucker, M. T. 2016. Thunder lizard song and dance: Skulls from *Brontosaurus* and *Apatosaurus* show marked divergence in vocalization and upright combat. *Geological Society of America Abstracts with Programs* 48(7), DOI: 10.1130/abs/2016AM-282844.

Bakker, R. T., Carpenter, K., Galton, P., Siegwarth, J., and Filla, J. 1990. A new latest Jurassic vertebrate fauna, from the highest levels of the Morrison Formation at Como Bluff, Wyoming, with comments on Morrison biochronology. *Hunteria* 2(6):1–19.

Bakker, R. T., Kralis, D., Siegwarth, J., and Filla, J. 1992. *Edmarka rex*, a new, gigantic theropod dinosaur from the middle Morrison Formation, Late Jurassic, of the Como Bluff outcrop region. *Hunteria* 2(9):1–24.

Barnes, F. A. 1988. *Canyonlands National Park Early History and First Descriptions*. Canyon Country Publications, Moab, UT, 160 pp.

———. 1989. *Hiking the Historic Route of the 1859 Macomb Expedition*. Canyon Country Publications, Moab, UT, 48 pp.

———. 2003. *The 1859 Macomb Expedition into Utah Territory*. Canyon Country Publications, Moab, UT, 144 pp.

Baron, M. G., Norman, D. B., and Barrett, P. M. 2017. A new hypothesis of dinosaur relationships and early dinosaur evolution. *Nature* 543:501–506.

Barthel, K. W., Swinburne, N. H. M., and Morris, S. C. 1990. *Solnhofen—a Study in Mesozoic Paleontology*. Cambridge University Press, New York, 236 pp.

Bartleson, B. L., and Jensen, J. A. 1988. The oldest (?) Morrison Formation dinosaur, Gunnison, Colorado. *Mountain Geologist* 25:129–139.

Behrensmeyer, A. K., Western, D., and Boaz, D. E. D. 1979. New perspectives in vertebrate paleoecology from a recent bone assemblage. *Paleobiology* 5:12–21.

Behrensmeyer, A. K., Damuth, J. D., DiMichele, W. A., Potts, R., Sues, H.-D., and Wing, S. L., eds. 1992. *Terrestrial Ecosystems through Time*. University of Chicago Press, Chicago, 568 pp.

Bennett, S. C. 2007. Reassessment of *Utahdactylus* from the Jurassic Morrison Formation of Utah. *Journal of Vertebrate Paleontology* 27:257–260.

Benson, R. B. J. 2008. New information on *Stokesosaurus*, a tyrannosauroid (Dinosauria: Theropoda) from North America and the United Kingdom. *Journal of Vertebrate Paleontology* 28:732–750.

Benson, R. B. J., Hunt, G., Carrano, M. T., and Campione, N. 2018. Cope's rule and the adaptive landscape of dinosaur body size evolution. *Palaeontology* 61:13–48.

Berman, D. S, and McIntosh, J. S. 1978. Skull and relationships of the Upper Jurassic sauropod *Apatosaurus* (Reptilia, Saurischia). *Bulletin of Carnegie Museum of Natural History*, no. 8, 35 pp.

Bertog, J. L. 2013. Updates and overview of the taphonomy of the Aaron Scott Quarry (Morrison Formation, Jurassic period), central Utah. *Geological Society of America Abstracts with Programs* 45(7):65.

Bertog, J., Jeffery, D. L., Coode, K., Hester, W. B., Robinson, R. R., and Bishop, J. 2014. Taphonomic patterns of a dinosaur accumulation in a lacustrine delta system in the Jurassic Morrison Formation, San Rafael Swell, Utah, USA. *Palaeontologia Electronica* 17(3):36A.

Bilbey, S. A. 1998. Cleveland-Lloyd Dinosaur Quarry—age, stratigraphy, and depositional environments. *Modern Geology* 22:87–120.

———. 1999. Taphonomy of the Cleveland-Lloyd Dinosaur Quarry in the Morrison Formation, central Utah—a lethal spring-fed pond. *In* Gillette, D. D., ed., *Vertebrate Paleontology in Utah*, Utah Geological Survey Miscellaneous Publication 99–1:121–133.

Bilbey, S. A., Hall, J. E., and Hall, D. A. 2000. Preliminary results on a new haplocanthosaurid sauropod dinosaur from the lower Morrison Formation of northeastern Utah. *Journal of Vertebrate Paleontology* 20(supp. 3):30A.

Bird, R. T. 1985. *Bones for Barnum Brown*. Texas Christian University Press, Fort Worth, 225 pp.

Bjork, P. R. 1983. The dinosaurs of South Dakota. *Proceedings of the South Dakota Academy of Sciences* 62:209–210.

Bonnan, M. F. 2004. Morphometric analysis of humerus and femur shape in Morrison sauropods: Implications for functional morphology and paleobiology. *Paleobiology* 30:444–470.

———. 2005. Pes anatomy in sauropod dinosaurs: Implications for functional morphology, evolution, and phylogeny. *In* Tidwell, V., and Carpenter, K., eds., *Thunder Lizards: The Sauropodomorph Dinosaurs*, Indiana University Press, Bloomington, 346–380.

———. 2007. Linear and geometric morphometric analysis of long bone scaling patterns in Jurassic neosauropod dinosaurs: Their functional and paleobiological implications. *Anatomical Record* 290:1089–1111.

Bonnan, M. F., and Wedel, M. J. 2004. First occurrence of *Brachiosaurus* (Dinosauria: Sauropoda) from the Upper Jurassic Morrison Formation of Oklahoma. *PaleoBios* 24:13–21.

Boyd, C. A. 2015. The systematic relationships and biogeographic history of ornithischian dinosaurs. *PeerJ* 3:e1523. DOI:10.7717/peerj.1523.

Bray, E. S., and Hirsch, K. F. 1998. Eggshell from the Upper Jurassic Morrison Formation. *Modern Geology* 23:219–240.

Breithaupt, B. H. 1990. Biography of William Harlow Reed: The story of a frontier fossil collector. *Earth Sciences History* 9:6–13.

———. 1998. Railroads, blizzards, and dinosaurs: A history of collecting in the Morrison Formation of Wyoming during the nineteenth century. *Modern Geology* 23:441–463.

Breithaupt, B. H., and Matthews, N. A. 2017. Precise 3D photogrammetry reveals new information on pterosaur ichnotaxonomy and terrestrial locomotion: Revisiting the ichnoholotype of *Pteraichnus saltwashensis*. *Journal of Vertebrate Paleontology*, Program and Abstracts, 2017, 86.

———. 2018. New visualizations of the three-dimensional, terrestrial world of the "dragon reptiles": Pterosaur tracks and photogrammetric ichnology. *Flugsaurier 2018: Los Angeles, Abstracts*, 19–22.

Brett-Surman, M., Jabo, S., Kroehler, P., Carrano, M., and Kvale, E. 2005. A new microvertebrate assemblage from the Upper Jurassic Morrison Formation, including mammals, theropods, and sphenodontians [abstract]. *Journal of Vertebrate Paleontology* 25(3):39A.

Brezinski, D. K., and Kollar, A. D. 2018. Origin of the Carnegie Quarry sandstone (Morrison Formation, Jurassic) at Dinosaur National Monument, Jensen, Utah. *Palaios* 33:94–105.

Briggs, D. E. G., Erwin, D. H., and Collier, F. J. 1994. *The Fossils of the Burgess Shale*. Smithsonian Institution Press, Washington, DC, 238 pp.

Brinkman, D. B., Stadtman, K., and Smith, D. 2000. New material of *Dinochelys whitei* Gaffney, 1979, from the Dry Mesa Quarry (Morrison Formation, Jurassic) of Colorado. *Journal of Vertebrate Paleontology* 20:269–274.

Brinkman, P. D. 2010. *The Second Jurassic Dinosaur Rush: Museums and Paleontology in America at the Turn of the Twentieth Century*. University of Chicago Press, Chicago, 345 pp.

Britt, B. B. 1991. Theropods of Dry Mesa Quarry (Morrison Formation, Late Jurassic), Colorado, with emphasis on the osteology of *Torvosaurus tanneri*. *Brigham Young University Geology Studies* 37:1–72.

———. 1997. Postcranial pneumaticity. *In* Currie, P. J., and Padian, K., eds., *Encyclopedia of Dinosaurs*, Academic Press, San Diego, 590–593.

Britt, B. B., and Naylor, B. G. 1994. An embryonic *Camarasaurus* (Dinosauria, Sauropoda) from the Upper Jurassic Morrison Formation (Dry Mesa Quarry, Colorado). *In* Carpenter, K., Hirsch, K. F., and Horner, J. R., eds., *Dinosaur Eggs and Babies*, Cambridge University Press, New York, 265–278.

Britt, B. B., Scheetz, R. D., and Dangerfield, A. 2008. A suite of dermestid beetle traces on dinosaur bone from the Upper Jurassic Morrison Formation, Wyoming, USA. *Ichnos* 15:59–71.

Bromley, R. G., Buatois, L. A., Genise, J. F., Labandeira, C. C., Mángano, M. G., Melchor, R. N., Schlirf, M., and Uchman, A. 2007. Comments on the paper "Reconnaissance of Upper Jurassic Morrison Formation ichnofossils, Rocky Mountain Region, USA: Paleoenvironmental, stratigraphic, and paleoclimatic significance of terrestrial and freshwater ichnocoenoses" by Stephen T. Hasiotis. *Sedimentary Geology* 200:141–150.

Bronzo, K. M., Fricke, H., Hoerner, M. E., Foster, J. R. and Lundstrom, C. 2017. Using oxygen, carbon and strontium isotope ratios of tooth enamel from dinosaurs to infer patterns of movement over the Late Jurassic landscape of CO, UT and WY. *Geological Society of America Abstracts with Programs* 49(6): DOI:10.1130/abs/2017AM-303693.

Brown, B. 1935. Sinclair dinosaur expedition, 1934. *Natural History* 36:2–15.

Brown, J. H. 1995. *Macroecology*. University of Chicago Press, Chicago, 284 pp.

Brown, J. T. 1972. The flora of the Morrison Formation (Upper Jurassic) of central Montana. PhD dissertation, University of Montana, Missoula.

———. 1975. Upper Jurassic and Lower Cretaceous ginkgophytes from Montana. *Journal of Paleontology* 49:724–730.

Brusatte, S. L., and Sereno, P. C. 2008. Phylogeny of Allosauroidea (Dinosauria: Theropoda): Comparative analysis and resolution. *Journal of Systematic Palaeontology* 6:155–182.

Buffetaut, E., Suteethorn, V., Cuny, G., Tong, H., Le Loeuff, J., Khansubha, S., and Jongautchariyakul, S. 2000. The earliest known sauropod dinosaur. *Nature* 407:72–74.

Buffetaut, E., Martill, D., and Escuillié, F. 2004. Pterosaurs as part of a spinosaur diet. *Nature* 430:33.

Buffrénil, V. de, Farlow, J. O., and de Ricqlés, A. 1986. Growth and function of *Stegosaurus* plates: Evidence from bone histology. *Paleobiology* 12:459–473.

Buscalioni, A. D. 2017. The Gobiosuchidae in the early evolution of Crocodyliformes. *Journal of Vertebrate Paleontology* 37(3):e1324459.

Butler, R. J., Upchurch, P., and Norman, D. B. 2008. The phylogeny of the ornithischian dinosaurs. *Journal of Systematic Palaeontology* 6:1–40.

Butler, R. J., Galton, P. M., Porro, L. B., Chiappe, L. M., Henderson, D. M., and Erickson, G. M. 2010. Lower limits of ornithischian dinosaur body size inferred from a new Upper Jurassic heterodontosaurid from North America. *Proceedings of the Royal Society B* 277:375–381.

Butler, R. J., Porro, L. B., Galton, P. M., and Chiappe, L. M. 2012. Anatomy and cranial functional morphology of the small-bodied dinosaur *Fruitadens haagarorum* from the Upper Jurassic of the USA. *PLoS ONE* 7(4):e31556.

Button, D. J., Rayfield, E. J., and Barrett, P. M. 2014. Cranial biomechanics underpins high sauropod diversity in resource-poor environments. *Proceedings of the Royal Society B* 281:20142114.

Button, D. J., Barrett, P. M., and Rayfield, E. J. 2016. Comparative cranial myology and biomechanics of *Plateosaurus* and *Camarasaurus* and evolution of the sauropod feeding apparatus. *Palaeontology* 59:887–913.

Bykowski, R. J. 2017. The evolution of carnivorous dinosaurs in response to changing prey body size. *Journal of Vertebrate Paleontology*, Program and Abstracts, 2017, 91.

Caldwell, M. W., Nydam, R. L., Palci, A., and Apesteguía, S. 2015. The oldest known snakes from the Middle Jurassic–Lower Cretaceous provide insights on snake evolution. *Nature Communications* 6:5996. DOI:10.1038/ncomms6996.

Callison, G. 1987. Fruita: A place for wee fossils. *In* Averett, W. R., ed., *Paleontology and Geology of the Dinosaur Triangle*, Museum of Western Colorado, Grand Junction, 91–96.

Callison, G., and Quimby, H. M. 1984. Tiny dinosaurs: Are they fully grown? *Journal of Vertebrate Paleontology* 3:200–209.

Calvo, J. O. 1994. Jaw mechanics in sauropod dinosaurs. *Gaia* 10:183–194.

Campbell, N. A. 1993. *Biology*. 3rd ed. Benjamin/Cummings Publishing, New York, 1,190 pp.

Campione, N. E., and Evans, D. C. 2012. A universal scaling relationship between body mass and proximal limb bone dimensions in quadrupedal terrestrial tetrapods. *BMC Biology* 10:60

Campione, N. E., Evans, D. C., Brown, C. M., and Carrano, M. T. 2014. Body mass estimation in non-avian bipeds using a theoretical conversion to quadrupedal stylopodial proportions. *Methods in Ecology and Evolution* 5:913–923.

Carballido, J. L., and Sander, P. M. 2014. Postcranial axial skeleton of *Europasaurus holgeri* (Dinosauria, Sauropoda) from the Upper Jurassic of Germany: Implications for sauropod ontogeny and phylogenetic relationships of basal Macronaria. *Journal of Systematic Palaeontology* 12:335–387.

Carballido, J. L., Marpmann, J. S., Schwarz-Wings, D., and Pabst, B. 2012. New information on a juvenile sauropod specimen from the Morrison Formation and reassessment of its systematic position. *Palaeontology* 55:567–582.

Carpenter, K. 1994. Baby *Dryosaurus* from the Upper Jurassic Morrison Formation of Dinosaur National Monument. *In* Carpenter, K., Hirsch, K. F., and Horner, J. R., eds., *Dinosaur Eggs and Babies*, Cambridge University Press, New York, 288–297.

———. 1997. Ankylosaurs. *In* Farlow, J. O., and Brett-Surman, M. K., eds., *The Complete Dinosaur*, Indiana University Press, Bloomington, 307–316.

———. 1998a. Armor of *Stegosaurus stenops*, and the taphonomic history of a new specimen from Garden Park, Colorado. *Modern Geology* 23:127–144.

———. 1998b. Redescription of the multituberculate *Zofiabaatar* and the paurodont *Foxraptor* from Pine Tree Ridge, Wyoming. *Modern Geology* 23:393–405.

———. 1999. *Eggs, Nests, and Baby Dinosaurs: A Look at Dinosaur Reproduction*. Indiana University Press, Bloomington, 336 pp.

———. 2006. Biggest of the big: A critical re-evaluation of the mega-sauropod *Amphicoelias fragillimus* Cope, 1878. *New Mexico Museum of Natural History and Science Bulletin* 36:131–137.

———. 2010. Variation in a population of Theropoda (Dinosauria): *Allosaurus* from the Cleveland-Lloyd Quarry (Upper Jurassic) USA. *Paleontological Research* 14:250–259.

———. 2013. History, sedimentology, and taphonomy of the Carnegie Quarry, Dinosaur National Monument, Utah. *Annals of Carnegie Museum* 81(3):153–232.

———. 2017. Comment (Case 3700)—opposition against the proposed designation of *Diplodocus carnegii* Hatcher, 1901 as the type species of *Diplodocus* Marsh, 1878 (Dinosauria, Sauropoda). *Bulletin of Zoological Nomenclature* 74:47–49.

———. 2018. *Maraapunisaurus fragillimus*, n.g. (formerly *Amphicoelias fragillimus*), a basal rebbachisaurid from the Morrison Formation (Upper Jurassic) of Colorado. *Geology of the Intermountain West* 5:227–244.

Carpenter, K., and Galton, P. M. 2001. Othniel Charles Marsh and the myth of the eight-spiked *Stegosaurus*. *In* Carpenter, K., ed., *The Armored Dinosaurs*, Indiana University Press, Bloomington, 76–102.

———. 2018. A photo documentation of bipedal ornithischian dinosaurs from the Upper Jurassic Morrison Formation, USA. *Geology of the Intermountain West* 5:167–207.

Carpenter, K., and Lamanna, M. C. 2015. The braincase assigned to the ornithopod dinosaur *Uteodon* McDonald, 2011, reassigned to *Dryosaurus* Marsh, 1894: Implications for iguanodontian morphology and taxonomy. *Annals of Carnegie Museum* 83:149–165.

Carpenter, K., and McIntosh, J. 1994. Upper Jurassic sauropod babies from the Morrison Formation. *In* Carpenter, K., Hirsch, K. F., and Horner, J. R., eds., *Dinosaur Eggs and Babies*, Cambridge University Press, New York, 265–278.

Carpenter, K., and Smith, M. 2001. Forelimb osteology and biomechanics of *Tyrannosaurus rex*. *In* Tanke, D. H., and Carpenter, K., eds., *Mesozoic Vertebrate Life*, Indiana University Press, Bloomington, 90–116.

Carpenter, K., and Tidwell, V. 1998. Preliminary description of a *Brachiosaurus* skull from Felch Quarry 1, Garden Park, Colorado. *Modern Geology* 23:69–84.

Carpenter, K., and Wilson, Y. 2008. New species of *Camptosaurus* (Ornithopoda: Dinosauria) from the Morrison Formation (Upper Jurassic) of Dinosaur National Monument, Utah, and a biomechanical analysis of its forelimb. *Annals of Carnegie Museum* 76(4):227–263.

Carpenter, K., Miles, C., and Cloward, K. 1998. Skull of a Jurassic ankylosaur (Dinosauria). *Nature* 393:782–783.

Carpenter, K., Miles, C. A., and Cloward, K. 2001. New primitive stegosaur from the Morrison Formation, Wyoming. *In* Carpenter, K., ed., *The Armored Dinosaurs*, Indiana University Press, Bloomington, 55–75.

Carpenter, K., Unwin, D., Cloward, K., Miles, C., and Miles, C. 2003. A new scaphognathine pterosaur from the Upper Jurassic Morrison Formation of Wyoming, USA. *In* Buffetaut, E., and Mazin, J.-M., eds., *Evolution and Palaeobiology of Pterosaurs*, Geological Society Special Publication 217:45–54.

Carpenter, K., Miles, C., and Cloward, K. 2005a. New small theropod from the Upper Jurassic Morrison Formation of Wyoming. *In* Carpenter, K., ed., *The Carnivorous Dinosaurs*, Indiana University Press, Bloomington, 23–48.

Carpenter, K., Miles, C., Ostrom, J. H., and Cloward, K. 2005b. Redescription of the small maniraptoran theropods *Ornitholestes* and *Coelurus* from the Upper Jurassic Morrison Formation of Wyoming. *In* Carpenter, K., ed., *The Carnivorous Dinosaurs*, Indiana University Press, Bloomington, 49–71.

Carpenter, K., Sanders, F., McWhinney, L. A., and Wood, L. 2005c. Evidence for predator-prey relationships: Examples for *Allosaurus* and *Stegosaurus*. *In* Carpenter, K., ed., *The Carnivorous Dinosaurs*, Indiana University Press, Bloomington, 325–350.

Carrano, M. T., and Choiniere, J. 2016. New information on the forearm and manus of *Ceratosaurus nasicornis* Marsh, 1884 (Dinosauria, Theropoda), with implications for theropod forelimb evolution. *Journal of Vertebrate Paleontology* DOI:10.1080/02724634.2015.1054497.

Carrano, M. T., and Sampson, S. D. 2008. The phylogeny of Ceratosauria (Dinosauria: Theropoda). *Journal of Systematic Palaeontology* 6:183–236.

Carrano, M. T., and Velez-Juarbe, J. 2006. Paleoecology of the Quarry 9 vertebrate assemblage from Como Bluff, Wyoming (Morrison Formation, Late Jurassic). *Palaeogeography, Palaeoclimatology, Palaeoecology* 237:147–159.

Carrano, M. T., Benson, R. B. J., and Sampson, S. D. 2012. The phylogeny of Tetanurae (Dinosauria: Theropoda). *Journal of Systematic Palaeontology* 10:211–300.

Carroll, R. L. 1997. *Patterns and Processes of Vertebrate Evolution*. Cambridge University Press, New York, 448 pp.

Cassiliano, M. L. 1997. Crocodiles, tortoises, and climate: A shibboleth reexamined. *Paleoclimates* 2:47–69.

Cassiliano, M. L., and Clemens, W. A. 1979. Symmetrodonta. *In* Lillegraven, J. A., Kielan-Jaworowska, Z., and Clemens, W. A., eds., *Mesozoic Mammals—the First Two-Thirds of Mammalian History*, University of California Press, Berkeley, 150–161.

Cawley, G. C., and Janacek, G. J. 2010. On allometric equations for predicting body mass of dinosaurs. *Journal of Zoology* 280:355–361.

Chatterjee, S., and Templin, R. J. 2004. Feathered coelurosaurs from China: New light on the arboreal origin of avian flight. In Currie, P. J., Koppelhus, E. B., Shugar, M. A., and Wright, J. L., eds., *Feathered Dragons*, Indiana University Press, Bloomington, 251–281.

Chenoweth, W. L. 1998. Uranium mining in the Morrison Formation. *Modern Geology* 23:427–439.

Chiappe, L. M., Salgado, L., and Coria, R. 2001. Embryonic skulls of titanosaur sauropod dinosaurs. *Science* 293:2444–2446.

Chiappe, L. M., Schmitt, J. G., Jackson, F. D., Garrido, A., Dingus, L., and Grellet-Tinner, G. 2004. Nest structure for sauropods: Sedimentary criteria for recognition of dinosaur nesting traces. *Palaios* 19:89–95.

Chin, K., and Kirkland, J. I. 1998. Probable herbivore coprolites from the Upper Jurassic Mygatt-Moore Quarry, western Colorado. *Modern Geology* 23:249–275.

Chinsamy, A. 1995. Ontogenetic changes in the bone histology of the Late Jurassic ornithopod *Dryosaurus lettowvorbecki*. *Journal of Vertebrate Paleontology* 15:96–104.

Chinsamy-Turan, A. 2005. *The Microstructure of Dinosaur Bone: Deciphering Biology with Fine-Scale Techniques*. Johns Hopkins University Press, Baltimore, 195 pp.

Christian, A., Müller, R. H. G., Christian, G., and Preuschoft, H. 1999. Limb swinging in elephants and giraffes and implications for the reconstruction of limb movements and speed estimates in large dinosaurs. *Mitteilungen aus dem Museum für Naturkunde in Berlin, Geowissenschaftliche Reihe* 2:81–90.

Christiansen, E. H., Kowallis, B. J., Dorais, M. J., Hart, G. L., Mills, C. N., Pickard, M., and Parks, E. 2015. The record of volcanism in the Brushy Basin Member of the Morrison Formation: Implications for the Late Jurassic of North America. *Geological Society of America Special Paper* 513.

Christiansen, N. A., and Tschopp, E. 2010. Exceptional stegosaur integument impressions from the Upper Jurassic Morrison Formation of Wyoming. *Swiss Journal of Geosciences* 103(2):163–171.

Christiansen, P. 2000. Feeding mechanisms of the sauropod dinosaurs *Brachiosaurus*, *Camarasaurus*, *Diplodocus*, and *Dicraeosaurus*. *Historical Biology* 14:137–152.

Christiansen, P., and Fariña, R. A. 2004. Mass prediction in theropod dinosaurs. *Historical Biology* 16:85–92.

Chure, D. J. 1994. *Koparion douglassi*, a new dinosaur from the Morrison Formation (Upper Jurassic) of Dinosaur National Monument; the oldest troodontid (Theropoda: Maniraptora). *Brigham Young University Geology Studies* 40:11–15.

———. 1995. A reassessment of the gigantic theropod *Saurophagus maximus* from the Morrison Formation (Upper Jurassic) of Oklahoma, USA. *Sixth Symposium on Mesozoic Terrestrial Ecosystems and Biota, Short Papers*, 103–106.

———. 1998. On the orbit of theropod dinosaurs. *Gaia* 15:233–240.

———. 2000. A new species of *Allosaurus* from the Morrison Formation of Dinosaur National Monument (UT-CO) and a revision of the theropod family Allosauridae. PhD dissertation, Columbia University, New York.

———. 2001. The second record of the African theropod *Elaphrosaurus* (Dinosauria, Ceratosauria) from the Western

Hemisphere. *Neues Jahrbuch für Geologie und Paläontologie Mh.* 2001(9):565–576.

Chure, D. J., and Engelmann, G. F. 1989. The fauna of the Morrison Formation in Dinosaur National Monument. In *Mesozoic/Cenozoic Paleontology: Classic Localities, Contemporary Approaches, 28th International Geologic Congress Field Trip Guidebook*, 8–14.

Chure, D. J., and Evans, S. E. 1998. A new occurrence of *Cteniogenys*, with comments on its distribution and abundance. *Modern Geology* 23:49–55.

Chure, D. J., and Loewen, M. A. 2020. Cranial anatomy of *Allosaurus jimmadseni*, a new species from the lower part of the Morrison Formation (Upper Jurassic) of western North America. *PeerJ* 8:e7803. DOI 10.7717/peerj.7803.

Chure, D. J., and Madsen, J. H. 1998. An unusual braincase (?*Stokesosaurus clevelandi*) from the Cleveland-Lloyd Dinosaur Quarry, Utah (Morrison Formation; Late Jurassic). *Journal of Vertebrate Paleontology* 18:115–125.

Chure, D. J., Turner, C., and Peterson, F. 1994. An embryo of *Camptosaurus* from the Morrison Formation (Jurassic, Middle Tithonian) in Dinosaur National Monument, Utah. In Carpenter, K., Hirsch, K. F., and Horner, J. R., eds., *Dinosaur Eggs and Babies*, Cambridge University Press, New York, 298–311.

Chure, D. J., Britt, B. B., and Madsen, J. H. 1997. A new specimen of *Marshosaurus bicentesimus* (Theropoda) from the Morrison Formation (Late Jurassic) of Dinosaur National Monument. *Journal of Vertebrate Paleontology* 17(suppl. 3):38A.

Chure, D. J., Litwin, R., Hasiotis, S., Evanoff, E., and Carpenter, K. 1998a. The fauna and flora of the Morrison Formation. *Modern Geology* 23:507–537.

Chure, D. J., Fiorillo, A. R., and Jacobsen, A. 1998b. Prey bone utilization by predatory dinosaurs in the Late Jurassic of North America, with comments on prey bone use by dinosaurs throughout the Mesozoic. *Gaia* 15:227–232.

Chure, D. J., Litwin, R., Hasiotis, S. T., Evanoff, E., and Carpenter, K. 2006. The fauna and flora of the Morrison Formation: 2006. *New Mexico Museum of Natural History and Science Bulletin* 36:233–249.

Cifelli, R. L., and Madsen, S. K. 1998. Triconodont mammals from the medial Cretaceous of Utah. *Journal of Vertebrate Paleontology* 18:403–411.

Clack, J. A. 2002. *Gaining Ground: The Origin and Evolution of Tetrapods*. Indiana University Press, Bloomington, 369 pp.

Clarac, F., De Buffrénil, V., Brochu, C., and Cubo, J. 2017. The evolution of bone ornamentation in Pseudosuchia: Morphological constraints versus ecological adaptation. *Biological Journal of the Linnean Society* 121:395–408.

Clarac, F., De Buffrénil, V., Cubo, J., and Quilhac, A. 2018. Vascularization in ornamented osteoderms: Physiological implications in ectothermy and amphibious lifestyle in the crocodylomorphs? *Anatomical Record* 301:175–183.

Clark, J. M. 1985. A new crocodilian from the Upper Jurassic Morrison Formation of western Colorado. Master's thesis, University of California, Berkeley.

———. 1994. Patterns of evolution in Mesozoic Crocodyliformes. In Fraser, N. C., and Sues, H.-D., eds., *In the Shadow of the Dinosaurs—Early Mesozoic Tetrapods*, Cambridge University Press, New York, 84–97.

———. 2011. A new shartegosuchid crocodyliform from the Upper Jurassic Morrison Formation of western Colorado. *Zoological Journal of the Linnean Society* 163:S152–S172.

Clemens, W. A., and Kielan-Jaworowska, Z. 1979. Multitu-
berculata. *In* Lillegraven, J. A., Kielan-Jaworowska, Z., and
Clemens, W. A., eds., *Mesozoic Mammals—the First Two-
Thirds of Mammalian History*, University of California Press,
Berkeley, 99–149.

Coe, M. J., Dilcher, D. L., Farlow, J. O., Jarzen, D. M., and
Russell, D. A. 1987. Dinosaurs and land plants. *In* Friis, E.
M., Chaloner, W. G., Crane, P. R., eds., *The Origins of An-
giosperms and Their Biological Consequences*, Cambridge
University Press, New York, 225–258.

Colbert, E. H. 1984. *The Great Dinosaur Hunters and Their
Discoveries*, Dover, New York, 283 pp.

———. 1993. Feeding strategies and metabolism in elephants
and sauropod dinosaurs. *American Journal of Science*
293A:1–19.

Condon, S. M., and Peterson, F. 1986. Stratigraphy of Middle
and Upper Jurassic rocks of the San Juan Basin: Historical perspective, current ideas, and remaining problems. *In*
Turner-Peterson, C. E., Santos, E. S., and Fishman, N. S.,
eds., *A Basin Analysis Case Study: The Morrison Formation,
Grants Uranium Region, New Mexico*. American Association
of Petroleum Geologists, *Studies in Geology* 22:7–26.

Connely, M. V. 2002. Stratigraphy and paleoecology of the
Morrison Formation, Como Bluff, Wyoming. Unpublished
master's thesis, Utah State University, Logan.

———. 2006. Paleoecology of pterosaur tracks from the Upper Jurassic Sundance and lower Morrison formations in
central Wyoming. *New Mexico Museum of Natural History
and Science Bulletin* 36:199–202.

Conway Morris, S., and Caron, J.-B. 2014. A primitive fish from
the Cambrian of North America. *Nature* 512:419–422.

Cooley, J. T., and Schmitt, J. G. 1998. An anastomosed fluvial
system in the Morrison Formation (Upper Jurassic) of southwest Montana. *Modern Geology* 22:171–208.

Coombs, W. P., Jr. 1975. Sauropod habits and habitats. *Palaeogeography, Palaeoclimatology, Palaeoecology* 17:1–33.

Cooper, W. E., Jr. 1994. Prey chemical discrimination, foraging
mode, and phylogeny. *In* Vitt, L. J., and Pianka, E. R., eds.,
Lizard Ecology, Princeton University Press, Princeton, NJ,
95–116.

Cope, E. D. 1877a. On reptilian remains from the Dakota beds
of Colorado. *Proceedings of the American Philosophical
Society* 17:193–196.

———. 1877b. On a gigantic saurian from the Dakota epoch
of Colorado. *Paleontological Bulletin* 25:5–10.

Costa, F., and Mateus, O. 2019. Dacentrurine stegosaurs (Dinosauria): A new specimen of *Miragaia longicollum* from the
Late Jurassic of Portugal resolves taxonomical validity and
shows the occurrence of the clade in North America. *PLoS
ONE* 14(11):e0224263.

Croft, D. A. 2001. Cenozoic environmental change in South
America as indicated by mammalian body size distributions
(cenograms). *Diversity and Distributions* 7:271–287.

Crompton, A. W., and Jenkins, F. A., Jr. 1967. American Jurassic symmetrodonts and Rhaetic "pantotheres." *Science*
155:1006–1009.

Cross, C. W. 1894. Pike's Peak Colorado. *Geological Atlas Folio
1894, United States Geological Survey*, 8 pp.

Curry, K. A. 1999. Ontogenetic histology of *Apatosaurus* (Dinosauria: Sauropoda): New insights on growth rates and
longevity. *Journal of Vertebrate Paleontology* 19:654–665.

Curtice, B. D. 1999. A report on the utility of sauropod caudal
vertebrae for generic identification and of the first occurrence of *Apatosaurus* in Arizona. *In* McCord, R. D., and

Boaz, D., eds., *Southwest Paleontological Symposium Proceedings, Mesa Southwest Museum Bulletin* 6:39–48.

Curtice, B. D., and Stadtman, K. 2001. The demise of *Dysty-
losaurus edwini* and a revision of *Supersaurus vivianae. In*
McCord, R. D., and Boaz, D., eds., Western Association of
Vertebrate Paleontologists and Southwest Paleontological
Symposium Proceedings, *Mesa Southwest Museum Bulletin*
8:33–40.

Curtice, B. D., and Wilhite, D. R. 1996. A re-evaluation of the
Dry Mesa dinosaur quarry sauropod fauna with a description of juvenile sauropod elements. *In* Huffman, A. C., Lund,
W. R., and Godwin, L. H., eds., *Geology and Resources of
the Paradox Basin, Utah Geological Association Guidebook*
25:325–338.

Curtice, B. D., Stadtman, K. L., and Curtice, L. J. 1996. A
reassessment of *Ultrasauros macintoshi* (Jensen, 1985).
In Morales, M., ed., *The Continental Jurassic, Museum of
Northern Arizona Bulletin* 60:87–95.

Curtice, B. D., Foster, J. R., and Wilhite, D. R. 1997. A statisti-
cal analysis of sauropod limb elements [abstract]. *Journal of
Vertebrate Paleontology* 17(suppl. 3):41A.

Czerkas, S. A. 1987. A reevaluation of the plate arrangement
on *Stegosaurus stenops. In* Czerkas, S. J., and Olson, E. C.,
eds., *Dinosaurs Past and Present*, vol. 2, University of Washington Press, Seattle, 83–99.

Czerkas, S. A. 1992. Discovery of dermal spines reveals a new
look for sauropod dinosaurs. *Geology* 20:1068–1070.

Czerkas, S. A., and Ford, T. 2018. Pterosaur or diapsid? The
search for the true *Utahdactylus. Flugsaurier 2018: Los Angeles, Abstracts*, 35–36.

Czerkas, S. A., and Mickelson, D. L. 2002. The first occurrence
of skeletal pterosaur remains in Utah. *In* Czerkas, S. J.,
Feathered Dinosaurs and the Origin of Flight, Dinosaur Museum, Blanding, UT, 3–13.

Dagg, A. I., and Foster, J. B. 1976. *The Giraffe: Its Biology, Behavior, and Ecology*. Robert E. Krieger, Malabar, FL, 232 pp.

Dalman, S. G. 2014a. New data on small theropod dinosaurs
from the Upper Jurassic Morrison Formation of Como Bluff,
Wyoming, USA. *Volumina Jurassica* 12(2):181–196.

———. 2014b. Osteology of a large allosauroid theropod from
the Upper Jurassic (Tithonian) Morrison Formation of Colorado, USA. *Volumina Jurassica* 12(2):159–180.

Davis, B. M., Cifelli, R. L., and Rougier, G. W. 2018. A preliminary report of the fossil mammals from a new microvertebrate locality in the Upper Jurassic Morrison Formation,
Grand County, Utah. *Geology of the Intermountain West*
5:1–8.

Dawkins, R. 2004. *The Ancestor's Tale*. Mariner Books, Boston,
673 pp.

De Andrade, M. B., Edmonds, R., Benton, M. J., and Schouten,
R. 2011. A new Berriasian species of *Goniopholis* (Mesoeucrocodylia, Neosuchia) from England, and a review
of the genus. *Zoological Journal of the Linnean Society*
163:S66–S108.

Delcourt, R. 2018. Ceratosaur palaeobiology: New insights
on evolution and ecology of the southern rulers. *Scientific
Reports* 8:9730.

Demar, D. G., and Carrano, M. T. 2018. New rhynchocephalian
(Reptilia, Lepidosauria) material from the Upper Jurassic
Morrison Formation, north-central Wyoming, U. S. A. con-
solidates a clade of American Sphenodontinae. *Journal of
Vertebrate Paleontology*, Program and Abstracts, 2018, 114.

Demko, T. M., and Parrish, J. T. 1998. Paleoclimatic setting of the Upper Jurassic Morrison Formation. *Modern Geology* 22:283–296.

Demko, T. M., Currie, B. S., and Nicoll, K. A. 2004. Regional paleoclimatic and stratigraphic implications of paleosols and fluvial/overbank architecture in the Morrison Formation (Upper Jurassic), Western Interior, USA. *Sedimentary Geology* 167:117–137.

Derr, M. E. 1974. Sedimentary structure and depositional environment of paleochannels in the Jurassic Morrison Formation near Green River, Utah. *Brigham Young University Geology Studies* 21:3–39.

Dodson, P. 1990. Sauropod paleoecology. *In* Weishampel, D. B., Dodson, P., and Osmólska, H., eds., *The Dinosauria*, University of California Press, Berkeley, 345–401.

Dodson, P., Behrensmeyer, A. K., Bakker, R. T., and McIntosh, J. S. 1980. Taphonomy and paleoecology of the dinosaur beds of the Jurassic Morrison Formation. *Paleobiology* 6:208–232.

Douglass, G. E. 2009. *Speak to the Earth and It Will Teach You: The Life and Times of Earl Douglass, 1862–1931.* www.booksurge.com, 449 pp.

Dunagan, S. P., and Turner, C. E. 2004. Regional paleohydrologic and paleoclimatic settings of wetland/lacustrine depositional systems in the Morrison Formation (Upper Jurassic), Western Interior, USA. *Sedimentary Geology* 167:269–296.

Dunham, A. E., Overall, K. L., Porter, W. P., and Forster, C. A. 1989. Implications of ecological energetics and biophysical and developmental constraints for life-history variation in dinosaurs. *Geological Society of America Special Paper* 238:1–19.

Eldridge, G. H. 1896. Geology of the Denver Basin in Colorado. *United States Geological Survey Monograph* 27:1–556.

ElShafie, S. J. 2017. Earliest evidence of tail regeneration in a fossil squamate. *Journal of Vertebrate Paleontology*, Program and Abstracts, 2017, 108.

Engelmann, G. F., and Callison, G. 1998. Mammalian faunas of the Morrison Formation. *Modern Geology* 23:343–379.

———. 1999. *Glirodon grandis*, a new multituberculate mammal from the Upper Jurassic Morrison Formation. *In* Gillette, D. D., ed., *Vertebrate Paleontology in Utah*, Utah Geological Survey Miscellaneous Publication 99–1:161–177.

Engelmann, G. F., Chure, D. J., and Fiorillo, A. R. 2004. The implications of a dry climate for the paleoecology of the fauna of the Upper Jurassic Morrison Formation. *Sedimentary Geology* 167:297–308.

Ensom, P., and Sigogneau-Russell, D. 2000. New symmetrodonts (Mammalia, Theria) from the Purbeck Limestone Group, Lower Cretaceous, southern England. *Cretaceous Research* 21:767–779.

Erickson, B. R. 1988. Notes on the postcranium of *Camptosaurus*. *Scientific Publications of the Science Museum of Minnesota* 6:1–13.

———. 2014. History of the Poison Creek Expeditions 1976–1990, with description of *Haplocanthosaurus* post cranials and a subadult diplodocid skull. *Science Museum of Minnesota Monograph* 8:1–29.

Escaso, F., Ortega, F., Dantas, P., Malafaia, E., Silva, B., and Gasulla, J. M. 2014. A new dryosaurid ornithopod (Dinosauria, Ornithischia) from the Late Jurassic of Portugal. *Journal of Vertebrate Paleontology* 34:1102–1112.

Evans, S. E. 1989. New material of *Cteniogenys* (Reptilia: Diapsida; Jurassic) and a reassessment of the phylogenetic position of the genus. *Neues Jarhbuch für Geologie und Paläontologie* 1989:577–589.

———. 1990. The skull of *Cteniogenys*, a choristodere (Reptilia: Archosauromorpha) from the Middle Jurassic of Oxfordshire. *Zoological Journal of the Linnean Society* 99:205–237.

———. 1991. The postcranial skeleton of the choristodere *Cteniogenys* (Reptilia: Diapsida) from the Middle Jurassic of England. *Geobios* 24(2):187–199.

———. 1992. A sphenodontian (Reptilia: Lepidosauria) from the Middle Jurassic of England. *Neues Jahrbuch für Geologie und Paläontologie Mh.* 1992:449–457.

———. 1996. *Parviraptor* (Squamata: Anguimorpha) and other lizards from the Morrison Formation at Fruita, Colorado. *In* Morales, M., ed., *The Continental Jurassic, Museum of Northern Arizona Bulletin* 60:243–248.

Evans, S. E., and Chure, D. J. 1998a. Paramacellodid lizard skulls from the Jurassic Morrison Formation at Dinosaur National Monument, Utah. *Journal of Vertebrate Paleontology* 18:99–114.

———. 1998b. Morrison lizards: Structure, relationships, and biogeography. *Modern Geology* 23:35–48.

———. 1999. Upper Jurassic lizards from the Morrison Formation of Dinosaur National Monument, Utah. *In* Gillette, D. D., ed., *Vertebrate Paleontology in Utah*, Utah Geological Survey Miscellaneous Publication 99–1:151–159.

Evans, S. E., and Milner, A. R. 1993. Frogs and salamanders from the Upper Jurassic Morrison Formation (Quarry Nine, Como Bluff) of North America. *Journal of Vertebrate Paleontology* 13:24–30.

Evans, S. E., Lally, C., Chure, D. J., Elder, A., and Maisano, J. A. 2005. A Late Jurassic salamander (Amphibia: Caudata) from the Morrison Formation of North America. *Zoological Journal of the Linnean Society* 143:599–616.

Farlow, J. O. 1976. Speculations about the diet and foraging behavior of large carnivorous dinosaurs. *American Midland Naturalist* 95:186–191.

———. 1980. Predator/prey biomass ratios, community food webs and dinosaur physiology. *In* Thomas, R. D. K., and Olson, E. C., eds., *A Cold Look at the Warm-Blooded Dinosaurs*, American Association for the Advancement of Science, Selected Symposium 28:55–84.

———. 1987. Speculations about the diet and digestive physiology of herbivorous dinosaurs. *Paleobiology* 13:60–72.

———. 1990. Dinosaur energetics and thermal biology. *In* Weishampel, D. B., Dodson, P., and Osmólska, H., eds., *The Dinosauria*, University of California Press, Berkeley, 43–55.

———. 1992. Sauropod tracks and trackmakers: Integrating the ichnological and skeletal records. *Zubia* 10:89–138.

———. 1993. On the rareness of big, fierce animals: Speculations about the body sizes, population densities, and geographic ranges of predatory mammals and large carnivorous dinosaurs. *American Journal of Science* 293A:167–199.

Farlow, J. O., and Holtz, T. R., Jr. 2002. The fossil record of predation in dinosaurs. *In* Kowalewski, M., and Kelley, P. H., eds., *The Fossil Record of Predation*, Paleontological Society Papers 8:251–265.

Farlow, J. O., and Pianka, E. R. 2003. Body size overlap, habitat partitioning, and living space requirements of terrestrial vertebrate predators: Implications for the paleoecology of large theropod dinosaurs. *Historical Biology* 16:21–40.

Farlow, J. O., Thompson, C. V., and Rosner, D. E. 1976. Plates of the dinosaur *Stegosaurus*: Forced convection heat loss fins? *Science* 192:1123–1125.

Farlow, J. O., Pittman, J. G., and Hawthorne, J. M. 1989. *Brontopodus birdi*, Lower Cretaceous sauropod footprints from the U. S. Gulf Coastal Plain. *In* Gillette, D. D., and Lockley, M. G., eds., *Dinosaur Tracks and Traces*, Cambridge University Press, New York, 371–394.

Farlow, J. O., Gatesy, S. M., Holtz, T. R., Hutchinson, J. R., and Robinson, J. M. 2000. Theropod locomotion. *American Zoologist* 40:640–663.

Farlow, J. O., Hurlburt, G. R., Elsey, R. M., Britton, A. R. C., and Langston, W., Jr. 2005. Femoral dimensions and body size of *Alligator mississippiensis*: Estimating the size of extinct mesoeucrocodylians. *Journal of Vertebrate Paleontology* 25:354–369.

Farlow, J. O., Coroian, I. D., and Foster, J. R. 2010. Giants on the landscape: Modelling the abundance of megaherbivorous dinosaurs of the Morrison Formation (Late Jurassic, western USA). *Historical Biology* 22:403–429.

Ferrusquía-Villafranca, I., Jiménez-Hidalgo, E., and Bravo-Cuevas, V. M. 1996. Footprints of small sauropods from the Middle Jurassic of Oaxaca, southeastern Mexico. *In* Morales, M., ed., *The Continental Jurassic, Museum of Northern Arizona Bulletin* 60:119–126.

Filla, B. J., and Redman, P. D. 1994. *Apatosaurus yahnahpin*: A preliminary description of a new species of diplodocid dinosaur from the Late Jurassic Morrison Formation of southern Wyoming, the first sauropod dinosaur found with a complete set of "belly ribs." *Wyoming Geological Association, 44th Annual Field Conference Guidebook*, 159–178.

Fillmore, R. 2011. *Geological Evolution of the Colorado Plateau of Eastern Utah and Western Colorado*. University of Utah Press, Salt Lake City, 496 pp.

Finke, D. L., and Denno, R. F. 2004. Predator diversity dampens trophic cascades. *Nature* 429:407–410.

Fiorillo, A. R. 1994. Time resolution at Carnegie Quarry (Morrison Formation: Dinosaur National Monument, Utah): Implications for dinosaur paleoecology. *University of Wyoming Contributions to Geology* 30:149–156.

———. 1998a. Bone modification features on sauropod remains (Dinosauria) from the Freezeout Hills Quarry N (Morrison Formation) of southeastern Wyoming and their contribution to fine-scale paleoenvironmental interpretation. *Modern Geology* 23:111–126.

———. 1998b. Dental microwear patterns of the sauropod dinosaurs *Camarasaurus* and *Diplodocus*: Evidence for resource partitioning in the Late Jurassic of North America. *Historical Biology* 13:1–16.

Foster, J. R. 1992. Fossil vertebrates of the Morrison Formation (Upper Jurassic) of the Black Hills, South Dakota and Wyoming. Unpublished master's thesis, South Dakota School of Mines and Technology, Rapid City.

———. 1995. Allometric and taxonomic limb bone robustness variability in some sauropod dinosaurs [abstract]. *Journal of Vertebrate Paleontology* 15(suppl. 3):29A.

———. 1996a. Sauropod dinosaurs of the Morrison Formation (Upper Jurassic), Black Hills, South Dakota and Wyoming. *University of Wyoming Contributions to Geology* 31:1–25.

———. 1996b. Fossil vertebrate localities in the Morrison Formation (Upper Jurassic) of western South Dakota. *In* Morales, M., ed., *The Continental Jurassic, Museum of Northern Arizona Bulletin* 60:255–263.

———. 2000. Paleobiogeographic homogeneity of dinosaur faunas during the Late Jurassic in western North America. New Mexico Museum of Natural History and Science Bulletin 17:47–50.

———. 2001a. Taphonomy and paleoecology of a microvertebrate assemblage from the Morrison Formation (Upper Jurassic) of the Black Hills, Crook County, Wyoming. *Brigham Young University Geology Studies* 46:13–33.

———. 2001b. Relative abundances of the Sauropoda (Dinosauria, Saurischia) of the Morrison Formation and implications for Late Jurassic paleoecology of North America. *Mesa Southwest Museum Bulletin* 8:47–60.

———. 2003a. Paleoecological analysis of the vertebrate fauna of the Morrison Formation (Upper Jurassic), Rocky Mountain region, USA. *New Mexico Museum of Natural History and Science Bulletin* 23:1–95.

———. 2003b. New specimens of *Eilenodon* (Reptilia, Sphenodontia) from the Morrison Formation (Upper Jurassic) of Colorado and Utah. *Brigham Young University Geology Studies* 47:17–22.

———. 2005a. Evidence of size-classes and scavenging in the theropod *Allosaurus fragilis* at the Mygatt-Moore Quarry (Late Jurassic), Rabbit Valley, Colorado [abstract]. *Journal of Vertebrate Paleontology* 25(suppl. 3):59A.

———. 2005b. New juvenile sauropod material from western Colorado, and the record of juvenile sauropods from the Upper Jurassic Morrison Formation. *In* Tidwell, V., and Carpenter, K., eds., *ThunderLizards: The Sauropodomorph Dinosaurs*. Indiana University Press, Bloomington, 141–153.

———. 2005c. New sauropod dinosaur specimens found near Moab, Utah, and the sauropod fauna of the Morrison Formation. *Canyon Legacy* 55:22–27.

———. 2006. The mandible of a juvenile goniopholidid (Crocodyliformes) from the Morrison Formation (Upper Jurassic) of Wyoming. *New Mexico Museum of Natural History and Science Bulletin* 36:101–105.

———. 2009. Preliminary body mass estimates for mammalian genera of the Morrison Formation (Upper Jurassic, North America). *PaleoBios* 28:114–122.

———. 2013. Ecological segregation of the Late Jurassic stegosaurian and iguanodontian dinosaurs of the Morrison Formation in North America: Pronounced or subtle? *PalArch's Journal of Vertebrate Palaeontology* 10(3):1–11.

———. 2015a. Dangers of low sample size in studies of sauropod dinosaur species diversity: A Morrison Formation case study. *Journal of Vertebrate Paleontology*, Program and Abstracts, 2015, 126.

———. 2015b. Theropod dinosaur ichnogenus *Hispanosauropus* identified from the Morrison Formation (Upper Jurassic), western North America. *Ichnos* 22:183–191.

———. 2018. A new atoposaurid crocodylomorph from the Morrison Formation (Upper Jurassic) of Wyoming, USA. *Geology of the Intermountain West* 5:287–295.

Foster, J. R., and Chure, D. J. 1998. Patterns of theropod diversity and distribution in the Late Jurassic Morrison Formation, western USA [abstract]. *5th International Symposium on the Jurassic System*, Vancouver, Abstracts and Program, 30–31.

———. 2000. An ilium of a juvenile *Stokesosaurus* (Dinosauria, Theropoda) from the Morrison Formation (Upper Jurassic: Kimmeridgian), Meade County, South Dakota. *Brigham Young University Geology Studies* 45:5–10.

———. 2006. Hindlimb allometry in the Late Jurassic theropod dinosaur *Allosaurus*, with comments on its abundance and distribution. *New Mexico Museum of Natural History and Science Bulletin* 36:119–122.

Foster, J. R., and Heckert, A. B. 2011. Ichthyoliths and other microvertebrate remains from the Morrison Formation (Upper Jurassic) of northeastern Wyoming: A screen-washed sample indicates a significant aquatic component to the fauna. *Palaeogeography, Palaeoclimatology, Palaeoecology* 305:264–279.

Foster, J. R., and Hunt-Foster, R. K. 2011. New occurrences of dinosaur skin of two types (Sauropoda? And Dinosauria indet.) from the Late Jurassic of North America (Mygatt-Moore Quarry, Morrison Formation). *Journal of Vertebrate Paleontology* 31:717–721.

Foster, J. R., and Lockley, M. G. 1995. Tridactyl dinosaur footprints from the Morrison Formation (Upper Jurassic) of northeast Wyoming. *Ichnos* 4:35–41.

Foster, J. R., and Lockley, M. G. 1997. Probable crocodilian tracks and traces from the Morrison Formation (Upper Jurassic) of eastern Utah. *Ichnos* 5:121–129.

———. 2006. The vertebrate ichnological record of the Morrison Formation (Upper Jurassic, North America). *New Mexico Museum of Natural History and Science Bulletin* 36:203–216.

Foster, J. R., and Martin, J. E. 1994. Late Jurassic dinosaur localities in the Morrison Formation of northeastern Wyoming. *Wyoming Geological Association, 44th Annual Field Conference Guidebook*, 115–126.

Foster, J. R., and McMullen, S. K. 2017. Paleobiogeographic distribution of Testudinata and neosuchian Crocodyliformes in the Morrison Formation (Upper Jurassic) of North America: Evidence of habitat zonation? *Palaeogeography, Palaeoclimatology, Palaeoecology* 468:208–215.

Foster, J. R., and Peterson, J. E. 2016. First report of *Apatosaurus* (Diplodocidae: Apatosaurinae) from the Cleveland-Lloyd Quarry in the Upper Jurassic Morrison Formation of Utah: Abundance, distribution, paleoecology, and taphonomy of an endemic North American sauropod clade. *Palaeoworld* 25:431–443.

Foster, J. R., and Trujillo, K. C. 2000. New occurrences of *Cteniogenys* (Reptilia, Choristodera) in the Late Jurassic of Wyoming and South Dakota. *Brigham Young University Geology Studies* 45:11–18.

———. 2004. The small-vertebrate sample of the Morrison Formation: Has collecting bias hidden a significant aquatic component in the Late Jurassic? [abstract]. *Journal of Vertebrate Paleontology* 24(suppl. 3):59A–60A.

Foster, J. R., and Wedel, M. J. 2014. *Haplocanthosaurus* (Saurischia: Sauropoda) from the lower Morrison Formation (Upper Jurassic) near Snowmass, Colorado. *Volumina Jurassica* 12(2):197–210.

Foster, J. R., Lockley, M. G., and Brockett, J. 1999. Possible turtle tracks from the Morrison Formation of southern Utah. *In* Gillette, D. D., ed., *Vertebrate Paleontology in Utah*, Utah Geological Survey Miscellaneous Publication 99–1:185–191.

Foster, J. R., Hamblin, A. H., and Lockley, M. G. 2000. The oldest evidence of a sauropod dinosaur in the western United States and other important vertebrate trackways from Grand Staircase-Escalante National Monument, Utah. *Ichnos* 7:169–181.

Foster, J. R., Holtz, T. R., Jr., and Chure, D. J. 2001. Contrasting patterns of diversity and community structure in the theropod faunas of the Late Jurassic and Late Cretaceous of western North America. *Journal of Vertebrate Paleontology* 21(supp. to n. 3):A51.

Foster, J. R., Trujillo, K. C., Madsen, S. K., and Martin, J. E. 2006. The Late Jurassic mammal *Docodon*, from the Morrison Formation of the Black Hills, Wyoming: Implications for abundance and biogeography of the genus. *New*

Mexico Museum of Natural History and Science Bulletin 36:165–169.

Foster, J. R., Hunt, R. K., and King, L. R. 2007. Taphonomy of the Mygatt-Moore Quarry, a large dinosaur bonebed in the Upper Jurassic Morrison Formation of western Colorado. *Geological Society of America Abstracts with Programs* 39(6):400.

Foster, J. R., Trujillo, K. C., Frost, F., and Mims, A. L. 2015. Summary of vertebrate fossils from the Morrison Formation (Upper Jurassic) at Curecanti National Recreation Area, central Colorado. *New Mexico Museum of Natural History and Science Bulletin* 67:43–50.

Foster, J. R., McHugh, J. B., Peterson, J. E., and Leschin, M. F. 2016a. Major bonebeds in mudrocks of the Morrison Formation (Upper Jurassic), northern Colorado Plateau of Utah and Colorado. *Geology of the Intermountain West* 3:33–66.

Foster, J. R., Irmis, R. B., Trujillo, K. C., McMullen, S. K., and Gillette, D. D. 2016b. *Dystrophaeus viaemalae* Cope from the basal Morrison Formation of Utah: Implications for the origin of eusauropods in North America. *Journal of Vertebrate Paleontology*, Program and Abstracts, 2016, 138.

Foster, J. R., Trujillo, K. C., and Chamberlain, K. R. 2017. A preliminary U-Pb zircon age for the Fruita Paleontological Area microvertebrate localities, Upper Jurassic Morrison Formation, Mesa County, CO. *Journal of Vertebrate Paleontology*, Program and Abstracts, 2017, 113.

Foster, J. R., Hunt-Foster, R. K., Gorman, M. A., II, Trujillo, K. C., Suarez, C. A., McHugh, J. B., Peterson, J. E., Warnock, J. P., and Schoenstein, H. E. 2018. Paleontology, taphonomy, and sedimentology of the Mygatt-Moore Quarry, a large dinosaur bonebed in the Morrison Formation, western Colorado—implications for Upper Jurassic dinosaur preservation modes. *Geology of the Intermountain West* 5:23–93.

Foster, J. R., Pagnac, D. C., and Hunt-Foster, R. K. 2020. An unusually diverse northern biota from the Morrison Formation (Upper Jurassic), Black Hills, Wyoming. *Geology of the Intermountain West* 7:29–67.

Foth, C., Evers, S. W., Pabst, B., Mateus, O., Flisch, A., Patthey, M., and Rauhut, O. W. M. 2015. New insights into the lifestyle of *Allosaurus* (Dinosauria: Theropoda) based on another specimen with multiple pathologies. *PeerJ* 3:e940. DOI:10.7717/peerj.940.

Fricke, H. C., Hencecroth, J., and Hoerner, M. E. 2011. Lowland-upland migration of sauropod dinosaurs during the Late Jurassic epoch. *Nature* 480:513–515.

Froese, R., and Pauly, D., eds. 2005. FishBase. http://www.fishbase.org.

Gaffney, E. S. 1979. The Jurassic turtles of North America. *Bulletin of the American Museum of Natural History* 162:95–135.

———. 1984. Historical analysis of theories of chelonian relationship. *Systematic Zoology* 33:283–301.

Galiano, H., and Albersdörfer, R. 2010. *Amphicoelias "brontodiplodocus"*: A new sauropod, from the Morrison Formation, Big Horn Basin, Wyoming, with taxonomic reevaluation of *Diplodocus, Apatosaurus, Barosaurus* and other genera. *Dinosauria International Ten Sleep Report Series*, no. 1, 50 pp.

Galli, K. G. 2014. Fluvial architecture element analysis of the Brushy Basin Member, Morrison Formation, western Colorado, USA. *Volumina Jurassica* 12(2):69–106.

Galton, P. M. 1981a. A rhamphorhyncoid pterosaur from the Upper Jurassic of North America. *Journal of Paleontology* 55:1117–1122.

———. 1981b. *Dryosaurus*, a hypsilophodontid dinosaur from the Upper Jurassic of North America and Africa: Postcranial skeleton. *Paläontographica Zeitschrift* 55:271–312.

———. 1982a. Juveniles of the stegosaurian dinosaur *Stegosaurus* from the Upper Jurassic of North America. *Journal of Vertebrate Paleontology* 2:47–62.

———. 1982b. *Elaphrosaurus*, an ornithomimid dinosaur from the Upper Jurassic of North America and Africa. *Paläontologische Zeitschrift* 56:265–275.

———. 1983. The cranial anatomy of *Dryosaurus*, a hypsilophodontid dinosaur from the Upper Jurassic of North America and east Africa, with a review of hypsilophodontids from the Upper Jurassic of North America. *Geologica et Palaeontologica* 17:207–243.

———. 1990. Stegosauria. *In* Weishampel, D. B., Dodson, P., and Osmólska, H., eds., *The Dinosauria*, University of California Press, Berkeley, 435–455.

———. 1997. Stegosaurs. *In* Farlow, J. O., and Brett-Surman, M. K., eds., *The Complete Dinosaur*, Indiana University Press, Bloomington, 291–306.

———. 2007. Teeth of ornithischian dinosaurs (mostly Ornithopoda) from the Morrison Formation (Upper Jurassic) of western USA. *In* Carpenter, K., ed., *Beaks and Horns: The Ornithopod and Ceratopsian Dinosaurs*, Indiana University Press, Bloomington, p. 17–47.

———. 2010. Species of plated dinosaur *Stegosaurus* (Morrison Formation, Late Jurassic) of western USA: New type species designation needed. *Swiss Journal of Geosciences* 103:187–198.

———. 2011. *Stegosaurus* Marsh, 1877 (Dinosauria, Ornithischia): Proposed replacement of the type species with *Stegosaurus stenops* Marsh, 1887. *Bulletin of Zoological Nomenclature* 68(2):127–133.

Galton, P. M., and Carpenter, K. 2016a. The plated dinosaur *Stegosaurus longispinus* Gilmore, 1914 (Dinosauria: Ornithischia; Upper Jurassic, western USA), type species of *Alcovasaurus* n. gen. *Neues Jahrbuch fur Geologie und Paläontologie, Abhandlungen* 279:185–208.

———. 2016b. Bipedal ornithischian dinosaurs (Heterodontosauridae and Ornithopoda) from the Morrison Formation (Upper Jurassic) of the western United States of America. *Journal of Vertebrate Paleontology*, Program and Abstracts, 2016, 141–142.

Galton, P. M., and Jensen, J. A. 1979. A new large theropod dinosaur from the Upper Jurassic of Colorado. *Brigham Young University Geology Studies* 26:1–12.

Gans, C. 1961. The feeding mechanism of snakes and its possible evolution. *American Zoologist* 1:217–227.

Gao K.-Q., and Shubin, N. H. 2003. Earliest known crown-group salamanders. *Nature* 422:424–428.

Gates, T. A. 2005. The Late Jurassic Cleveland-Lloyd Dinosaur Quarry as a drought-induced assemblage. *Palaios* 20:363–375.

Gauthier, J. A. 1986. Saurischian monophyly and the origin of birds. *Memoirs of the California Academy of Sciences* 8:1–40.

Gee, C. T. 2011. Dietary options for the sauropod dinosaurs from and integrated botanical and paleobotanical perspective. *In* Klein, N., Remes, K., Gee, C. T., and Sander, P. M., eds., *Biology of the Sauropod Dinosaurs: Understanding the Life of Giants*, Indiana University Press, Bloomington, 34–56.

Gee, C. T. 2016. Emerging data on the Morrison flora: Opulent conifer forests or a hinterland xeriscape for the sauropods? *Journal of Vertebrate Paleontology*, Program and Abstracts, 2016, 143–144.

Gee, C. T., Dayvault, R. D., Stockey, R. A., and Tidwell, W. D. 2014. Greater palaeobiodiversity in conifer seed cones in the Upper Jurassic Morrison Formation of Utah, USA. *Palaeobiodiversity and Palaeoenvironments* 94:363–375.

Gee, C. T., Sprinkel, D. A., Bennis, M. B., and Gray, D. E. 2019. Silicified logs of *Agathoxylon hoodii* (Tidwell et Medlyn) comb. nov. from Rainbow Draw, near Dinosaur National Monument, Uintah County, Utah, USA, and their implications for araucariaceous conifer forests in the Upper Jurassic Morrison Formation. *Geology of the Intermountain West* 6:77–92.

Gerkema, M. P., Davies, W. I. L., Foster, R. G., Menaker, M., and Hut, R. A. 2013. The nocturnal bottleneck and the evolution of activity patterns in mammals. *Proceedings of the Royal Society B* 280:20130508.

Gierliński, G. D., and Sabath, K. 2008. Stegosaurian footprints from the Morrison Formation of Utah and their implications for interpreting other ornithischian tracks. *Oryctos* 8:29–46.

Gill, F. L., Hummel, J., Reza Sharifi, A., Lee, A. P., and Lomax, B. H. 2018. Diets of giants: The nutritional value of sauropod diet during the Mesozoic. *Palaeontology* 61:647–658.

Gillette, D. D. 1991. *Seismosaurus halli*, gen. et sp. nov., a new sauropod dinosaur from the Morrison Formation (Upper Jurassic/Lower Cretaceous) of New Mexico, USA. *Journal of Vertebrate Paleontology* 11:417–433.

———. 1994. *Seismosaurus: The Earth Shaker*. Columbia University Press, New York, 205 pp.

———. 1996a. Stratigraphic position of the sauropod *Dystrophaeus viaemalae* Cope 1877 and its evolutionary implications. *In* Morales, M., ed., *The Continental Jurassic, Museum of Northern Arizona Bulletin* 60:59–68.

———. 1996b. Origin and early evolution of the sauropod dinosaurs of North America: The type locality and stratigraphic position of *Dystrophaeus viaemalae* Cope 1877. *In* Huffman, A. C., Jr., Lund, W. R., and Godwin, L. H., eds., *Geology and Resources of the Paradox Basin*, Utah Geological Association Guidebook 25:313–324.

Gilmore, C. W. 1910. A new rhynchocephalian reptile from the Jurassic of Wyoming, with notes on the fauna of "Quarry 9." *Proceedings of the United States National Museum* 37:35–42.

———. 1914. Osteology of the armored Dinosauria in the United States National Museum, with special reference to the genus *Stegosaurus*. *United States National Museum Bulletin* 89, 136 pp.

———. 1920. Osteology of the carnivorous Dinosauria in the United States National Museum, with special reference to the genera *Antrodemus* (*Allosaurus*) and *Ceratosaurus*. *United States National Museum Bulletin* 110, 159 pp.

———. 1925. A nearly complete articulated skeleton of *Camarasaurus*, a saurischian dinosaur from the Dinosaur National Monument, Utah. *Memoirs of the Carnegie Museum* 10(3):347–384.

———. 1926. A new Aëtosaurian reptile from the Morrison Formation of Utah. *Annals of the Carnegie Museum* 16:325–342.

———. 1928. Fossil lizards of North America. *Memoirs of the National Academy of Sciences* 22(3):1–169.

———. 1936. Osteology of *Apatosaurus*, with special reference to specimens in the Carnegie Museum. *Memoirs of the Carnegie Museum* 11:175–300.

Givnish, T. J., Wong, S. C., Stuart-Williams, H., Holloway-Phillips, M., and Farquhar, G. D. 2014. Determinants of maximum tree height in *Eucalyptus* species along a rainfall gradient in Victoria, Australia. *Ecology* 95:2991–3007.

Godefroit, P., Sinitsa, S. M., McNamara, M. E., Cincotta, A. C., Dhouailly, D., and Reshetova, S. 2017. Integumentary structures in *Kulindadromeus zabaikalicus*, a basal neornithischian dinosaur from the Jurassic of Siberia. *Journal of Vertebrate Paleontology*, Program and Abstracts, 2017, 119.

Göhlich, U. B., Chiappe, L. M., Clark, J. M., and Sues, H.-D. 2005. The systematic position of the Late Jurassic alleged dinosaur *Macelognathus* (Crocodylomorpha: Sphenosuchia). *Canadian Journal of Earth Sciences* 42:307–321.

Good, S. C. 2004. Paleoenvironmental and paleoclimatic significance of freshwater bivalves in the Upper Jurassic Morrison Formation, Western Interior, USA. *Sedimentary Geology* 167:163–176.

Goodwin, M. B., Clemens, W. A., Hutchison, J. H., Wood, C. B., Zavada, M. S., Kemp, A., Duffin, C. J., and Schaff, C. R. 1999. Mesozoic continental vertebrates with associated palynostratigraphic dates from the northwestern Ethiopian plateau. *Journal of Vertebrate Paleontology* 19:728–741.

Gorman, M. A., Miller, I. M., Pardo, J. D., and Small, B. J. 2008. Plants, fish, turtles, and insects from the Morrison Formation: A Late Jurassic ecosystem near Cañon City, Colorado. *In* Raynolds, R. G., ed., *Roaming the Rocky Mountains and Environs: Geological Field Trips*, Geological Society of America Field Guide 10:295–310.

Gotelli, N. J., and Graves, G. R. 1996. *Null Models in Ecology*. Smithsonian Institution Press, Washington, DC, 368 pp.

Grande, L., and Bemis, W. E. 1998. A comprehensive phylogenetic study of amiid fishes (Amiidae) based on comparative skeletal anatomy: An empirical search for interconnected patterns of natural history. *Society of Vertebrate Paleontology Memoir* 4, 690 pp.

Gunga, H.-C., Kirsh, K., Rittweger, J., Röcker, L., Clarke, A., Albertz, J., Wiedemann, A., Mokry, S., Suthau, T., Wehr, A., Heinrich, W.-D., and Schultze, H.-P. 1999. Body size and body volume distribution in two sauropods from the Upper Jurassic of Tendaguru (Tanzania). *Mitteilungen aus dem Museum für Naturkunde in Berlin, Geowissenschaftliche Reihe* 2:31–102.

Gunga, H.-C., Suthau, T., Bellmann, A., Stoinski, S., Friedrich, A., Trippel, T., Kirsch, K., and Hellwich, O. 2008. A new body mass estimation of *Brachiosaurus brancai* Janensch, 1914 mounted and exhibited at the Museum of Natural History (Berlin, Germany). *Fossil Record* 11:33–38.

Hahn, G., and Hahn, R. 2004. The dentition of the Plagiaulacida (Multituberculata, Late Jurassic to Early Cretaceous). *Geologica et Palaeontologica* 38:119–159.

Hairston, N. G., Sr. 1987. *Community Ecology and Salamander Guilds*. Cambridge University Press, New York, 230 pp.

Hallam, A. 1994. Jurassic climates as inferred from the sedimentary and fossil record. *In* Allen, J. R. L., Hoskins, B. J., Sellwood, B. W., Spicer, R. A., and Valdes, P. J., eds., *Palaeoclimates and Their Modelling with Special Reference to the Mesozoic Era*, Chapman and Hall, London, 79–88.

Hallett, M., and Wedel, M. J. 2016. *The Sauropod Dinosaurs: Life in the Age of Giants*. Johns Hopkins University Press, Baltimore, 320 pp.

Hanna, R. R. 2002. Multiple injury and infection in a sub-adult theropod dinosaur *Allosaurus fragilis* with comparisons to allosaur pathology in the Cleveland-Lloyd Dinosaur Quarry collection. *Journal of Vertebrate Paleontology* 22:76–90.

Haq, B. U., Hardenbol, J., and Vail, P. R. 1987. Chronology of fluctuating sea levels since the Triassic. *Science* 235:1156–1167.

Harris, J. D. 2006. The significance of *Suuwassea emilieae* (Dinosauria: Sauropoda) for flagellicaudatan intrarelationships and evolution. *Journal of Systematic Palaeontology* 4:185–198.

Harris, J. D., and Carpenter, K. 1996. A large pterodactyloid from the Morrison Formation (Late Jurassic) of Garden Park, Colorado. *Neues Jahrbuch für Geologie und Paläontologie Mh.* 1996:473–484.

Harris, J. D., and Dodson, P. 2004. A new diplodocoid sauropod dinosaur from the Upper Jurassic Morrison Formation of Montana, USA. *Acta Palaeontologica Polonica* 49:197–210.

Hartman, S. 1996. Biomechanics of the arms of allosaurids and the implications for behavior. *In* Paleoenvironments of the Jurassic, *Tate Museum Guidebook* no. 1, Casper, WY, 88–98.

Hartman, S., Mortimer, M., and Lovelace, D. 2017. New information on a paravian theropod from the Morrison Formation. *Journal of Vertebrate Paleontology*, Program and Abstracts, 2017, 126.

Hartman, S., Mortimer, M., Wahl, W. R., Lomax, D., Lippincott, J., and Lovelace, D. M. 2019. A new paravian dinosaur from the Late Jurassic of North America supports a late acquisition of avian flight. *PeerJ* 7:e7247. DOI:10.7717/peerj.7247.

Hasiotis, S. T. 2004. Reconnaissance of Upper Jurassic Morrison Formation ichnofossils, Rocky Mountain Region, USA: Paleoenvironmental, stratigraphic, and paleoclimatic significance of terrestrial and freshwater ichnocoenoses. *Sedimentary Geology* 167:177–268.

Hasiotis, S. T., and Demko, T. M. 1996a. Terrestrial and freshwater trace fossils, Upper Jurassic Morrison Formation, Colorado Plateau. *In* Morales, M., ed., *The Continental Jurassic, Museum of Northern Arizona Bulletin* 60:355–370.

———. 1996b. Ant (Hymenoptera: Formicidae) nest ichnofossils, Upper Jurassic Morrison Formation, Colorado Plateau: Evolutionary and ecologic implications [abstract]. *Geological Society of America Abstracts with Programs* 28(7):A106.

Hatcher, J. B. 1901. *Diplodocus* (Marsh): Its osteology, taxonomy, and probable habits, with a restoration of the skeleton. *Memoirs of the Carnegie Museum* 1(1):1–63.

———. 1903. Osteology of *Haplocanthosaurus*, with description of a new species, and remarks on the probable habits of the Sauropoda and the age and origin of the Atlantosaurus beds. *Memoirs of the Carnegie Museum* 2(1):1–72.

Hay, O. P. 1908. The fossil turtles of North America. *Carnegie Institution of Washington Publication* 75:1–586.

Hecht, M. K. 1970. The morphology of *Eodiscoglossus*, a complete Jurassic frog. *American Museum Novitates* 2424:1–17.

Hecht, M. K., and Estes, R. 1960. Fossil amphibians from Quarry Nine. *Postilla* 46:1–19.

Heck, K. L., Jr., van Belle, G., and Simberloff, D. 1975. Explicit calculation of the rarefaction diversity measurement and the determination of sufficient sample size. *Ecology* 56:1459–1461.

Heckert, A. B., Zeigler, K. E., Lucas, S. G., Spielmann, J. A., Hester, P. M., Peterson, Ronald E., Peterson, Rodney E., and D'Andrea, N. V. 2003a. Geology and paleontology of the Upper Jurassic (Morrison Formation: Brushy Basin Member) Peterson Quarry, central New Mexico. *New Mexico Geological Society Guidebook, 54th Field Conference, Geology of the Zuni Plateau*, 315–324.

Heckert, A. B., Spielmann, J. A., Lucas, S. G., Altenberg, R., and Russell, D. M. 2003b. An Upper Jurassic theropod dinosaur from the Section 19 Mine, Morrison Formation, Grants Uranium District. *New Mexico Geological Society Guidebook, 54th Field Conference, Geology of the Zuni Plateau*, 309–314.

Hedrick, B. P., Tumarkin-Deratzian, A. R., and Dodson, P. 2014. Bone microstructure and relative age of the holotype specimen of the diplodocoid sauropod dinosaur *Suuwassea emilieae*. *Acta Palaeontologica Polonica* 59:295–304.

Heesy, C. P., and Hall, M. I. 2010. The nocturnal bottleneck and the evolution of mammalian vision. *Brain, Behavior and Evolution* 75:195–203.

Heller, P. L., Ratigan, D., Trampush, S., Noda, A., McElroy, B., Drever, J., Huzurbazar, S. 2015. Origins of bimodal stratigraphy in fluvial deposits: An example from the Morrison Formation (Upper Jurassic), western U. S. A. *Journal of Sedimentary Research* 85:1466–1477.

Henderson, D. M. 1998. Skull and tooth morphology as indicators of niche partitioning in sympatric Morrison Formation theropods. *Gaia* 15:219–226.

Henrici, A. C. 1997. The frog fauna from the Late Jurassic Morrison Formation at the Rainbow Park microsite, Utah [abstract]. *Journal of Vertebrate Paleontology* 17(suppl. 3):52A.

———. 1998a. New anurans from the Rainbow Park microsite, Dinosaur National Monument, Utah. *Modern Geology* 23:1–16.

———. 1998b. A new pipoid anuran from the Late Jurassic Morrison Formation at Dinosaur National Monument, Utah. *Journal of Vertebrate Paleontology* 18:321–332.

Herne, M. C., and Lucas, S. G. 2006. *Seismosaurus hallorum*: Osteological reconstruction from the holotype. *New Mexico Museum of Natural History and Science Bulletin* 36:139–148.

Hildebrand, M. 1995. *Analysis of Vertebrate Structure*. John Wiley, New York, 657 pp.

Hirsch, K. F. 1994. Upper Jurassic eggshells from the Western Interior of North America. *In* Carpenter, K., Hirsch, K. F., and Horner, J. R., eds., *Dinosaur Eggs and Babies*, Cambridge University Press, New York, 137–150.

Holland, W. J. 1915. A new species of *Apatosaurus*. *Annals of the Carnegie Museum* 10:143–145.

———. 1924. The skull of *Diplodocus*. *Memoirs of the Carnegie Museum* 9:379–403.

Holman, J. A. 2003. *Fossil Frogs and Toads of North America*. Indiana University Press, Bloomington, 246 pp.

Holtz, T. R., Jr. 1994. The phylogenetic position of the Tyrannosauridae: Implications for theropod systematics. *Journal of Paleontology* 68:1100–1117.

———. 1995. Adaptive trends in major subgroups of theropods and related taxa. *Journal of Vertebrate Paleontology* 15(suppl. 3):35A.

———. 1998. A new phylogeny of the carnivorous dinosaurs. *Gaia* 15:5–61.

———. 2003. Dinosaur predation, evidence and ecomorphology. *In* Kelley, P. H., Kowalewski, M., and Hansen, T. A., eds., *Predator-Prey Interactions in the Fossil Record*, Kluwer Academic/Plenum, New York, 325–340.

Holtz, T. R., Jr., Brinkman, D. L., and Chandler, C. L. 1994. Denticle morphometrics and a possibly different life habit for the theropod dinosaur *Troodon*. *Journal of Vertebrate Paleontology* 14(3):30A.

Holtz, T. R., Jr., Molnar, R. E., and Currie, P. J. 2004. Basal Tetanurae. *In* Weishampel, D. B., Dodson, P., and Osmólska, H., eds., *The Dinosauria*, 2nd ed., University of California Press, Berkeley, 71–110.

Hopp, T. P., and Orsen, M. J. 2004. Dinosaur brooding behavior and the origin of flight feathers. *In* Currie, P. J., Koppelhus, E. B., Shugar, M. A., and Wright, J. L., eds., *Feathered Dragons*, Indiana University Press, Bloomington, 234–250.

Hotton, C. L., and Baghai-Riding, N. L. 2010. Palynological evidence for conifer dominance within a heterogeneous landscape in the Late Jurassic Morrison Formation, U. S. A. *In* Gee, C. T., ed., *Plants in Mesozoic Time: Morphological Innovations, Phylogeny, Ecosystems*, Indiana University Press, Bloomington, 295–328.

———. 2016. Palynology of the Late Jurassic Morrison Formation: New insights into floristics, paleoclimate, phytogeography, and tetrapod herbivory. *Journal of Vertebrate Paleontology*, Program and Abstracts, 2016, 157.

Hu Y., Wang Y., Luo Z., and Li C. 1997. A new symmetrodont mammal from China and its implications for mammalian evolution. *Nature* 390:137–142.

Huene, F. von. 1904. *Dystrophaeus viaemalae* Cope in neuer Beleuchtung. *Neues Jahrbuch fur Mineralogie, Geologie, und Palaeontologie, Beil-Bd* 19:319–333.

Hunt, A. P., and Lucas, S. G. 1987. J. W. Stovall and the Mesozoic of the Cimarron Valley, Oklahoma and New Mexico. *New Mexico Geological Society Guidebook, 38th Field Conference*, 139–151.

———. 2006. A small theropod dinosaur from the Upper Jurassic of eastern New Mexico with a checklist of small theropods from the Morrison Formation of western North America. *New Mexico Museum of Natural History and Science Bulletin* 36:115–118.

Hunt, T. C., and Richmond, D. R. 2018. The aquatic vertebrate community of a bone-dry pond: The historic Stovall Quarry 8, Morrison Formation in the panhandle of Oklahoma. *Journal of Vertebrate Paleontology*, Program and Abstracts, 2018, 152–153.

Hunt, A. P., Lucas, S. G., Krainer, K., and Spielmann, J. 2006. The taphonomy of the Cleveland-Lloyd Dinosaur Quarry, Upper Jurassic Morrison Formation, Utah: A re-evaluation. *New Mexico Museum of Natural History and Science Bulletin* 36:57–65.

Hurlbert, S. H. 1971. The nonconcept of species diversity: A critique and alternative parameters. *Ecology* 52:577–586.

Huttenlocker, A. K., Grossnickle, D. M., Kirkland, J. I., Schultz, J. A., and Luo Z.-X. 2018. Late-surviving stem mammal links the lowermost Cretaceous of North America and Gondwana. *Nature* DOI:10.1038/s41586–018–0126-y.

ICZN. 2013. Opinion 2320 (Case 3536): *Stegosaurus* Marsh, 1877 (Dinosauria, Ornithischia): Type species replaced with *Stegosaurus stenops* Marsh, 1887. *Bulletin of Zoological Nomenclature* 70(2):129–130.

Ikejiri, T. 2005. Distribution and biochronology of *Camarasaurus* (Dinosauria, Sauropoda) from the Jurassic Morrison Formation of the Rocky Mountain region. *New Mexico Geological Society, 56th Field Conference Guidebook, Geology of the Chama Basin*, 367–379.

Ikejiri, T., Watkins, P. S., and Gray, D. J. 2006. Stratigraphy, sedimentology, and taphonomy of a sauropod quarry from the upper Morrison Formation of Thermopolis, central Wyoming. *New Mexico Museum of Natural History and Science Bulletin* 36:39–46.

Imlay, R. W. 1980. Jurassic paleobiogeography of the conterminous United States in its continental setting. *United States Geological Survey Professional Paper* 1062, 134 pp.

Irmis, R. B., Nesbitt, S. J., Padian, K., Smith, N. D., Turner, A. H., Woody, D., and Downs, A. 2007. A Late Triassic dinosauromorph assemblage from New Mexico and the rise of dinosaurs. *Science* 317:358–361.

Janensch, W. 1920. Über *Elaphrosaurus bambergi* und die Megalosaurier aus den Tendaguru-Schichten Deutsch-Ostafrikas. *Sitzungsber. Ges. Naturforsch*. Freunde Berlin 1920:225–235.

Janzen, D. H. 1976. The depression of reptile biomass by large herbivores. *American Naturalist* 110:371–400.

Jenkins, F. A., Jr. 1969. Occlusion in *Docodon* (Mammalia, Docodonta). *Postilla* 139:1–24.

Jenkins, F. A., Jr., and Crompton, A. W. 1979. Triconodonta. *In* Lillegraven, J. A., Kielan-Jaworowska, Z., and Clemens, W. A., eds., *Mesozoic Mammals—the First Two-Thirds of Mammalian History*, University of California Press, Berkeley, 74–90.

Jennings, D. S., Platt, B. F., and Hasiotis, S. T. 2006. Distribution of vertebrate trace fossils, Upper Jurassic Morrison Formation, Bighorn Basin, Wyoming, USA: Implications for differentiating paleoecological and preservational bias. *New Mexico Museum of Natural History and Science Bulletin* 36:183–192.

Jensen, J. A. 1985. Three new sauropod dinosaurs from the Upper Jurassic of Colorado. *Great Basin Naturalist* 45:697–709.

———. 1987. New brachiosaur material from the Late Jurassic of Utah and Colorado. *Great Basin Naturalist* 47:592–608.

———. 1988. A fourth new sauropod dinosaur from the Upper Jurassic of the Colorado Plateau and sauropod bipedalism. *Great Basin Naturalist* 48:121–145.

Jensen, J. A., and Padian, K. 1989. Small pterosaurs and dinosaurs from the Uncompahgre Fauna (Brushy Basin Member, Morrison Formation: ?Tithonian), Late Jurassic, western Colorado. *Journal of Paleontology* 63:364–373.

Ji Q., Luo Z., and Ji S. 1999. A Chinese triconodont mammal and mosaic evolution of the mammalian skeleton. *Nature* 398:326–330.

Ji Q., Luo Z., Yuan C., Wible, J. R., Zhang J., and Georgi, J. A. 2002. The earliest known eutherian mammal. *Nature* 416:816–822.

Johnstone, W. 2017. *Hey! What Do You Do with A Dinosaur? The San Ysidro Dinosaur*. www.willjohnstone.com.

Jones, M. E. H. 2009. Dentary tooth shape in *Sphenodon* and its fossil relatives (Diapsida: Lepidosauria: Rhynchocephalia). *Comparative Dental Morphology* 13:9–15.

Jones, M. E. H., Bever, G. S., Foster, J. R., Evans, S. E., Sertich, J. J. W., and Carrano, M. T. 2014. A new look at rhynchocephalian reptiles from the Late Jurassic Morrison Formation. *Mid-Mesozoic: The Age of Dinosaurs in Transition*, abstracts, 61.

Jones, M. E. H., Watson, A. P., Sertich, J. W., Foster, J. R., Garbe, U., and Salvemini, F. 2016. Neutron computed tomography succeeds where x-ray computed tomography fails: Enamel thickness in the herbivorous rhynchocephalian *Eilenodon* from the Late Jurassic Morrison Formation of USA. *Journal of Vertebrate Paleontology*, Program and Abstracts, 2016, 163.

Jones, M. E. H., Lucas, P. W., Tucker, A. S., Watson, A. P., Sertich, J. J. W., Foster, J. R., Williams, R., Garbe, U., Bevitt, J. J., and Salvemini, F. 2018. Neutron scanning reveals unexpected complexity in the enamel thickness of an herbivorous Jurassic reptile. *Journal of the Royal Society Interface* 15:20180039.

Jones, T. D., Farlow, J. O., Ruben, J. A., Henderson, D. M., and Hillenius, W. J. 2000. Cursoriality in bipedal archosaurs. *Nature* 406:716–718.

Joyce, W. G. 2007. Phylogenetic relationships of Mesozoic turtles. *Bulletin of the Peabody Museum of Natural History* 48(1):3–102.

Joyce, W. G., Lucas, S. G., Scheyer, T. M., Heckert, A. B., and Hunt, A. P. 2008. A thin-shelled reptile from the Late Triassic of North America and the origin of the turtle shell. *Proceedings of the Royal Society B* DOI:10.1098/rspb.2008.1196.

Kielan-Jaworowska, Z., and Hurum, J. H. 2001. Phylogeny and systematics of multituberculate mammals. *Palaeontology* 44:389–429.

Kielan-Jaworowska, Z., Cifelli, R. L., and Luo Z.-X. 2004. *Mammals from the Age of Dinosaurs*. Columbia University Press, New York, 630 pp.

Kilbourne, B., and Carpenter, K. 2005. Redescription of *Gargoyleosaurus parkpinorum*, a polacanthid ankylosaur from the Upper Jurassic of Albany County, Wyoming. *Neues Jahrbuch für Geologie und Paläontologie Abh.* 237:111–160.

King, L. R., and Heckert, A. B. 2018. A preliminary analysis of the Mogan Site, a new Late Jurassic vertebrate microfossil site in the Morrison Formation in northeastern Wyoming. *Journal of Vertebrate Paleontology*, Program and Abstracts, 2018, 159.

King, L. R., Foster, J. R., and Scheetz, R. D. 2006. New pterosaur specimens from the Morrison Formation and a summary of the Late Jurassic pterosaur record of the Rocky Mountain region. *New Mexico Museum of Natural History and Science Bulletin* 36:109–113.

Kirkland, J. I. 1987. Upper Jurassic and Cretaceous lungfish tooth plates from the Western Interior, the last dipnoan faunas of North America. *Hunteria* 2(2):1–16.

———. 1994. Predation of dinosaur nests by terrestrial crocodilians. *In* Carpenter, K., Hirsch, K. F., and Horner, J. R., eds., *Dinosaur Eggs and Babies*, Cambridge University Press, New York, p. 124–133.

———. 1998. Morrison fishes. *Modern Geology* 22:503–533.

———. 2006. Fruita Paleontological Area (Upper Jurassic, Morrison Formation), western Colorado: An example of terrestrial taphofacies analysis. *New Mexico Museum of Natural History and Science Bulletin* 36:67–95.

Kirkland, J. I., and Carpenter, K. 1994. North America's first pre-Cretaceous ankylosaur (Dinosauria) from the Upper Jurassic Morrison Formation of western Colorado. *Brigham Young University Geology Studies* 40:25–42.

Kirkland, J. I., Carpenter, K., Hunt, A. P., and Scheetz, R. D. 1998. Ankylosaur (Dinosauria) specimens from the Upper Jurassic Morrison Formation. *Modern Geology* 23:145–177.

Kirkland, J. I., Scheetz, R. D., and Foster, J. R. 2005. Jurassic and Lower Cretaceous dinosaur quarries of western Colorado and eastern Utah. *In* Richard, G., comp., *2005 Rocky Mountain Section of the Geological Society of America Field Trip Guidebook*, Grand Junction Geological Society, Field Trip 402:1–26.

Kirkland, J. I., Carpenter, K., and Morgan, K. 2016. The Morrison Formation's ankylosaurs. *Journal of Vertebrate Paleontology*, Program and Abstracts, 2016, 167.

Kirkland, J. I., DeBlieux, D. D., Hunt-Foster, R. K., Foster, J. R., Trujillo, K. C., and Finzel, E. 2020. The Morrison Formation and its bounding strata on the western side of the Blanding basin, San Juan County, Utah. *Geology of the Intermountain West* 7:137195.

Koch, A. L., Frost, F., and Trujillo, K. C. 2006. Paleontological discoveries at Curecanti National Recreations Area and Black Canyon of the Gunnison National Park, Upper Jurassic Morrison Formation, Colorado. *New Mexico Museum of Natural History and Science Bulletin* 36:35–38.

Kohl, M. F., and McIntosh, J. S. 1997. *Discovering Dinosaurs in the Old West—the Field Journals of Arthur Lakes*. Smithsonian Institution Press, Washington, DC, 198 pp.

Kowallis, B. J., Christiansen, E. H., Deino, A. L., Peterson, F., Turner, C. E., Kunk, M. J., and Obradovich, J. D. 1998. The age of the Morrison Formation. *Modern Geology* 22:235–260.

Kraus, M. J. 1979. Eupantotheria. *In* Lillegraven, J. A., Kielan-Jaworowska, Z., and Clemens, W. A., eds., *Mesozoic Mammals—the First Two-Thirds of Mammalian History*, University of California Press, Berkeley, 162–171.

Krause, D. W. 1982. Jaw movement, dental function, and diet in the Paleocene multituberculate *Ptilodus. Paleobiology* 8:265–281.

Krebs, B. 1971. Evolution of the mandible and lower dentition in dryolestids (Pantotheria, Mammalia). *In* Kermack, D. M., and Kermack, K. A., eds., *Early Mammals, Zoological Journal of the Linnean Society* 50(suppl. 1):89–102.

———. 1991. Das Skelett von *Henkelotherium guimarotae* gen. et sp. nov. (Eupantotheria, Mammalia) aus dem Oberen Jura von Portugal. *Berliner Geowissenschaftliche Abhandlungen* 133, 110 pp.

Kron, D. G. 1979. Docodonta. *In* Lillegraven, J. A., Kielan-Jaworowska, Z., and Clemens, W. A., eds., *Mesozoic Mammals—the First Two-Thirds of Mammalian History*, University of California Press, Berkeley, 91–98.

Langer, M. C., Ezcurra, M. D., Rauhut, O. W. M., Benton, M. J., Knoll, F., McPhee, B. W., Novas, F. E., Pol, D., and Brusatte, S. L. 2017. Untangling the dinosaur family tree. *Nature* DOI:10.1038/nature24011.

Lara, M. B., Foster, J. R., Kirkland, J. I., and Howells, T. F. 2020. First fossil true water bugs (Heteroptera, Nepomorpha) from Upper Jurassic strata of North America (Morrison Formation, southeastern Utah). *Historical Biology* doi.org/10.1080/08912963.2020.1755283.

Larson, G. 2003. *The Complete Far Side: Volume One, 1980–1986*. Andrews McMeel, Kansas City, MO, 644 pp.

Leardi, J. M., Pol, D., and Clark, J. M. 2017. Detailed anatomy of the braincase of *Macelognathus vagans* Marsh, 1884 (Archosauria, Crocodylomorpha) using high resolution tomography and new insights on basal crocodylomorph phylogeny. *PeerJ* 5:e2801. DOI:10.7717/peerj.2801.

Legendre, S. 1986. Analysis of mammalian communities from the late Eocene and Oligocene of southern France. *Palaeovertebrata* 16(4):191–212.

Lillegraven, J. A., and Krusat, G. 1991. Cranio-mandibular anatomy of *Haldanodon exspectatus* (Docodonta; Mammalia) from the Late Jurassic of Portugal and its implications to the evolution of mammalian characters. *University of Wyoming Contributions to Geology* 28:39–138.

Lockley, M. G. 1991. *Tracking Dinosaurs—a New Look at an Ancient World*. Cambridge University Press, New York, 238 pp.

Lockley, M. G., and Foster, J. R. 2006. Dinosaur and turtle tracks from the Morrison Formation (Upper Jurassic) of Colorado National Monument, with observations on the taxonomy of vertebrate swim tracks. *New Mexico Museum of Natural History and Science Bulletin* 36:193–198.

———. 2010. An assemblage of probable crocodylian traces and associated dinosaur tracks from the lower Morrison Formation (Upper Jurassic) of eastern Utah. New Mexico Museum of Natural History and Science Bulletin 51:93–97.

———. 2017. A review of vertebrate tracks from the Morrison Formation, western USA. *Journey to the Jurassic: Exploring the Morrison Formation*, 10th Founders Symposium Program and Abstracts, Western Interior Paleontological Society, 38–40.

Lockley, M. G., and Hunt, A. P. 1995. *Dinosaur Tracks and Other Fossil Footprints of the Western United States*. Columbia University Press, New York, 338 pp.

———. 1998. A probable stegosaur track from the Morrison Formation of Utah. *Modern Geology* 23:331–342.

Lockley, M. G., Houck, K. J., and Prince, N. K. 1986. North America's largest dinosaur trackway site: Implications for Morrison Formation paleoecology. *Geological Society of America Bulletin* 97:1163–1176.

Lockley, M. G., Farlow, J. O., and Meyer, C. A. 1994. *Brontopodus* and *Parabrontopodus* ichnogen. nov. and the significance of wide and narrow-gauge sauropod trackways. *Gaia* 10:135–146.

Lockley, M. G., Logue, T. J., Moratalla, J. J., Hunt, A. P., Schultz, R. J., and Robinson, J. W. 1995. The fossil trackway *Pteraichnus* is pterosaurian, not crocodilian: Implications for the global distribution of pterosaur tracks. *Ichnos* 4:7–20.

Lockley, M. G., Santos, V. F. D., Meyer, C., and Hunt, A. P. 1998a. A new dinosaur tracksite in the Morrison Formation, Boundary Butte, southeastern Utah. *Modern Geology* 23:317–330.

Lockley, M. G., Foster, J. R., and Hunt, A. P. 1998b. A short summary of dinosaur tracks and other fossil footprints from the Morrison Formation. *Modern Geology* 23:277–290.

Lockley, M. G., Wright, J. L., Langston, W., and West, E. S. 2001. New pterosaur track specimens and tracksites in the Late Jurassic of Oklahoma and Colorado: Their paleobiological significance and regional ichnological context. *Modern Geology* 24:179–203.

Lockley, M. G., Gierlinski, G., Matthews, N., Xing, L., Foster, J. R., and Cart, K. 2017. New dinosaur track occurrences from the Upper Jurassic Salt Wash Member (Morrison Formation) of southeastern Utah: Implications for thyreophoran trackmaker distribution and diversity. *Palaeogeography, Palaeoclimatology, Palaeoecology* 470:116–121.

Lockley, M. G., Foster, J. R., and Hunt-Foster, R. K. 2018. The first North American *Deltapodus* trackway in a diverse *Anomoepus*, theropod, sauropod and turtle track assemblage from the Upper Jurassic Salt Wash Member (Morrison Formation) of eastern Utah. *New Mexico Museum of Natural History and Science Bulletin* 79:407–416.

Loewen, M. A. 2004. Variation and stratigraphic distribution of *Allosaurus* within the Late Jurassic Morrison Formation. *Geological Society of America, Abstracts with Programs* 36(5):524.

López-Arbarello, A., Sferco, E., and Rauhut, O. W. M. 2013. A new genus of coccolepidid fishes (Actinopterygii, Chondrostei) from the continental Jurassic of Patagonia. *Palaeontologia Electronica* 16(1):1–23.

Lovelace, D. M. 2006. An Upper Jurassic Morrison Formation fire-induced debris flow: Taphonomy and paleoenvironments of a sauropod (Sauropoda: *Supersaurus vivianae*) locality, east-central Wyoming. *New Mexico Museum of Natural History and Science Bulletin* 36:47–56.

Lovelace, D. M., Hartman, S. A., Wahl, W. R. 2007. Morphology of a specimen of *Supersaurus* (Dinosauria, Sauropoda) from the Morrison Formation of Wyoming, and a re-evaluation of diplodocid phylogeny. *Arquivos do Museu Nacional, Rio de Janeiro* 65:527–544.

Lucas, S. G. 2018. The Upper Jurassic Morrison Formation in north-central New Mexico—linking Colorado Plateau stratigraphy to the stratigraphy of the High Plains. *Geology of the Intermountain West* 5:117–129.

Lucas, S. G., and Heckert, A. B. 2015. New Mexico's record of Jurassic fossil vertebrates. *New Mexico Museum of Natural History and Science Bulletin* 68:97–104.

Lucas, S. G., Spielmann, J. A., Rinehart, L. F., Heckert, A. B., Herne, M. C., Hunt, A. P., Foster, J. R., and Sullivan, R. M. 2006a. Taxonomic status of *Seismosaurus hallorum*, a Late Jurassic sauropod dinosaur from New Mexico. *New Mexico Museum of Natural History and Science Bulletin* 36:149–161.

Lucas, S. G., Rinehart, L. F., and Heckert, A. B. 2006b. *Glyptops* (Testudines, Pleurosternidae) from the Upper Jurassic Morrison Formation, New Mexico. *New Mexico Museum of Natural History and Science Bulletin* 36:97–99.

Lull, R. S. 1919. The sauropod dinosaur *Barosaurus* Marsh. *Memoirs of the Connecticut Academy of Arts and Sciences* 6:1–42.

Luo Z.-X., and Wible, J. R. 2005. A Late Jurassic digging mammal and early mammalian diversification. *Science* 308:103–107.

Luo Z.-X., Kielan-Jaworowska, Z., and Cifelli, R. L. 2002. In quest for a phylogeny of Mesozoic mammals. *Acta Palaeontologica Polonica* 47:1–78.

Luo, Z.-X., Meng, Q.-J., Ji, Q., Liu, D., Zhang, Y.-G., and Neander, A. I. 2015. Evolutionary development in basal mammaliaforms as revealed by a docodontan. *Science* 347:760–764.

Macdonald, D. 2001. *The Encyclopedia of Mammals*. Brown Reference Group, London, 930 pp.

Madsen, J. H. 1974. A new theropod dinosaur from the Upper Jurassic of Utah. *Journal of Paleontology* 48:27–31.

———. 1976a. *Allosaurus fragilis*: A revised osteology. *Utah Geological Survey Bulletin* 109, 163 pp.

———. 1976b. A second new theropod dinosaur from the Late Jurassic of east central Utah. *Utah Geology* 3:51–60.

Madsen, J. H., and Welles, S. P. 2000. *Ceratosaurus* (Dinosauria, Theropoda), a revised osteology. *Utah Geological Survey Miscellaneous Publication* 00–2, 80 pp.

Madsen, J. H., McIntosh, J. S., and Berman, D. S. 1995. Skull and atlas-axis complex of the Upper Jurassic sauropod *Camarasaurus* Cope (Reptilia: Saurischia). *Bulletin of Carnegie Museum of Natural History* 31:1–115.

Madsen, S. K. 2010. *Exploring Desert Stone: John N. Macomb's 1859 Expedition to the Cayonlands of the Colorado*. Utah State University Press, Logan, 273 pp.

Maidment, S. C. R. 2010. Stegosauria: A historical review of the body fossil record and phylogenetic relationships. *Swiss Journal of Geosciences* 103:199–210.

Maidment, S. C. R., and Muxworthy, A. 2019. A chronostratigraphic framework for the Upper Jurassic Morrison Formation, western U. S. A. *Journal of Sedimentary Research* 89:1017–1038.

Maidment, S. C. R., Norman, D. B., Barrett, P. M., and Upchurch, P. 2007. Systematics and phylogeny of Stegosauria (Dinosauria: Ornithischia). *Journal of Systematic Palaeontology* 6:367–407.

Maidment, S. C. R., Brassey, C., and Barrett, P. M. 2015. The postcranial skeleton of an exceptionally complete individual of the plated dinosaur *Stegosaurus stenops* (Dinosauria: Thyreophora) from the Upper Jurassic Morrison Formation of Wyoming, U. S. A. *PLoS ONE* 10(10):e0138352. DOI:10.1371/journal.pone.0138352.

Maidment, S. C. R., Balikova, E., and Muxworthy, A. R. 2017. Magnetostratigraphy of the Upper Jurassic Morrison Formation at Dinosaur National Monument, Utah, and prospects for using magnetostratigraphy as a correlative tool in the Morrison Formation. *In* Zeigler, K. E., and Parker, W. G., eds., *Terrestrial Depositional Systems: Deciphering Complexities through Multiple Stratigraphic Methods*, Elsevier, Amsterdam, 279–302.

Maidment, S. C. R., Woodruff, D. C., and Horner, J. R. 2018. A new specimen of the ornithischian dinosaur *Hesperosaurus mjosi* from the Upper Jurassic Morrison Formation of Montana, USA, and implications for growth and size in Morrison stegosaurs. *Journal of Vertebrate Paleontology* DOI:10.1080/02724634.2017.1406366.

Maier, G. 2003. *African Dinosaurs Unearthed: The Tendaguru Expeditions*. Indiana University Press, Bloomington, 380 pp.

Main, R. P., de Ricqlès, A., Horner, J. R., and Padian, K. 2005. The evolution and function of thyreophoran dinosaur scutes: Implications for plate function in stegosaurs. *Paleobiology* 31:291–314.

Maltese, A., Tschopp, E., Holwerda, F., and Burnham, D. 2018. The real Bigfoot: A pes from Wyoming, USA is the largest sauropod pes ever reported and the northern-most occurrence of brachiosaurids in the Upper Jurassic Morrison Formation. *PeerJ* 6:e5250. DOI:10.7717/peerj.5250.

Mannion, P. D., Upchurch, P., Mateus, O., Barnes, R. N., and Jones, M. E. H. 2012. New information on the anatomy and systematic position of *Dinheirosaurus lourinhanensis* (Sauropoda: Diplodocoidea) from the Late Jurassic of Portugal, with a review of European diplodocoids. *Journal of Systematic Palaeontology* 10:521–551.

Mannion, P. D., Upchurch, P., Barnes, R. N., and Mateus, O. 2013. Osteology of the Late Jurassic Portuguese sauropod dinosaur *Lusotitan atalaiensis* (Macronaria) and the evolutionary history of basal titanosauriforms. *Zoological Journal of the Linnean Society* 168:98–206.

Maor, R., Dayan, T., Ferguson-Gow, H., and Jones, K. E. 2017. Temporal niche expansion in mammals from a nocturnal ancestor after dinosaur extinction. *Ecology and Evolution* DOI:10.1038/s41559–017–0366–5.

Marsh, O. C. 1877. Notice of some new dinosaurian reptiles from the Jurassic Formation. *American Journal of Science* (ser. 3) 14:514–516.

———. 1878a. New pterodactyl from the Jurassic of the Rocky Mountains. *American Journal of Science* (ser. 3) 16:233–234.

———. 1878b. Principal characters of American Jurassic dinosaurs—part 1. *American Journal of Science* (ser. 3) 16:411–416.

———. 1878c. Fossil mammal from the Jurassic of the Rocky Mountains. *American Journal of Science* (ser. 3) 15:459.

———. 1879a. Principal characters of American Jurassic dinosaurs—part 2. *American Journal of Science* (ser. 3) 17:86–92.

———. 1879b. Notice of new Jurassic reptiles. *American Journal of Science* (ser. 3) 18:501–505.

———. 1879c. Additional remains of Jurassic mammals. *American Journal of Science* (ser. 3) 18:215–216.

————. 1880. Notice of Jurassic mammals representing two new orders. *American Journal of Science* (ser. 3) 20:235–239.

————. 1881. Principal characters of American Jurassic dinosaurs—part 5. *American Journal of Science* (ser. 3) 21:417–423.

————. 1884. Principal characters of American Jurassic dinosaurs—part 8, the Order Theropoda. *American Journal of Science* (ser. 3) 27:329–340.

————. 1887a. American Jurassic mammals. *American Journal of Science* (ser. 3) 33:327–348.

————. 1887b. Principal characters of American Jurassic dinosaurs—part 9, the skull and dermal armor of *Stegosaurus*. *American Journal of Science* (ser. 3) 34:413–417.

————. 1890. Description of new dinosaurian reptiles. *American Journal of Science* (ser. 3) 39:81–86.

————. 1899. Footprints of Jurassic dinosaurs. *American Journal of Science* (ser. 4) 7:227–232.

Martin, James E., and Foster, J. R. 1998. First Jurassic mammals from the Black Hills, northeastern Wyoming. *Modern Geology* 23:381–392.

Martin, Jeremy E., Rabi, M., and Csiki, Z. 2010. Survival of *Theriosuchus* (Mesoeucrocodylia: Atoposauridae) in a Late Cretaceous archipelago: A new species from the Maastrictian of Romania. *Naturwissenschaften* 97:845–854.

Martin, Jeremy E., Rabi, M., Csiki-Sava, Z., and Vasile, S. 2014. Cranial morphology of *Theriosuchus sympiestodon* (Mesoeucrocodylia, Atoposauridae) and the widespread occurrence of *Theriosuchus* in the Late Cretaceous of Europe. *Journal of Paleontology* 88:444–456.

Martin, Jeremy E., Delfino, M., and Smith, T. 2016. Osteology and affinities of Dollo's goniopholidid (Mesoeucrocodylia) from the Early Cretaceous of Bernissart, Belgium. *Journal of Vertebrate Paleontology* 36(6):e1222534.

Martin, T. 1999. Dryolestidae (Dryolestoidea, Mammalia) aus dem Oberen Jura von Portugal. *Abhandlungen der Senckenbergischen Naturforschenden Gesellschaft* 550:1–119.

Martin, T. 2005. Postcranial anatomy of *Haldanodon exspectatus* (Mammalia, Docodonta) from the Late Jurassic (Kimmeridgian) of Portugal and its bearing for mammalian evolution. *Zoological Journal of the Linnean Society* 145:219–248.

Martin, T., and Krebs, B. 2000. *Guimarota: A Jurassic Ecosystem*. Dr. Friedrich Pfeil, Munich, 155 pp.

Mateus, O., and Antunes, M. T. 2000a. *Torvosaurus* sp. (Dinosauria: Theropoda) in the Late Jurassic of Portugal. *Livro de Resumos do I Congresso Ibérico de Paleontologia*, 115–117.

————. 2000b. *Ceratosaurus* (Dinosauria: Theropoda) in the Late Jurassic of Portugal. *Abstract Volume of the 31st International Geological Congress, Rio de Janeiro, Brazil*.

————. 2001. *Draconyx loureiroi*, a new Camptosauridae (Dinosauria, Ornithopoda) from the Late Jurassic of Lourinhã, Portugal. *Annales de Paléontologie* 87:61–73.

Mateus, O., and Tschopp, E. 2013. *Cathetosaurus* as a valid sauropod genus and comparisons with *Camarasaurus*. *Journal of Vertebrate Paleontology*, Program and Abstracts, 2013, 173.

Mateus, O., Walen, A., and Atunes, M. T. 2006. The large theropod fauna of the Lourinhã Formation (Portugal) and its similarity to that of the Morrison Formation, with a description of a new species of *Allosaurus*. *New Mexico Museum of Natural History and Science Bulletin* 36:123–129.

Mateus, O., Maidment, S. C. R., and Christiansen, N. A. 2009. A new long-necked "sauropod-mimic" stegosaur and the evolution of the plated dinosaurs. *Proceedings of the Royal Society B* 276:1815–1821.

Mateus, O., Dinis, J., and Cunha, P. P. 2017. The Lourinhã Formation: The Upper Jurassic to lowermost Cretaceous of the Lusitanian Basin, Portugal—landscapes where dinosaurs walked. *Ciências da Terra Earth Sciences Journal* 19:75–97.

Mathews, J. C., Large, D. S., Williams, S. A., and Tremaine, K. 2018. Ten years of excavation at the Late Jurassic (Tithonian) Hanksville-Burpee Dinosaur Quarry (Morrison Formation, Brushy Basin Member), and a possible new specimen of *Torvosaurus tanneri*. *Journal of Vertebrate Paleontology*, Program and Abstracts, 2018, 177.

Matsumoto, R., and Evans, S. E. 2010. Choristoderes and the freshwater assemblages of Laurasia. *Journal of Iberian Geology* 36:253–274.

Mattison, C. 1987. *Frogs and Toads of the World*. Facts on File, New York, 191 pp.

————. 1989. *Lizards of the World*. Facts on File, New York, 192 pp.

Mayor, A. 2005. *Fossil Legends of the First Americans*. Princeton University Press, Princeton, NJ, 446 pp.

Mazin, J., Hantzpergue, P., Lafaurie, G., and Vignaud, P. 1995. Des pistes de ptérosaures dans le Tithonien de Crayssac (Quercy, France). *Comptes Rendus Académie des Sciences Paris* 321(2a):417–424.

McDonald, A. T. 2011. The taxonomy of species assigned to *Camptosaurus* (Dinosauria: Ornithopoda). *Zootaxa* 2783:52–68.

McHugh, J. B. 2018. Evidence for niche partitioning among ground-height browsing sauropods from the Upper Jurassic Morrison Formation of North America. *Geology of the Intermountain West* 5:95–103.

McIntosh, J. S. 1981. Annotated catalogue of the dinosaurs (Reptilia, Archosauria) in the collections of Carnegie Museum of Natural History. *Bulletin of Carnegie Museum of Natural History*, no. 18, 67 pp.

————. 1990a. Sauropoda. *In* Weishampel, D. B., Dodson, P., and Osmólska, H., eds., *The Dinosauria*, University of California Press, Berkeley, 345–401.

————. 1990b. Species determination in sauropod dinosaurs with tentative suggestions for their classification. *In* Carpenter, K., and Currie, P. J., eds., *Dinosaur Systematics: Perspectives and Approaches*, Cambridge University Press, New York, 53–69.

————. 1990c. The second Jurassic dinosaur rush. *Earth Sciences History* 9:22–27.

————. 1997. The saga of a forgotten sauropod dinosaur. *In* Wolberg, D. L., Stump, E., and Rosenberg, G. D., eds., *Dinofest International Proceedings*, National Academy of Sciences, Philadelphia, 7–12.

————. 2005. The genus *Barosaurus* Marsh (Sauropoda, Diplodocidae). *In* Tidwell, V., and Carpenter, K., eds., *Thunder-Lizards: The Sauropodomorph Dinosaurs*, Indiana University Press, Bloomington, 38–77.

McIntosh, J. S., and Carpenter, K. 1998. The holotype of *Diplodocus longus*, with comments on other specimens of the genus. *Modern Geology* 23:85–110.

McIntosh, J. S., and Williams, M. E. 1988. A new species of sauropod dinosaur, *Haplocanthosaurus delfsi* sp. nov., from the Upper Jurassic Morrison Fm. of Colorado. *Kirtlandia* 43:3–26.

McIntosh, J. S., Coombs, W. P., and Russell, D. A. 1992. A new diplodocid sauropod (Dinosauria) from Wyoming, USA. *Journal of Vertebrate Paleontology* 12:158–167.

McIntosh, J. S., Miller, W. E., Stadtman, K. L., and Gillette, D. D. 1996. The osteology of *Camarasaurus lewisi* (Jensen, 1988). *Brigham Young University Geology Studies* 41:73–115.

McIntosh, J. S., Brett-Surman, M. K., and Farlow, J. O. 1997. Sauropods. *In* Farlow, J. O., and Brett-Surman, M. K., eds., *The Complete Dinosaur*, Indiana University Press, Bloomington, 264–290.

McKenna, M. C., and Bell, S. K. 1997. *Classification of Mammals above the Species Level.* Columbia University Press, New York, 631 pp.

McMullen, S. K. 2016. Controls on the stratigraphic distribution of non-marine fossils: A case study in the Jurassic Morrison Formation, western USA. *Geological Society of America Abstracts with Programs* 48(7): DOI: 10.1130/abs/2016AM-287727.

McWhinney, L. A., Rothschild, B. M., and Carpenter, K. 2001. Posttraumatic chronic osteomyelitis in *Stegosaurus* dermal spikes. *In* Carpenter, K., ed., *The Armored Dinosaurs*, Indiana University Press, Bloomington, 141–156.

Meiri, S. 2010. Length-weight allometries in lizards. *Journal of Zoology* 281:218–226.

Melstrom, K. M., D'Emic, M. D., Chure, D. J., and Wilson, J. A. 2016. A juvenile sauropod dinosaur from the Late Jurassic of Utah, U.S.A., presents further evidence of an avian style air sac system. *Journal of Vertebrate Paleontology* DOI:10.1080/02724634.2016.1111898.

Meng, J.-Q., Ji, Q., Zhang, Y.-G., Liu, D., Grossnickle, D. M., and Luo, Z.-X. 2015. An arboreal docodont from the Jurassic and mammaliaform ecological diversification. *Science* 347:764–768.

Miall, A. D. 2010. Alluvial deposits. In James, N. P., and Dalrymple, R. W., eds. *Facies Models* 4, Geological Association of Canada, St. John's, 105-137.

Michelis, I. 2004. Taphonomie des Howe Quarry (Morrison Formation, Oberer Jura), Bighorn County, Wyoming, USA. Unpublished PhD dissertation, Rheinischen Friedrich-Wilhelms-Universität Bonn.

Miller, C. N. 1987. Terrestrial vegetation in the northern Rocky Mountains before the appearance of flowering plants. *Annals of Missouri Botanical Garden* 74:692–706.

Miller, W. E., Horrocks, R. D., and Madsen, J. H. 1996. The Cleveland-Lloyd dinosaur quarry, Emery County, Utah: A U. S. natural landmark (including history and quarry map). *Brigham Young University Geology Studies* 41:3–24.

Mocho, P., and Chiappe, L. M. 2018. New camarasaurid specimen from the Gnatalie Quarry in southern Utah (Morrison Formation, U.S.A.). *Journal of Vertebrate Paleontology*, Program and Abstracts, 2018, 184.

Mocho, P., Royo-Torres, R., and Ortega, F. 2014a. Phylogenetic reassessment of *Lourinhasaurus alenquerensis*, a basal Macronaria (Sauropoda) from the Upper Jurassic of Portugal. *Zoological Journal of the Linnean Society* 170:875–916.

Mocho, P., Royo-Torres, R., Malafaia, E., Escaso, F., and Ortega, F. 2014b. A preliminary evaluation of Diplodocidae record from the Upper Jurassic of Lusitanian Basin (W, Portugal). *In* Pereira, I., Amaral, F., and Vinhas, A., eds., *IV CJIG, LEG 2014, Livro de Actas.* Pólo de Estremoz de UÉvora, Évora, p. 85–88.

Molnar, R. E., and Clifford, H. T. 2000. Gut contents of a small ankylosaur. *Journal of Vertebrate Paleontology* 20:194–196.

Monaco, P. E. 1998. A short history of dinosaur collecting in the Garden Park Fossil Area, Cañon City, Colorado. *Modern Geology* 23:465–480.

Mook, C. C. 1967. Preliminary description of a new Goniopholid crocodilian. *Kirtlandia* 2:1–10.

Mosbrugger, V., Gee, C. T., Belz, G., and Ashraf, A. R. 1994. Three-dimensional reconstruction of an in-situ Miocene peat forest from the Lower Rhine Embayment, northwestern Germany—new methods in palaeovegetation analysis. *Palaeogeography, Palaeoclimatology, Palaeoecology* 110:295–317.

Myers, T. S., and Fiorillo, A. R. 2009. Evidence for gregarious behavior and age segregation in sauropod dinosaurs. *Palaeogeography, Palaeoclimatology, Palaeoecology* 274:96–104.

Myers, T. S., and Storrs, G. W. 2007. Taphonomy of the Mother's Day Quarry, Upper Jurassic Morrison Formation, south-central Montana, USA. *Palaios* 22:652–666.

Myers, T. S., Tabor, N. J., and Rosenau, N. A. 2014. Multiproxy approach reveals evidence of highly variable paleoprecipitation in the Upper Jurassic Morrison Formation (western United States). *Geological Society of America Bulletin* 126:1105–1116.

Myhrvold, N. P. 2016. Dinosaur metabolism and the allometry of maximum growth rate. *PLoS ONE* 11(11):e0163205.

Norman, D. B. 2004. Basal Iguanodontia. *In* Weishampel, D. B., Dodson, P., and Osmólska, H., eds., *The Dinosauria.* 2nd ed., University of California Press, Berkeley, 413–437.

Noto, C. R., and Grossman, A. 2010. Broad-scale patterns of Late Jurassic dinosaur paleoecology. *PLoS ONE* 5(9):e12553. DOI:10.1371/journal.pone.0012553.

Novas, F. E. 2009. *The Age of Dinosaurs in South America.* Indiana University Press, Bloomington, 452 pp.

Nydam, R. L., Chure, D. J., and Evans, S. E. 2013. *Schillerosaurus* gen. nov., a replacement name for the lizard genus *Shilleria* Evans and Chure, 1999 a junior homonym of *Schilleria* Dahl, 1907. *Zootaxa* 3734:99–100.

Nydam, R. L., Caldwell, M. W., Apesteguía, S., and Palci, A. 2016. Lizards and snakes of the Morrison Formation (Upper Jurassic, USA) and the early global distribution and diversity of squamates. *Journal of Vertebrate Paleontology*, Program and Abstracts, 2016, 197.

Olszewski, T. D. 2010. Diversity partitioning using Shannon's entropy and its relationship to rarefaction. *Paleontological Society Papers* 16:95–116.

Osborn, H. F. 1903. *Ornitholestes hermanni*, a new compsognathoid dinosaur from the Upper Jurassic. *Bulletin of the American Museum of Natural History* 19:459–464.

Osborn, H. F., and Mook, C. C. 1921. *Camarasaurus, Amphicoelias*, and other sauropods of Cope. *Memoirs of the American Museum of Natural History* 3:247–287.

Ostrom, J. H. 1971. On the systematic position of *Macelognathus vagans. Postilla* 153:1–10.

———. 1980. *Coelurus* and *Ornitholestes*: Are they the same? *In* Jacobs, L. L., ed., *Aspects of Vertebrate History*, Museum of Northern Arizona Press, Flagstaff, 245–256.

Ostrom, J. H., and McIntosh, J. S. 1966. *Marsh's Dinosaurs—the Collections from Como Bluff.* Yale University Press, New Haven, CT, 388 pp.

O'Sullivan, R. B. 2010. The lower and upper contacts of the Upper Jurassic Bluff Sandstone Member of the Morrison Formation in southeastern Utah. *New Mexico Geological Society Guidebook, 61st Field Conference, Four Corners Country*, 101–106.

Owen, D. E., Walters, L. J., Jr., and Beck, R. G. 1984. The Jackpile Sandstone Member of the Morrison Formation in west-central New Mexico—a formal definition. *New Mexico Geology* 6:45–52.

Owen, R. 1879. On the association of dwarf crocodiles (*Nannosuchus* and *Theriosuchus pusillus*, e.g.) with the diminutive mammals of the Purbeck Shales. *Quarterly Journal of the Geological Society* 35:148–155.

Owen-Smith, R. N. 1988. *Megaherbivores: The Influence of Very Large Body Size on Ecology*. Cambridge University Press, New York, 369 pp.

Packard, G. C., Boardman, T. J., and Birchard, G. F. 2009. Allometric equations for predicting body mass of dinosaurs. *Journal of Zoology* 279:102–110.

———. 2010. Allometric equations for predicting body mass of dinosaurs: A comment on Cawley & Janacek (2010). *Journal of Zoology* 282:221–222.

Padian, K. 1998. Pterosaurians and ?avians from the Morrison Formation (Upper Jurassic, Western U.S.). *Modern Geology* 23:57–68.

Padian, K., and Olsen, P. E. 1984. The fossil trackway *Pteraichnus*: Not pterosaurian but crocodilian. *Journal of Paleontology* 58:178–184.

Pardo, J. D., Huttenlocker, A. K., Small, B. J., and Gorman, M. A. 2010. The cranial morphology of a new genus of lungfish (Osteichthyes: Dipnoi) from the Upper Jurassic Morrison Formation of North America. *Journal of Vertebrate Paleontology* 30:1352–1359.

Parrish, J. T. 1993. Mesozoic climates of the Colorado Plateau. *Museum of Northern Arizona Bulletin* 59:1–11.

Parrish, J. T., Peterson, F., and Turner, C. E. 2004. Jurassic "savannah"—plant taphonomy and climate of the Morrison Formation (Upper Jurassic, western USA). *Sedimentary Geology* 167:137–162.

Parry, L. A., Baron, M. G., and Vinther, J. 2017. Multiple optimality criteria support Ornithoscelida. *Royal Society Open Science* 4:170833.

Paul, G. S. 1988a. *Predatory Dinosaurs of the World*. Simon and Schuster, New York, 464 pp.

———. 1988b. The brachiosaur giants of the Morrison and Tendaguru with a description of a new subgenus, *Giraffatitan*, and a comparison of the world's largest dinosaurs. *Hunteria* 2(3):1–14.

———. 1998. Terramegathermy and Cope's rule in the land of titans. *Modern Geology* 23:179–217.

Peirce, R. J. 2006. A nearly complete dentary of the ornithopod dinosaur *Othnielia rex* from the Morrison Formation of Wyoming. *New Mexico Museum of Natural History and Science Bulletin* 36:163–164.

Pérez-Moreno, B. P., Chure, D. J., Pires, C., Marques Da Silva, C., Dos Santos, V., Dantas, P., Póvoas, L., Cachão, M., Sanz, J. L., and Galopim de Carvalho, A. M. 1999. On the presence of *Allosaurus fragilis* (Theropoda: Carnosauria) in the Upper Jurassic of Portugal: First evidence of an intercontinental dinosaur species. *Journal of the Geological Society, London* 156:449–452.

Peterson, F., and Turner, C. E. 1998. Stratigraphy of the Ralston Creek and Morrison Formations (Upper Jurassic) near Denver, Colorado. *Modern Geology* 22:3–38.

Peterson, F., and Turner-Peterson, C. E. 1987. The Morrison Formation of the Colorado Plateau: Recent advances in sedimentology, stratigraphy, and paleotectonics. *Hunteria* 2(1):1–18.

Peterson, J. E., Warnock, J. P., Ebhart, S. L., Clawson, S. R., and Noto, C. R. 2017. New data towards the development of a comprehensive taphonomic framework for the Late Jurassic Cleveland-Lloyd Dinosaur Quarry, central Utah. *PeerJ* 5:e3368.

Peterson, J. E., Lovelace, D. M., Connely, M. V., McHugh, J. B., Hartman, S., Hayes, L., and Strey, F. 2018. Computed tomography, segmentation, and rapid prototyping of the dental batteries of an apatosaurine (Sauropoda, Diplodocidae) specimen from the Upper Jurassic Morrison Formation, Como Bluff, Wyoming. *Journal of Vertebrate Paleontology*, Program and Abstracts, 2018, 198.

Peterson, O. A., and Gilmore, C. W. 1902. *Elosaurus parvus*, a new genus and species of the Sauropoda. *Carnegie Museum Annals* 1:490–499.

Pianka, E. R., and Vitt, L. J. 2003. *Lizards: Windows to the Evolution of Diversity*. University of California Press, Berkeley, 333 pp.

Pinsof, J. D. 1983. A Jurassic lungfish (Dipnoi: Ceratodontidae) from western South Dakota. *Proceedings of the South Dakota Academy of Sciences* 62:75–79.

Poyato-Ariza, F. J. 2003. Dental characters and phylogeny of pycnodontiform fishes. *Journal of Vertebrate Paleontology* 23:937–940.

Pritchard, A. C., Turner, A. H., Allen, E. R., and Norell, M. A. 2013. Osteology of a North American goniopholidid (*Eutretauranosuchus delfsi*) and palate evolution in Neosuchia. *American Museum Novitates* 3783:1–56.

Prothero, D. R. 1981. New Jurassic mammals from Como Bluff, Wyoming, and the interrelationships of non-tribosphenic Theria. *Bulletin of the American Museum of Natural History* 167:281–325.

Prothero, D. R., and Estes, R. 1980. Late Jurassic lizards from Como Bluff, Wyoming and their palaeobiogeographic significance. *Nature* 286:484–486.

Puétolas-Pascual, E., Canudo, J. I., and Sender, L. M. 2015. New material from a huge specimen of *Anteophthalmosuchus* cf. *escuchae* (Goniopholididae) from the Albian of Andorra (Teruel, Spain): Phylogenetic implications. *Journal of Iberian Geology* 41:41–56.

Raath, M. A., and McIntosh, J. S. 1987. Sauropod dinosaurs from the central Zambezi Valley, Zimbabwe, and the age of the Kadzi Formation. *South African Journal of Geology* 90:107–119.

Raisanen, D. C. W., and Hasiotis, S. T. 2018. New ichnotaxa of vertebrate burrows from the Salt Wash Member, Upper Jurassic Morrison Formation, south-eastern Utah (USA). *Annales Societatis Geologorum Poloniae* 88:181–202.

Rasmussen, T. E., and Callison, G. 1981a. A new species of triconodont mammal from the Upper Jurassic of Colorado. *Journal of Paleontology* 55:628–634.

———. 1981b. A new herbivorous sphenodontid (Rhynchocephalia: Reptilia) from the Jurassic of Colorado. *Journal of Paleontology* 55:1109–1116.

Rauhut, O. W. M. 2001. Herbivorous dinosaurs from the Late Jurassic (Kimmeridgian) of Guimarota, Portugal. *Proceedings of the Geologists' Association* 112:275–283.

———. 2003. A tyrannosauroid dinosaur from the Upper Jurassic of Portugal. *Palaeontology* 46:903–310.

———. 2005. Post-cranial remains of "coelurosaurs" (Dinosauria, Theropoda) from the Late Jurassic of Tanzania. *Geological Magazine* 142:97–107.

———. 2006. A brachiosaurid sauropod from the Late Jurassic Cañadón Calcáreo Formation of Chubut, Argentina. *Fossil Record* 9:226–237.

———. 2011. Theropod dinosaurs from the Late Jurassic of Tendaguru (Tanzania). *In* Barrett, P. M., and Milner, A. R., eds., *Studies on Fossil Tetrapods, Palaeontological Association Special Papers in Palaeontology* 86:195–239.

Rauhut, O. W. M., and Carrano, M. T. 2016. The theropod dinosaur *Elaphrosaurus bambergi* Janensch, 1920, from the Late Jurassic of Tendaguru, Tanzania. *Zoological Journal of the Linnean Society* 178:546–610.

Rauhut, O. W. M., Remes, K., Fechner, R., Cladera, G., and Puerta, P. 2005. Discovery of a short-necked sauropod dinosaur from the Late Jurassic period of Patagonia. *Nature* 435:670–672.

Rauhut, O. W. M., Heyng, A. M., López-Arbarello, A., and Hecker, A. 2012a. A new rhynchocephalian from the Late Jurassic of Germany with a dentition that is unique amongst tetrapods. *PLoS ONE* 7(10): e46839.

Rauhut, O. W. M., Foth, C., Tischlinger, H., and Norell, M. A. 2012b. Exceptionally preserved juvenile megalosauroid theropod dinosaur with filamentous integument from the Late Jurassic of Germany. *Proceedings of the National Academy of Sciences* 109:11746–11751.

Raven, T. J., and Maidment, S. C. R. 2017. A new phylogeny of Stegosauria (Dinosauria, Ornithischia). *Palaeontology* DOI:10.1111/pala.12291.

Ray, G. R. 1941. Big for his day. *Natural History* 48:36–39.

Rayfield, E. J., Norman, D. B., Horner, C. C., Horner, J. R., Smith, P. M., Thomason, J. J., and Upchurch, P. 2001. Cranial design and function in a large theropod dinosaur. *Nature* 409:1033–1037.

Rea, T. 2001. *Bone Wars: The Excavation and Celebrity of Andrew Carnegie's Dinosaur*. University of Pittsburgh Press, Pittsburgh, 276 pp.

Rees, P. M., Noto, C. R., Parrish, J. M., and Parrish, J. T. 2004. Late Jurassic climates, vegetation, and dinosaur distributions. *Journal of Geology* 112:643–653.

Reid, R. E. H. 1997. Dinosaurian physiology: The case for "intermediate" dinosaurs. *In* Farlow, J. O., and Brett-Surman, M. K., eds., *The Complete Dinosaur*, Indiana University Press, Bloomington, 449–473.

———. 2012. "Intermediate" dinosaurs: The case updated. *In* Brett-Surman, M. K., Holtz, T. R., and Farlow, J. O., eds., *The Complete Dinosaur*, 2nd ed., Indiana University Press, Bloomington, 873–921.

Remes, K. 2007. A second Gondwanan diplodocid dinosaur from the Upper Jurassic Tendaguru Beds of Tanzania, east Africa. *Palaeontology* 50:653–667.

Retallack, G. J. 1997. Dinosaurs and dirt. *In* Wolberg, D. L., Stump, E., and Rosenberg, G. D., eds., *Dinofest International Proceedings*, National Academy of Sciences, Philadelphia, 345–359.

Richmond, D. R., and Morris, T. H. 1996. The dinosaur death-trap of the Cleveland-Lloyd Quarry, Emery County, Utah. *Museum of Northern Arizona Bulletin* 60:533–545.

Rigby, J. K., Jr. 1982. *Camarasaurus* cf. *supremus* from the Morrison Formation near San Ysidro, New Mexico—The San Ysidro dinosaur. *New Mexico Geological Society Guidebook* 33:271–272.

Riggs, E. S. 1903. *Brachiosaurus altithorax*, the largest known dinosaur. *American Journal of Science* (ser. 4) 15:299–306.

———. 1904. Structure and relationships of opisthocoelian dinosaurs, part II: The Brachiosauridae. *Field Columbian Museum Publication 94, Geological Series* 2(6):229–247.

Robb, J. 1986. Tuatara. *In* Halliday, T. R., and Adler, K., eds., *The Encyclopaedia of Reptiles and Amphibians*, Facts on File, New York, 134–135.

Robinson, J. W., and McCabe, P. J. 1997. Sandstone-body and shale-body dimensions in a braided fluvial system: Salt Wash Sandstone Member (Morrison Formation), Garfield County,

Utah. *American Association of Petroleum Geologists Bulletin* 81:1267–1291.

Roderick, W. R. T., Cutkosky, M. R., and Lentink, D. 2017. Touchdown to take-off: At the interface of flight and surface locomotion. *Interface Focus* 7:20160094. DOI:10.1098/rsfs.2016.0094.

Rodríguez, J. 1999. Use of cenograms in mammalian paleoecology: A critical review. *Lethaia* 32:331–347.

Rogers, R. R., Krause, D. W., and Curry Rogers, K. 2003. Cannibalism in the Madagascan dinosaur *Majungatholus atopus*. *Nature* 422:515–518.

Rosenzweig, M. L. 1995. *Species Diversity in Space and Time*. Cambridge University Press, New York, 436 pp.

Rotatori, F. M., Moreno-Azanza, M., and Mateus, O. 2020. New information on ornithopod dinosaurs from the Late Jurassic of Portugal. *Acta Palaeontologica Polonica* 65: DOI. org/10.4202/app.00661.2019.

Rougier, G. W., Wible, J. R., and Hopson, J. A. 1996. Basicranial anatomy of *Priacodon fruitaensis* (Triconodontidae, Mammalia) from the Late Jurassic of Colorado, and a reappraisal of Mammaliaform interrelationships. *American Museum Novitates* 3183:1–38.

Rougier, G. W., Spurlin, B. K., and Kik, P. K. 2003. A new specimen of *Eurylambda aequicrurius* and considerations on "symmetrodont" dentition and relationships. *American Museum Novitates* 3398:1–15.

Rougier, G. W., Sheth, A. S., Carpenter, K., Appella-Guiscafre, L., and Davis, B. M. 2015. A new species of *Docodon* (Mammaliaformes: Docodonta) from the Upper Jurassic Morrison Formation and a reassessment of selected craniodental characters in basal mammaliaforms. *Journal of Mammalian Evolution* 22:1–16.

Royo-Torres, R., Upchurch, P., Kirkland, J. I., DeBlieux, D. D., Foster, J. R., Cobos, A., and Alcalá, L. 2017. Descendants of the Jurassic turiasaurs from Iberia found refuge in the Early Cretaceous of western USA. *Scientific Reports* 7:14311.

Russell, D. A. 1989. *The Dinosaurs of North America*. North-Word, Minocqua, WI, 240 pp.

Russell, D., Béland, P., and McIntosh, J. S. 1980. Paleoecology of the dinosaurs of Tendaguru (Tanzania). *Mémoire Société Géologique de France* 139:169–175.

Salisbury, S. W. 2002. Crocodilians from the Lower Cretaceous (Berriasian) Purbeck Limestone Group of Dorset, southern England. *Special Papers in Palaeontology* 68:121–144.

Salisbury, S. W., and Naish, D. 2011. Crocodilians. *In* Batten, D. J., ed., *English Wealden Fossils*, Palaeontological Association Field Guide to Fossils, no. 14, 305–369.

Salisbury, S. W., Willis, P. M. A., Peitz, S., and Sander, P. M. 1999. The crocodilian *Goniopholis simus* from the Lower Cretaceous of North-Western Germany. *In* Unwin, D. M., ed., Cretaceous Fossil Vertebrates, *Special Papers in Palaeontology* 60:121–148.

Sander, P. M. 1999. Life history of Tendaguru sauropods as inferred from long bone histology. *Mitteilungen aus dem Museum für Naturkunde in Berlin, Geowissenschaftliche Reihe* 2:103–112.

Sander, P. M., Christian, A., Clauss, M., Fechner, R., Gee, C. T., Griebeler, E.-M., Gunga, H.-C., Hummel, J., Mallison, H., Perry, S. F., Preuschoft, H., Rauhut, O. W. M., Remes, K., Tütken, T., Wings, O., and Witzel, U. 2011. Biology of the sauropod dinosaurs: The evolution of gigantism. *Biological Reviews* 86:117–155.

Santos, E. S., and Turner-Peterson, C. E. 1986. Tectonic setting of the San Juan Basin in the Jurassic. *In* Turner-Peterson, C.

E., Santos, E. S., and Fishman, N. S., eds., *A Basin Analysis Case Study: The Morrison Formation, Grants Uranium Region, New Mexico*. American Association of Petroleum Geologists, *Studies in Geology* 22:27–34.

Scheetz, R. 1991. Progress report of juvenile and embryonic *Dryosaurus* remains from the Upper Jurassic Morrison Formation of Colorado. *In* Averett, W. R., ed., *Guidebook for Dinosaur Quarries and Tracksites Tour: Western Colorado and Eastern Utah*, Grand Junction Geological Society, 27–29.

Scheffer, M., Xu, C., Hantson, S., Holmgren, M., Los, S. O., and van Nes, E. H. 2018. A global climate niche for giant trees. *Global Change Biology* 24:2875–2883.

Schmitz, L., and Motani, R. 2011. Nocturnality in dinosaurs inferred from scleral ring and orbit morphology. *Science* 332:705–708.

Schultz, J. A., Bhullar, B.-A. S., and Luo, Z.-X. 2017. Re-examination of the Jurassic mammaliaform *Docodon victor* by computed tomography and occlusal functional analysis. *Journal of Mammalian Evolution* DOI:10.1007/s10914-017-9418-5.

Schumacher, B. A. 2008. The Last Chance Site, a new sauropod quarry from the Upper Jurassic Morrison Formation, southeastern Colorado. *In* Farley, G. H., and Choate, J. R., eds., *Fort Hays Studies, Special Issue No. 2, Unlocking the Unknown: Papers Honoring Dr. Richard J. Zakrzewski*, Fort Hays State University, Hays, KS, 77–88.

———. 2018. Extensive new trackways of *Megalosauripus* and *Parabrontopodus* at "Dinosaur Lake," the Purgatoire Valley Dinosaur Tracksite, Morrison Formation, southeastern Colorado. *Journal of Vertebrate Paleontology*, Program and Abstracts, 2018, 212.

Schwarz, D., and Salisbury, S. W. 2005. A new species of *Theriosuchus* (Atoposauridae, Crocodylomorpha) from the Late Jurassic (Kimmeridgian) of Guimarota, Portugal. *Geobios* 38:779–802.

Schwarz, D., Ikejiri, T., Breithaupt, B. H., Sander, P. M., and Klein, N. 2007. A nearly complete skeleton of an early juvenile diplodocid (Dinosauria: Sauropoda) from the lower Morrison Formation (Late Jurassic) of north central Wyoming and its implications for early ontogeny and pneumaticity in sauropods. *Historical Biology* 19:225–253.

Schwarz, D., Raddatz, M., and Wings, O. 2017. *Knoetschkesuchus langenbergensis* gen. nov. sp. nov., a new atoposaurid crocodyliform from the Upper Jurassic Langenberg Quarry (Lower Saxony, northwestern Germany), and its relationships to *Theriosuchus*. *PLoS ONE* 12(2):e0160617.

Senter, P. 2006. Forelimb function in *Ornitholestes hermanni* Osborn (Dinosauria, Theropoda). *Palaeontology* 49:1029–1034.

Sereno, P. C. 1986. Phylogeny of the bird-hipped dinosaurs (Order Ornithischia). *National Geographic Research* 2:234–256.

———. 1999. The evolution of dinosaurs. *Science* 284:2137–2147.

Sertich, J. 2017. Extreme theropod richness of the Morrison Formation and the unusual theropod specimens of the Denver Museum of Nature and Science. *Journey to the Jurassic: Exploring the Morrison Formation*, 10th Founders Symposium Program and Abstracts, Western Interior Paleontological Society, 33–35.

Seymour, R. S., and Lillywhite, H. B. 2000. Hearts, neck posture and metabolic intensity of sauropod dinosaurs. *Proceedings of the Royal Society of London B* 267:1883–1887.

Shipman, P. 1998. *Taking Wing: Archaeopteryx and the Evolution of Bird Flight*. Simon and Schuster, New York, 336 pp.

Shu D.-G., Luo H.-L., Conway Morris, S., Zhang X.-L., Hu S.-X., Chen L., Han J., Zhu M., Li Y., and Chen L.-Z. 1999. Lower Cambrian vertebrates from south China. *Nature* 402:42–46.

Shubin, N. H., Daeschler, E. B., and Jenkins, F. A., Jr. 2014. Pelvic girdle and fin of *Tiktaalik roseae*. *Proceedings of the National Academy of Sciences* 111:893–899.

Simpson, G. G. 1926a. The fauna of Quarry Nine. *American Journal of Science* (ser. 5) 12:1–11.

———. 1926b. American terrestrial Rhynchocephalia. *American Journal of Science* 5:12–16.

———. 1926c. Mesozoic Mammalia—IV: The multituberculates as living animals. *American Journal of Science* (ser. 5) 11:228–250.

———. 1927. Mesozoic Mammalia—VI: Genera of Morrison Pantotheres. *American Journal of Science* (ser. 5) 13:411–416.

———. 1928. *A Catalogue of the Mesozoic Mammalia in the Geological Department of the British Museum*. British Museum, London, 215 pp.

———. 1929. American Mesozoic Mammalia. *Memoirs of the Peabody Museum of Yale University* 3(1):1–171.

———. 1933. Paleobiology of Jurassic mammals. *Palaeobiologica* 5:127–158.

———. 1937. A new Jurassic mammal. *American Museum Novitates* 943:1–6.

Smith, D. K. 1998. A morphometric analysis of *Allosaurus*. *Journal of Vertebrate Paleontology* 18:126–142.

Smith, D. K., Allen, E. R., Sanders, R. K., and Stadtman, K. L. 2010. A new specimen of *Eutretauranosuchus* (Crocodyliformes; Goniopholididae) from Dry Mesa, Colorado. *Journal of Vertebrate Paleontology* 30:1466–1477.

Smith, D. M., Gorman, M. A., II, Pardo, J. D., and Small, B. J. 2011. First fossil Orthoptera from the Jurassic of North America. *Journal of Paleontology* 85:102–105.

Smith, E. 2017. A new trace fossil assemblage interpreted to be produced by social insects from the Upper Jurassic Morrison Formation of eastern Utah. *Journey to the Jurassic: Exploring the Morrison Formation*, 10th Founders Symposium Program and Abstracts, Western Interior Paleontological Society, 30–31.

Smith, E., Loewen, M., and Kirkland, J. I. 2016. Trace fossil evidence for the presence of social insects in the Late Jurassic Morrison Formation of eastern Utah. *Geological Society of America Abstracts with Programs* 48(7): DOI: 1130/abs/2016AM-278863.

Smith, L. M. 2003. *Playas of the Great Plains*. University of Texas Press, Austin, 257 pp.

Snively, E., Cotton, J. R., Ridgely, R., and Witmer, L. M. 2013. Multibody dynamics model of head and neck function in *Allosaurus* (Dinosauria, Theropoda). *Palaeontologia Electronica* 16(2):11A.

Spotila, J. R. 1980. Constraints of body size and environment on the temperature regulation of dinosaurs. *In* Thomas, R. D. K., and Olson, E. C., eds., *A Cold Look at the Warm-Blooded Dinosaurs*, American Association for the Advancement of Science, Selected Symposium 28:233–252.

Sprague, M., and McLain, M. A. 2018. Resolving the *Mesadactylus* complex of Dry Mesa Quarry, Morrison Formation, Colorado. *Journal of Vertebrate Paleontology*, Program and Abstracts, 2018, 220.

Steel, R. 1973. Crocodylia. *In* Kuhn, O., ed., *Handbuch der Paläoherpetologie* 16:1–116.

Stevens, K. A. 2013. The articulation of sauropod necks: Methodology and mythology. *PLoS ONE* 8(10):e78572.

Stevens, K. A., and Parrish, J. M. 1996. Articulating three-dimensional computer models of sauropod cervical vertebrae [abstract]. *Journal of Vertebrate Paleontology* 16(suppl. 3):67A.

———. 1997. Comparisons of neck form and function in the Diplodocidae [abstract]. *Journal of Vertebrate Paleontology* 17(suppl. 3):79A.

———. 1999. Neck posture and feeding habits of two Jurassic sauropod dinosaurs. *Science* 284:798–800.

———. 2005a. Neck posture, dentition, and feeding strategies in Jurassic sauropod dinosaurs. *In* Tidwell, V., and Carpenter, K., eds., *Thunder-lizards: The sauropodomorph dinosaurs*, Indiana University Press, Bloomington, p. 212–232.

———. 2005b. Digital reconstructions of sauropod dinosaurs and implications for feeding. *In* Curry, K. A., and Wilson, J. A., eds., *The sauropods: Evolution and paleobiology*, University of California Press, Berkeley, p. 178–200.

Stokes, W. L. 1957. Pterodactyl tracks from the Morrison Formation. *Journal of Paleontology* 31:952–954.

Stokes, W. L. 1985. The Cleveland-Lloyd dinosaur quarry—Window to the Past. *United States Government Printing Office*, Washington, DC, 27 pp.

Storrs, G. W., and Gower, D. J. 1993. The earliest possible choristodere (Diapsida) and gaps in the fossil record of semi-aquatic reptiles. *Journal of the Geological Society, London* 150:1103–1107.

Storrs, G. W., Oser, S. E., and Aull, M. 2013. Further analysis of a Late Jurassic dinosaur bone-bed from the Morrison Formation of Montana, USA, with a computed three-dimensional reconstruction. *Earth and Environmental Science Transactions of the Royal Society of Edinburgh* 103:443–458.

Stovall, J. W. 1938. The Morrison of Oklahoma and its dinosaurs. *Journal of Geology* 46:583–600.

Sun, C., Wang, H., Dilcher, D. L., Li, T., Li, Y., and Na, Y. 2015. A new species of *Czekanowskia* (Czekanowskiales) from the Middle Jurassic of Ordos Basin, China. *Botanica Pacifica* 4:149–155.

Sundell, C. A. 2005. Recent discoveries from the Morrison Formation of north east Wyoming. *Geological Society of America 57th Annual Meeting, Rocky Mountain Section, Abstracts with Programs* 37(6):12.

Swanson, B. A., Santucci, V. L., Madsen, S. K., Elder, A. S., and Kenworthy, J. P. 2005. Arches National Park Paleontological Survey. *Geologic Resources Division Technical Report* NPS/NRGRD/GRDTR-05/01, 36 pp.

Szigeti, G. J., and Fox, J. E. 1981. Unkpapa Sandstone (Jurassic), Black Hills, South Dakota, and eolian facies of the Morrison Formation. *Society of Economic Paleontologists and Mineralogists, Special Publication* 31:331–349.

Tanner, L. H., Galli, K. G., and Lucas, S. G. 2014. Pedogenic and lacustrine features of the Brushy Basin Member of the Upper Jurassic Morrison Formation in western Colorado: Reassessing the paleoclimatic interpretations. *Volumina Jurassica* 12(2):115–130.

Taylor, M. P. 2009. A re-evaluation of *Brachiosaurus altithorax* Riggs 1902 (Dinosauria, Sauropoda) and its generic separation from *Giraffatitan brancai* (Janensch, 1914). *Journal of Vertebrate Paleontology* 29:787–806.

Taylor, M. P., and Naish, D. 2005. The phylogenetic taxonomy of Diplodocoidea (Dinosauria: Sauropoda). *PaleoBios* 25:1–7.

Taylor, M. P., and Wedel, M. J. 2013a. The effect of intervertebral cartilage on neutral posture and range of motion in the necks of sauropod dinosaurs. *PLoS ONE* 8(10):e78214. DOI:10.1371/journal.pone.0078214.

———. 2013b. Why sauropods had long necks, and why giraffes have short necks. *PeerJ* 1:e36. DOI:10.7717/peerj.36.

Taylor, M. P., Wedel, M. J., and Naish, D. 2009. Head and neck posture in sauropod dinosaurs inferred from extant animals. *Acta Palaeontologica Polonica* 54:213–220.

Tennant, J. P., and Mannion, P. D. 2014. Revision of the Late Jurassic crocodyliform *Alligatorellus*, and evidence for allopatric speciation driving high diversity in western European atoposaurids. *PeerJ* 2:e599. DOI:10.7717/peerj.599.

Tennant, J. P., Mannion, P. D., and Upchurch, P. 2016. Evolutionary relationships and systematics of Atoposauridae (Crocodylomorpha: Neosuchia): Implications for the rise of Eusuchia. *Zoological Journal of the Linnean Society* 177:854–936.

Thulborn, R. A. 1990. *Dinosaur Tracks*. Chapman and Hall, London, 410 pp.

Tidwell, W. D. 1990. Preliminary report on the megafossil flora of the Upper Jurassic Morrison Formation. *Hunteria* 2(8):1–11.

Tidwell, W. D., and Medlyn, D. A. 1993. Conifer wood from the Upper Jurassic of Utah, USA—part II: *Araucarioxylon hoodii* sp. nov. *Palaeobotanist* 42:70–77.

Tidwell, W. D., Britt, B. B., and Ash, S. R. 1998. Preliminary floral analysis of the Mygatt-Moore Quarry in the Jurassic Morrison Formation, west-central Colorado. *Modern Geology* 22:341–378.

Tidwell, W. D., Connely, M., and Britt, B. B. 2006. A flora from the base of the Upper Jurassic Morrison Formation near Como Bluff, Wyoming, USA. *New Mexico Museum of Natural History and Science Bulletin* 36:171–181.

Tipper, J. C. 1979. Rarefaction and rarefiction—the use and abuse of a method in paleoecology. *Paleobiology* 5:423–434.

Travouillon, K. J., and Legendre, S. 2009. Using cenograms to investigate gaps in mammalian body mass distributions in Australian mammals. *Palaeogeography, Palaeoclimatology, Palaeoecology* 272:69–84.

Tremaine, K., D'Emic, M., Williams, S., Hunt-Foster, R., Foster, J., and Mathews, J. 2015. Paleoecological implications of a new specimen of the ankylosaur *Mymoorapelta maysi* from the Hanksville-Burpee Quarry, latest Jurassic (Tithonian) Morrison Formation (Brushy Basin Member). *Journal of Vertebrate Paleontology*, Program and Abstracts, 2015, 226.

Trenham, P. C. 2001. Terrestrial habitat use by adult California tiger salamanders. *Journal of Herpetology* 35:343–346.

Trujillo, K. C. 1999. Vertebrate paleontology, stratigraphy, and sedimentology of new microvertebrate localities in the Morrison (Upper Jurassic) and Cloverly (Lower Cretaceous) formations, Ninemile Hill, Carbon County, Wyoming. Unpublished master's thesis, University of Wyoming, Laramie.

———. 2006. Clay mineralogy of the Morrison Formation (Upper Jurassic–?Lower Cretaceous), and its use in long distance correlation and paleoenvironmental analyses. *New Mexico Museum of Natural History and Science Bulletin* 36:17–23.

Trujillo, K. C., and Kowallis, B. J. 2015. Recalibrated legacy ^{40}Ar/^{39}Ar ages for the Upper Jurassic Morrison Formation, Western Interior, U. S. A. *Geology of the Intermountain West* 2:1–8.

Trujillo, K. C., Foster, J. R., Hunt-Foster, R. K., and Chamberlain, K. R. 2014. A U/Pb age for the Mygatt-Moore Quarry, Upper Jurassic Morrison Formation, Mesa County, Colorado. *Volumina Jurassica* 12(2):107–114.

Tschopp, E., and Mateus, O. 2013. The skull and neck of a new flagellicaudatan sauropod from the Morrison Formation and its implication for the evolution and ontogeny of diplodocid dinosaurs. *Journal of Systematic Palaeontology* 11:853–888.

———. 2016. Case 3700 *Diplodocus* Marsh, 1878 (Dinosauria, Sauropoda)—proposed designation of *D. carnegii* Hatcher, 1901 as the type species. *Bulletin of Zoological Nomenclature* 73:17–24.

———. 2017. Osteology of *Galeamopus pabsti* sp. nov. (Sauropoda: Diplodocidae), with implications for neurocentral closure timing, and the cervico-dorsal transition in diplodocids. *PeerJ* 5:e3179. DOI:10.7717/peerj.3179.

Tschopp, E., Mateus, O., Kosma, R., Sander, M., Joger, U., and Wings, O. 2014. A specimen-level cladistics analysis of *Camarasaurus* (Dinosauria, Sauropoda) and a revision of camarasaurid taxonomy. *Journal of Vertebrate Paleontology*, Program and Abstracts, 2014.

Tschopp, E., Mateus, O., and Benson, R. B. J. 2015. A specimen-level phylogenetic analysis and taxonomic revision of Diplodocidae (Dinosauria, Sauropoda). *PeerJ* 3:e857. DOI:10.7717/peerj.857.

Tschopp, E., Mateus, O., and Norell, M. 2018a. Complex overlapping joints between facial bones allowing limited anterior sliding movements of the snout in diplodocid sauropods. *American Museum Novitates* 3911:1–16.

Tschopp, E., Mateus, O., Marzola, M., and Norell, M. 2018b. Indications for a horny beak and extensive supraorbital connective tissue in diplodocid sauropods. *Journal of Vertebrate Paleontology*, Program and Abstracts, 2018, 229.

Tschopp, E., Brinkman, D., Henderson, J., Turner, M. A., and Mateus, O. 2018c. Considerations of the replacement of a type species in the case of the sauropod dinosaur *Diplodocus* Marsh, 1878. *Geology of the Intermountain West* 5: 245–262.

Tucker, R. T. 2011. Taphonomy of Sheridan College Quarry 1, Buffalo, Wyoming: Implications for reconstructing historic dinosaur localities including Utterback's 1902–1910 Morrison dinosaur expeditions. *Geobios* 44:527–541.

Turner, C. E., and Fishman, N. S. 1991. Jurassic Lake T'oo'dichi': A large alkaline, saline lake, Morrison Formation, eastern Colorado Plateau. *Geological Society of America Bulletin* 103:538–558.

Turner, C. E., and Peterson, F. 1999. Biostratigraphy of dinosaurs in the Upper Jurassic Morrison Formation of the Western Interior, USA. *In* Gillette, D. D., ed., *Vertebrate Paleontology in Utah*, Utah Geological Survey Miscellaneous Publication 99–1:77–114.

———. 2004. Reconstruction of the Upper Jurassic Morrison Formation extinct ecosystem—a synthesis. *Sedimentary Geology* 167:309–355.

Turner-Peterson, C. E. 1986. Fluvial sedimentology of a major uranium-bearing sandstone—a study of the Westwater Canyon Member of the Morrison Formation, San Juan Basin, New Mexico. *In* Turner-Peterson, C. E., Santos, E. S., and Fishman, N. S., eds., *A Basin Analysis Case Study: The Morrison Formation, Grants Uranium Region, New Mexico.* American Association of Petroleum Geologists, *Studies in Geology* 22:47–75.

Tyler, N., and Ethridge, F. G. 1983. Depositional setting of the Salt Wash Member of the Morrison Formation, southwest Colorado. *Journal of Sedimentary Petrology* 53:67–82.

Upchurch, P. 1998. The phylogenetic relationships of sauropod dinosaurs. *Zoological Journal of the Linnean Society* 124:43–103.

Upchurch, P., and Mannion, P. D. 2009. The first diplodocid from Asia and its implications for the evolutionary history of sauropod dinosaurs. *Palaeontology* 52:1195–1207.

Upchurch, P., Barrett, P. M., and Dodson, P. 2004a. Sauropoda. *In* Weishampel, D. B., Dodson, P., and Osmólska, H., eds., *The Dinosauria*, 2nd ed., University of California Press, Berkeley, 259–322.

Upchurch, P., Tomida, Y., and Barrett, P. M. 2004b. A new specimen of *Apatosaurus ajax* (Sauropoda: Diplodocidae) from the Morrison Formation (Upper Jurassic) of Wyoming, USA. *National Science Museum Tokyo, Monographs* 26:1–108.

Valdes, P. 1994. Atmospheric general circulation models of the Jurassic. *In* Allen, J. R. L., Hoskins, B. J., Sellwood, B. W., Spicer, R. A., and Valdes, P. J., eds., *Palaeoclimates and Their Modelling, with Special Reference to the Mesozoic Era*, Chapman and Hall, London, 79–88.

Valdes, P. J., and Sellwood, B. W. 1992. A palaeoclimate model for the Kimmeridgian. *Palaeogeography, Palaeoclimatology, Palaeoecology* 95(1–2):47–72.

Varricchio, D. J., Martin, A. J., and Katsura, Y. 2007. First trace and body fossil evidence of a burrowing, denning dinosaur. *Proceedings of the Royal Society B* 274:1361–1368.

Walker, A. D. 1970. A revision of the Jurassic reptile *Hallopus victor* (Marsh), with remarks on the classification of crocodiles. *Philosophical Transactions of the Royal Society of London* 257:323–372.

Wang X., and Zhou Z. 2004. Pterosaur embryo from the Early Cretaceous. *Nature* 429:621.

Wang, Y., Dong, L., and Evans, S. E. 2016. Polydactyly and other limb abnormalities in the Jurassic salamander *Chunerpeton* from China. *Palaeobiodiversity and Palaeoenvironments* 96:49–59.

Warnock, J. P., Peterson, J. E., Clawson, S. R., Matthews, N. A., and Breithaupt, B. H. 2018. Close-range photogrammetry of the Cleveland-Lloyd Dinosaur Quarry, Upper Jurassic Morrison Formation, Emery County, Utah. *Geology of the Intermountain West* 5: 271–285.

Waskow, K. 2017. Growth rates of giants: Histological evidence for size related differences in growth models between normal sized diplodocoids and a unique assemblage of dwarfed Late Jurassic diplodocoids from the Mother's Day Quarry (Morrison Formation, Montana, USA). *Journal of Vertebrate Paleontology*, Program and Abstracts, 2017, 210–211.

Weaver, J. C. 1983. The improbable endotherm: The energetics of the sauropod dinosaur *Brachiosaurus*. *Paleobiology* 9:173–182.

Wedel, M. J. 2003a. Vertebral pneumaticity, air sacs, and the physiology of sauropod dinosaurs. *Paleobiology* 29:243–255.

———. 2003b. The evolution of vertebral pneumaticity in sauropod dinosaurs. *Journal of Vertebrate Paleontology* 23:344–357.

———. 2005. Postcranial skeletal pneumaticity in sauropods and its implications for mass estimates. *In* Curry Rogers, K. A., and Wilson, J. A., eds., *The Sauropods: Evolution* and *Paleobiology*, University of California Press, Berkeley, 201–228.

———. 2009. Evidence for bird-like air sacs in saurischian dinosaurs. *Journal of Experimental Zoology* 311A:611–628.

Wedel, M. J., and Taylor, M. P. 2013. Caudal pneumaticity and pneumatic hiatuses in the sauropod dinosaurs *Giraffatitan* and *Apatosaurus*. *PLoS ONE* 8(10):e78213. DOI:10.1371/journal.pone.0078213.

Weissmann, G. S., Hartley, A. J., Scuderi, L. A., Nichols, G. J., Davidson, S. K., Owen, A., Atchley, S. C., Bhattacharya, P., Chakraborty, T., Ghosh, P., Nordt, L. C., Michel, L., and Tabor, N. J. 2013. Prograding distributive fluvial systems—geomorphic models and ancient examples. *SEPM Special Publications* 104:131–147.

Wellnhofer, P. 1991. *The Illustrated Encyclopedia of Pterosaurs*. Crescent Books, New York, 192 pp.

———. 2004. The plumage of *Archaeopteryx*: Feathers of a dinosaur? *In* Currie, P. J., Koppelhus, E. B., Shugar, M. A., and Wright, J. L., eds., *Feathered Dragons*, Indiana University Press, Bloomington, 282–300.

Werner, J., and Griebeler, E. M. 2014. Allometries of maximum growth rate versus body mass at maximum growth indicate that non-avian dinosaurs had growth rates typical of fast growing ectothermic sauropsids. *PLoS ONE* 9(2):e88834.

Whitlock, J. A. 2011a. Re-evaluation of *Australodocus bohetii*, a putative diplodocoid sauropod from the Tendaguru Formation of Tanzania, with comment on Late Jurassic sauropod faunal diversity and palaeoecology. *Palaeogeography, Palaeoclimatology, Palaeoecology* 309:333–341.

Whitlock, J. A. 2011b. A phylogenetic analysis of Diplodocoidea (Saurischia: Sauropoda). *Zoological Journal of the Linnean Society* 161:872–915.

Whitlock, J. A. 2011c. Inferences of diplodocoid (Sauropoda: Dinosauria) feeding behavior from snout shape and microwear analyses. *PLoS ONE* 6(4):e18304. DOI:10.1371/journal.pone.0018304.

Whitlock, J. A., and Wilson, J. A. 2018. The Late Jurassic sauropod dinosaur "*Morosaurus*" *agilis* Marsh 1889 reexamined and reinterpreted as a dicraeosaurid. *Journal of Vertebrate Paleontology*, Program and Abstracts, 2018, 238.

Whitlock, J. A., D'Emic, M. D., and Wilson, J. A. 2011. Cretaceous diplodocids in Asia? Re-evaluating the phylogenetic affinities of a fragmentary specimen. *Palaeontology* 54:351–364.

Whitlock, J. A., Trujillo, K. C., and Hanik, G. M. 2018. Assemblage-level structure in Morrison Formation dinosaurs, Western Interior, USA. *Geology of the Intermountain West* 5:9–22.

Wieland, G. R. 1920. The longneck sauropod *Barosaurus*. *Science* 51:528–530.

Wiersma, K., and Sander, P. M. 2017. The dentition of a well-preserved specimen of *Camarasaurus* sp.: Implications for function, tooth replacement, soft part reconstruction, and food intake. *Paläontologische Zeitschrift* 91:145–161.

Wilhite, D. R., and Curtice, B. D. 1998. Ontogenetic variation in sauropod dinosaurs. *Journal of Vertebrate Paleontology* 18(suppl. 3):86A.

Williams, R. M. E., Irwin, R. P., III, Zimbelman, J. R., Chidsey, T. C., and Eby, D. E. 2011. Field guide to exhumed paleochannels near Green River, Utah: Terrestrial analogs for sinuous ridges on Mars. *Geological Society of America Special Paper* 483:483–505.

Williamson, T. E., and Chure, D. J. 1996. A large allosaurid from the Upper Jurassic Morrison Formation, (Brushy Basin Member), west central New Mexico. *In* Morales, M., ed., *The Continental Jurassic, Museum of Northern Arizona Bulletin* 60:73–80.

Williston, S. W. 1901. The dinosaurian genus *Creosaurus*, Marsh. *American Journal of Science* (ser. 4) 11(62):111–114.

Wilson, J. A. 2002. Sauropod dinosaur phylogeny: Critique and cladistics analysis. *Zoological Journal of the Linnean Society* 136:217–276.

Wilson, J. A., and Sereno, P. C. 1998. Early evolution and higher-level phylogeny of sauropod dinosaurs. *Society of Vertebrate Paleontology Memoir 5, Journal of Vertebrate Paleontology* 18(Suppl. 2):1–68.

Wings, O. 2015. The rarity of gastroliths in sauropod dinosaurs—a case study in the Late Jurassic Morrison Formation, western USA. *Fossil Record* 18:1–16.

Witmer, L. M., Chatterjee, S., Franzosa, J., and Rowe, T. 2003. Neuroanatomy of flying reptiles and implications for flight, posture and behaviour. *Nature* 425:950–953.

Witton, M. P. 2013. *Pterosaurs: Natural History, Evolution, Anatomy*. Princeton University Press, Princeton, NJ, 291 pp.

Woodruff, D. C., and Foster, J. R. 2014. The fragile legacy of *Amphicoelias fragillimus* (Dinosauria: Sauropoda; Morrison Formation—latest Jurassic). *Volumina Jurassica* 12(2):211–220.

———. 2017. The first specimen of *Camarasaurus* (Dinosauria: Sauropoda) from Montana: The northernmost occurrence of the genus. *PLoS ONE* 12(5):e0177423.

Woodruff, D. C., Storrs, G. W., Curry Rogers, K., Carr, T. D., and Wilson, J. 2015. The smallest known diplodocid skull: New insights into sauropod cranial development. *Journal of Vertebrate Paleontology*, Program and Abstracts, 2015, 241.

Woodruff, D. C., Carr, T. D., Storrs, G. W., Waskow, K., Scannella, J. B., Nordén, K. K., and Wilson, J. P. 2018. The smallest diplodocid skull reveals cranial ontogeny and growth-related dietary changes in the largest dinosaurs. *Scientific Reports* 8:14341. DOI:10.1038/s41598–018–32620-x.

Xing, L., Miyashita, T., Zhang, J., Li, D., Ye, Y., Skiya, T., Wang, F., and Currie, P. J. 2015a. A new sauropod dinosaur from the Late Jurassic of China and the diversity, distribution, and relationships of mamenchisaurids. *Journal of Vertebrate Paleontology* DOI:10.1080/02724634.2014.889701.

Xing, L., Miyashita, T., Currie, P. J., You, H., Zhang, J., and Dong, Z. 2015b. A new basal eusauropod from the Middle Jurassic of Yunnan, China, and faunal compositions and transitions of Asian sauropodomorph dinosaurs. *Acta Palaeontologica Polonica* 60:145–154.

Young, M. T., Rayfield, E. J., Holliday, C. M., Witmer, L. M., Button, D. J., Upchurch, P., and Barrett, P. M. 2012. Cranial biomechanics of *Diplodocus* (Dinosauria, Sauropoda): Testing hypotheses of feeding behavior in an extinct megaherbivore. *Naturwissenschaften* 99:637–643.

Young, M. T., Tennant, J. P., Brusatte, S. L., Challands, T. J., Fraser, N. C., Clark, N. D. L., and Ross, D. A. 2016. The first definitive Middle Jurassic atoposaurid (Crocodylomorpha, Neosuchia), and a discussion on the genus *Theriosuchus*. *Zoological Journal of the Linnean Society* 176:443–462.

Young, R. G. 1991. A dinosaur nest in the Jurassic Morrison Formation, western Colorado. *In* Averett, W. R., ed., *Guidebook for Dinosaur Quarries and Tracksites Tour: Western Colorado and Eastern Utah*, Grand Junction Geological Society, 1–15.

Yuan, C.-X., Ji, Q., Meng, Q.-J., Tabrum, A. R., and Luo, Z.-X. 2013. Earliest evolution of multituberculate mammals revealed by a new Jurassic fossil. *Science* 341:779–783.

Index

Italicized page numbers indicate occurrences in figures or tables.

JOHN FOSTER is a paleontologist at the Utah Field House of Natural History State Park Museum in Vernal, Utah. He has conducted fieldwork and research in the Morrison Formation throughout the Rocky Mountain region for 30 years, starting while a graduate student at the South Dakota School of Mines and Technology and the University of Colorado, Boulder. He previously worked as executive director of the Moab Museum in Moab, Utah, and curator of paleontology at the Museums of Western Colorado in Fruita, Colorado. He is also author of *Cambrian Ocean World: Ancient Sea Life of North America*.